New Spaces in Physics

After the development of manifolds and algebraic varieties in the previous century, mathematicians and physicists have continued to advance concepts of space. This book and its companion explore various new notions of space, including both formal and conceptual points of view, as presented by leading experts at the *New Spaces in Mathematics and Physics* workshop held at the Institut Henri Poincaré in 2015.

This volume covers a broad range of topics in mathematical physics, including noncommutative geometry, supergeometry, derived symplectic geometry, higher geometric quantization, intuitionistic quantum logic, problems with the continuum description of spacetime, twistor theory, loop quantum gravity, and geometry in string theory. It is addressed primarily to mathematical physicists and mathematicians, but also to historians and philosophers of these disciplines.

MATHIEU ANEL is a Visiting Assistant Professor at Carnegie Mellon University. His research interests include higher category theory, algebraic topology, and topos theory.

GABRIEL CATREN is Permanent Researcher in philosophy of physics at the French National Centre for Scientific Research (CNRS). His research interests include the foundations of classical and quantum mechanics, and the foundations of gauge theories.

New Spaces in Physics
Formal and Conceptual Reflections

Edited by

MATHIEU ANEL
Carnegie Mellon University

GABRIEL CATREN
CNRS - Université de Paris

CAMBRIDGE
UNIVERSITY PRESS

University Printing House, Cambridge CB2 8BS, United Kingdom

One Liberty Plaza, 20th Floor, New York, NY 10006, USA

477 Williamstown Road, Port Melbourne, VIC 3207, Australia

314–321, 3rd Floor, Plot 3, Splendor Forum, Jasola District Centre,
New Delhi – 110025, India

103 Penang Road, #05-06/07, Visioncrest Commercial, Singapore 238467

Cambridge University Press is part of the University of Cambridge.

It furthers the University's mission by disseminating knowledge in the pursuit of
education, learning, and research at the highest international levels of excellence.

www.cambridge.org
Information on this title: www.cambridge.org/9781108490627
DOI: 10.1017/9781108854399

© Cambridge University Press 2021

This publication is in copyright. Subject to statutory exception
and to the provisions of relevant collective licensing agreements,
no reproduction of any part may take place without the written
permission of Cambridge University Press.

First published 2021
3rd printing 2021

Printed in the United Kingdom by TJ Books Limited, Padstow Cornwall

A catalogue record for this publication is available from the British Library.

Library of Congress Cataloging-in-Publication Data
Names: Anel, Mathieu, editor. | Catren, Gabriel, 1973– editor.
Title: New spaces in physics : formal and conceptual reflections /
edited by Mathieu Anel, Carnegie Mellon University, Pennsylvania,
Gabriel Catren, Centre National de la Recherche Scientifique (CNRS), Paris.
Description: New York : Cambridge University Press, 2020. | Includes
bibliographical references and index.
Identifiers: LCCN 2020006655 (print) | LCCN 2020006656 (ebook) | ISBN
9781108490627 (hardback) | ISBN 9781108854429 (epub)
Subjects: LCSH: Mathematics–Research.
Classification: LCC QA11.2 .N525 2020 (print) | LCC QA11.2 (ebook) |
DDC 516–dc23
LC record available at https://lccn.loc.gov/2020006655
LC ebook record available at https://lccn.loc.gov/2020006656

ISBN - 2 Volume Set 978-1-108-85436-8 Hardback
ISBN - Volume I 978-1-108-49063-4 Hardback
ISBN - Volume II 978-1-108-49062-7 Hardback

Cambridge University Press has no responsibility for the persistence or accuracy of URLs
for external or third-party internet websites referred to in this publication and does not
guarantee that any content on such websites is, or will remain, accurate or appropriate.

Contents

Contents for New Spaces in Mathematics   *page* vii

Introduction   1
Mathieu Anel and Gabriel Catren

PART I   NONCOMMUTATIVE AND SUPERCOMMUTATIVE GEOMETRIES   21

1 **Noncommutative Geometry, the Spectral Standpoint**   23
Alain Connes

2 **The Logic of Quantum Mechanics (Revisited)**   85
Klaas Landsman

3 **Supergeometry in Mathematics and Physics**   114
Mikhail Kapranov

PART II   SYMPLECTIC GEOMETRY   153

4 **Derived Stacks in Symplectic Geometry**   155
Damien Calaque

5 **Higher Prequantum Geometry**   202
Urs Schreiber

PART III   SPACETIME   279

6 **Struggles with the Continuum**   281
John C. Baez

7	**Twistor Theory: A Geometric Perspective for Describing the Physical World**	327
	Roger Penrose	
8	**Quantum Geometry of Space**	373
	Muxin Han	
9	**Stringy Geometry and Emergent Space**	407
	Marcos Mariño	

Contents for *New Spaces in Mathematics*

	Contents for New Spaces in Physics	*page* vii
	Introduction	1
	Mathieu Anel and Gabriel Catren	
	PART I DIFFERENTIAL GEOMETRY	29
1	**An Introduction to Diffeology**	31
	Patrick Iglesias-Zemmour	
2	**New Methods for Old Spaces: Synthetic Differential Geometry**	83
	Anders Kock	
3	**Microlocal Analysis and Beyond**	117
	Pierre Schapira	
	PART II TOPOLOGY AND ALGEBRAIC TOPOLOGY	153
4	**Topo-logie**	155
	Mathieu Anel and André Joyal	
5	**Spaces as Infinity-Groupoids**	258
	Timothy Porter	
6	**Homotopy Type Theory: The Logic of Space**	322
	Michael Shulman	
	PART III ALGEBRAIC GEOMETRY	405
7	**Sheaves and Functors of Points**	407
	Michel Vaquié	
8	**Stacks**	462
	Nicole Mestrano and Carlos Simpson	

9	**The Geometry of Ambiguity: An Introduction to the Ideas of Derived Geometry**	505
	Mathieu Anel	
10	**Geometry in dg-Categories**	554
	Maxim Kontsevich	

Introduction

Mathieu Anel and Gabriel Catren

Contents

New Spaces in Physics	1
1 Summaries of the Chapters	8
2 Acknowledgments	18
References	18

New Spaces in Physics

Two fundamental scientific revolutions took place in physics during the first decades of the 20th century: Einstein's geometric description of the gravitational interaction by means of the *general theory of relativity* and the development of *quantum mechanics*. These two revolutions radically modified our understanding of the laws that rule the physical phenomena taking place at opposite (astrophysical and microscopic) spatiotemporal scales.

On one hand, general relativity introduced into physics essential geometric ideas and tools mainly developed during the 19th century in pure mathematics, notably differential geometry, Riemannian geometry, and tensor calculus. Moreover, general relativity provided the motivating example of the general program – launched by H. Weyl around 1918 – intended to provide similar geometric descriptions of the other fundamental (electromagnetic and nuclear) interactions. This "geometrization program" was finally achieved in the 1950s in the framework of the *Yang–Mills theories* and acquired a solid mathematical foundation and geometric interpretation with the theory of Cartan (for general relativity) and Ehresmann (for Yang–Mills theories) connections on principal fiber bundles.[1] Both general relativity and Yang–Mills theories define the so-called *gauge theories of fundamental interactions*, where the term *gauge* refers to the fact that these theories are endowed with a *local symmetry* associated to the possibility to choose different coordinate systems ("gauges") at each spatiotemporal location. Moreover, this geometrization of the fundamental interactions provided the cornerstone of the so-called *standard model* of elementary particles and the associated attempts to unify the four fundamental interactions (where the most celebrated success of this program up to now was the Glashow–Weinberg–Salam unification of the electromagnetic and weak interactions).

On the other hand, quantum mechanics – with its utilization of noncommutative operator algebras on Hilbert spaces – has a strong algebraic flavor that has obstructed to a certain extent the construction of a conceptual interpretation based on a geometric intuition. The main obstacle to the comprehension of quantum mechanics in geometric terms is given by the *noncommutative* character of the algebras of quantum observables. Indeed, this central feature of the quantum formalism has as a consequence that – differently from the *commutative* algebras of classical observables – the quantum observables cannot be understood as functions on an "ordinary" space. This essential feature of quantum mechanics introduces a sort of discontinuity between this theory on one hand and both classical mechanics (which relies on a solid geometric intuition) and the gauge theories of the fundamental interactions on the other.

Roughly speaking, the main lines of research leading to new notions of space in physics after the quantum and the relativistic revolutions can be understood as attempts to understand quantum mechanics in more geometric terms on one hand and to quantize general relativity on the other. Let us consider first the "geometrization" of quantum mechanics. Is it possible to construct noncommutative quantum algebras out of geometric structures? What would be gained by doing so? First, it is worth stressing that quantum mechanics is a formalism that – up to now – could not be endowed with a unanimously accepted conceptual interpretation, being the landscape of competing interpretations populated with radically different conceptual schemes. Now, casting quantum mechanics in more geometric terms redounds in a gain of a conceptual and more intuitive understanding that might pave the road for solving this interpretative conundrum. For instance, the *geometric quantization* formalism developed by Kirillov, Kostant, and Souriau presents quantum mechanics in the same geometric formalism – the theory of connections on fiber bundles – used in gauge theories (see [5, 18, 19, 31]). Now, since gauge

---

[1] For a history of the path that led from general relativity to Yang–Mills theories (and a collection of some of the corresponding seminal papers), see [27].

theories are better understood than quantum mechanics from a conceptual standpoint, geometric quantization provides a useful bridge to transport this conceptual clarity to quantum mechanics. Second, since classical mechanics relies on a clear geometric basis, the geometrization of quantum mechanics might improve the comprehension of the relationship between quantum mechanics and classical mechanics.

Among the different ways according to which mathematicians can construct noncommutative algebras from geometry, three constructions became relevant in physics, namely,

- the deformation of a ring of functions (giving rise in particular to the *deformation quantization* of a Poisson manifold; see, for instance, [32] and references therein);
- the endomorphisms of a fiber bundle (giving rise in particular to the *geometric quantization* of a symplectic manifold [5, 18, 19, 31]);
- the convolution algebra of a groupoid (giving rise to *noncommutative methods* [6]).

Now, both deformation quantization and geometric quantization strongly rely on the *symplectic formulation of classical mechanics*. Here a main player for the development of physical geometry during the 20th century enters the scene: *symplectic geometry*. Thanks to the work of mathematicians like Arnold, Maslov, Souriau, and Weinstein, among others, the explosion of research in symplectic geometry during the 20th century led to a deep transformation of our comprehension of classical mechanics (see, for instance, [1, 2, 10, 21, 22, 31]). In the framework of this symplectic geometrization of classical mechanics, fundamental new notions and theories were introduced, such as Souriau's *moment map* [25, 31], the Marsden–Weinstein *symplectic reduction* [23], and Weinstein's *symplectic "category"* and *Lagrangian correspondences* [33]. In the wake of this symplectic refoundation of classical mechanics, it is also worth mentioning the development of the theory of variational calculus on jet bundles and the development of multisymplectic geometry launched by De Donder and Weyl and continued – more recently – by Kijowski, among others. In this extended context, important new notions were introduced, such as the *covariant phase space*, the *Peierls bracket*, and the *variational bicomplex* (see, for instance, [7, 8, 17, 26, 31, 34]).

From a conceptual standpoint, the great importance of symplectic (and Poisson) geometry is that it encodes what we could call the *classical seeds* of quantum mechanics. By doing so, the development of symplectic geometry allowed us to significantly reduce the gap between classical and quantum mechanics. It could even be argued that symplectic geometry opened the path to the comprehension of quantum mechanics as a continuous extension of classical mechanics and no longer as a sort of "new paradigm" discontinuously separated from the classical one (see Schreiber's contribution in Chapter 5). For instance, both in *deformation quantization* and in *geometric quantization*, classical structures (namely, the Poisson structure and the symplectic structure, respectively) encode fundamental quantum features. While in deformation quantization the Poisson structure provides the first term of the "quantum" deformation (in the formal parameter $\hbar$) of the commutative algebra of functions on a phase space, in geometric quantization the symplectic structure defines the curvature (on the prequantization fiber bundle) that explains the noncommutativity of quantum operators.[2] Moreover, one of the central facts of symplectic geometry is the existence of a correspondence defined by the symplectic structure between observables (functions on a phase space) and what could be called *classical operators* (Hamiltonian vector fields). In this way, the fundamental role played by operators in mechanics – far from being a quantum innovation – is already a central feature of classical mechanics.[3] It is also worth mentioning that the category-theoretic "points" of a symplectic manifold are given by its Lagrangian submanifolds.[4] According to Guillemin and Sternberg, the notion of Lagrangian submanifold encodes the classical seeds of the quantum indeterminacies:

> *The Heisenberg uncertainty principle says that it is impossible to determine simultaneously the position and momentum of a quantum-mechanical particle. This can be rephrased as follows: the smallest subsets of classical phase space in which the presence of a quantum-mechanical particle can be detected are its Lagrangian submanifolds. For this reason it makes sense to regard the Lagrangian submanifolds of phase space* [rather than its set-theoretic points] *as being its true 'points'* [11].

In this way, it could be argued that if the notion of localization in phase space (in the sense of "being at a certain point" of phase space) is not defined with respect to its set-theoretic points but rather with respect to the Lagrangian "points," then the Heisenberg indeterminacy principle does not forbid a localization of a quantum particle in phase space. All in all, these

---

[2] It is worth noting that this is in complete analogy to the fact that in general relativity and Yang–Mills theories, the noncommutativity of parallel transports results from the presence of a nontrivial curvature.

[3] In the framework of geometric quantization, quantum operators are in fact defined by means of a vertical extension (where *vertical* means in the direction of the fibers of the corresponding prequantization fiber bundle) of these classical operators (see, for instance, [5]).

[4] Considered from the standpoint of category theory, the Lagrangian submanifolds of a symplectic manifold (M, ω) are the $(*, 0)$-points of M in Weinstein's symplectic "category" (where $(*, 0)$ is the trivial symplectic manifold, that is, the morphisms (Lagrangian correspondences) $(*, 0) \to (M, \omega)$.

different insights brought forward by the development of symplectic geometry are permitting us to progressively sublate the simplistic opposition between the supposedly stable and well-understood realm of classical mechanics and the still-unsolved conceptual problems posited by quantum mechanics. By pushing this line of thought to its limit, it could even be argued that the missing insights permitting us to construct a satisfactory conceptual interpretation of quantum mechanics might stem from a better comprehension of classical mechanics and its symplectic foundations. In this sense, the explosion of research in symplectic geometry is pulling back the problem of interpreting quantum mechanics to an unexpected problem: the problem of reinterpreting classical mechanics.

Another direct repercussion on geometry elicited by the development of quantum mechanics is given by the study of hypothetical "spaces" supporting (or dual to) *noncommutative* "algebras of functions." The new branch of geometry known as *noncommutative geometry* might have been inspired by the capacity to generate new notions of space associated to the *geometry–algebra dualities*, that is, to the dualities between spaces and the algebras of "functions" on them (for instance, the duality between affine schemes and commutative rings or the Gelfand–Naimark duality between compact Hausdorff topological spaces and commutative unital C^*-algebras). Indeed, the geometry–algebra dualities naturally lead to the introduction of new spaces by means of the following pattern: given a particular instantiation of a geometry–algebra duality, one can generalize the corresponding algebra of functions – by passing, for instance, to noncommutative algebras – and try to interpret the new algebra as an "algebra of functions" on a generalized space. However, it is not clear to what extent the noncommutative approaches to geometry do really produce "noncommutative spaces" dual to the corresponding algebras. An alternative way to understand noncommutative geometry could be the following. Given "ordinary" (commutative) spaces, one can define noncommutative invariants. These invariants do not always allow us to reconstruct the space, but they encode nonetheless certain important geometric aspects like *properness* or *smoothness* (see, for instance, Chapter 10 of *New Spaces in Mathematics*). The important fact is that these noncommutative invariants endowed with a geometric meaning permit us to introduce certain geometric concepts and intuitions into the realm of noncommutative algebra.

The formulation of quantum mechanics and general relativity naturally leads to the *quantum gravity* program, that is, to the different research programs intended to quantize general relativity (for instance, superstring theory, loop quantum gravity, semiclassical quantum gravity, causal sets, dynamical triangulations, lattice quantum gravity, and the asymptotic safety program, among others[5]). The general expression *quantize general relativity* denotes here both the application of standard quantization methods (e.g., canonical quantization, path integral) to general relativity in its Lagrangian or Hamiltonian formulation and the direct construction of a theory out of which general relativity and the continuum description of spacetime is supposed to emerge in some "classical" approximation.

The supposed necessity to quantize general relativity can be justified on different grounds, for instance,

- the idea that quantum gravity is required to deal with spacetime singularities taking place at very high energies and very small scales (such as the big bang and black hole singularities);
- the fact that while general relativity describes (by means of the Einstein field equations) the coupling between *classical* matter and the geometry of spacetime, all matter is currently described in the framework of *quantum* field theory;
- the idea that the unification between gravity and the other *quantum* gauge fields carrying the electromagnetic and nuclear interactions requires us also to describe gravity in quantum terms – by taking into account that the nongravitational interactions are mediated by the so-called *gauge bosons* (like the photon for the electromagnetic interaction), this argumentative line led (mainly in the framework of perturbative string theory) to the postulation of a hypothetical massless spin-2 particle that mediates the gravitational interaction, the *graviton*;[6]
- the arguments based on the finite character of black hole entropy (see, for instance, [29]).

Besides these particular motivations, a more straightforward argument is the following. Since

1. general relativity is already a *classical* theory in the sense that it can be cast in terms of classical (Hamiltonian or Lagrangian) mechanics (e.g., ADM formalism, Einstein–Hilbert action); and
2. classical mechanics has been superseded by (or extended to) quantum mechanics,

then general relativity has to be recast in quantum-mechanical terms.

---

[5] For an overview of different approaches to quantum gravity, see, for instance, [24] and references therein.

[6] It is worth noting that a straightforward application of the perturbative methods of quantum field theory to the gravitational interaction leads to a perturbative nonrenormalizability. This obstacle has been the main motivation for the development of *nonperturbative approaches* to quantum gravity.

Despite the still highly speculative nature of the field, research in quantum gravity has already had a significant impact on mathematical geometry. First, string theory already had important repercussions on research in pure geometry (e.g., mirror symmetry, Gromov–Witten invariants, and enumerative geometry; see, for instance, [3, 15, 16]). Second, research in quantum gravity opened the field of *quantum geometry*, that is, the study of different geometric structures, out of which the classical and continuum spacetime geometry described by general relativity can be reobtained in some form of "classical" limit. In very general terms, the field of quantum geometry explores ideas such as

- a fundamental discretization of spacetime (an idea that goes back to Riemann [28] and reappears in almost every approach to quantum gravity);
- spaces described by noncommutative coordinates (e.g., noncommutative geometry);
- quantum indeterminacies and fluctuations of geometric quantities;
- linear superpositions of geometries.

For instance (as Mariño explains in Chapter 9), string theory addresses different forms of deformation (stringy, quantum) of classical Riemannian geometry resulting from the quantum description of dynamical extended objects (strings and eventually p-branes). In turn, loop quantum gravity studies certain geometric structures – the canonical spin-networks and the covariant spinfoams – arising from a more or less direct quantization of general relativity (see Han's contribution in Chapter 8). Other approaches explore the possibility of understanding the classical and continuum description of spacetime geometry – as well as geometric notions like *dimension* and *locality* – as an emergent description arising from *nongeometric* or *pregeometric* (a term introduced by Wheeler [34]) degrees of freedom. Examples of these supposed pregeometric structures are the *causal sets*, that is, sets representing spacetime events endowed with an order relation encoding the causal structure [9, 30], or combinatorial structures like simplicial objects and graphs (e.g., *quantum graphity* [14]). However, the characterization of these structures as non- or pregeometric is problematic (do they really "*break loose at the start from all mention of geometry and distance?*" [34]), and it might seem more appropriate to state that the different "pregeometric" scenarios proposed thus far remove certain geometric features of the classical and continuum description of spacetime conveyed by general relativity (e.g., continuity, differential structure, distance, dimensionality, or locality).

Let us consider now in some detail the different chapters of this volume.

1 Summaries of the Chapters

1.1 Part I Noncommutative and Supercommutative Geometries

1.1.1 Noncommutative Geometry, the Spectral Standpoint
(Alain Connes)

The construction of quotients of spaces has been an important source of definitions of new notions of space. The space of leaves of a dense foliation does not have enough open subsets to be described as a manifold or even as a topological space. The spaces of orbits of group actions that are not free have singularities that a topology or a differential structure cannot encode. Several methods have been invented to work with these objects, some using category theory (e.g., sheaves and stacks, topoi, diffeologies), others algebra. The noncommutative geometry of A. Connes belongs to this latter class. The basic idea is to replace the commutative ring of observable functions on the quotient by the noncommutative convolution algebra of the foliation or the group action. This construction is justified by the fact that, when the quotient exists, the categories of modules over the function ring or over the convolution algebra coincide.[7] However, the latter construction is better behaved than the former.

From a more conceptual standpoint, the basic principle of Connes's noncommutative geometry is to substitute the equivalence relation associated to a quotient operation by the corresponding *action groupoid* of identifications. The main difference between an equivalence relation and a groupoid is that the latter keeps track of the fact that different points might be identified in many different ways (which includes *a fortiori* the particular case of possible nontrivial stabilizers). In this sense, an equivalence relation can be understood as a truncated groupoid where the possibly multiple concrete identifications between two elements are collapsed to the abstract fact that they are equivalent. This transition from equivalence relations to groupoids leads to the consideration of a particular noncommutative algebraic structure, namely, the *convolution algebra on the action groupoid* (where the noncommutativity is a direct consequence of the noncommutativity of compositions in the groupoid). As it was stressed by Connes in [6, §1.1, pp. 40–45], this kind of noncommutative algebra was implicitly discovered by Heisenberg in the seminal 1925 article in which he proposed the matrix formulation of quantum mechanics [12].[8]

---

[7] Technically, they are Morita-equivalent algebras.

[8] In Heisenberg's matrix formulation, the relations between physical quantities are governed by the noncommutative algebra of matrices that represent these quantities. Connes argued that the Ritz–Rydberg combination principle that models the experimental results provided by atomic

Noncommutative geometry consists in defining a certain number of geometric notions (infinitesimal calculus, integration and measure theory, metric, etc.) in terms of algebras that are not necessarily commutative. The central notion is that of *spectral triplets* $(\mathcal{A}, \mathcal{H}, D)$ encoding a "space" with a metric and a measure theory.[9] The commutative algebra of functions on a Riemannian (spin^c) manifold is reinterpreted by Connes as an algebra of operators acting on a Hilbert space of spinors, and the inverse line element of the Riemannian structure is encoded (in Connes's *distance formula*) by the corresponding Dirac operator. Now, the central insight is that this setting remains valid when we substitute the commutative algebra functions by a noncommutative algebra of operators acting on a Hilbert space.

It is also worth noting that Connes's version of noncommutative geometry is motivated by the problem of quantizing gravity and unifying the four fundamental interactions. The inverse line element defined by the Dirac operator D encodes not only the gravitational interaction (associated as usual to the metric) but also the electromagnetic, and nuclear – weak and strong – interactions (which are associated to the inner fluctuations of the metric). This results in a successful derivation of the Lagrangian of the standard model from a Lorentzian spacetime crossed with a specific finite noncommutative space. Interestingly enough, the different physical forces are unified by means of the metric structure of the noncommutative space, thereby giving rise to a sort of generalized gravity theory.

1.1.2 The Logic of Quantum Mechanics (Revisited) (Klaas Landsman)

Landsman's contribution can be inscribed among the attempts to generalize the classical notions of space by using the framework provided by the geometry–algebra duality. Starting with

- the (constructive versions of the) Gelfand–Naimark duality between *commutative* unital C^*-algebras and compact Hausdorff topological spaces; and
- the Stone duality between the category of boolean lattices (with homomorphisms of orthocomplemented lattices as arrows) and totally disconnected compact Hausdorff spaces (Stone spaces),

---

spectroscopy (which were incompatible with the classical predictions) can be encoded in a groupoid of frequencies whose convolution algebra is nothing but the algebra of matrices discovered by Heisenberg.

[9] More precisely, a general spectral triplet $(\mathcal{A}, \mathcal{H}, D)$ is given by a $*$-algebra $\mathcal{A}$ endowed with a representation by bounded operators on a Hilbert space $\mathcal{H}$ and an unbounded self-adjoint Dirac operator D acting on $\mathcal{H}$ and encoding a generalized notion of distance that extends the Riemannian notion of distance to the noncommutative realm.

Landsman moves forward to the intuitionistic/noncommutative realm by addressing

- the Priestley duality between bounded distributive lattices and Priestley spaces; and
- the Esakia duality between Heyting algebras and Esakia spaces.

The ultimate goal of this progression is a conjectured duality between arbitrary unital C^*-algebras and some Heyting algebras. The result of this work in progress would be the construction of a model of an *intuitionistic quantum logic* that has the opposite features from Birkhoff and von Neumann's quantum logic [4]. This means that such an intuitionistic quantum logic is distributive (which paves the way to an interpretation of the logical operations $\wedge$ and $\vee$ as a disjunction and a conjunction, respectively) but does not keep the law of the excluded middle (which, according to Landsman, matches quantum features such as Schrödinger cat situations).

Interestingly enough, this construction of an intuitionistic quantum logic can be related to topos theory. Briefly, we can associate to any unital C^*-algebra A the topos of covariant functors $\mathcal{C}(A) \to Set$ on the posetal category $\mathcal{C}(A)$ of all unital commutative subalgebras of A.

1.1.3 Supergeometry in Mathematics and Physics (Mikhail Kapranov)

Kapranov's contribution addresses the quandaries of supergeometry in mathematics and supersymmetry in physics from an original homotopical perspective. According to Kapranov, the challenge posited by supergeometry and supersymmetry is to understand the formal and conceptual structures underlying the $\pm$ sign rules that govern the supercommutation structures in both mathematics and physics. These structures involve vector spaces with a $\mathbb{Z}/2\mathbb{Z}$-grading together with a monoidal structure involving Koszul's sign rule. Now, an important caveat is here necessary: the similarities between formalisms discovered by physicists and mathematicians might sometimes be misleading. According to Kapranov, an instance of this danger is provided by these supercommutative structures. Indeed, a careful comparative study of supercommutative structures in mathematics and physics leads Kapranov to conclude that the formal similarity should not lead to an identification: the $\mathbb{Z}/2\mathbb{Z}$ of mathematicians is not the same as the $\mathbb{Z}/2\mathbb{Z}$ of physicists.

From a mathematical standpoint, supergeometry is the study of geometric objects whose rings of functions are commutative superalgebras $A = A^{\bar{0}} \oplus A^{\bar{1}}$ composed of even and odd elements subjected to the corresponding supercommutation rules. In this way, supergeometry can be added to the list of

Introduction 11

new geometries (along noncommutative spaces, topoi, and so forth) associated to the attempts to generalize the standard commutative algebras of functions of the geometry–algebra dualities. What Kapranov calls the *principle of naturality of supers* states that supercommutative algebras – rather than being a noncommutative generalization – can be understood as a natural "super" extension of commutative algebra itself (which is implicit in the term *super*-commutative rather than *non*-commutative).[10] From a physical standpoint, Kapranov understands supersymmetry as a particular case of the heuristic *principle of square roots* according to which certain quantities of "immediate physical interest" (e.g., real quantities like the probability density in quantum mechanics) are bilinear combinations of more fundamental quantities (e.g., complex quantities like the wave function in quantum mechanics).

The original idea addressed by Kapranov in his contribution is that supercommutative structures are nothing but the tip of an homotopical iceberg associated to the first level of the *sphere spectrum* $\mathbb{S}$ (where the group $\{\pm\}$ associated to the sign rules is the first homotopy group of $\mathbb{S}$). To explain this, M. Kapranov proposes to push further the comprehension of the group of integers. Classically, $\mathbb{Z}$ is the free commutative group on one generator. However, this is no longer the case in a homotopical or higher categorical context. If sets are replaced by ∞-groupoids (see Chapters 5, 6 and 9 of *New Spaces in Mathematics*), the free commutative group on one generator is no longer $\mathbb{Z}$ but the *sphere spectrum* $\mathbb{S}$.[11] As Kapranov writes in the wake of Grothendieck, the sphere spectrum – being the homotopic version of the ring $\mathbb{Z}$ of entire numbers – "is the most fundamental commutative object," that is (we could say), the "supercommutative" object. By doing so, Kapranov establishes an unexpected link between supercommutative structures in mathematics and physics on one hand and the homotopical reconceptualization of the abstract notion of identity in terms of concrete (and possibly multiple) identifications on the other hand.

The sphere spectrum is a nontrivial homotopy type whose homotopy invariants are the so-called *stable homotopy groups of spheres*. The π_0 of this homotopy type recovers $\mathbb{Z}$, but its π_1 and π_2 are both $\mathbb{Z}/2\mathbb{Z}$. This last feature of the sphere spectrum allows Kapranov to explain the similarity

---

[10] For instance, given a supercommutative algebra $A = A^{\overline{0}} \oplus A^{\overline{1}}$, the even part $A^{\overline{0}}$ defines an ordinary affine scheme $Spec(A^{\overline{0}})$ and the odd part (being nilpotent) enriches this scheme by adding an "infinitesimal neighborhood" to it.

[11] Intuitively, the difference is that the products ab and ba are equal in $\mathbb{Z}$, whereas they are only homotopic in $\mathbb{S}$. It follows that a square a^2 inherits a nontrivial loop in the space $\mathbb{S}$. In this regard, $\mathbb{Z}$ is constructed from the space $\mathbb{S}$ by contracting these loops into trivial loops. But by doing so, $\mathbb{Z}$ is no longer described as a free object but rather as an object with relations.

between the supercommutative structures of mathematicians and physicists. Both are working with a sign rule, but these are controlled by different levels of the homotopy of the sphere spectrum (the 1-truncation for mathematicians and 2-truncation for physicists). Kapranov's conclusion is that the practice of mathematicians and physicists can be unified by considering the new notion of vector spaces graded by $\mathbb{S}$ (rather than $\mathbb{Z}$), a notion that does not yet formally exist and would "open a fantastic possibility of higher super-mathematics." The chapter finishes with a sketch of what this theory could be.

1.2 Part II Symplectic Geometry

1.2.1 Derived Stacks in Symplectic Geometry (Damien Calaque)

An important recent development in the history of the relations between symplectic geometry and mechanics is provided by the "derived" enhancement of symplectic geometry discussed in Chapter 4.

The reformulation of mechanics in terms of symplectic geometry led to a number of problems related to the limitations of the notion of manifold. For example, a fundamental operation in symplectic geometry is the *symplectic reduction* of a Hamiltonian group action on a symplectic manifold. This process combines a *restriction* to a subspace of phase space and a *projection* to a quotient space (called, in physics jargon, *constraint surface* and *reduced phase space*, respectively) [13, 23]. Now, these two operations might produce singularities. The result of this symplectic reduction is again a symplectic manifold when there are no singularities, but the definition of a symplectic structure around a singular point becomes problematic when singularities are present.

The recent development of derived geometry (see Chapter 9 of *New Spaces in Mathematics*) has succeeded in defining a general notion of singular symplectic space in the context of algebraic geometry.[12] The introduction of these new symplectic spaces allowed us to regularize important features of the theory: Lagrangian correspondences (i.e., the morphisms in Weinstein's symplectic category) can always be composed, symplectic reductions are

---

[12] It is worth noting that methods developed for dealing with constrained Hamiltonian systems (like the BRST cohomological reformulation of symplectic reduction or the BV formalism [13, 20]) already implicitly encode ideas coming from the domain of "derived mathematics." In the framework of derived geometry, the ad hoc (co)homological methods used in physics are interpreted as a way to deal with singular points: degenerated systems of constraint equations produce nontransverse intersections and degenerated symmetries produce nonfree group actions and singular quotients. The (co)homological structure generated by the so-called *ghosts* and *antighosts* of the BRST formalism serves to regularize these possible pathological situations.

always symplectic, and the symplectic structure built from the transgression construction always exists. The "derivation" of symplectic geometry has also led to some new features. First, the extension of the notion of symplectic structure to the notion of *shifted symplectic structure* allowed us to realize that certain important spaces (like the intersection of any two Lagrangian correspondences, the quotient stack $\mathfrak{g}^*//G$ of a coadjoint action, and the classifying space $BG = *//G$) are in fact endowed with shifted symplectic structures. Moreover, Lagrangian correspondences are no longer given by subspaces but rather by general maps that are not necessarily injective.[13]

1.2.2 Higher Prequantum Geometry (Urs Schreiber)

As we have said before, the standard (and too simplistic) demarcation line between a supposedly well-understood realm of classical mechanics and the quandaries of quantum physics has been blurred by both the explosion in the second half of the 20th century of research on *symplectic* and *Poisson geometries* and the development of quantization formalisms that strongly rely on this geometric description of classical mechanics, notably, *geometric quantization* and *deformation quantization*.

Schreiber's Chapter 5 moves forward in the direction of sublating the opposition between classical and quantum mechanics

- by showing that the prequantization construction in geometric quantization (i.e., the definition of a linear fiber bundle on the phase space with a connection whose curvature is defined by the symplectic form) can be understood as a *global lifting of local data*; and
- by developing this understanding of prequantum geometry for *covariant* field theories, that is, for theories that are local in spacetime M rather than in space.

Briefly, the main idea is to generalize the definition of the Lagrangian (of a theory that describes physical fields given by sections of a bundle $\varphi \colon E \to M$ over spacetime) from *globally defined Lagrangian* to *families of local Lagrangians endowed with gluing data*. Technically, this amounts to defining a Lagrangian as a function on E with values in the *moduli space of Čech–Deligne cocycles*, which classify *gerbes with connections*. Roughly speaking, a *gerbe* is a generalization of a G-principal fiber bundle where the fibers are isomorphic not to G but rather to the classifying spaces $B^n G$ for some n.

---

[13] Interestingly enough, moment maps are particular instances of these generalized Lagrangian correspondences.

Geometrically, this means that the topological twists are not introduced at the level of the identifications between the fibers (given by G-valued transition functions on the twofold intersections $U_{i_1} \cap U_{i_2}$, where U_{i_n} are open sets of a covering) but rather at the level of the higher cocycle consistency conditions defined on the higher n-fold intersections $U_{i_1} \cap ... \cap U_{i_n}$.

The resulting *higher prequantum geometry* allows us to prequantize local field theory in an explicitly *local* (or *covariant*) and *gauge-invariant* manner. It is a major conceptual step for the comprehension of mechanics to understand the prequantization construction (which yields the noncommutative algebra of quantum operators) in terms of what we could characterize as *global classical mechanics*. A related important feature of Schreiber's formalism is that the global degrees of freedom given by the topology of the bundle (e.g., the instanton sector) – rather than being fixed in an ad hoc manner – are incorporated in the definition of the generalized Lagrangians as functions with values in a classifying space. In this sense, Schreiber's formalism gives a further step toward fulfilling the heuristic principle of "background independence," that is, the principle according to which physical theories have to be as free as possible from the presupposition of ad hoc geometric background structures.

1.3 Part III Spacetime

1.3.1 Struggles with the Continuum (John C. Baez)

An important motivation for constructing a mathematical description of physical nature is given by the possibility of making predictions of future states by means of computations. In general, such computations require us to introduce idealizations or approximations (for instance, it is easier to represent planets as points rather than as three-dimensional objects). In particular, the representation of spacetime by means of the continuum of real numbers seems to be one of these idealizations. Even if spacetime might well be discrete at small scales, it is helpful to assume that it is continuous in almost all branches of physics. In particular, this assumption permits us to use differential calculus and approximations by series expansions. Now, while making idealizations is an essential component of scientific activity, idealizations might lead to inconsistencies when pushed beyond their limits (e.g., forces or speeds can become infinite, series expansion can diverge) or become an obstruction to the construction of a rigorous formulation of the theories at stake.

In Chapter 6, Baez analyzes a number of problems posited in different branches of physics by the assumption of spacetime continuity, for instance, the problems related to the collision and noncollision singularities in Newtonian point particles interacting gravitationally, the problem of self-forces in electromagnetic theory (exerted by the field created by a particle on itself), the renormalization of infinities in quantum field theory, and the questions and problems posited by (black hole and cosmological) singularities in general relativity. All of these problems are related to the possibility of having arbitrarily small or large real numbers (typically, small distances creating infinite forces).

Overall, Baez's chapter shows that the interactions between mathematical theories, physical theories, and reality are more complex than what is commonly assumed. Theories can be incomplete, be inconsistent, or fail to make predictions in certain regimes, even if they are in agreement (at least to a certain extent) with experiments. Some of the problems posited by the assumption of continuity have led to important solutions (e.g., the discretization of energy in quantum mechanics), but most of them are still open. While in pure mathematics, certain research programs address the problems posited by the mathematical continuum (e.g., nonstandard analysis, synthetic differential geometry, constructivism, finitism, ultrafinitism), Baez and many physicists believe that a successful theory of quantum gravity might shed new light on the problems associated with the assumption of spacetime continuity.

1.3.2 Twistor Theory: A Geometric Perspective for Describing the Physical World (Roger Penrose)

Penrose's twistor theory provides a new approach to fundamental physics in which quantum mechanics and relativity theory are combined in an original way that departs from the more established quantum gravity approaches. Rather than combining them by applying (for instance) quantization techniques to general relativity, the communicating vessel between the two theories is surprisingly provided by complex analysis and holomorphic geometry.

Twistor theory provides a framework in which spacetime is a derived notion with respect to an underlying arena provided by complex *twistor geometry*. One of the central intuitions of twistor theory is that ray-lights (endowed with an angular momentum twist) should be considered more fundamental than spacetime points. In a sense, this amounts to taking seriously the fact that spacetime events connected by a ray-light are "separated" by a null spatiotemporal interval. In turn, this projective stance amounts to ascribing a more fundamental importance to the physics of massless particles and the corresponding conformal invariance.

A central feature of twistor theory is that it is specifically adapted to the fact that the physical spacetime *"that macroscopically presents to us"* has $3+1$

dimensions. This feature of the theory contrasts both with approaches like string theory that require extra space dimensions and with the mathematical drive toward full generality. In particular, the central role played by complex numbers in quantum mechanics is directly related by Penrose to the fact that physical space is of dimension 3. In turn, the *Riemann sphere* that provides this relation also shows up in a relativistic context, now as the *celestial sphere* surrounding an observer. Twistor theory is based on the intuition according to which this dual role played by the Riemann sphere – far from being a mere coincidence – provides an essential hint toward a unification of both theories based on 2-spinor calculus.

Besides unfolding the different trends of thought that led to twistor theory, Penrose also introduces in Chapter 7 a recent attempt (called *palatial twistor theory*) intended to overcome the *googly problem* related to the left-handed (or anti-self-dual) character of the gravitational fields in Penrose's nonlinear graviton construction.

1.3.3 Quantum Geometry of Space (Muxin Han)

One possible strategy to construct a *quantum theory of gravity* is to apply a quantization formalism to the canonical (constrained) formulation of general relativity (for instance, the ADM Hamiltonian formalism). Differently from Yang–Mills theories (where the dynamical variable is a connection), the fundamental dynamical variable in the standard Einstein–Hilbert formulation of general relativity is the 3-metric of the corresponding spatial hypersurfaces of a four-dimensional spacetime manifold M_4 (with respect to a chosen foliation of M_4 in three-dimensional hypersurfaces M_3^t). An important step forward intended to overcome the impasses of the original quantum geometrodynamics program (impasses that are mainly associated with the nonlinear nature of the corresponding constraints) was given by Ashtekar's reformulation of general relativity as a dynamical theory of $SU(2)$ connections. The main step of this program is the definition of the *quantum configuration space* $\overline{\mathcal{A}}$ of gravity as a space of $SU(2)$-holonomies around spatial "loops" defined by "generalized" or "distributional connections."[14] This means that the wave functions of the quantum theory will be given by functions of these $SU(2)$-holonomies

$$\psi_\gamma(A) = \psi(h_{e_1}(A),...,h_{e_n}(A)),$$

where γ is a graph (or "network") immersed in M_3, $(e_1,...,e_n)$ are the n edges of γ, and $h_{e_i}(A)$ is the holonomy of A along the edge e_i. The main

---

[14] The distributional nature of the connections means that the usual smoothness assumption on the connections was dropped, that is, the map $e \mapsto h_e(A)$ can be discontinuous, in order to cope with the problem of defining a diffeomorphism invariant measure on the configuration space.

advantage of this connection-theoretical reformulation of general relativity is that it brings Einstein's theory closer to the Yang–Mills theories that describe the other fundamental interactions and whose quantization is well understood.

Along these lines, *loop quantum gravity* was developed – thanks to the efforts of physicists like Ashtekar, Rovelli, Smolin, Thiemann, and Lewandowski, among others – as a *background independent* and *nonperturbative* approach to quantum gravity out of which a well-defined picture of quantum space and quantum spacetime emerged in the form of the *spin-networks* and *spinfoams* states, respectively. Since the spin-network states are common eigenvectors of the area and the volume operators with discrete spectra, loop quantum gravity provides a concrete description of a discrete quantum geometry obtained by quantizing a quantum configuration space of distributional $SU(2)$-connections.

1.3.4 Stringy Geometry and Emergent Space (Marcos Mariño)

The problem of quantizing gravity has also triggered research in theoretical physics aiming to extend the physics of fundamental point-particles to higher-dimensional objects such as strings, membranes, and p-branes. In particular, string theory provides a unified theory of the fundamental interactions with propagating gravitons (and black hole solutions), that is to say, a unified theory of the four fundamental interactions including a perturbative theory of quantum gravity. As Mariño explains in his contribution, this perturbative quantum theory of gravity can be geometrically understood in terms of a 2-parameter *stringy* and *quantum* deformation of classical Riemannian geometry (where the deformation parameters are the string length l_s and the string coupling constant g_{st}, respectively). In concrete examples, classical Riemannian geometry "emerges" as an effective description in the limit of a point-particle approximation ($l_s \to 0$) of noninteracting ($g_{st} \to 0$) strings.

Regarding the stringy deformation, a manifold X – rather than being "probed" by points (limit case in which $l_s \to 0$) – is endowed with all possible maps from Riemann surfaces to X, where these maps describe the possible worldsheets of the string. In the simpler case given by *topological string theory*, the quantum theory of strings embedded in a Calabi–Yau manifold X gave rise to the celebrated results in enumerative geometry associated to the Gromov–Witten invariants.

In the case of a nonzero string coupling constant g_{st}, quantum corrections associated to Riemann surfaces with a nonzero genus have to be considered. Since the corresponding genus expansion is divergent, a nonperturbative formulation of string theory is needed (the formulation of which remains problematic). Now, it was understood that a nonperturbative formulation of string theory requires us to consider higher-dimensional objects (p-branes) that are "invisible" from the standpoint of the perturbative theory. One of the most fruitful avenues of research toward a nonperturbative formulation of string theory is given by the so-called *holographic dualities*, notably Maldacena's AdS/CFT correspondence (where AdS/CFT stands for *anti–de Sitter/conformal field theory*). These are dualities between string theories (and the corresponding spacetime gravitational physics) on one hand and quantum gauge theories in lower dimension on the other.

2 Acknowledgments

The material for this book and its companion (*New Spaces in Mathematics: Formal and Conceptual Reflections*) grew up from the conference New Spaces in Mathematics and Physics organized by the editors in 2015 at the Institut Henri Poincaré in Paris. Both the conference and the book project have been realized in the framework of the ERC project "Philosophy of Canonical Quantum Gravity" piloted by G. Catren. Information on the conference (including videos of the talks) is available at https://www.youtube.com/playlist?list=PLRxtDuSeiaXYy17D56Era8ns3V4vmWmqE.

The history of the evolution of space in mathematics and physics attempted here does not pretend to be comprehensive. For various reasons, important chapters are missing (orbifolds; tropical geometry; Berkovich spaces; $\mathbb{F}_1$-geometry; motives; quantales; constructive analysis and its real numbers; formal topology; or other approaches to quantum gravity, such as causal sets, group field theory, or dynamical triangulations), and we apologize to the reader who might have expected to find these in this work.

This project has received funding from the European Research Council under the European Community's Seventh Framework Programme (FP7/2007-2013 Grant Agreement 263523, project Philosophy of Canonical Quantum Gravity).

References

[1] R. Abraham and J. E. Marsden, *Foundations of Mechanics*, 2nd ed., Addison-Wesley (1978)

[2] V. I. Arnold, *Mathematical Methods of Classical Mechanics*, Springer (1989)

[3] P. Aspinwall, B. Greene, and D. Morrison, *Calabi–Yau moduli space, mirror manifolds and spacetime topology change in string theory*, Preprint, arXiv:hep-th/9309097

[4] G. Birkhoff and J. von Neumann, *The logic of quantum mechanics*, Ann. Math. 37 (1936) 823–43

[5] J.-L. Brylinski, *Loop spaces, characteristic classes, and geometric quantization*, Progress in Mathematics 107, Birkhäuser (1993)

[6] A. Connes, *Noncommutative Geometry*, Academic Press (1994)

[7] C. Crnković and E. Witten, *Covariant description of canonical formalism in geometrical theories*, in S. W. Hawking and W. Israel, eds., *Three Hundred Years of Gravitation,* Cambridge University Press (1987) 676–84

[8] P. Deligne and D. Freed, *Classical field theory*, in P. Deligne, P. Etingof, D. S. Freed, L. C. Jeffrey, D. Kazhdan, J. W. Morgan, D. R. Morrison, and E. Witten, eds., *Quantum Fields and Strings: A Course for Mathematicians*, American Mathematical Society and the Institute for Advanced Studies (1999) 137–225

[9] F. Dowker, *Introduction to causal sets and their phenomenology*, General Relativity Gravitation 45 (2013) 1651–67

[10] V. Guillemin and S. Sternberg, *Symplectic Techniques in Physics*, Cambridge University Press (1984)

[11] V. Guillemin and S. Sternberg, *Geometric quantization and multiplicities of group representations*, Invent. Math. 67 (1982) 515–38

[12] W. Heisenberg, *Quantum-theoretical re-interpretation of kinematic and mechanical relations*, Z. Phys. 33 (1925) 879–93

[13] M. Henneaux and C. Teitelboim, *Quantization of Gauge Systems*, Princeton University Press (1994)

[14] T. Konopka, F. Markopoulou, and L. Smolin, *Quantum graphity*, Phys. Rev. D 77 (2006) 104029

[15] M. Kontsevich, *Homological algebra of mirror symmetry*, Proc. ICM Zürich (1994), arXiv:alg-geom/9411018

[16] M. Kontsevich and Y. Manin, *Gromov–Witten classes, quantum cohomology, and enumerative geometry*, Comm. Math. Phys. 164 (1994) 525–62

[17] J. Kijowski and W. M. Tulczyjew, *A Symplectic Framework for Field Theories*, Lecture Notes in Physics 107, Springer (1979)

[18] A. A. Kirillov, *Unitary representations of nilpotent Lie groups*, Russian Math. Surveys 17 (1962) 53–104

[19] B. Kostant, *Quantization and Unitary Representations*, Lecture Notes in Mathematics 170, Springer (1970) 87–207

[20] B. Kostant and S. Sternberg, *Symplectic reduction, BRS cohomology, and infinite dimensional Clifford algebras*, Ann. Phys. 176 (1987) 49–113

[21] J.-L. Koszul and Y. Zou, *Introduction to Symplectic Geometry*, Springer (2019)

[22] P. Libermann and C. M. Marle, *Symplectic Geometry and Analytical Mechanics*, Springer (1987)

[23] J. E. Marsden and A. Weinstein, *Reduction of symplectic manifolds with symmetry*, Rept. Math. Phys. 5 (1974) 121–30

[24] D. Oriti, *Approaches to Quantum Gravity*, Cambridge University Press (2009)

[25] J.-P. Ortega and T. S. Ratiu, *Momentum Maps and Hamiltonian Reduction*, Birkhäuser (2004)

[26] R. E. Peierls, *The commutation laws of relativistic field theory*, Proc. R. Soc. London, Ser. A 214 (1952) 143–57

[27] L. O'Raifeartaigh, *The Dawning of Gauge Theory*, Princeton University Press (1997)
[28] B. Riemann, *Uber die Hypothesen, welche der Geometrie zu grunde liegen*, W. K. Clifford (trans.), Nature VIII (1873) 14–17, 36, 37. (Original work published 1867)
[29] R. Sorkin, *Ten theses on black hole entropy*, Stud. Hist. Philos. Mod. Phys. 36 (2005) 291–301
[30] R. D. Sorkin, *Causal sets: Discrete gravity* (Notes for the Valdivia Summer School), in A. Gomberoff and D. Marolf, eds., *Lectures on Quantum Gravity*, Series of the Centro De Estudios Científicos, Plenum (2005)
[31] J.-M. Souriau, *Structure of Dynamical Systems: A Symplectic View of Physics*, Birkhäuser (1997)
[32] D. Sternheimer, *Deformation quantization: Twenty years after*, Preprint, arXiv: math/9809056
[33] A. Weinstein, *Symplectic geometry*, Bull. AMS 5 (1981) 1–13
[34] J. A. Wheeler, *Pregeometry: Motivations and prospects*, in A. R. Marlov, ed., *Quantum Theory and Gravitation*, Academic Press (1980) 1–11
[34] G. J. Zuckerman, *Action principles and global geometry*, in S. T. Yau, ed., *Mathematical Aspects of String Theory,* World Scientific (1987) 259–84

Mathieu Anel
Department of Philosophy, Carnegie Mellon University
mathieu.anel@gmail.com

Gabriel Catren
Laboratoire SPHERE – Sciences, Philosophie, Histoire
(UMR 7219, CNRS, Université de Paris)
gabrielcatren@gmail.com

PART I

Noncommutative and Supercommutative
Geometries

1
Noncommutative Geometry, the Spectral Standpoint

Alain Connes

In memory of John Roe, and in recognition of his pioneering achievements in coarse geometry and index theory.

Contents

1	Introduction, Two Key Examples	23
2	New Paradigm: Spectral Geometry	30
3	New Tool: Quantized Calculus	43
4	Cyclic Cohomology	50
5	Overall Picture, Interaction with Other Fields	53
6	Spectral Realization of Zeros of Zeta and the Scaling Site	64
References		71

1 Introduction, Two Key Examples

Noncommutative geometry has its roots both in quantum physics and in pure mathematics. One of its main themes is to explore and investigate "new spaces" that lie beyond the scope of standard tools. In physics the discovery of such new spaces goes back to Heisenberg's matrix mechanics, which revealed the noncommutative nature of the phase space of quantum systems. The appearance of a whole class of "new spaces" in mathematics and the need to treat them with new tools came later. It is worth explaining in simple terms when one can recognize that a given space such as the space of leaves of a foliation, the space of irreducible representations of a discrete group, the space of Penrose tilings, the space of lines[1] in a negatively curved compact Riemann surface, the space of rank one subgroups of $\mathbb{R}$, and so on is "noncommutative"

---

[1] Infinite geodesics.

in the sense that classical tools are inoperative. The simple characteristic of such spaces is visible even at the level of the underlying "sets": for such spaces their cardinality is the same as the continuum, but nevertheless, it is not possible to put them constructively in bijection with the continuum. More precisely, any explicitly constructed map from such a set to the real line fails to be injective! It should then be pretty obvious to the reader that there is a real problem when one wants to treat such spaces in the usual manner. One may of course dismiss them as "pathological," but this is ignoring their abundance since they appear typically when one defines a space as an inductive limit. Moreover, extremely simple toposes admit such spaces as their spaces of points, and discarding them is an act of blindness. The reason to call such spaces "noncommutative" is that if one accepts to use noncommuting coordinates to encode them, and one extends the traditional tools to this larger framework, everything falls in place. The basic principle is that one should take advantage of the presentation of the space as a quotient of an ordinary space by an equivalence relation, and instead of effecting the quotient in one stroke, one should associate to the equivalence relation its convolution algebra over the complex numbers, thus keeping track of the equivalence relation itself. The noncommutativity of the algebra betrays the identification of points[2] of the ordinary space, and when the quotient happens to exist as an ordinary space, the algebra is Morita-equivalent to the commutative algebra of complex valued functions on the quotient. One needs of course to refine the perception of the space one is handling according to the natural hierarchy of geometric points of view:

- measure theory;
- topology;
- differential geometry;
- metric geometry.

It is important to stress from the start that noncommutative geometry is not just a "generalization" of ordinary geometry, inasmuch as while it covers as a special case the known spaces, it handles them from a completely different perspective, which is inspired by quantum mechanics rather than by its classical limit. Moreover, there are many surprises, and the new spaces admit features that have no counterpart in the ordinary case, such as their time evolution, which appears already from their measure theory and turns them into thermodynamical objects.

---

[2] The groupoid identifying two points gives the noncommutative algebra of matrices $M_2(\mathbb{C})$.

The obvious question, then, from a conservative standpoint is why one needs to explore such new mathematical entities, rather than staying in the well-paved classical realm. The simple answer is that noncommutative geometry allows one to comprehend in a completely new way the following two spaces whose relevance can hardly be questioned:

> **Spacetime and NCG**
>
> The simplest reason why noncommutative geometry is relevant for the understanding of the geometry of spacetime is the key role of the non-abelian gauge theories in the standard model of particles and forces. The gauge theories enhance the symmetry group of gravity, that is, the group of diffeomorphisms Diff(M) of spacetime, to a larger group which is a semidirect product with the group of gauge transformations of the second kind. Searching for a geometric interpretation of the larger group as a group of diffeomorphisms of a higher-dimensional manifold is the essence of the Kaluza–Klein idea. Noncommutative geometry gives another track. Indeed, the group of diffeomorphisms Diff(M) is the group of automorphisms of the star algebra $\mathcal{A}$ of smooth functions on M, and if one replaces $\mathcal{A}$ by the (Morita-equivalent) noncommutative algebra $M_n(\mathcal{A})$ of matrices over $\mathcal{A}$, one enhances the group of automorphisms of the algebra in exactly the way required by the non-abelian gauge theory with gauge group SU(n). The Riemannian geometric paradigm is extended to the noncommutative world in an operator-theoretic and spectral manner. A geometric space is encoded by its algebra of coordinates $\mathcal{A}$ and its "line element," which specifies the metric. The new geometric paradigm of spectral triples (see Section 2) encodes the discrete and the continuum on the same stage, which is Hilbert space. The Yukawa coupling matrix D_F of the standard model provides the inverse line element for the finite geometry $(\mathcal{A}_F, \mathcal{H}_F, D_F)$, which displays the fine structure of spacetime detected by the particles and forces discovered so far. For a long time the structure of the finite geometry was introduced "by hand" following the trip inward bound [227] in our understanding of matter and forces as foreseen by Newton –
>
> *The Attractions of Gravity, Magnetism and Electricity, reach to very sensible distances, and so have been observed by vulgar Eyes, but there may be others which reach to so small distances as hitherto escape Observation.*
>
> – and adapting by a bottom-up process the finite geometry to the particles and forces, with the perfect fitting of the Higgs phenomenon

and the seesaw mechanism with the geometric interpretation. There was, however, no sign of an end to this quest nor any sensible justification for the presence of the noncommutative finite structure. This state of affairs changed recently [30, 31] with the simultaneous quantization of the fundamental class in K-homology and in K-theory. The K-homology fundamental class is represented by the Dirac operator. Representing the K-theory fundamental class, by requiring the use of the Feynman slash of the coordinates, explains the slight amount of noncommutativity of the finite algebra $\mathcal{A}_F$ from Clifford algebras. From a purely geometric problem emerged the very same finite algebra $\mathcal{A}_F = M_2(\mathbb{H}) \oplus M_4(\mathbb{C})$ that was the outcome of the bottom-up approach.

One big plus of the noncommutative presentation is the economy in encoding. This economy is familiar in written language, where the order of letters is so important. Apart from the finite algebra $\mathcal{A}_F$, the algebra of coordinates is generated by adjoining a "punctuation symbol" Y of K-theoretic nature, which is a unitary with $Y^4 = 1$. It is the noncommutativity of the obtained system that gets one outside the finite-dimensional algebraic framework and generates the continuum. It allows one to write a higher analogue of the commutation relations that quantizes the volume and encodes all four-dimensional spin geometries.

Zeta and NCG

Let us explain in a simple manner why the problem of locating the zeros of the Riemann zeta function is intimately related to the adele class space. In fact, what matters for the zeros is not the Riemann zeta function itself but rather the ideal it generates among holomorphic functions of a suitable class. A key role as a generator of this ideal is played by the operation on functions $f(u)$ of a real positive variable u defined by

$$f \mapsto E(f), \quad E(f)(v) := \sum_{\mathbb{N}^\times} f(nv). \tag{1.1}$$

Indeed this operation is, in the variable $\log v$, a sum of translations $\log v \mapsto \log v + \log n$ by $\log n$, and thus after a suitable Fourier transform it becomes a product by the Fourier transform of the sum of the Dirac masses $\delta_{\log n}$, that is, by the Riemann zeta function $\sum e^{-is \log n} = \zeta(is)$. It is thus by no means mysterious that one obtains a spectral realization by considering the cokernel of the map E. But what is interesting is the geometric meaning of this fact and the link with the explicit formulas.

Noncommutative Geometry, the Spectral Standpoint

Thanks to the use of adeles in the thesis of Tate, the above summation over $\mathbb{N}^\times$ relates to a summation over the group $\mathbb{Q}^*$ of nonzero rational numbers. The natural adelic space on which the group $\mathbb{Q}^*$ is acting is simply the space $\mathbb{A}_\mathbb{Q}$ of adeles over $\mathbb{Q}$, and in order to focus on the Riemann zeta function, one needs to restrict oneself to functions invariant under the maximal compact subgroup $\hat{\mathbb{Z}}^*$ of the group of idèle classes that acts canonically on the noncommutative space of adele classes: $\mathbb{Q}^*\backslash\mathbb{A}_\mathbb{Q}$. In fact, let Y be the quotient $\mathbb{A}_f/\hat{\mathbb{Z}}^*$ of the locally compact space of finite adeles by $\hat{\mathbb{Z}}^*$ and $y \in Y$ the element whose components are all 1. The closure F of the orbit $\mathbb{N}^\times y$ is compact, and one has for $q \in \mathbb{Q}^*$ the equivalence $qy \in F \iff q \in \pm\mathbb{N}^\times$. It is this property that allows one to replace the sum over $\mathbb{N}^\times$ involved in (1.1) by the summation over the associated group $\mathbb{Q}^*$, simply by considering the function $1_F \otimes f$ on the product $Y \times \mathbb{R}$. The space of adele classes $X = \mathbb{Q}^*\backslash\mathbb{A}_\mathbb{Q}/\hat{\mathbb{Z}}^* = \mathbb{Q}^*\backslash(Y \times \mathbb{R})$ is noncommutative because the action of $\mathbb{Q}^*$ on $\mathbb{A}_\mathbb{Q}$ is ergodic for the Haar measure. This geometrization gives a trace formula interpretation of the explicit formulas of Riemann–Weil (see Section 6).

The same space X appears naturally from a totally different perspective, which is that of Grothendieck toposes. Indeed, the above action of $\mathbb{N}^\times$ by multiplication on the half line $[0, \infty)$ naturally gives rise to a Grothendieck topos, and it turns out that X is the space of points of this topos as shown in our recent work with C. Consani [85, 86]. Moreover, the trace formula interpretation of the explicit formulas of Riemann–Weil allows one to rewrite the Riemann zeta function in the Hasse–Weil form, where the role of the Frobenius is played by its analogue in characteristic 1. The space X is the space of points of the arithmetic site defined over the semifield $\mathbb{R}_+^{\max}$, and the Galois group of $\mathbb{R}_+^{\max}$ over the boolean semifield $\mathbb{B}$ is the multiplicative group $\mathbb{R}_+^*$, which acts by the automorphisms $x \mapsto x^\lambda$, and its action on X replaces the action of the Frobenius on the points over $\bar{\mathbb{F}}_q$ in finite characteristics. This topos-theoretic interpretation yields a natural structure sheaf of tropical nature on the above scaling site, which becomes a curve in a tropical sense, and the two sides – (topos theoretic, tropical) on one hand and (adelic, noncommutative) on the other hand – are in the same relative position as the two sides of the class field theory isomorphism.

The theory of Grothendieck toposes is another crucial extension of the notion of space that plays a major role in algebraic geometry and is liable to treat quotient spaces in a successful manner. The main difference of point of view between noncommutative geometry and the theory of toposes is that noncommutative geometry makes fundamental use of the complex numbers as coefficients and thus has a very close relation with the formalism of quantum mechanics. It takes complex numbers as the preferred coefficient field and Hilbert space as the main stage. The birth of noncommutative geometry can be traced back to a night in June 1925 when, around three in the morning, W. Heisenberg, while working alone on the island of Helgoland in the North Sea, discovered matrix mechanics and the noncommutativity of the phase space of microscopic mechanical systems. After the formulation of Heisenberg's discovery as matrix mechanics, von Neumann reformulated quantum mechanics using Hilbert space operators and went much further with Murray in identifying "subsystems" of a quantum system as "factorizations" of the underlying Hilbert space $\mathcal{H}$. They discovered unexpected factorizations that did not correspond to tensor product decompositions $\mathcal{H} = \mathcal{H}_1 \otimes \mathcal{H}_2$ and developed the theory of factors, which they classified into three types. In my thesis I showed that factors $\mathcal{M}$ admit a canonical time evolution, that is, a canonical homomorphism

$$\mathbb{R} \xrightarrow{\delta} \text{Out}(\mathcal{M}) = \text{Aut}(\mathcal{M})/\text{Int}(\mathcal{M}),$$

by showing that the class of the modular automorphism σ_t^φ in $\text{Out}(\mathcal{M})$ does not depend on the choice of the faithful normal state φ. This of course relied crucially on the Tomita–Takesaki theory [261], which constructs the map $\varphi \mapsto \sigma_t^\varphi$. The above uniqueness of the class of the modular automorphism [40] drastically changed the status of the two invariants, which I had previously introduced in [38, 39] by making them computable. The kernel of δ, $T(M) = \text{Ker}\delta$, forms a subgroup of $\mathbb{R}$, the *periods* of M, and many nontrivial, nonclosed subgroups appear in this way. The fundamental invariant of factors is the *modular spectrum* $S(M)$ of [38]. Its intersection $S(M) \cap \mathbb{R}_+^*$ is a closed subgroup of $\mathbb{R}_+^*$ [43], and this gave the subdivision of type III into type III$_\lambda$ $\iff$ $S(M) \cap \mathbb{R}_+^* = \lambda^{\mathbb{Z}}$. I showed moreover that the classification of factors of type III$_\lambda$, $\lambda \in [0, 1)$ is reduced to that of type II and automorphisms

$$M = N \rtimes_\theta \mathbb{Z}, \ N \ \textbf{type} \ \text{II}_\infty, \ \theta \in \text{Aut}(N).$$

In the case III$_\lambda$, $\lambda \in (0, 1)$, N is a factor and the automorphism $\theta \in \text{Aut}(N)$ is of module λ, that is, it scales the trace by the factor λ. In the III$_0$ case, N has a nontrivial center, and, using the restriction of θ to the center, this gave a very rich invariant, a flow, and I used it in 1972 [41] to show the existence of hyperfinite non-ITPFI factor. Only the case III$_1$ remained open in my thesis [42] and was solved later by Takesaki [260] using crossed product by $\mathbb{R}_+^*$.

At that point one could fear that the theory of factors would remain foreign from mainstream mathematics. This is not the case thanks to the following construction [45] of the von Neumann algebra $W(V, F)$ canonically associated to a foliated manifold V with foliation F. It is the von Neumann algebra of random operators, that is, of bounded families (T_ℓ) parameterized by leaves ℓ and where each (T_ℓ) acts in the Hilbert space $L^2(\ell)$ of square integrable half-densities on the leaf. Modulo equality almost everywhere, one gets a von Neumann algebra for the algebraic operations given by pointwise addition and multiplication:

$$W(V, F) := \{(T_\ell) \mid \text{acting on } L^2(\ell)\}.$$

The geometric examples lead naturally to the most exotic factors, the prototype being the Anosov foliation of the sphere bundle of Riemann surfaces (see Section 6.2), whose von Neumann algebra is the hyperfinite factor of type III$_1$.

The above construction of the von Neumann algebra of a foliation only captures the "measure theory" of the space of leaves. But such spaces inherit a much richer structure from the topological and differential geometric structure of the ambient manifold V. This fact served as a fundamental motivation for the development of noncommutative geometry, and Figure 1.1 gives an overall picture of how the refined properties of leaf spaces and more general noncommutative spaces are treated in noncommutative geometry. One of the

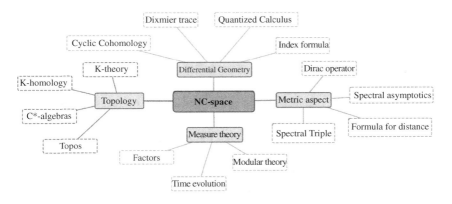

Figure 1.1 Aspects of a noncommutative space.

striking new features is the richness of the measure theory and how it interacts with a priori foreign features such as the secondary characteristic classes of foliations. Cyclic cohomology plays a central role in noncommutative geometry as the replacement of de Rham cohomology [51, 52]. One of its early striking applications obtained in [50] (following previous work of S. Hurder) using as a key tool the time derivative of the cyclic cocycle, which is the transverse fundamental class, is that the flow of weights[3] of the von Neumann algebra of a foliation with nonvanishing Godbillon–Vey class admits an invariant probability measure. This implies immediately that the von Neumann algebra is of type III and exhibits the deep interplay between characteristic classes and ergodic theory of foliations.

2 New Paradigm: Spectral Geometry

In the name "noncommutative geometry" the adjective *noncommutative* has an obvious algebraic meaning if one recalls that *commutative* is one of the leitmotivs of algebraic geometry. As mentioned earlier, reducing the new theory to the extension of previously known notions beyond the commutative case misses the point. In fact, we shall now justify the name "geometry" given to the field. The new paradigm for a geometric space is spectral. A spectral geometry is given by a triple $(\mathcal{A}, \mathcal{H}, D)$, where $\mathcal{A}$ is an algebra of operators in the Hilbert space $\mathcal{H}$ and where D is a self-adjoint operator, generally unbounded, acting in the same Hilbert space $\mathcal{H}$. Intuitively, the algebra $\mathcal{A}$ represents the functions on the geometric space, and the operator D plays the role of the inverse line element. It specifies how to measure distances by the formula

$$d(a,b) = \operatorname{Sup} |f(a) - f(b)|, f \text{ such that } \|[D, f]\| \leq 1. \qquad (2.1)$$

Given a spin compact Riemannian manifold M, the associated spectral triple $(\mathcal{A}, \mathcal{H}, D)$ is given by the algebra $\mathcal{A}$ of functions on M acting in the Hilbert space $\mathcal{H}$ of L^2 spinors and the Dirac operator D. The formula (2.1) gives back the geodesic distance in the usual Riemannian sense but measures distances in a Kantorovich dual manner. The word *spectral* in *spectral triple* (or *spectral geometry*) has two origins: the straightforward one is the well-known Gelfand duality between a (locally compact) space and its algebra of complex valued continuous functions (vanishing at infinity); the deeper one is the full reconstruction of the geometry from the spectrum of the Dirac operator

---

[3] The fundamental invariant of [42, 44].

Noncommutative Geometry, the Spectral Standpoint 31

(the D in spectral triples) and the relative position in Hilbert space (the $\mathcal{H}$ of the spectral triple) of the two von Neumann algebras given on one hand by the weak closure (double commutant) of the algebra $\mathcal{A}$ (the $\mathcal{A}$ of the spectral triple) and on the other hand by the functions $f(D)$ of the operator D. We refer to [75] for this point.[4] The great advantage of the new paradigm is that (2.1) continues to make sense for spaces that are no longer arcwise connected and works equally well for discrete or fractal spaces. Moreover, the paradigm does not make use of the commutativity of $\mathcal{A}$. When $\mathcal{A}$ is noncommutative, the evaluation $f(a)$ of $f \in \mathcal{A}$ at a point a no longer makes sense, but the formula (2.1) still makes sense and measures the distance between states[5] on $\mathcal{A}$, that is, positive linear forms $\phi \colon \mathcal{A} \to \mathbb{C}$, normalized by $\phi(1) = 1$. The formula measures the distance between states as follows:

$$d(\phi, \psi) = \operatorname{Sup} |\phi(f) - \psi(f)|, f \text{ such that } \|[D, f]\| \leqslant 1. \qquad (2.2)$$

The new spectral paradigm of geometry extends the Riemannian paradigm, and we explained in [94] the extent to which it allows one to go further in the query of Riemann on the geometric paradigm of space in the infinitely small. Here *space* is the space in which we live and is, after all, the only one that truly deserves the name of "geometry", independently of all the later developments that somehow usurp this name. The new paradigm ensures that the line element does encapsulate all the binding forces known so far and is flexible enough to follow the inward bound quest of the geometric structure of space for microscopic distances [227]. The words of Riemann are so carefully chosen and his vision is so far-seeing that the reader will, we hope, excuse our repeated use of this quotation:

> *Nun scheinen aber die empirischen Begriffe, in welchen die räumlichen Massbestimmungen gegründet sind, der Begriff des festen Körpers und des Lichtstrahls, im Unendlichkleinen ihre Gültigkeit zu verlieren; es ist also sehr wohl denkbar, dass die Massverhältnisse des Raumes im Unendlichkleinen den Voraussetzungen der Geometrie nicht gemäss sind, und dies würde man in der That annehmen müssen, sobald sich dadurch die Erscheinungen auf einfachere Weise erklären liessen. Es muss also entweder das dem Raume zu Grunde liegende Wirkliche eine discrete*

---

[4] One recovers the algebra of smooth functions from the von Neumann algebra of measurable ones as those measurable functions that preserve the intersection of domains of powers of D. One then gets the topological space as the Gelfand spectrum of the norm closure of the algebra of smooth functions, and one also gets its smooth structure, its Riemannian metric, etc.
[5] The pure states are the extreme points of the space of states on $\mathcal{A}$. In quantum mechanics they play the role of the points of the classical phase space. They are endowed with the natural equivalence relation defined by the unitary equivalence of the associated irreducible representations. The space of equivalence classes of pure states is typically a noncommutative space, except in the easy type I situation.

Mannigfaltigkeit bilden, oder der Grund der Massverhältnisse ausserhalb, in darauf wirkenden bindenen Kräften, gesucht werden.[6]

We also explained in [94] the deep roots of the notion of spectral geometry in pure mathematics from the conceptual understanding of the notion of (smooth, compact, oriented, and spin) manifold. The fundamental property of a "manifold" is not Poincaré duality in ordinary homology but is Poincaré duality in the finer theory called KO-homology. The key result behind this, due to D. Sullivan (see [208, Epilogue]), is that a PL-bundle is the same thing[7] as a spherical fibration together with a KO-orientation. The fundamental class in KO-homology contains all the information about the Pontrjagin classes of the manifold, and these are not at all determined by its homotopy type: in the simply connected case, only the signature class is fixed by the homotopy type.

The second key step toward the notion of spectral geometry is the link with the Hilbert space and operator formalism, which came as a byproduct of the work of Atiyah and Singer on the index theorem. They understood that operators in Hilbert space provide the right realization for KO-homology cycles [2, 252]. Their original idea was developed by Brown–Douglas–Fillmore [14], Voiculescu [270], and Mishchenko [210] and acquired its definitive form in the work of Kasparov at the end of the 1970s. The great new tool is bivariant Kasparov theory [172, 173], and as far as K-homology cycles are concerned, the right notion is already in Atiyah's paper [2]: a K-homology cycle on a compact space X is given by a representation of the algebra $C(X)$ (of continuous functions on X) in a Hilbert space $\mathcal{H}$, together with a Fredholm operator F acting in the same Hilbert space fulfilling some simple compatibility condition (of commutation modulo compact operators) with the action of $C(X)$. One striking feature of this representation of K-homology cycles is that the definition does not make any use of the commutativity of the algebra $C(X)$.

At the beginning of the 1980s, motivated by numerous examples of noncommutative spaces arising naturally in geometry from foliations and in particular the noncommutative torus $\mathbb{T}^2_\theta$ described in Section 2.1, I realized that specifying an unbounded representative of the Fredholm operator gave the

---

[6] Now it seems that the empirical notions on which the metric determinations of space are founded, the notion of a solid body and of a ray of light, cease to be valid in the infinitely small. It is therefore quite conceivable that the metric relations of space in the infinitely small do not conform to the hypotheses of geometry; and we ought in fact to assume this, if we can thereby obtain a simpler explanation of phenomena. Either therefore the reality that underlies space must be discrete or we must seek the foundation of its metric relations outside it, in binding forces that act upon it.

[7] Modulo the usual "small-print" qualifications at the prime 2 [250].

right framework for spectral geometry. The corresponding K-homology cycle[8] only retains the stable information and is insensitive to deformations, while the unbounded representative encodes the metric aspect. These are the deep mathematical reasons at the root of the notion of spectral triple. Since the year 2000 [67], important progress has been made in the reconstruction theorem [78], which gives an abstract characterization of the spectral triples associated to ordinary spin geometries. Moreover, a whole subject has developed concerning noncommutative manifolds, and for lack of space, we shall not cover it here but refer to the papers [68, 69, 70, 71, 98, 110, 111, 188]. We apologize for the brevity of our account of the topics covered in the present survey; the reader will find a more complete treatment and bibliography in the collection of surveys [34, 89, 103, 113, 116, 123, 133, 194, 198, 271, 275, 277].

2.1 Geometry and the Modular Theory

Foliated manifolds provide a very rich source of examples of noncommutative spaces, and one is quite far from a full understanding of these spaces. In this section we discuss in an example the interaction of the new geometric paradigm with the modular theory (type III). The example is familiar from [67], it is the noncommutative torus $\mathbb{T}^2_\theta$ whose K-theory was computed in the breakthrough paper of Pimsner and Voiculescu [231]. We first briefly recall the geometric picture underlying the noncommutative torus $\mathbb{T}^2_\theta$ and then describe the new development since year 2000, which displays a highly nontrivial interaction between the new paradigm of geometry and the modular type III theory. The underlying geometric picture is the Kronecker foliation $dy = \theta dx$ of the unit torus $V = \mathbb{R}^2/\mathbb{Z}^2$ as shown in Figure 1.2. If one considers the restriction of the foliation to a neighborhood of a transversal as shown in Figure 1.2, the structure is that of the product of a circle S^1 by an interval, and the leaf space is S^1. But for the whole foliation of the torus and irrational θ, the leaf space is a noncommutative space. It corresponds to the identification of those points on S^1 that are on the same orbit of the irrational rotation of angle θ. At the algebraic level the foliation algebra is the same[9] as the crossed product $C(\mathbb{T}^2_\theta) := C(S^1) \rtimes_\theta \mathbb{Z}$ of the transversal by the irrational rotation of angle θ.

---

[8] In [4], S. Baaj and P. Julg have extended the unbounded construction to the bivariant case. The most efficient axioms have been found by M. Hilsum in [165], which allows us to handle the case of symmetric non-self-adjoint operators.
[9] Up to Morita-equivalence, the algebraic notion has been adapted to C^*-algebras by M. Rieffel in [241].

The differential geometry of the noncommutative torus $\mathbb{T}_\theta^2$ was first defined and investigated in *C. R. Acad. Sci. Paris* (*cf.* [47]), "C^* algèbres et géométrie différentielle." While at the C^*-algebra level the presentation of the algebra $C(\mathbb{T}_\theta^2)$ of continuous functions on $\mathbb{T}_\theta^2$ is by means of two unitaries U, V such that

$$VU = e^{2\pi i \theta} UV,$$

the smooth functions on $\mathbb{T}_\theta^2$ are given by the dense subalgebra $C^\infty(\mathbb{T}_\theta^2) \subset C(\mathbb{T}_\theta^2)$ of all series

$$x = \sum_{m,n \in \mathbb{Z}} a_{m,n} U^m V^n,$$

such that the sequence of complex coefficients $(a_{m,n})$ is rapidly decaying in the sense that

$$\sup_{m,n \in \mathbb{Z}} |a_{m,n}|(1 + |n| + |m|)^k < \infty$$

for any nonnegative integer k. Thus one has a simple explicit description of the generic element of $C^\infty(\mathbb{T}_\theta^2)$ (while there is no such explicit description for $C(\mathbb{T}_\theta^2)$). At the geometric level one has the smooth groupoid $G = S^1 \rtimes_\theta \mathbb{Z}$, which is the reduction of the holonomy groupoid of the foliation to the transversal. The elements of G are described equivalently as pairs of points of S^1 that are on the same orbit of the irrational rotation or as pairs $(u, n) \in S^1 \times \mathbb{Z}$. The algebra is the convolution algebra of G. Any other transversal L gives canonically a module $C^\infty(\mathbb{T}_\theta^2, L)$ over the convolution algebra of G, and the main discovery of [47] can be summarized as follows:

1. The Schwartz space $\mathcal{S}(\mathbb{R})$ is a finite projective module over $C^\infty(\mathbb{T}_\theta^2)$ for the action of the generators U, V on Schwartz functions $\xi(s)$ given by translation of the variable $s \mapsto s - \theta$ and multiplication by $e^{2\pi i s}$.

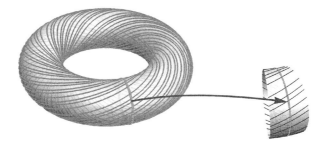

Figure 1.2 Kronecker foliation $dy = \theta dx$.

2. The Murray–von Neumann dimension of $\mathcal{S}(\mathbb{R})$ is θ.
3. The module $\mathcal{S}(\mathbb{R})$ is naturally endowed with a connection of constant curvature, and the product of the constant curvature by the Murray–von Neumann dimension is an integer.

It is this integrality of the product of the constant curvature by the (irrational) Murray–von Neumann dimension that was the starting point of noncommutative differential geometry (cf. [51, 52]). It holds irrespective of the transversal, and the description of the general finite projective modules thus obtained was done in [47].[10] The differential geometry as well as pseudo-differential operator calculus of the noncommutative torus $\mathbb{T}^2_\theta$ were first developed in [47]. As in the case of an ordinary torus, the conformal structure is specified by a complex modulus $\tau \in \mathbb{C}$ with $\Im(\tau) > 0$. The flat spectral geometry $(\mathcal{A}, \mathcal{H}, D_0)$ is constructed using the two derivations δ_1, δ_2 which generate the action by rotation of the generators U, V of the algebra $\mathcal{A} = C^\infty(\mathbb{T}^2_\theta)$. To obtain a curved geometry $(\mathcal{A}, \mathcal{H}, D)$ from the flat one $(\mathcal{A}, \mathcal{H}, D_0)$, one introduces (cf. [36, 77]) a noncommutative Weyl conformal factor (or dilaton), which changes the metric by modifying the noncommutative volume form while keeping the same conformal structure. Both notions of volume form and of conformal structure are well understood in the general case (cf. [59, Section VI]). The new and crucial ingredient, which has no classical analog, is the modular operator Δ of the nontracial weight $\varphi(a) = \varphi_0(ae^{-h})$ associated to the dilaton h.

In noncommutative geometry the local geometric invariants such as the Riemannian curvature are extracted from the spectral triple using the functionals defined by the coefficients of heat kernel expansion

$$\mathrm{Tr}(ae^{-tD^2}) \sim_{t \searrow 0} \sum_{n \geq 0} a_n(a, D^2) t^{\frac{-d+n}{2}}, \quad a \in \mathcal{A},$$

where d is the dimension of the geometry. Equivalently, one may consider special values of the corresponding zeta functions. Thus the local curvature of the geometry is detected by the high-frequency behavior of the spectrum of D coupled with the action of the algebra $\mathcal{A}$ in $\mathcal{H}$. In the case of $\mathbb{T}^2_\theta$ the computation of the value at $s = 0$ of the zeta function $\mathrm{Tr}(a|D|^{-2s})$ for the two-dimensional curved geometry associated to the dilaton h, or equivalently of the coefficient $a_2(a, D^2)$ of the heat expansion, was started in the late 1980s (cf. [36]), and the proof of the analogue of the Gauss–Bonnet formula was published in [77]. It was subsequently extended in [118] to the case of arbitrary values of the complex modulus τ (one had $\tau = i$ in [77]). In these papers, only

---

[10] And was often misattributed as in [201].

the total integral of the curvature was needed, and this allowed one to make simplifications under the trace that are no longer possible when one wants to fully compute the local expression for the functional $a \in \mathcal{A} \mapsto a_2(a, D^2)$.

The complete calculation of $a_2(a, D^2)$ was actually performed in our joint work with H. Moscovici in [79] and independently by F. Fathizadeh and M. Khalkhali in [119]. The spectral geometry of noncommutative tori is a key testing ground for new ideas, and basic contributions have been made in [77, 79, 120, 121, 123, 125, 126, 167, 176, 192, 196, 234], with beautiful surveys by M. Lesch and H. Moscovici [194] and by F. Fathizadeh and M. Khalkhali [123], to which we refer for a more complete account and bibliography. In this topic the hard direct calculations alternate with their conceptual understanding, and the main result of [79] is the conceptual explanation for the complicated formulas. The explanation is obtained by expressing in terms of a closed formula the Ray–Singer log-determinant of D^2. The gradient of the log-determinant functional yields in turn a local curvature formula, which arises as a sum of two terms, each involving a function in the modular operator, of one and two variables, respectively. Computing the gradient in two different ways leads to the proof of a deep internal consistency relation between these two distinct constituents. This relation was checked successfully and gave confidence in the pertinence of the whole framework. Moreover, it elucidates the meaning of the intricate two operator–variable function, which we now briefly describe. As in the case of the standard torus viewed as a complex curve, the total Laplacian associated to such a spectral triple splits into two components, one $\triangle_\varphi$ on functions and the other $\triangle_\varphi^{(0,1)}$ on $(0,1)$-forms, the two operators being isospectral outside zero. The corresponding curvature formulas involve second-order (outer) derivatives of the Weyl factor. For the Laplacian $\triangle_\varphi$ the result is of the form

$$a_2(a, \triangle_\varphi) = -\frac{\pi}{2\tau_2} \varphi_0 \left(a \left(K_0(\nabla)(\triangle(h)) + \frac{1}{2} H_0(\nabla_1, \nabla_2)(\square_\Re(h)) \right) \right), \quad (2.3)$$

where $\nabla = \log \triangle$ is the inner derivation implemented by $-h$,

$$\triangle(h) = \delta_1^2(h) + 2\Re(\tau)\delta_1\delta_2(h) + |\tau|^2 \delta_2^2(h);$$

$\square_\Re$ is the Dirichlet quadratic form

$$\square_\Re(\ell) := (\delta_1(\ell))^2 + \Re(\tau)\left(\delta_1(\ell)\delta_2(\ell) + \delta_2(\ell)\delta_1(\ell)\right) + |\tau|^2 (\delta_2(\ell))^2;$$

and ∇_i, $i = 1, 2$, signifies that ∇ is acting on the ith factor. The operators $K_0(\nabla)$ and $H_0(\nabla_1, \nabla_2)$ are new ingredients, whose occurrence has no classical analogue and is a vivid manifestation of the nonunimodular nature of the curved geometry of the noncommutative 2-torus with Weyl factor. The functions $K_0(u)$ and $H_0(u,v)$ by which the modular derivatives act seem at first very complicated and of course beg for a conceptual understanding. This was obtained in [79], where, denoting $\tilde{K}_0(s) = 4\frac{\sinh(s/2)}{s}K_0(s)$ and $\tilde{H}_0(s,t) = 4\frac{\sinh((s+t)/2)}{s+t}H_0(s,t)$, we give an abstract proof of the functional relation

$$-\frac{1}{2}\tilde{H}_0(s_1,s_2) = \frac{\tilde{K}_0(s_2) - \tilde{K}_0(s_1)}{s_1+s_2} + \frac{\tilde{K}_0(s_1+s_2) - \tilde{K}_0(s_2)}{s_1} - \frac{\tilde{K}_0(s_1+s_2) - \tilde{K}_0(s_1)}{s_2}, \tag{2.4}$$

which determines the whole structure in terms of the function $\tilde{K}_0$, which turns out to be (up to the factor $\frac{1}{8}$) the generating function of the Bernoulli numbers, that is, one has $\frac{1}{8}\tilde{K}_0(u) = \sum_1^\infty \frac{B_{2n}}{(2n)!} u^{2n-2}$.

Among the recent developments are the full understanding of the interplay of the geometry with Morita-equivalence (a purely noncommutative feature) in [192] and a new hard computation in [91]. This latter result gives the formulas for the a_4 term for the noncommutative torus and, as a consequence, allows one to compute this term for the noncommutative 4-tori that are products of two 2-tori. We managed in [91] to derive abstractly and check the analogs of the above functional relations, but the formulas remain very complicated and mysterious, thus begging for a better understanding. One remarkable fact is that the noncommutative 4-tori that are products of two 2-tori are no longer conformally flat so that the computation of the a_4 term for them involves a more complicated modular structure that is given by a two-parameter group of automorphisms. This provides a first hint in the following program:

> The long-range goal in exploring the spectral geometry of noncommutative tori is to arrive at the formulation of the full-fledged modular theory involving the full Jacobian matrix rather than its determinant, as suggested long ago by the transverse geometry of foliations in [50], where the reduction from type III to type II was done for the full Jacobian. The hypoelliptic theory was used in [64] to obtain the transverse geometry at the type II level. The first step of the general theory is to adapt the twist defined in [76] to a notion involving the full Jacobian.

2.2 Inner Fluctuations of the Metric

In the same way as inner automorphisms enhance the group of diffeomorphisms, they provide special deformations of the metric for a spectral geometry

($\mathcal{A}, \mathcal{H}, D$) that correspond to the gauge fields of non-abelian gauge theories. In our joint work with A. Chamseddine and W. van Suijlekom [29], we obtained a conceptual understanding of the role of the gauge bosons in physics as the inner fluctuations of the metric. I will describe this result here in a nontechnical manner.

Ignoring the important nuance coming from the real structure J, the inner fluctuations of the metric were first defined as the transformation

$$D \mapsto D + A, \quad A = \sum a_j[D, b_j], \quad a_j, b_j \in \mathcal{A}, \quad A = A^*,$$

which imitates the way classical gauge bosons appear as matrix-valued one-forms in the usual framework. The really important facts were that the spectral action applied to $D + A$ delivers the Einstein–Yang–Mills action, which combines gravity with matter in a natural manner, and that the gauge invariance becomes transparent at this level since an inner fluctuation coming from a gauge potential of the form $A = u[D, u^*]$ where u is a unitary element (i.e., $uu^* = u^*u = 1$) simply results in a unitary conjugation $D \mapsto uDu^*$, which does not change the spectral action. What we discovered in our joint work with A. Chamseddine and W. van Suijlekom [29] is that the inner fluctuations arise in fact from the action on metrics (i.e., the D) of a canonical *semigroup* Pert($\mathcal{A}$) that only depends on the algebra $\mathcal{A}$ and extends the unitary group. The semigroup is defined as the self-conjugate elements

$$\mathrm{Pert}(\mathcal{A}) := \{A = \sum a_j \otimes b_j^{\mathrm{op}} \in \mathcal{A} \otimes \mathcal{A}^{\mathrm{op}} \mid \sum a_j b_j = 1, \, \theta(A) = A\},$$

where θ is the antilinear automorphism of the algebra $\mathcal{A} \otimes \mathcal{A}^{\mathrm{op}}$ given by

$$\theta : \sum a_j \otimes b_j^{\mathrm{op}} \mapsto \sum b_j^* \otimes a_j^{*\mathrm{op}}.$$

The composition law in Pert($\mathcal{A}$) is the product in the algebra $\mathcal{A} \otimes \mathcal{A}^{\mathrm{op}}$. The action of this semigroup Pert($\mathcal{A}$) on the metrics is given, for $A = \sum a_j \otimes b_j^{\mathrm{op}}$, by

$$D \mapsto D' = {}^A D = \sum a_j D b_j.$$

Moreover, the transitivity of inner fluctuations, which is a key feature, since one wants an inner fluctuation of an inner fluctuation to be itself an inner fluctuation, now results from the semigroup structure according to the equality ${}^{A'}({}^A D) = {}^{(A'A)} D$. We refer to [29] for the full treatment. Inner fluctuations have been adapted in the twisted case in [186, 187].

2.3 The Spectral Action and Standard Model Coupled to Gravity

The new spectral paradigm of geometry, because of its flexibility, provides the needed tool to refine our understanding of the structure of physical space in the small and to "seek the foundation of its metric relations outside it, in binding forces which act upon it".

The main idea, described in detail in [94], is that the line element now embodies not only the force of gravity but all the known forces, electroweak and strong, appearing from the spectral action and the inner fluctuations of the metric. This provides a completely new perspective on the geometric interpretation of the detailed structure of the standard model and of the Brout–Englert–Higgs mechanism. One gets the following simple mental picture for the appearance of the scalar field: imagine that the space under consideration is two-sided like a sheet S of paper in two dimensions. Then, when differentiating a function on such a space, one may restrict the function to either side $S_\pm$ of the sheet and thus obtain two spin-one fields. But one may also take the finite difference $f(s_+) - f(s_-)$ of the function at the related points of the two sides. The corresponding field is clearly insensitive to local rotations and is a scalar spin-zero field. This, in a nutshell, is how the Brout–Englert–Higgs field appears geometrically once one accepts that there is a "fine structure" that is revealed by the detailed structure of the standard model of matter and forces. This allows one to uncover the geometric meaning of the Lagrangian of gravity coupled to the standard model. This extremely complicated Lagrangian is obtained from the spectral action developed in our joint work with A. Chamseddine [18], which is the only natural additive spectral invariant of a noncommutative geometry. To comply with Riemann's requirement that the inverse line element D embodies the forces of nature, it is evidently important that we do not separate artificially the gravitational part from the gauge part and that D encapsulates both forces in a unified manner. In the traditional geometrization of physics, the gravitational part specifies the metric while the gauge part corresponds to a connection on a principal bundle. In the NCG framework, D encapsulates both forces in a unified manner, and the gauge bosons appear as inner fluctuations of the metric but form an inseparable part of the latter. The noncommutative geometry dictated by physics is the product of the ordinary four-dimensional continuum by a finite noncommutative geometry $(\mathcal{A}_F, \mathcal{H}_F, D_F)$ that appears naturally from the classification of finite geometries of KO-dimension equal to 6 modulo 8 (cf. [20, 22]). The finite-dimensional algebra $\mathcal{A}_F$ that appeared is of the form

$$\mathcal{A}_F = C_+ \oplus C_-, \quad C_+ = M_2(\mathbb{H}), \quad C_- = M_4(\mathbb{C}).$$

The agreement of the mathematical formalism of spectral geometry and all its subtleties, such as the periodicity of period 8 of the KO-theory, might still be accidental, but I personally became convinced of the pertinence of this approach when recovering [20] the seesaw mechanism (which was dictated by the pure math calculation of the model) while I was unaware of its key physics role to provide very small nonzero masses to the neutrinos, and of how it is "put by hand" in the standard model. The low Higgs mass then came in 2012 as a possible flaw of the model, but in [27] we proved the compatibility of the model with the measured value of the Higgs mass, due to the role in the renormalization of the scalar field that was already present in [24] but had been ignored thinking that it would not affect the running of the self-coupling of the Higgs.

In all the previous developments, we had followed the "bottom-up" approach, that is, we uncovered the details of the finite noncommutative geometry $(\mathcal{A}_F, \mathcal{H}_F, D_F)$ from the experimental information contained in the standard model coupled to gravitation. In 2014, in collaboration with A. Chamseddine and S. Mukhanov [30, 31], we were investigating the purely geometric problem of encoding 4-manifolds in the most economical manner in the spectral formalism. Our problem had no a priori link with the standard model of particle and forces, and the idea was to treat the coordinates in the same way as the momenta are assembled together in a single operator using the gamma matrices. The great surprise was that this investigation gave the conceptual explanation of the finite noncommutative geometry from Clifford algebras! This is described in detail in [94], to which we refer. What we obtained is a higher form of the Heisenberg commutation relations between p and q, whose irreducible Hilbert space representations correspond to four-dimensional spin geometries. The role of p is played by the Dirac operator and the role of q by the Feynman slash of coordinates using Clifford algebras. The proof that all spin geometries are obtained relies on deep results of immersion theory and ramified coverings of the sphere. The volume of the four-dimensional geometry is automatically quantized by the index theorem, and the spectral model, taking into account the inner automorphisms due to the noncommutative nature of the Clifford algebras, gives Einstein gravity coupled with the slight extension of the standard model, which is a Pati–Salam model. This model was shown in [28, 32] to yield unification of coupling constants. We refer to the survey [34] by A. Chamseddine and W. van Suijlekom for an excellent account of the whole story of the evolution of this theory from the early days to now.

The dictionary between the physics terminology and the fine structure of the geometry is of the following form:

Standard Model	Spectral Model
Higgs boson	inner metric$^{(0,1)}$
gauge bosons	inner metric$^{(1,0)}$
fermion masses u, ν	Dirac$^{(0,1)}$ in $\uparrow$
CKM matrix, masses down	Dirac$^{(0,1)}$ in $(\downarrow 3)$
lepton mixing, masses leptons e	Dirac$^{(0,1)}$ in $(\downarrow 1)$
Majorana mass matrix	Dirac$^{(0,1)}$ on $E_R \oplus J_F E_R$
gauge couplings	fixed at unification
Higgs scattering parameter	fixed at unification
tadpole constant	$-\mu_0^2 |\mathbf{H}|^2$

2.4 Dimension 4

In our work with A. Chamseddine and S. Mukhanov [30, 31], the dimension 4 plays a special role for the following reason. In order to encode a manifold M, one needs to construct a pair of maps ϕ, ψ from M to the sphere (of the same dimension d) in such a way that the sum of the pullbacks of the volume form of the (round) sphere vanishes nowhere on M. This problem is easy to solve in dimensions 2 and 3 because one first writes M as a ramified cover $\phi : M \to S^d$ of the sphere and one precomposes ϕ with a diffeomorphism f of M such that $\Sigma \cap f(\Sigma) = \emptyset$, where Σ is the subset where ϕ is ramified. This subset is of codimension 2 in M, and there is no difficulty to find f because $(d-2) + (d-2) < d$ for $d < 4$. It is worth mentioning that the 2 for the codimension of Σ is easy to understand from complex analysis: for an arbitrary smooth map $\phi : M \to S^d$, the Jacobian will vanish on a codimension 1 subset, but in one-dimensional complex analysis, the Jacobian is a sum of squares and its vanishing means the vanishing of the derivative which gives two conditions rather than one. Note also in this respect that quaternions do not help.[11] Thus dimension $d = 4$ is the critical dimension for the above existence problem of the pair ϕ, ψ. Such a pair does not always exist,[12] but as shown in [30, 31], it always exists for spin manifolds, which is the relevant case for us.

The higher form of the Heisenberg commutation relation mentioned above involves in dimension d the power d of the commutator $[D, Z]$ of the Dirac operator with the operator Z which is constructed (using the real structure J; see [94]) from the coordinates. We shall now explain briefly how this fits perfectly with the framework of D. Sullivan on Sobolev manifolds, that is, of manifolds of dimension d where the pseudo-group underlying the atlas preserves continuous functions with one derivative in L^d. He discovered the intriguing special role of dimension 4 in this respect. He showed in [259] that topological manifolds in dimensions > 5 admit bi-Lipschitz coordinates, and these are unique up to small perturbations; moreover, existence and uniqueness also hold for Sobolev structures: one derivative in L^d. A stronger result was known classically for dimensions 1, 2, and 3. There the topology controls the smooth structure up to small deformation. In dimension 4, he proved with S. Donaldson in [108] that for manifolds with coordinate atlases related by the pseudo-group preserving continuous functions with one derivative in L^4, it is possible to develop the $SU(2)$ gauge theory and the famous Donaldson invariants. Thus in dimension 4 the Sobolev manifolds behave like the smooth ones as opposed to Freedman's abundant topological manifolds.[13] The obvious question, then, is to what extent the higher Heisenberg equation of [30, 31] singles out the Sobolev manifolds as the relevant ones for the functional integral involving the spectral action.

2.5 Relation with Quantum Gravity

Our approach to the geometry of spacetime is not concerned with quantum gravity but it addresses a more basic related question which is to understand "why gravity coupled to the standard model," which we view as a preliminary. The starting point of our approach is an extension of Riemannian geometry beyond its classical domain, which provides the needed flexibility to answer the query of Riemann in his inaugural lecture. The modification of the

---

[11] As an exercise, one can compute the Jacobian of the power map $q \mapsto q^n$, $q \in \mathbb{H}$ and show that it vanishes on a codimension 1 subset. For instance, for $n = 3$, the Jacobian is given by $9\left(3a^2 - b^2 - c^2 - d^2\right)^2 \left(a^2 + b^2 + c^2 + d^2\right)^2$.
[12] It does not exist for $M = P_2(\mathbb{C})$.
[13] For which any modulus of continuity whatsoever is not known.

geometric paradigm comes from the confluence of the abstract understanding of the notion of manifold from its fundamental class in KO-homology with the advent of the formalism of quantum mechanics. This confluence comes from the realization of cycles in KO-homology from representations in Hilbert space. The final touch on the understanding of the geometric reason behind gravity coupled to the standard model came from the simultaneous quantization of the fundamental class in KO-homology and its dual in KO-theory, which gave rise to the higher Heisenberg relation. So far, we remain at the level of first quantization, and the problem of second quantization is open. A reason why this issue cannot be ignored is that the quantum corrections to the line element as explained in Section 3.2.3 (and shown in Figure 1.3) are ony the tip of the iceberg since the dressing also occurs for all the n-point functions for fermions, while Figure 1.3 only involves the two point function. One possible strategy to pass to this second quantized higher level of geometry is to try to give substance to the proposal we did in [94]:

> *The duality between KO-homology and KO-theory is the origin of the higher Heisenberg relation. As already mentioned in [54], algebraic K-theory, which is a vast refinement of topological K-theory, is begging for the development of a dual theory and one should expect profound relations between this dual theory and the theory of interacting quanta of geometry. As a concrete point of departure, note that the deepest results on the topology of diffeomorphism groups of manifolds are given by the Waldhausen algebraic K-theory of spaces and we refer to [112] for a unifying picture of algebraic K-theory.*

> The role of the second quantization of fermions in "Entropy and the Spectral Action" [33], which interprets the spectral action as an entropy, is a first step toward building from quantum field theory a *second quantized* version of spectral geometry.

Note also that an analogous result to [33] has been obtained in [15] for the bosonic case.

3 New Tool: Quantized Calculus

A key tool of noncommutative geometry is the quantized calculus. Its origin is the notion of real variable provided by the formalism of quantum mechanics in terms of self-adjoint operators in Hilbert space. One very striking basic fact is that if $\phi \colon \mathbb{R} \to \mathbb{R}$ is an arbitrary Borel function and H is a self-adjoint operator in Hilbert space, then $\phi(H)$ makes sense even though ϕ can have a completely different definition in various parts of $\mathbb{R}$. It is this fact that reveals in which sense the formalism of quantum mechanics is compatible with the naive notion of real variable. It is superior to the classical notion of random variable because it models discrete variables with finite multiplicity as coexisting with continuous variables, while a discrete random variable with finite multiplicity only exists on a countable sample space. The price one pays for this coexistence is noncommutativity since the commutant of a self-adjoint operator H with countable spectrum of multiplicity 1 is the algebra of Borel functions $\phi(H)$, and they all have countable range and atomic spectral measure. This implies in particular that the infinitesimal variables of the theory, which are modeled by compact operators, cannot commute with the continuous variables. We refer to [94] for a detailed discussion of this point. We now display the dictionary that translates the classical geometric notions into their noncommutative analogs.

3.1 Quantum Formalism and Variables, the Dictionary

We refer to Chapter 4 of [59] for a detailed description of the calculus and many concrete examples.

While the first lines of the dictionary in Table 1.1 are standard in quantum mechanics, they become more and more involved as one goes down the list. We refer to the survey of Dan Voiculescu [271] for his deep work on commutants modulo normed ideals and the modulus of quasicentral approximation for an n-tuple of operators with respect to a normed ideal, which provides a fundamental tool, used crucially in the reconstruction theorem [78]. We did include the inner fluctuations in the dictionary as corresponding to the gauge theories, and the precise relation is explained in Section 2.2. Note also in this respect that the alteration of the metric by a Weyl factor is of a similar nature as the inner fluctuations. One understands from [59] how the Beltrami differentials act on conformal structures when the latter are encoded in bounded form, but the problem of encoding a general change of metric is yet open. After listing below the various entries of the dictionary, we shall concentrate on the highly nontrivial part, which is integration in the form of the Dixmier trace and the Wodzicki residue.

3.2 The Principle of Locality in NCG

It is worth describing a fundamental principle that has emerged over the years. It concerns the notion of "locality" in noncommutative geometry. The notion

Noncommutative Geometry, the Spectral Standpoint

of locality is built in for topological spaces or more generally for toposes. It is also an essential notion in modern physics.

Table 1.1.

real variable $f : X \to \mathbb{R}$	self-adjoint operator H in Hilbert space
range $f(X) \subset \mathbb{R}$ of the variable	spectrum of the operator H
composition $\phi \circ f$, ϕ measurable	measurable functions $\phi(H)$ of self-adjoint operators
bounded complex variable Z	bounded operator A in Hilbert space
infinitesimal variable dx	compact operator T
infinitesimal of order $\alpha > 0$	characteristic values $\mu_n(T) = O(n^{-\alpha})$ for $n \to \infty$
algebraic operations on functions	algebra of operators in Hilbert space
integral of function $\int f(x)dx$	$\!\!\!\!-\!\!\!\!\!\int T =$ coefficient of $\log(\Lambda)$ in $\mathrm{Tr}_\Lambda(T)$
line element $ds^2 = g_{\mu\nu} dx^\mu dx^\nu$	$ds = \bullet\!\!-\!\!\!-\!\!\bullet$: fermion propagator D^{-1}
$d(a,b) = \mathrm{Inf} \int_\gamma \sqrt{g_{\mu\nu} dx^\mu dx^\nu}$	$d(\mu,\nu) = \mathrm{Sup} \lvert \mu(A) - \nu(A)\rvert, \lvert\lVert [D,A] \rVert \leq 1.$
Riemannian geometry (X, ds^2)	spectral geometry $(\mathcal{A}, \mathcal{H}, D)$
curvature invariants	asymptotic expansion of spectral action
gauge theory	inner fluctuations of the metric
Weyl factor perturbation	$D \mapsto \rho D \rho$
conformal geometry	Fredholm module $(\mathcal{A}, \mathcal{H}, F)$, $F^2 = 1$.
perturbation by Beltrami differential	$F \mapsto (aF + b)(bF + a)^{-1}$, $a = (1 - \mu^2)^{-1/2}, b = \mu a$
distributional derivative	quantized differential $dZ := [F, Z]$
measure of conformal weight p	$f \mapsto \!\!\!\!-\!\!\!\!\!\int f(Z) \lvert dZ \rvert^p$

Locality also plays a key role in noncommutative geometry and makes sense in the above framework of the quantized calculus, but it acquires another meaning, more subtle than the straightforward topological one. The idea comes from the way the Fourier transform translates the local behavior of functions on a space in terms of the decay at ∞ of their Fourier coefficients. The relevant properties of the "local functionals" that makes them "local" is translated in Fourier by their dependence on the asymptotic behavior of the Fourier coefficients.

This is familiar in particle physics, where it is the ultraviolet behavior that betrays the fine local features, in particular through the ultraviolet divergencies. More specifically, in quantum field theory, the counterterms that need to be added along the way in the renormalization process as corrections to the initial Lagrangian appear from the ultraviolet divergencies and are automatically "local." One merit of the divergencies is to be insensitive to bounded perturbations of the momentum space behavior. The fundamental ones appear either as the coefficient of a logarithmic divergency when a cutoff parameter Λ (with the dimension of an energy) is moved to ∞ or, in the Dim-Reg regularization scheme, as a residue.

In noncommutative geometry, the above two ways of isolating the relevant term in divergencies have been turned into two key tools that allow one to construct "local functionals." These are

- the Dixmier trace;
- the Wodzicki residue.

These tools are functionals on suitable classes of operators, and their role is to filter out the irrelevant details. In some sense these filters wipe out the quantum details and give a semiclassical picture.

3.2.1 Dixmier Trace and Fractals

The integration in noncommutative geometry needs to combine two properties that seem contradictory: being a positive trace (to allow for cyclic permutations under the functional) and vanishing on infinitesimals of high enough order (order > 1). This latter property together with the approximation from below of positive compact operators by finite rank ones entails that the integration cannot commute with increasing limits and is nonnormal as a functional. Such a functional was discovered by J. Dixmier in 1966 [107]. One might be skeptical, at first, on the use of such functionals for the reason invoked in the introduction concerning the nonconstructive aspect of some proofs for the cardinality of natural noncommutative spaces. Fortunately, there is a perfect answer to this objection both at the abstract general level and in practice. The ingredient needed to construct the nonnormal functional is a limiting process $\lim_\omega$ on bounded sequences of real numbers. It is required to be linear and positive (i.e., $\lim_\omega((a_n)) \geq 0$ if $a_n \geq 0$ $\forall n$) and to vanish for sequences that tend to 0 at infinity. The book of Hardy [148] on divergent series shows the role of iterated Cesaro means and of such limiting processes in many questions of number theory related to Tauberian theorems. If one would require the limiting process to be multiplicative, it would correspond to an ultrafilter and would easily be shown to break the rule of measurability that we imposed from the start in the introduction. At this point a truly remarkable result of G. Mokobodzki [206] saves the day. He managed to show, using the continuum hypothesis, the existence of a limiting process that, besides the above conditions, is **universally measurable** and fulfills

$$\lim_\omega \left(\int a(\alpha) d\mu(\alpha) \right) = \int \lim_\omega (a(\alpha)) d\mu(\alpha) \tag{3.1}$$

for any bounded measurable family $a(\alpha)$ of sequences of real numbers. In other words, if one uses this medial limit of Mokobodzki, one no longer need worry about the measurability problems of the limit, and one can permute it freely with integrals! One might object that the medial limit is not explicitly constructed since its existence relies on the continuum hypothesis but the medial limit is to be used as a tool, and there are very few places in mathematics where one can use the power of subtle independent axioms such as the continuum hypothesis. We refer to the article of Paul Cohen [37] for the interplay between axioms of set theory and mathematical proof. He explains how certain proofs might require the use of stronger axioms even though the original question is formulated in simple number-theoretic language. But the objection that the Dixmier trace is "nonconstructively defined" does not apply in practice, the main point being that this issue does not arise when there is convergence. Since year 2000, a lot of progess has been made, and for instance, the construction of Dixmier traces and the proof of its important properties have been extended to the type II situation in [10]. Moreover, a very powerful school with great analytical skill has developed around F. Sukochev and D. Zanin and their book with S. Lord [199]. With them we have undertaken the task of giving complete proofs of several important results of [59] which were only briefly sketched there. In [92] we give complete proofs of Theorem 17 of Chapter 4 of [59] and in [93], with E. McDonald, the proof of the result on Julia sets announced in [59, p. 23]. At the technical level, an important hypothesis plays a role concerning the invariance of the limiting process under rescaling. It is simpler to state it for a limiting process ω on bounded continuous functions

$f(t)$ of $t \in [0, \infty]$, where $\lim_\omega f$ neglects all functions that tend to 0 at ∞. One lets $\alpha > 0$ and defines the notation

$$\lim_{\omega^\alpha} f := \lim_\omega g, \quad g(u) := f(u^\alpha), \quad \forall u \geqslant 0. \tag{3.2}$$

Then the relevant additional requirement is power invariance, that is, the equality of $\lim_{\omega^\alpha}$ with $\lim_\omega$. It can always be achieved due to the amenability of solvable groups and is technically very useful, as shown in Section 7 of [92]. We refer to [198] for a survey of the many key developments around the singular traces whose prototype examples were uncovered by Dixmier.

3.2.2 Wodzicki Residue

The Wodzicki residue [274] is a remarkable discovery that considerably enhances the power of the quantized calculus because it allows one, under suitable assumptions, to give sound meaning to the integral of infinitesimals whose order is strictly less than 1, while it agrees (up to normalization) with the Dixmier trace for infinitesimals of order $\geqslant 1$. Just as the Dixmier trace, it vanishes for infinitesimals of order > 1, but the extension of its domain is a considerable improvement that has no classical meaning. A striking example is given by the following new line in the dictionary (up to normalization):

scalar curvature of manifold of dimension 4	functional $\mathcal{A} \ni f \mapsto \int f ds^2 = \int R f ds^4$

The Wodzicki residue was extended to the framework of spectral triples with simple dimension spectrum in our joint work [60] with Henri Moscovici, where we obtained a local expression (local in the above sense) for the cyclic cocycle that computes the index pairing (see Section 4). There is a revealing example [72] where the computation of this local formula displays well the simplifications that occur when one manipulates infinitesimals (in the operator-theoretic sense) modulo those of high order. The explicit form of the operators provided by the quantum group is quite involved, but when one works modulo higher order, all sorts of irrelevant details are wiped out. The noncommutative residue was extended by R. Ponge to Heisenberg manifolds in [232], and new invariants of CR-geometry were obtained in a series of papers [233].

3.2.3 Locality and Spectral Action

Another striking result since the year 2000 was obtained in our joint work with Ali Chamseddine [19]. It gives a general local formula for the scale-independent terms $\zeta_{D+A}(0) - \zeta_D(0)$ in the spectral action, under the perturbation of D given by the addition of a gauge potential A. Here $(\mathcal{A}, \mathcal{H}, D)$ is a spectral triple with simple dimension spectrum consisting of positive integers, and the general formula has finitely many terms and gives

$$\zeta_{D+A}(0) - \zeta_D(0) = -\!\!\int \log(1 + A\,D^{-1}) = \sum_n \frac{(-1)^n}{n} \!\!\int (A\,D^{-1})^n. \tag{3.3}$$

In dimension 4 this formula reduces to

$$\begin{aligned}\zeta_{D+A}(0) - \zeta_D(0) = &-\!\!\int AD^{-1} + \frac{1}{2}\!\!\int (AD^{-1})^2 \\ &- \frac{1}{3}\!\!\int (AD^{-1})^3 + \frac{1}{4}\!\!\int (AD^{-1})^4,\end{aligned} \tag{3.4}$$

and under the hypothesis of vanishing of the tadpole term, the variation (3.4) is the sum of a Yang–Mills action and a Chern–Simons action relative to a cyclic 3-cocycle on the algebra $\mathcal{A}$. This result should be seen as a first step in the long-term program of performing the renormalization technique entirely in the framework of spectral triples. We refer to [255, 256, 257] for fundamental steps in this direction. After the initial breakthrough discovery in our joint work with D. Kreimer of the conceptual meaning of perturbative renormalization as the Birkhoff decomposition, we unveiled in our joint work with M. Marcolli [74, Theorem 1.106 and Corollary 1.107], an affine group scheme that acts on the coupling constants of physical theories as the fundamental ambiguity inherent to the renormalization process.

At the conceptual level one can expect that since the local functionals filter out the quantum details, and give a semiclassical picture of the quantum reality, the knowledge of this semiclassical picture only allows one to guess the quantum reality. In fact, the interpretation of the line element ds as the fermion propagator (in the Euclidean signature) already gives a hint of the modifications of geometry due to the quantum corrections. Indeed, these corrections "dress" the fermion propagator as shown in Figure 1.3.

This dressing of the propagator should be interpreted physically as quantum corrections to the geometry, that is, as the replacement of the initial "classical" spectral triple by a new one in which the inverse line element undergoes the same dressing. In perturbative renormalization, this can only be expressed

Figure 1.3 Dressed propagator.

in the form of an asymptotic series in powers of $\hbar$, which modifies D by multiplication by such a series of functions of the operator D^2/Λ^2. This suggests that in reality the geometry of spacetime is far more interrelated to the quantum and should not be fixed and frozen to its semiclassical limit when one performs renormalization. The geometric paradigm of noncommutative geometry is well suited for this interaction with the quantum world. We refer to the books [258] of W. van Suijlekom and [114] of M. Eckstein and B. Iochum for complete accounts of the above topics on the spectral action.

4 Cyclic Cohomology

The quantized calculus briefly presented in the dictionary replaces the differential df of a function by the commutator $[F, f]$, which is an operator in Hilbert space where the self-adjoint operator F acts and fulfills $F^2 = 1$. One can then develop the calculus using products of k one forms as forms of degree k and the graded commutator with F as the differential. One has $d^2 = 0$ because $F^2 = 1$. Moreover, under good summability conditions, the higher forms will be trace class operators, and putting things together, one will get an abstract cycle of degree n on the algebra $\mathcal{A}$, that is, a graded differential algebra Ω that has $\mathcal{A}$ in degree zero and is equipped with a closed graded trace $\tau \colon \Omega^n \to \mathbb{C}$. The integer n is only relevant by its parity and by being large enough so that all n forms are trace class operators. In 1981 (see the Oberwolfach 1981 report [48] on "Spectral Sequence and Homology of Currents for Operator Algebras") I discovered cyclic cohomology, the SBI sequence, and the related spectral sequence from the above framework. In particular, the periodicity operator S in cyclic cohomology was imposed by the ability to replace n by $n + 2$ in this framework. Since then, cyclic cohomology has played a central role in noncommutative geometry as the replacement of de Rham cohomology. Moreover, and parallel to what happens in differential geometry, the integral pairing between K-homology and K-theory is computed by the pairing of the Chern characters in cyclic theory according to the diagram

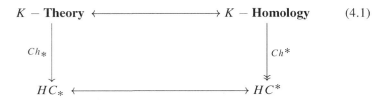
(4.1)

The Chern character in K-homology is defined as explained above from the quantized calculus associated to a Fredholm module. However, the obtained formula is nonlocal and difficult to compute explicitly. In [59, Theorem 8], I stated an explicit formula using the Dixmier trace for the Hochschild class of the Chern character, that is, its image under the forgetful map I from cyclic to Hochschild cohomology. Following my unpublished notes, a proof was included in the book [124]. The result was then refined and improved in [16].

The Hochschild class of the character only gives partial information on the Chern character; a complete local formula, involving the Wodzicki residue in place of the Dixmier trace, was obtained in 1995 in our joint work with H. Moscovici [60]. In fact, index theory is an essential application of the cyclic theory, and we shall briefly describe its role in Section 5.1.

de Rham homology of manifold	periodic cyclic cohomology
Atiyah–Singer index theorem	local index formula in NCG [60]
characteristic classes and Lie algebra cohomology	cyclic cohomology of Hopf algebras

4.1 Transverse Elliptic Theory and Cyclic Cohomology of Hopf Algebras

In our work with H. Moscovici on the transverse elliptic operators for foliations, we were led to develop the theory of characteristic classes in the framework of noncommutative geometry as a replacement of its classical differential geometric ancestor. We also developed another key ingredient of the transverse geometry, which is the geometric analogue of the reduction from type III to type II and automorphisms, and we used the theory of hypoelliptic operators to obtain a spectral triple describing the type II geometry. In fact, the explicit computations involved in the local index formula for this spectral triple dictated two essential new ingredients: first, a Hopf algebra $\mathcal{H}_n$ that governs transverse geometry in codimension n, and second, a general construction of the cyclic cohomology of Hopf algebras as a far-reaching generalization of Lie algebra cohomology [1, 64, 65, 66, 146, 147]. This theory has been vigorously developed by H. Moscovici and B. Rangipour [216, 217, 218, 219]; the DGA version of the Hopf cyclic cohomology and the characteristic map from Hopf cyclic to cyclic were investigated by S. Gorokhovsky [135].

4.2 Cyclic Cohomology and Archimedean Cohomology

In [82] with C. Consani, we showed that cyclic homology which was invented for the needs of noncommutative geometry is the right theory to obtain, using the λ-operations, Serre's Archimedean local factors of the L-function of an arithmetic variety as regularized determinants. Cyclic homology provides a conceptual general construction of the sought for "Archimedean cohomology." This resurgence of cyclic homology in the area of number theory was totally unexpected and is a witness of the coherence of the general line of thought. In fact the recent work [95, 96] of G. Cortinas, J. Cuntz, R. Meyer, and G. Tamme relates rigid cohomology to cyclic homology.

4.3 Topological Cyclic Homology

We refer to the book [112] for a complete view of the relation of topological Hochschild and cyclic theories with algebraic K-theory. At the conceptual level one may think of topological cyclic homology as cyclic homology performed over the smallest possible ground ring, which in algebraic topology is not the familiar ring $\mathbb{Z}$ of integers but the far more absolute sphere spectrum. The great advantage of working over this base is that the analogue of the tensor product becomes much more manageable as far as the action of permutations on the factors as well as their fixed points are concerned. This fact gives rise to new operations that do not exist in the general framework. These operations include an analogue of the Frobenius operator, and in his work with I. Madsen [151, 152], and then in [150, 153, 154], L. Hesselholt has shown that the local and global Witt construction, the de Rham–Witt complex, and the Fontaine theory all emerge naturally in this framework. All the

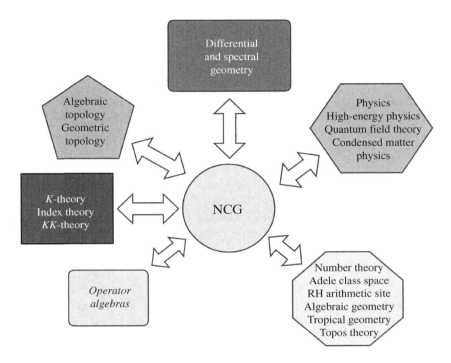

Figure 1.4 Interactions with noncommutative geometry.

more since he showed recently how topological periodic cyclic homology with its inverse Frobenius operator may be used to give a cohomological interpretation of the Hasse–Weil zeta function of a scheme smooth and proper over a finite field in the form entirely similar to [82]. Finally, the paper [87] shows that topological cyclic homology is cyclic homology when applied to the gamma rings of G. Segal and that this framework, based on the sphere spectrum, subsumes all previous attempts to develop "absolute algebra." We refer to the work of B. Dundas in [113] for a survey of the topological cyclic theory and an attempt to extend the construction to incorporate the unstable information.

5 Overall Picture, Interaction with Other Fields

We describe in this section a few of the interactions of noncommutative geometry with other fields with no claim of being exhaustive.

5.1 Index Theory

5.1.1 Type II Index Theory

Atiyah's index theorem for covering spaces [3] gave a striking application of the type II index theory that he used to define the index of an elliptic operator acting on the noncompact Galois covering of a smooth compact manifold. In Atiyah's theorem the type II index takes the same value as on the compact quotient and is hence integer valued. In [46] the type II index theorem for foliations displayed the role of the Ruelle–Sullivan current associated to a transverse measure, and the index as well as the dimensions of the solution spaces are real valued and give striking examples of real dimensions as discovered by Murray and von Neumann. These real dimensions are best understood at the intuitive level in the same way as densities of infinite sets along the leaves relate to cardinalities of finite sets, and this point of view is extensively developed in [46].

5.1.2 Higher Index Theory

The need for a higher index theory is already visible starting with the longitudinal index theorem for foliations, which made use of a transverse measure as explained in Section 5.1.1. Such transverse measures do not always exist, and the theory would seem to be limited to the type II situation as opposed to the type III that often occurs in geometry. Fortunately, the need of a transverse measure for the formulation of the theorem can be overcome when the result is formulated in terms of K-theory, as was done in our joint work with G. Skandalis [53]. But in order to obtain numeric indices, one still needs to define natural numerical invariants of K-theory. This is obtained using the pairing of K-theory with cyclic cohomology, and the early applications of the theory go back to the work on the transverse fundamental class [50]. The strategy can be divided in to several steps:

1. Prove analytic invariance or vanishing properties of the analytic index as an element of the K-theory of a suitable algebra (most often a C^*-algebra associated to the geometric situation).
2. Compute the pairing of the analytic index with cyclic cocycles by explicit geometric formulas, in the spirit of the original result of Atiyah and Singer.
3. Show that the algebraic pairing with the cyclic cocycle in fact makes sense at the level of the K-theory of the C^*-algebra.

Note that steps 1 and 3 are of a hard analytic nature, while step 2 is of a more algebraic and computational nature. The third step is usually the most difficult. The early applications of the higher index theory included the vanishing of $\hat{A}$-genus for a foliated compact manifold with leafwise positive scalar curvature [9, 50, 283], the Novikov conjecture for Gelfand–Fuchs classes [50], and the Novikov conjecture for hyperbolic groups [57, Section 5.3]. The proof of the Novikov conjecture for Gelfand–Fuchs classes illustrates the power of cyclic cohomology theory, and there is no other approach available to prove this result, which inspired the recent proof by S. Gong, J. Wu, and G. Yu of the Novikov conjecture for discrete subgroups of the group of volume-preserving diffeomorphisms in [134]. M. Puschnigg has made key contributions [238, 239, 240] to step 3 and was able to use it to prove the Kadison–Kaplansky conjecture (which states that the reduced group C^*-algebra does not contain nontrivial idempotents) for hyperbolic groups. Moreover, R. Meyer fully developed in [205] the cyclic theory in the context of bornological algebras. A basic new idea is the notion of analytic nilpotence for bornological algebras, which allows him to use the Cuntz–Quillen approach to cyclic theories. In [214, 215], H. Moriyoshi and P. Piazza proved a higher index analogue of the Atiyah–Patodi–Singer formula for foliated manifolds using the pairing with the Godbillon–Vey cocycle. In [158, 159, 160], N. Higson and J. Roe construct an analytical analog of surgery theory and connect this theory with the usual surgery theory for manifolds. They define an analytic analog of algebraic Poincaré complexes and show that such a complex has a signature, lying in the K-theory of the appropriate C^*-algebra, and that this signature is invariant under bordism and homotopy. They construct a natural commutative diagram of exact sequences, sending the surgery exact sequence of Wall to an analytic surgery sequence improving on previous results of J. Rosenberg [248]. The above gives only a brief glance at the primary higher index theory, and we refer to the survey of van Erp and Gorokhovsky [116] for an excellent introduction to the subject as well as a detailed account of the local index formula in NCG.

We now turn to the recent developments [135, 136, 214, 230, 276, 277, 278] in the secondary higher index theory. When the primary index vanishes, as in the case of positive scalar curvature, or in the relative situation of homotopic manifolds for the signature, the secondary theory comes into play, and one obtains a secondary invariant called the rho-invariant. This theory was first introduced by Higson and Roe [160, 162]. We refer to the splendid survey by Zhizhang Xie and Guoliang Yu [277] for the remarkable recent developments in this domain. We shall not attempt (for lack of space) to do justice to these considerable advances but briefly mention the papers of P. Piazza and T. Schick [230] and of G. Yu and Z. Xie [276, 277, 278] who develop sophisticated secondary invariants involving the group C^*-algebra of Galois covers to analyze the connected components of the space of metrics of positive scalar curvature. They proved in full generality the connection of the higher index of the Dirac operator on a spin manifold with boundary to the higher rho-invariant of the Dirac operator on the boundary, where the boundary is endowed with a positive scalar curvature metric. In [273], with S. Weinberger, G. Yu and Z. Xie settled a long-standing question by showing that the higher rho-invariant is a group homomorphism. They then apply this result to study nonrigidity of topological manifolds by giving a lower bound for the size of the reduced structure group of a closed oriented topological manifold, by the number of torsion elements in the fundamental group of the manifold. Note that step 3 above comes into play when one uses cyclic cohomology to obtain numerical invariants by pairing with the rho-invariant. In [35] the pairing of delocalized cyclic cocycles with the higher rho-invariants is computed in terms of the delocalized eta invariants, which were shown in [279] to be algebraic if the Baum–Connes conjecture holds, thus giving a new test of the conjecture. We refer to [277] for an account of this theory.

5.1.3 Algebraic Index Theory

The algebraic index theory is a vast subject that has been successfully developed by R. Nest and B. Tsygan (see [220, 221, 222, 223, 224]) and with Bressler and Gorokhovsky in [13]. It is an approach to the Atiyah–Singer index formula using deformation quantizations. It gives a powerful way of understanding conceptually and proving the algebraic index formulas that play, for instance, a key role in step 2 of the higher index theory (Section 5.1.2). The algebra of pseudo-differential operators provides a deformation quantization of the algebra of smooth functions on the cotangent space of a manifold. The index map is formulated using the trace density, which gives a map from the Hochschild or negative cyclic homology of the deformation quantization to the de Rham cohomology of the manifold. We refer to [13, 220] for this line of development as well as to [228, 229] for index theory in the framework of Lie groupoids, and to [211, 212, 213, 225] for the analysis on singular spaces consisting of a smooth interior part and a tower of foliated structures at the boundary, including corner and cone points and edges of all possible dimensions. This leads us to the next topic, which is the geometrization of the pseudodifferential calculus and of deformations.

5.2 Smooth Groupoids and Their C^*-Algebras in Geometry

In [59] the construction of the tangent groupoid of a manifold was presented as a geometric framework for the theory of pseudo-differential operators.

The tangent groupoid embodies in a geometric manner the deformation from symbols to operators. The main tool there is the functorial association of a C^*-algebra $C^*(G)$ to a smooth groupoid G. As in the case of discrete groups, it admits a reduced and unreduced version, a distinction that disappears in the amenable case. This geometrization of operators allows one to give a completely geometric proof of the Atiyah–Singer index theorem as sketched in [59] and shown in [164]. In a remarkable series of papers C. Debord and G. Skandalis [99, 100, 101] have considerably extended this idea and advanced a general program of geometrization of the existing pseudo-differential calculi. In her work C. Debord had shown how to construct the holonomy groupoid for a singular foliation defined by a Lie algebroid when the fiber dimension is the same as the leaf dimension. The general construction of the holonomy groupoid for a singular foliation was done by Androulidakis and Skandalis, and this was the starting point of the index theory in this framework. In [102], C. Debord and G. Skandalis construct Lie groupoids by using the classical procedures of blowups and of deformations to the normal cone. They show that the blowup of a manifold sitting in a transverse way in the space of objects of a Lie groupoid leads to a calculus similar to the Boutet de Monvel calculus for manifolds with boundary. Their geometric construction gives rise to extensions of C^*-algebras, and they compute the corresponding K-theory maps and the associated index. Smooth groupoids and the associated C^*-algebras thus provide a powerful tool in differential geometry. We refer to [103] for an excellent survey of this theory. This tool also applies successfully to representation theory, as shown by N. Higson [161], who gave substance to the idea of G. Mackey on the analogy between complex semisimple groups and their Cartan motion groups.

5.3 Novikov Conjecture for Hyperbolic Groups

In the *C. R. Acad. Sc. Paris* Note of 1988 "Conjecture de Novikov et groupes hyperboliques" [56] and in [57], Henri Moscovici and I proved the Novikov conjecture for hyperbolic groups using cyclic cohomology combined with analytic techniques. The basic observation, which was part of the origin of cyclic cohomology, is that, given a discrete group Γ, a group cocycle $c \in Z^n(\Gamma, \mathbb{C})$ gives a cyclic cocycle τ_c on the noncommutative group ring $\mathbb{C}[\Gamma]$. Moreover, an index formula [56, 57] shows that the pairing of the cyclic cocycle gives the higher signature. The main difficulty is to pass from the very small domain of definition of the cocycle provided by $\mathbb{C}[\Gamma]$ to a larger domain having the same K-theory as the C^*-algebra $C_r^*(\Gamma)$ of the group, in which the signature is known to be homotopy invariant. The key technical tool we used was the Haagerup property, which provides a good notion of Schwartz rapid decay elements in $C_r^*(\Gamma)$ for hyperbolic groups. This, together with the existence of bounded group cocycles representing cohomology classes of dimension > 1 (which is dual to the nonvanishing of the Gromov norm), gave the solution of step 3 of the higher index technique of Section 5.1.2 and proved the Novikov conjecture for hyperbolic groups [57]. Subsequently, and together with M. Gromov and H. Moscovici, we produced another proof based on geometry. The original proof based on analysis and subalgebras of C^*-algebras stable under holomorphic functional calculus is still a main tool for the Baum–Connes conjecture, as in the work of V. Lafforgue, who proved the latter conjecture for all hyperbolic groups (see Section 5.4).

5.4 The Baum–Connes Conjecture

In this section we give a short overview of the Baum–Connes conjecture and refer to [133] for a remarkable survey of this topic. At the beginning of 1980, I had shown that closed transversals of foliations provide idempotents in the C^*-algebra of the foliation. This geometric construction played a key role in the explicit construction of the finite projective modules for the noncommutative torus [47] as described in Section 2.1. It was, however, quite clear that this geometric construction could not describe all of the K-theory of $C^*(V, F)$ even in the simplest case of a fibration. I met Paul Baum at the Kingston conference of 1980, and he explained his work with Ron Douglas describing the K-theory of a manifold W using geometric cycles given by triples (M, E, f), where M is a compact manifold, E is a vector bundle over M, and $f: M \to W$ is a continuous map, while one makes a suitable K-orientation hypothesis. This construction immediately gave a clue about how to extend the construction of K-theory classes from transversals to a much more general and flexible one; the principle at work there is a general principle of noncommutative geometry, which is worth emphasizing because of its potential use in other circumstances. It can be stated as follows:

> While it is in general difficult to map a noncommutative space X (such as a leaf space) to an ordinary space, it is quite easy to construct maps the other way and use such maps to test homological properties of X.

In fact, in the case of foliations (V, F), to get a map $f: M \to V/F$ from a manifold to a leaf space, it is enough to cover M by open sets and maps $f_j: \Omega_j \to V$ and to provide a one-cocycle ensuring that the maps match in the leaf space. The Baum–Connes conjecture was formulated by P. Baum and me in 1981 and had great impact on the K-theory of operator algebras (see [133]). It was first stated for discrete groups and foliations and for Lie groups (as the Connes–Kasparov conjecture) and then was generalized (see Baum–Connes–Higson [6], and J.-L. Tu [265]) to the framework of crossed products by locally compact groupoids. It has been a major application of the Kasparov bivariant theory. By its relevance for representation theory of real or p-adic Lie groups, the analysis of discrete groups, quantum groups, group actions, and so on it is one of the most powerful and unifying principles emerging as a link between algebraic topology and analysis. It holds in full generality for Lie groups and in the amenable framework, and very powerful results have been obtained (see, e.g., [155]), but its limits were found in [163], where counterexamples were obtained.

Among the most remarkable achievements since 2000 were the breakthrough contributions of Vincent Lafforgue, who developed the analogue of the Kasparov bivariant theory in the context of Banach algebras [184] and successfully applied this new tool combined with his work on the comparison of K-theory in the Banach and C^*-contexts. He was able to cross the obstruction given, for discrete groups, by the property T of Kazhdan, which had from the beginning in the 1980s forced one to use the reduced C^*-algebra of the group as the natural one for the general formulation but seemed an insurmountable obstacle in general. In [183] V. Lafforgue establishes the Baum–Connes conjecture for many groups with property T. In [209] I. Mineyev and G. Yu showed that hyperbolic groups are strongly bolic in the sense of V. Lafforgue and therefore satisfy the Baum–Connes conjecture. The results of Lafforgue cover not only reductive real and p-adic Lie groups but also discrete cocompact subgroups of real Lie groups of real rank one and discrete cocompact subgroups of real Lie groups SL_3 over a local field. The conjecture with coefficients was proved by P. Julg for the property T group $Sp(n, 1)$ in [169]. Finally, in [185], V. Lafforgue proved the Baum–Connes conjecture with coefficients for hyperbolic groups – a stunning positive result.

5.5 Coarse Geometry and the Coarse BC-Conjecture

Coarse geometry is the study of spaces from a "large-scale" point of view. Its link with C^*-algebras and index theory is due to John Roe [243, 244, 245, 247]. The book [245] gives a self-contained and excellent introduction to the topic. In his thesis John Roe managed to develop the index theory for noncompact spaces such as the leaves of a foliation of a compact manifold without a need for the ambient compact manifold (which was needed for the L^2 index theorem for foliations), and he associated a C^*-algebra to a coarse geometry, thus allowing one to apply the operator technique in this new context. His construction is obtained by abstracting the properties that in the foliation context hold for operators along the leaves coming from the holonomy groupoid. Let (X, d) be a metric space in which every closed ball is compact. Let $\mathcal{H}$ be a separable Hilbert space equipped with a faithful and nondegenerate representation of $C_0(X)$ whose range contains no nonzero compact operator. The Roe algebra associated to a coarse geometry is the C^*-algebra $C^*(X, d)$, which is the norm closure of operators T in $\mathcal{H}$ that are locally compact,[14] and of finite propagation,[15] two notions that have a clear geometric meaning using the metric d. This C^*-algebra is independent of the choice of the representation of $C_0(X)$. The BC-conjecture is adapted to the coarse context under the hypothesis that (X, d) has bounded geometry and using a discrete model.

In [281] Guoliang Yu proved the Novikov conjecture for groups with finite asymptotic dimension, and in [282] he proved the coarse BC-conjecture for metric spaces that admit a uniform embedding into Hilbert space, and as a corollary the Novikov conjecture for discrete groups whose coarse geometry is embeddable in Hilbert space. Moreover, he introduced Property A for metric spaces and showed that it implies uniform embedding into Hilbert space. We refer to the textbook [226] for large-scale geometry and the many other applications of Property A.

At the conceptual level, the coarse point of view is best expressed in Gromov's words:

> To regain the geometric perspective one has to change one's position and move the observation point far away from X. Then the metric in X seen from the distance δ becomes the original distance divided by δ and as δ tends to infinity the points in X coalesce into a connected continuous solid unity which occupies the visual horizon without any gaps or holes and fills our geometers heart with joy.

This point of view is clearly fitting with the principle of locality in noncommutative geometry explained in Section 3.2, provided one works "in Fourier," that is, one deals with momentum space. The large-scale geometry of momentum space corresponds to the local structure of space. Thus the momentum space of a spectral geometry $(\mathcal{A}, \mathcal{H}, D)$ should be viewed as a coarse space in a suitable manner. A first attempt is to consider the distance $d(A, A') := \|A - A'\|$ defined by the operator norm on the gauge potentials, that is, the self-adjoint elements of the $\mathcal{A}$-bimodule of formal one forms

---

[14] That is, such that fT and Tf are compact operators for any $f \in C_0(X)$.
[15] That is, their support stays at a finite distance from the diagonal. A pair $(x, y) \in X \times X$ is not in the support of T iff there are $f, g \in C_0(X)$, $f(x) \neq 0$, $g(y) \neq 0$, $fTg = 0$.

$\Omega^1 := \{\sum a_j db_j \mid a_j, b_j \in \mathcal{A}\}$, which one endows with

$$d(A, A') = \|A - A'\|_D, \text{ where } \|a_j db_j\|_D := \|\sum a_j[D, b_j]\|. \tag{5.1}$$

The $\mathcal{A}$-bimodule structure should be combined with the metric structure to investigate this coarse space. Note that the "zooming out" for the coarse metric corresponds to the rescaling $D \mapsto D/\Lambda$ for large Λ, which governs the spectral action. Under this rescaling $D \mapsto D/\Lambda$, the metric $d(\phi, \psi)$ on states of (2.2) is multiplied by Λ, that is, one zooms in, which corresponds to the duality between the two metric structures, the first given by (2.2) on the state space and the second given by (5.1) on gauge potentials.

5.6 Comparison with Toposes

Both noncommutative geometry and the theory of toposes provide a solution to the problem of the coexistence of the continuous and the discrete – the first by the use of the quantum formalism for real variables as self-adjoint operators and the second by a far-reaching generalization of the notion of space which embodies usual topological spaces on the same footing as combinatorical data such as small categories. The relation between noncommutative geometry and the theory of toposes belongs to the same principle as the relation established by the Langlands correspondence between analysis of complex representations on one hand and Galois representations on the other.

The functorial construction of the C^*-algebra associated to a smooth groupoid (Section 5.2) is the prototype of the correspondence between noncommutative geometry and the theory of toposes. The role of the operator algebra side is to obtain global results such as index theory, homotopy invariance of the signature, or vanishing theorems of the index of Dirac operators in the presence of positive scalar curvature. Such results would be inaccessible in a purely local theory.

A systematic relation between the two theories has been developed by S. Henry in [149] and is based on the use of groupoids on both sides.

But as explained in [15], the theory of Grothendieck toposes goes very far beyond its geometric role as a generalization of the notion of space. The new input comes from its relation with logics and the key notion of the classifying topos for a geometric theory of first order. We refer to [15] for an introduction to this fundamental aspect of the theory and the role of the duality between a topos and its presentation from a specific site.

From my own point of view, the great surprise was that very simple toposes provide, through their sets of points, very natural examples of noncommutative spaces. The origin of this finding came from realizing that the classifying space of a small category $\mathcal{C}$ gets much improved when it is replaced by the topos $\hat{\mathcal{C}}$ of contravariant functors from $\mathcal{C}$ to the category of sets. In the *C. R. Acad. Sc. Paris* Note of 1983 "Cohomologie cyclique et foncteurs Ext^n" (*cf.* [49]) I introduced the cyclic category Λ and showed that cyclic cohomology is a derived ext-functor after embedding the nonadditive category of algebras in the abelian category of Λ-modules. I also showed that the classifying space (in Quillen's sense) of the small category Λ is the same as for the compact group $U(1)$, thus exhibiting the strong relation of Λ with the circle. But since the classifying space of the ordinal category (of totally ordered finite sets and nondecreasing maps) is contractible, the information captured by the classifying space is only partial. This loss of information is completely repaired by the use of the topos $\hat{\mathcal{C}}$ associated to a small category $\mathcal{C}$. For instance, the topos thus associated to the ordinal category classifies the abstract intervals (totally ordered sets with a minimal and a maximal element) and encodes Hochschild cohomology as a derived functor. The lambda operations in cyclic homology [197] enrich the cyclic category Λ to the epicyclic category and the latter fibers over the category $(\star, \mathbb{N}^\times)$ with a single object and morphisms given by the semigroup $\mathbb{N}^\times$ of nonzero positive integers under multiplication. The topos $\widehat{\mathbb{N}^\times}$ associated as above to this category underlies the arithmetic site (whose structure also involves a structure sheaf), and it was a great surprise to realize [83] that the points of this topos form a noncommutative space intimately related to the adele class space of Section 6.

One is surely quite far from a complete understanding of the interrelations between topos theory and noncommutative geometry, and as another instance of an unexpected relation, we mention that the above topos $\widehat{\mathbb{N}^\times}$ underlying the arithmetic site qualifies for representing the point in noncommutative geometry. More precisely, noncommutative geometry works with (separable) C^*-algebras up to Morita-equivalence, and in each class, there is a unique (up to isomorphism) A that is stable, that is, such that $A \simeq A \otimes \mathcal{K}$, where $\mathcal{K}$ is the algebra of compact operators. The key fact then [84, Section 8] is that the algebra $\mathcal{K}$ of compact operators is naturally a C^*-algebra in the above topos, that is, it admits a natural action (unique up to inner) of the semigroup $\mathbb{N}^\times$ by endomorphisms, which is hence inherited by any stable C^*-algebra. Moreover, the classification of matroids by J. Dixmier [106] is given by the same space as the space of points of the topos, and the corresponding algebras are obtained as stalks of the sheaf associated to $\mathcal{K}$.

5.7 Quantum Field Theory on Noncommutative Spaces

Noncommutative tori appear naturally in compactifications of string theory, as shown in [140] (based on [141]), where the need for developing quantum field theory in such noncommutative spaces became apparent. The simplest case to consider is the Moyal deformation of Euclidean space. What was quickly discovered was that there is a mixing between ultraviolet and infrared divergencies so that the lack of compactness of the resolvent of the Laplacian becomes a real problem. In a remarkable series of papers H. Grosse and R. Wulkenhaar [138] together with V. Rivasseau and his collaborators [242] managed to overcome this difficulty and to renormalize the theory by adding a quadratic term that replaces the Laplacian by the harmonic oscillator. They then developed Euclidean ϕ^4-quantum field theory on four-dimensional Moyal space with harmonic propagation even further than its commutative analog. For instance, they showed that in contrast with the commutative case, there is no Landau ghost in this theory: it is asymptotically safe [105], leading to the tantalizing prospect of a full nonperturbative construction, including sectors with Feynman graphs of all genera.

Grosse and Wulkenhaar went on to extensively study and fully solve the planar sector of the theory [139] (certainly its most interesting part since it contains all the ultraviolet divergencies). This sector also identifies with an infinite volume limit of the model in a certain sense. Very surprisingly, both Euclidean invariance and translation invariance are restored in this limit. Building on the identities introduced in [105], they were able to fully solve all correlation functions of a model in terms of a single sector of the two-point function. They went even further and gave strong numerical evidence [140] (and a recent proof in the case of the two-point function of the ϕ^3 theory, in collaboration with A. Sako [141]) that Osterwalder–Schrader positivity also holds for such models. In particular, given the specific problems (such as Gribov ambiguities) that have plagued and delayed the construction of ordinary non-abelian gauge theories, the planar sector of the Grosse–Wulkenhaar theory has become our best candidate for a rigorous reconstruction of a nontrivial Wightman theory in four dimensions.

The study of the Grosse–Wulkenhaar model has also been inspirational for promising generalizations of matrix models and noncommutative field theories, namely, the tensor models and tensor field theories discovered and developed by R. Gurau, V. Rivasseau, and collaborators [144]. We refer to [275] for an extensive survey of the remarkable developments of quantum field theory in noncommutative spaces.

5.8 Noncommutative Geometry and Solid State Physics

Since the pioneering work of Jean Bellissard on the quantum Hall effect and the deep relation he unveiled as the noncommutative nature of the Brillouin zone (see [59, Chapter 4, p. 6] the relation between noncommutative geometry and solid state physics has been vigorously pursued. We refer to the book [235] and papers [236, 237] for recent progress as well as to the papers of T. Loring on the finite-dimensional manifestations of the K-theory obstructions and of C. Bourne, A. Carey, and A. Rennie [11] for the bulk-edge correspondence.

6 Spectral Realization of Zeros of Zeta and the Scaling Site

We describe in this section the main steps from quantum statistical mechanical models arising in number theory, the adele class space as a noncommutative space, to the scaling site, an object of algebraic geometry underlying the spectral realization of the zeros of the Riemann zeta function.

6.1 Quantum Statistical Mechanics and Number Theory

Discrete groups Γ provide very nontrivial examples of factors of type II_1, and their left regular representation given by the left action of Γ in the Hilbert space $\ell^2(\Gamma)$ of square integrable functions on Γ generates a type II_1 factor $R(\Gamma)$ as long as the nontrivial conjugacy classes of elements of Γ are infinite. It was shown by Atiyah that the type II index theory relative to $R(\Gamma)$ can be used very successfully for Galois coverings of compact manifolds. To obtain type III factors from discrete groups, one considers the relative situation of pairs (Γ, Γ_0), where $\Gamma_0 \subset \Gamma$ is a subgroup that is almost normal inasmuch as the left action of Γ_0 on the coset space Γ/Γ_0 only has finite orbits. The left action of Γ in the Hilbert space $\ell^2(\Gamma/\Gamma_0)$ then generates a von Neumann algebra that is no longer of type II in general and whose modular theory depends on the integer valued function $L(\gamma) :=$ cardinality of the orbit $\Gamma_0(/\gamma\Gamma_0)$ in the coset space Γ/Γ_0. In fact, the commutant von Neumann algebra is generated by the Hecke algebra $\mathcal{H}(\Gamma, \Gamma_0)$ of convolution of functions f of double cosets which have finite support in $\Gamma_0\backslash\Gamma/\Gamma_0$. The modular automorphism of the canonical state is given by [12]

$$\sigma_t(f) = \left(\frac{L(\gamma)}{R(\gamma)}\right)^{it} f(\gamma), \ R(\gamma) := L(\gamma^{-1}).$$

Almost normal subgroups $\Gamma_0 \subset \Gamma$ arise naturally by considering the inclusion of points over $\mathbb{Z}$ inside points over $\mathbb{Q}$ for algebraic groups as in the construction of Hecke algebras. The BC system arises from the "$ax + b$" algebraic group P. By construction, $P_\mathbb{Z}^+ \subset P_\mathbb{Q}^+$ is an inclusion $\Gamma_0 \subset \Gamma$ of countable groups, where $P_\mathbb{Z}^+$ and $P_\mathbb{Q}^+$ denote the restrictions to $a > 0$. This inclusion fulfills the above commensurability condition, and the associated Hecke algebra with its dynamics is the BC system. Its main interest is that it exhibits spontaneous symmetry breaking [12] when it is cooled down using the thermodynamics of noncommutative spaces (see [74] for this general notion) and that its partition function is the Riemann zeta function. Considerable progress has been done since 2000 on extending the BC system to number fields. For lack of space, we shall not describe these developments but refer to [74] and to [181, 182]. The paper [181] elucidates the functoriality of the construction in Theorem 4.4, where the authors show that the construction of Bost–Connes-type systems extends to a functor which to an inclusion of number fields $K \subset L$ assigns a C^*-correspondence equivariant with respect to their suitably rescaled natural dynamics. In the paper [182] M. Laca, N. Larsen, and S. Neshveyev succeeded to extend the properties of fabulous states of the BC system to arbitrary number fields. We refer to the introduction of their paper for a historical account of the various steps that led to the solution after the initial steps of Ha and Paugam. They consider the Hecke pair consisting of the group P_K^+ of affine transformations of a number field K that preserves the orientation in every real embedding and the subgroup P_O^+ of transformations with algebraic integer coefficients. They then show that the Hecke algebra $\mathcal{H}$ associated to the pair (P_K^+, P_O^+) fulfills perfect analogs of the properties of the BC system, including phase transitions and Galois action. More precisely, they obtain an arithmetic subalgebra (through the isomorphism of $\mathcal{H}$ to a corner in the larger system established in [181]) on which ground states exhibit the "fabulous states" property with respect to an action of the Galois group $\text{Gal}(K_{\text{ab}} \colon H_+(K))$, where $H_+(K)$ is the narrow Hilbert class field and K_{ab} is the maximal abelian extension of K.

We refer to [97] for a survey of the other important lines of development coming from the exploration by Joachim Cuntz and his collaborators of the C^*-algebra associated to the action of the multiplicative semigroup of a Dedekind ring on its additive group. Representations of such actions give rise to particularly intriguing problems, and the study of the corresponding C^*-algebras has motivated many of the new methods and general results obtained in this area.

Another very interesting broad generalization of the BC system has been developed by M. Marcolli and G. Tabuada [27]. Using the Tannakian formalism, they categorify the algebraic data used in constructing the system such as roots of unity, algebraic numbers, and Weil numbers. They study the partition function, low-temperature Gibbs states, and Galois action on zero-temperature states for the associated quantum statistical system and show that in the particular case of the Weil numbers, the partition function and the low-temperature Gibbs states can be described as series of polylogarithms.

6.2 Anosov Foliation

The factor of type III_1 generated by the regular representation of the BC system is the same (unique injective factor of type III_1 [145]) as the factor associated to the Anosov foliation of the sphere bundle of a Riemann surface of genus > 1. This fact suggested an analogy between the geometry of the continuous decomposition (as a crossed product: type $II_\infty \rtimes \mathbb{R}_+^*$) in both cases, which led me in [62], based on the paper of Guillemin [143], to the adele class space as a natural candidate for the trace formula interpretation of the explicit formulas of Riemann–Weil. It is worth explaining briefly the framework[16] of [143]. Let $\Gamma \subset SL(2,\mathbb{R})$ be a discrete cocompact subgroup, and $V = SL(2,\mathbb{R})/\Gamma$. Let F be the foliation of V whose leaves are the orbits of the action on the left of the subgroup $P \subset SL(2,\mathbb{R})$ of upper triangular matrices. At the Lie algebra level, P is generated by the elements

$$E^+ = \begin{pmatrix} 0 & 1 \\ 0 & 0 \end{pmatrix}, \quad H = \frac{1}{2}\begin{pmatrix} 1 & 0 \\ 0 & -1 \end{pmatrix}.$$

One denotes by η and ξ the corresponding vector fields. The associated flows are the horocycle and geodesic flows. They correspond to the one-parameter subgroups of $P \subset SL(2,\mathbb{R})$ given by

$$n(a) = \begin{pmatrix} 1 & a \\ 0 & 1 \end{pmatrix}, \quad g(t) = \begin{pmatrix} e^{t/2} & 0 \\ 0 & e^{-\frac{t}{2}} \end{pmatrix}.$$

The horocycle flow is normalized by the geodesic flow; more precisely, one has $g(t)n(a)g(-t) = n(ae^t)$. The von Neumann algebra of the foliation, $M = W(V, F)$, is a factor of type III_1, and its continuous decomposition is visible at the geometric level: the associated factor of type II_∞ is the von Neumann algebra $N = W(V, \eta)$ of the horocycle foliation. The one-parameter

---

[16] The same paper [143] was used two years later by Deninger [104] to motivate his search for a hypothetical cohomology theory using foliations.

group of automorphisms θ_λ scaling the trace is given by the action of the geodesic flow by automorphisms of $N = W(V, \eta)$ which has clear geometric meaning by the naturality of the construction of the von Neumann algebra of a foliation. The trace on $N = W(V, \eta)$ corresponds to the Ruelle Sullivan current, which is the contraction of the $SL(2, \mathbb{R})$-invariant volume form of M by the vector field η. The rescaling of the trace by the automorphisms θ_λ follows from the rescaling of η by the geodesic flow. In [143] a heuristic proof of the Selberg trace formula is given using the action of the geodesic flow on the horocycle foliation. Our interpretation of the explicit formulas as a trace formula involves in a similar manner the action of $\mathbb{R}_+^*$ on the geometric space associated to the type II_∞ factor of the continuous decomposition for the BC system. This geometric space is the adele class space (for the precise formulation, we refer to [74]).

6.3 The Adele Class Space

In [63] we gave a spectral realization of the zeros of the Riemann zeta function and of L-functions based on the above action of the idele class group on the noncommutative space of adele classes which is the type II space associated to the BC system by the reduction from type III to type II and automorphisms. This result determined a geometric interpretation of the Riemann–Weil explicit formulas as a Lefschetz formula and also a reformulation of the Riemann hypothesis in terms of the validity of a trace formula. The understanding of the geometric side of the trace formula is simpler when one works with the full adele class space, that is, when one does not further divide by $\hat{\mathbb{Z}}^*$. The local contribution from a place v of a global field K is obtained as the distributional trace of an integral $\int f(\lambda^{-1}) T_\lambda d^*\lambda$ of operators of the form, with $\lambda \in K_v^*$,

$$(T_\lambda \xi)(x) := \xi(\lambda x), \quad \forall x \in K_v,$$

where K_v is the local completion of K at the place v. The distributional trace of T_λ is the integral $\int k(x, x) dx$ on the diagonal of the Schwartz kernel that represents T_λ, that is, by

$$(T_\lambda \xi)(x) = \int k(x, y) \xi(y) dy \Rightarrow k(x, y) = \delta(y - \lambda x).$$

This gives for the trace the formula $\int k(x, x) dx = \int \delta(x - \lambda x) dx = |1 - \lambda|^{-1}$ using the change of variables $u = (1 - \lambda)x$ and the local definition of the module, which implies $dx = |1 - \lambda|^{-1} du$. For the convolution operator $\int f(\lambda^{-1}) T_\lambda d^*\lambda$ this delivers the local contribution to the Riemann–Weil explicit formula as $\int_{K_v^*} f(\lambda^{-1}) |1 - \lambda|^{-1} d^*\lambda$. The above computation is formal, but as shown in [63], the more precise calculation even delivers the subtle finite parts involved in the Riemann–Weil explicit formula. In [204], R. Meyer showed how, by relaxing the Sobolev condition, one can effectively re-prove the explicit formulas as a trace formula on the adele class space.

6.4 The Scaling Site

6.4.1 Hasse–Weil Form of the Riemann Zeta Function

After a few years of direct attack using only analysis, I came to the conclusion that a much better understanding of the geometry underlying the adele class space was required to make a vigorous advance toward the solution of this problem. This work was undertaken in the ongoing collaboration with C. Consani. Among the major results obtained in these past years is a geometric framework in which one can transpose to the case of RH many of the ingredients of the Weil proof for function fields as reformulated by Mattuck, Tate, and Grothendieck. The starting point [80, 81] is the determination and interpretation as intersection number using the geometry of the adele class space of the real counting function $N(q)$, which gives the complete Riemann zeta function by a Hasse–Weil formula in the limit $q \to 1$, in the line of the limit geometry on $\mathbb{F}_q$ for $q = 1$ proposed by J. Tits and C. Soulé.

6.4.2 Characteristic 1

The limit $q \to 1$ is taken analytically in the above development, and this begs for an algebraic understanding of the meaning of "characteristic 1." Characteristic p is defined by the congruence $p \simeq 0$. While the congruence $1 \simeq 0$ is fruitless, its variant given by $1 + 1 \simeq 1$ opens the door to a whole field that we discuss in [90] as the "world of characteristic 1," whose historical origin goes quite far back. It has strong connections with the field of optimization [128, 129], with tropical geometry, and with lattice theory, and, very importantly, it appeared independently with the school of "dequantization," of V. P. Maslov and his collaborators [178, 195]. They developed a satisfactory algebraic framework that encodes the semiclassical limit of quantum mechanics and called it idempotent analysis. The starting observation is that one can encode the limit $\hbar \to 0$ simply by conjugating the addition of numbers by the power operation $x \mapsto x^\epsilon$ and passing to the limit when $\epsilon \to 0$. The new addition of positive real numbers is

$$\lim_{\epsilon \to 0} \left(x^{\frac{1}{\epsilon}} + y^{\frac{1}{\epsilon}} \right)^\epsilon = \max\{x, y\} = x \vee y.$$

When endowed with this operation as addition and with the usual multiplication, the positive real numbers become a semifield $\mathbb{R}_+^{\max}$. It is of characteristic 1, that is, $1 \vee 1 = 1$, and contains the smallest semifield of characteristic 1, namely, the boolean semifield $\mathbb{B} = \{0, 1\}$. Moreover, $\mathbb{R}_+^{\max}$ admits nontrivial automorphisms, and one has

$$\mathrm{Gal}_{\mathbb{B}}(\mathbb{R}_+^{\max}) := \mathrm{Aut}_{\mathbb{B}}(\mathbb{R}_+^{\max}) = \mathbb{R}_+^*, \quad \mathrm{Fr}_\lambda(x) = x^\lambda, \quad \forall x \in \mathbb{R}_+^{\max}, \lambda \in \mathbb{R}_+^*,$$

thus providing a first glimpse at an answer to Weil's query in [272] of an algebraic framework in which the connected component of the idele class group would appear as a Galois group. The most striking discovery of this school of Maslov, Kolokoltsov, and Litvinov [178, 195] is that the Legendre transform, which plays a fundamental role in all of physics and, in particular, in thermodynamics in the 19th century, is simply the Fourier transform in the framework of idempotent analysis.

6.4.3 Tropical Geometry and Riemann–Roch Theorems

The tropical semi-ring $\mathbb{N}_{\min} = \mathbb{N} \cup \{\infty\}$ with the operations min and $+$ was introduced by Imre Simon in [251] to solve a decidability problem in rational language theory. His work is at the origin of the term tropical, which was coined by Marco Schutzenberger. Tropical geometry is a vast subject (see, e.g., [130, 200, 207]). We refer to [268] for an excellent introduction starting from the sixteenth Hilbert problem. In its simplest form [23] a tropical curve is given by a graph with a usual line metric on its edges. The natural structure sheaf on the graph is the sheaf of real-valued functions that are continuous, convex, and piecewise affine with integral slopes. The operations on such functions are given by $(f \vee g)(x) = f(x) \vee g(x)$ for all $x \in \Gamma$, and the product is given by pointwise addition. One also adjoins the constant $-\infty$, which plays the role of the zero element in the semi-rings of sections. One proceeds as in the classical case with the construction of the sheaf of semifields of quotients and finds the same type of functions as above, but no longer convex. Cartier divisors make sense, and one finds that the order of a section f at a point $x \in \Gamma$ is given by the sum of the (integer-valued) outgoing slopes. The conceptual explanation of why the discontinuities of the derivative should be interpreted as zeros or poles is due to Viro [269], who showed that it follows automatically if one understands that, as seen when dealing with valuations, the sum $x \vee x$ of two equal terms should be viewed as ambiguous with all values in the interval $[-\infty, x]$ on equal footing. In their work Baker and Norine [5] proved in the discrete setup of graphs an analog of the Riemann–Roch theorem whose essence is that the inequality $\mathrm{Deg}(D) \geqslant g$

(where g is the genus of the graph) for a divisor implies that the divisor is equivalent to an effective divisor. Once translated into language of the chip firing game [5], this fact is equivalent to the existence of a winning strategy if one assumes that the total sum of dollars attributed to the vertices of the graph is $\geqslant g$. We refer to [23] for the tropical curve version of the Baker and Norine theorem.

6.4.4 The Topos, Its Points, and Structure Sheaf

The major discovery [83, 84, 85] is that of the scaling site, the topos of $\mathbb{N}^\times$-equivariant sheaves on the Euclidean half-line, which is obtained by extension of scalars from the arithmetic site. The points of this topos form exactly the sector of the adele class space involved in RH, and the action of $\mathbb{R}_+^*$ on the adele class space corresponds to the action of the Frobenius. This provides in full the missing geometric structure on the adele class space, which becomes a tropical curve since the topos inherits, from its construction by extension of scalars, a natural sheaf of regular functions as piecewise affine convex functions. This structure is central in the well-known results on the localization of zeros of analytic functions that involve Newton polygons in the non-Archimedean case and Jensen's formula in the complex case. The new feature given by the action of $\mathbb{N}^\times$ corresponds to the transformation $f(z) \mapsto f(z^n)$ on analytic functions. The Newton polygons play a key role in the construction of the analogue of the Frobenius correspondences as a one-parameter semigroup of correspondences already defined at the level of the arithmetic site. Finally, the restriction to the periodic orbit of the scaling flow associated to each prime p gives a quasi-tropical structure that turns this orbit into a variant $C_p = \mathbb{R}_+^*/p^{\mathbb{Z}}$ of the classical Jacobi description $\mathbb{C}^*/q^{\mathbb{Z}}$ of an elliptic curve. On C_p, [86] develops the theory of Cartier divisors, determines the structure of the quotient of the abelian group of divisors by the subgroup of principal divisors, develops the theory of theta functions, and proves the Riemann–Roch formula, which involves real-valued dimensions, as in the type II index theory. The current situation concerning the evolution of this strategy toward RH is summarized in the essay [90] in the volume on open problems in mathematics, initiated by John Nash. After developing homological algebra in characteristic 1 in [88] we understood in [89] how to go back and forth to the complex situation using the above Jensen formula so that the fundamental structure takes place over $\mathbb{C}$ and fits well with noncommutative geometry.

References

[1] R. Akbarpour and M. Khalkhali, *Hopf algebra equivariant cyclic homology and cyclic homology of crossed product algebras*, J. Reine Angew. Math. 559 (2003) 137–52

[2] M. Atiyah, *Global theory of elliptic operators*, in *1970 Proc. Internat. Conf. on Functional Analysis and Related Topics (Tokyo, 1969)*, Univ. of Tokyo Press (1969) 21–30

[3] M. F. Atiyah, *Elliptic operators, discrete groups and von Neumann algebras*, in *Analyse et topologie*, Astérisque 32/33, Soc. Math. France (1976) 43–72

[4] S. Baaj and P. Julg, *Théorie bivariante de Kasparov et opérateurs non bornés dans les C^*-modules hilbertiens*, C. R. Acad. Sci. Paris Sér. I 296 (1983) 875–78

[5] M. Baker and S. Norine, *Riemann–Roch and Abel–Jacobi theory on a finite graph*, Adv. Math. 215 (2007) 766–88

[6] P. Baum, A. Connes, and N. Higson, *Classifying space for proper actions and K-theory of group C^*-algebras*, in *C^*-algebras: 1943–1993 (San Antonio, TX, 1993)*, Contemp. Math. 167, AMS (1994) 240–91

[7] P. F. Baum and E. van Erp, *K-homology and index theory on contact manifolds*, Acta Math. 213 (2014) 1–48

[8] M. Benameur, A. Gorokhowsky, and E. Leichtnam, *The higher twisted index theorem for foliations*, J. Funct. Anal. 273 (2017) 496–558

[9] M.-T. Benameur and J. Heitsch, *Enlargeability, foliations, and positive scalar curvature*, Invent. Math. 215 (2019) 367–82

[10] M.-T. Benameur and T. Fack, *Type II non-commutative geometry. I. Dixmier trace in von Neumann algebras*, Adv. Math. 199 (2006) 29–87

[11] C. Bourne, A. Carey, and A. Rennie, *A non-commutative framework for topological insulators*, Rev. Math. Phys. 28 (2016) 1650004

[12] J. B. Bost and A. Connes, *Hecke algebras, Type III factors and phase transitions with spontaneous symmetry breaking in number theory*, Selecta Math. (New Series) 1 (1995) 411–57

[13] P. Bressler, A. Gorokhovsky, R. Nest, and B. Tsygan, *Algebraic index theorem for symplectic deformations of gerbes*, in A. Connes, A. Gorokhovsky, M. Lesch, M. Pflaum, and B. Rangipour, eds., *Noncommutative Geometry and Global Analysis*, Contemp. Math. 546, AMS (2011) 23–38

[14] L. G. Brown, R. G. Douglas, and P. A. Fillmore, *Extensions of C^*-algebras and K-homology*, Ann. Math. 105 (1977) 265–324

[15] O. Caramello and L. Lafforgue, *Sur la dualité des topos et de leurs présentations et ses applications: une introduction*, Preprint IHES (2016)

[16] A. Carey, A. Rennie, F. Sukochev, and D. Zanin, *Universal measurability and the Hochschild class of the Chern character*, J. Spectr. Theory 6 (2016) 1–41

[17] A. Carey and F. Sukochev, *Dixmier traces and some applications in noncommutative geometry*, Russ. Math. Surveys 61 (2006) 1039–99

[18] A. Chamseddine and A. Connes, *The spectral action principle*, Comm. Math. Phys. 186 (1997) 731–50

[19] A. Chamseddine and A. Connes, *Inner fluctuations of the spectral action*, J. Geom. Phys. 57 (2006) 1–21
[20] A. Chamseddine, A. Connes, and M. Marcolli, *Gravity and the standard model with neutrino mixing*, Adv. Theo. Math. Phys. 11 (2007) 991–1089
[21] A. Chamseddine and A. Connes, *Quantum gravity boundary terms from the spectral action of noncommutative space*, Phys. Rev. Lett. 99 (2007) 071302
[22] A. Chamseddine and A. Connes, *Why the standard model?*, J. Geom. Phys. 58 (2008) 38–47
[23] A. Chamseddine and A. Connes, *The uncanny precision of the spectral action*, Comm. Math. Phys. 293 (2010) 867–97
[24] A. Chamseddine and A. Connes, *Noncommutative geometry as a framework for unification of all fundamental interactions including gravity*, Fortsch. Phys. 58 (2010) 553
[25] A. Chamseddine and A. Connes, *Noncommutative geometric spaces with boundary: spectral action*, J. Geom. Phys. 61 (2011) 317–32
[26] A. Chamseddine and A. Connes, *Spectral action for Robertson–Walker metrics*, J. High Energy Phys. 10 (2012) 101
[27] A. Chamseddine and A. Connes, *Resilience of the spectral standard model*, JHEP 1209 (2012) 104
[28] A. Chamseddine, A. Connes, and W. D. van Suijlekom, *Beyond the spectral standard model: Emergence of Pati–Salam unification*, JHEP 11 (2013) 132
[29] A. Chamseddine, A. Connes, and W. D. van Suijlekom, *Inner fluctuations in noncommutative geometry without the first order condition*, J. Geom. Phys. 73 (2013) 222
[30] A. Chamseddine, A. Connes, and V. Mukhanov, *Geometry and the quantum: Basics*, JHEP 12 (2014) 098
[31] A. Chamseddine, A. Connes, and V. Mukhanov, *Quanta of geometry: Noncommutative aspects*, Phys. Rev. Lett. 114 (2015)
[32] A. Chamseddine, A. Connes, and W. D. van Suijlekom, *Grand unification in the spectral Pati–Salam models*, JHEP 2511 (2015) 011
[33] A. Chamseddine, A. Connes, and W. D. van Suijlekom, *Entropy and the spectral action*, arXiv:1809.02944
[34] A. Chamseddine and W. D. van Suijlekom, *A survey of spectral models of gravity coupled to matter*, Springer Surveys (forthcoming)
[35] X. Chen, J. Wang, Z. Xie, and G. Yu, *Delocalized eta invariants, cyclic cohomology and higher rho invariants*, Preprint, arXiv:1901.02378
[36] P. B. Cohen and A. Connes, *Conformal geometry of the irrational rotation algebra*, Preprint, MPI/92-93
[37] P. Cohen, *Skolem and pessimism about proofs in mathematics*, Phil. Trans. R. Soc. A 363 (2005) 2407–18
[38] A. Connes, *Un nouvel invariant pour les algèbres de von Neumann*, C. R. Acad. Sci. Paris Ser. A-B 273 (1971) A900–A903
[39] A. Connes, *Calcul des deux invariants d'Araki et Woods par la théorie de Tomita et Takesaki*, C. R. Acad. Sci. Paris Ser. A-B 274 (1972) A175–77
[40] A. Connes, *Groupe modulaire d'une algèbre de von Neumann*, C. R. Acad. Sci. Paris Sér. A-B 274 (1972) 1923–26

[41] A. Connes, *Une classification des facteurs de type III*, C. R. Acad. Sci. Paris Ser. A-B 275 (1972) A523–25
[42] A. Connes, *Une classification des facteurs de type III*, Ann. Sci. École Norm. Sup. 6 (1973) 133–252
[43] A. Connes and A. van Daele, *The group property of the invariant S of von Neumann algebras*, Math. Scand. 32 (1973) 187–92
[44] A. Connes and M. Takesaki, *The flow of weights on factors of type III*, Tôhoku Math. J. 29 (1977) 473–575
[45] A. Connes, *The von Neumann algebra of a foliation*, in G. Dell'Antonio, S. Doplicher, and G. Jona-Lasinio, eds., *Mathematical Problems in Theoretical Physics*, Lecture Notes in Physics 80, Springer (1978) 145–51
[46] A. Connes, *Sur la théorie non commutative de l'intégration*, in P. de la Harpe, ed., *Algèbres d'opérateurs*, Lecture Notes in Mathematics 725, Springer (1979) 19–143
[47] A. Connes, C^*-*algèbres et géométrie différentielle*, C. R. Acad. Sci. Paris Sér. A-B 290 (1980) A599–A604
[48] A. Connes, *Spectral sequence and homology of currents for operator algebras*, Math. Forschungsinst. berwolfach Tagungsber. 41/81, Funktionalanalysis und C^*-Algebren (1981) 27–9/3–10
[49] A. Connes, *Cohomologie cyclique et foncteur Ext^n*, C.R. Acad. Sci. Paris Ser. I Math. 296 (1983) 953–58
[50] A. Connes, *Cyclic cohomology and the transverse fundamental class of a foliation*, in Geometric Methods in H. Araki and E. G. Effros, eds., *Operator Algebras (Kyoto, 1983)*, Pitman Research Notes in Mathematics 123, Longman (1986) 52–144
[51] A. Connes, *Noncommutative differential geometry. Part I: The Chern character in K-homology*, Preprint I.H.E.S. (M/82/53) (1982); *Part II: de Rham homology and noncommutative algebra*, Preprint IHES (M/83/19) (1983)
[52] A. Connes, *Noncommutative differential geometry*, Inst. Hautes Études Sci. Publ. Math. 62 (1985) 257–336
[53] A. Connes and G. Skandalis, *The longitudinal index theorem for foliations*, Publ. Res. Inst. Math. Sci. Kyoto 20 (1984) 1139–83
[54] A. Connes, *Leçon inaugurale au Collège de France* (January 11, 1985), www.alainconnes.org/docs/lecollege.pdf
[55] A. Connes and M. Rieffel, *Yang–Mills for noncommutative two tori*, in *Operator Algebras and Mathematical Physics (Iowa City, Iowa, 1985)*, Contemp. Math. Oper. Algebra. Math. Phys. 62, AMS (1987) 237–66
[56] A. Connes and H. Moscovici, *Conjecture de Novikov et groupes hyperboliques*, C.R. Acad. Sci. Paris Ser. I Math. 307 (1988) 475–80
[57] A. Connes and H. Moscovici, *Cyclic cohomology, the Novikov conjecture and hyperbolic groups*, Topology 29 (1990) 345–88
[58] A. Connes, M. Gromov, and H. Moscovici, *Group cohomology with Lipschitz control and higher signatures*, Geom. Functional Anal. 3 (1993) 1–78
[59] A. Connes, *Noncommutative Geometry*, Academic Press (1994)
[60] A. Connes and H. Moscovici, *The local index formula in noncommutative geometry*, GAFA 5 (1995) 174–243

[61] A. Connes, M. Douglas, and A. Schwarz, *Noncommutative geometry and matrix theory: Compactification on tori*, J. High Energy Physics 2 (1998)

[62] A. Connes, *Formule de trace en géométrie non-commutative et hypothèse de Riemann*, C. R. Acad. Sci. Paris Sér. I Math. 323 (1996) 1231–36

[63] A. Connes, *Trace formula in noncommutative geometry and the zeros of the Riemann zeta function*, Selecta Math. (N.S.) 5 (1999) 29–106

[64] A. Connes and H. Moscovici, *Hopf algebras, cyclic cohomology and the transverse index theorem*, Commun. Math. Phys. 198 (1998) 199–246

[65] A. Connes and H. Moscovici, *Cyclic cohomology and Hopf algebras*, Letters Math. Phys. 48 (1999) 97–108

[66] A. Connes and H. Moscovici, *Differentiable cyclic cohomology and Hopf algebraic structures in transverse geometry*, in E. Ghys, P. de la Harpe, V. F. R. Jones, V. Sergiescu, and T. Tsuboi, eds., *Essays on Geometry and Related Topics: Mémoires dédiés à André Haefliger*, vol. 1, Monogr. Enseign. Math. 38, Enseignement Math. (2001) 217–55

[67] A. Connes, *Noncommutative geometry – year 2000*, in *GAFA 2000 (Tel Aviv, 1999)*, Geom. Funct. Anal., Special Volume, Part II (2000) 481–559

[68] A. Connes and G. Landi, *Noncommutative manifolds, the instanton algebra and isospectral deformations*, Comm. Math. Phys. 221 (2001) 141–59

[69] A. Connes and M. Dubois-Violette, *Noncommutative finite-dimensional manifolds. I. Spherical manifolds and related examples*, Comm. Math. Phys. 230 (2002) 539–79

[70] A. Connes and M. Dubois-Violette, *Moduli space and structure of noncommutative 3-spheres*, Lett. Math. Phys. 66 (2003) 91–121

[71] A. Connes and M. Dubois-Violette, *Non commutative finite-dimensional manifolds II. Moduli space and structure of noncommutative 3-spheres*, Comm. Math. Phys. 281 (2008) 23–127

[72] A. Connes, *Cyclic cohomology, quantum group symmetries and the local index formula for $SU_q(2)$*, J. Inst. Math. Jussieu 3 (2004) 17–68

[73] A. Connes and M. Marcolli, *Renormalization and motivic Galois theory*, Int. Math. Res. Notices 76 (2004) 4073–91

[74] A. Connes and M. Marcolli, *Noncommutative Geometry, Quantum Fields, and Motives*, Colloquium Publications 55, AMS (2008)

[75] A. Connes, *A unitary invariant in Riemannian geometry*, Int. J. Geom. Methods Mod. Phys. 5 (2008) 1215–42

[76] A. Connes and H. Moscovici, *Type III and spectral triples*, in *Traces in Number Theory, Geometry and Quantum Fields*, Aspects Math. E38, Friedr. Vieweg (2008) 57–71

[77] A. Connes and P. B. Tretkoff, *The Gauss–Bonnet theorem for the noncommutative two torus*, in C. Consani and A. Connes, eds., *Noncommutative Geometry, Arithmetic, and Related Topics*, Johns Hopkins University Press (2011)

[78] A. Connes, *On the spectral characterization of manifolds*, J. Noncommut. Geom. 7 (2013) 1–82

[79] A. Connes and H. Moscovici, *Modular curvature for noncommutative two-tori*, J. Amer. Math. Soc. 27 (2014) 639–84

[80] A. Connes and C. Consani, *Schemes over* $\mathbb{F}_1$ *and zeta functions*, Compositio Math. 146 (2010) 1383–1415

[81] A. Connes and C. Consani, *From monoids to hyperstructures: In search of an absolute arithmetic*, in G. Van Dijk and M. Wakayama, eds., *Casimir Force, Casimir Operators and the Riemann Hypothesis*, de Gruyter (2010) 147–98

[82] A. Connes and C. Consani, *Cyclic homology, Serre's local factors and the* λ-*operations*, J. K-Theory 14 (2014) 1–45

[83] A. Connes and C. Consani, *The arithmetic site*, Comptes Rendus Math. Ser. I 352 (2014) 971–75

[84] A. Connes and C. Consani, *Geometry of the arithmetic site*, Adv. Math. 291 (2016) 274–329

[85] A. Connes and C. Consani, *The scaling site*, C. R. Math. Acad. Sci. Paris 354 (2016) 1–6

[86] A. Connes and C. Consani, *Geometry of the scaling site*, C. Sel. Math. New Ser. (2017)

[87] A. Connes and C. Consani, *Absolute algebra and Segal's* Γ-*rings: au dessous de* $\overline{Spec(\mathbb{Z})}$, J. Number Theory 162 (2016) 518–51

[88] A. Connes and C. Consani, *Homological algebra in characteristic one*, Preprint, arXiv:1703.02325

[89] A. Connes and C. Consani, *The Riemann-Roch strategy, complex lift of the scaling site*, Springer Surveys (forthcoming)

[90] A. Connes, *An essay on the Riemann hypothesis*, in J. Forbes Nash, Jr. and M. Th. Rassias, eds., *Open Problems in Mathematics*, Springer (2016)

[91] A. Connes and F. Fathizadeh, *The term* a_4 *in the heat kernel expansion of noncommutative tori*, Preprint, arXiv:1611.09815

[92] A. Connes, F. Sukochev, and D. Zanin, *Trace theorems for quasi-Fuchsian groups*, Mat. Sb. 208 (2017) 59–90

[93] A. Connes, F. Sukochev, and D. Zanin, *Conformal trace theorem for Julia sets of quadratic polynomials*, Ergodic Theory Dyn. Sys. (forthcoming)

[94] A. Connes, *Geometry and the quantum*, in J. Kouneiher, ed., *Foundations of Mathematics and Physics One Century after Hilbert*, Springer (2018) 159–96

[95] G. Cortinas, J. Cuntz, R. Meyer, and G. Tamme, *Nonarchimedean bornologies, cyclic homology and rigid cohomology*, Doc. Math. 23 (2018) 1197–1245

[96] G. Cortinas, J. Cuntz, R. Meyer, and G. Tamme, *Weak completions, bornologies and rigid cohomology*, J. Geom. Phys. 129 (2018) 192–99

[97] J. Cuntz, C^*-*algebras associated with algebraic actions*, in T. M. Carlsen, N. S. Larsen, S. Neshveyev, and C. Skau, eds., *Operator Algebras and Applications – the Abel Symposium*, Springer (2017) 151–65

[98] L. Dabrowski, G. Landi, A. Sitarz, W. Suijlekom, and J. Varilly, *The Dirac operator on* $SU_q(2)$, Comm. Math. Phys. 259 (2005) 729–59

[99] C. Debord and G. Skandalis, *Pseudodifferential extensions and adiabatic deformation of smooth groupoid actions*, Bull. Sci. Math. 139 (2015) 750–76

[100] C. Debord and G. Skandalis, *Adiabatic groupoid, crossed product by* $\mathbb{R}_+^*$ *and pseudodifferential calculus*, Adv. Math. 257 (2014) 66–91

[101] C. Debord and G. Skandalis, *Stability of Lie groupoid* C^*-*algebras*, J. Geom. Phys. 105 (2016) 66–74

[102] C. Debord and G. Skandalis, *Lie groupoids, exact sequences, Connes–Thom elements, connecting maps and index maps*, J. Geom. Phys. 129 (2018) 255–68
[103] C. Debord and G. Skandalis, *Lie groupoids, pseudodifferential calculus and index theory*, Springer Surveys (forthcoming)
[104] C. Deninger, *Some analogies between number theory and dynamical systems on foliated spaces*, Documenta Mathematical, ICM (1998) 163–86
[105] M. Disertori, R. Gurau, J. Magnen, and V. Rivasseau, *Vanishing of beta function of non commutative Phi**4(4) theory to all orders*, Phys. Lett. B 649 (2007) 95
[106] J. Dixmier, *On some C^*-algebras considered by Glimm*, J. Funct. Anal. 1 (1967) 182–203
[107] J. Dixmier, *Existence de traces non normales*, C. R. Acad. Sci. Paris Ser. A-B 262 (1966) A1107–8
[108] S. Donaldson and D. Sullivan, *Quasiconformal 4-manifolds*, Acta Math. 163 (1989) 181–252
[109] R. Dong and M. Khalkhali, *Second quantization and the spectral action*, Preprint, arXiv:1903.09624
[110] M. Dubois-Violette and G. Landi, *Noncommutative products of Euclidean spaces*, Lett. Math. Phys. 108 (2018) 2491–2513
[111] M. Dubois-Violette and G. Landi, *Noncommutative Euclidean spaces*, J. Geom. Phys. 130 (2018) 315–30
[112] B. Dundas, T. Goodwillie, and R. McCarthy, *The Local Structure of Algebraic K-Theory*, Algebra and Applications 18, Springer (2013)
[113] B. Dundas, *Cyclic homology in a special world*, Springer Surveys (forthcoming)
[114] M. Eckstein and B. Iochum, *Spectral Action in Noncommutative Geometry*, SpringerBriefs in Mathematical Physics 27, Springer, (2018)
[115] E. van Erp, *Noncommutative topology and the world's simplest index theorem*, Proc. Natl. Acad. Sci. USA 107 (2010) 8549–56
[116] E. van Erp and A. Gorokhovsky, *Index theory and noncommutative geometry: A survey*, Springer Surveys (forthcoming)
[117] W. Fan, F. Fathizadeh, and M. Marcolli, *Modular forms in the spectral action of Bianchi IX gravitational instantons*, J. High Energy Phys. 234 (2019) 37 pp.
[118] F. Fathizadeh and M. Khalkhali, *The Gauss–Bonnet theorem for noncommutative two tori with a general conformal structure*, J. Noncommut. Geom. 6 (2012) 457–80
[119] F. Fathizadeh and M. Khalkhali, *Scalar curvature for the noncommutative two torus*, J. Noncommut. Geom. 7 (2013) 1145–83
[120] F. Fathizadeh and M. Khalkhali, *Scalar curvature for noncommutative four-tori*, J. Noncommut. Geom. 9 (2015) 473–503
[121] F. Fathizadeh, A. Ghorbanpour, and M. Khalkhali, *Rationality of spectral action for Robertson–Walker metrics*, J. High Energy Phys. 12 (2014) 64
[122] F. Fathizadeh and M. Marcolli, *Periods and motives in the spectral action of Robertson–Walker spacetimes*, Comm. Math. Phys. 356 (2017) 641–71
[123] F. Fathizadeh and M. Khalkhali, *Curvature in noncommutative geometry*, Springer Surveys (forthcoming)
[124] H. Figueroa, J. M. Gracia-Bondía, and J. Varilly, *Elements of Noncommutative Geometry*, Birkhäuser (2000)

[125] R. Floricel, A. Ghorbanpour, and M. Khalkhali, *The Ricci curvature in noncommutative geometry*, J. Noncommut. Geom. 13 (2019) 269–96

[126] A. Ghorbanpour and M. Khalkhali, *Spectral geometry of functional metrics on noncommutative tori*, Preprint, arXiv:1811.04004

[127] A. Gathmann and M. Kerber, *A Riemann-Roch theorem in tropical geometry*, Math. Z. 259 (2008) 217–30

[128] S. Gaubert, *Methods and Applications of (max, +) Linear Algebra*, Lecture Notes in Computer Science 1200, Springer (1997) 261–82

[129] S. Gaubert, *Two lectures on the max-plus algebra*, in *Proceedings of the 26th Spring School of Theoretical Computer Science* (1998) 83–147

[130] I. Gelfand, M. Kapranov, and A. Zelevinsky, *Discriminants, Resultants, and Multidimensional Determinants*, Mathematics: Theory and Applications, Birkhauser (1994)

[131] M. P. Gomez-Aparicio, *Représentations non unitaires, morphisme de Baum-Connes et complétions inconditionnelles*, J. Noncommut. Geom. 3 (2009) 419–46

[132] M. P. Gomez-Aparicio, *Morphisme de Baum-Connes tordu par une représentation non unitaire*, J. K-Theory 6 (2010) 23–68

[133] M. P. Gomez-Aparicio, P. Julg, and A. Valette, *The Baum–Connes conjecture, an extended survey*, Springer Surveys (forthcoming)

[134] S. Gong, J. Wu, and G. Yu, *The Novikov conjecture, the group of volume preserving diffeomorphisms, and Hilbert–Hadamard spaces*, Preprint, arXiv:1811.02086

[135] A. Gorokhovsky, *Secondary characteristic classes and cyclic cohomology of Hopf algebras*, Topology 41 (2002) 993–1016

[136] A. Gorokhovsky, H. Moriyoshi, and P. Piazza, *A note on the higher Atiyah–Patodi–Singer index theorem on Galois coverings*, J. Noncommut. Geom. 10 (2016) 265–306

[137] A. Gorokhovsky and H. Moscovici, *Index pairing with Alexander–Spanier cocycles*, J. Geom. Phys. 133 (2018) 195–209.

[138] H. Grosse and R. Wulkenhaar, *Renormalization of phi**4 theory on noncommutative R**4 in the matrix base*, Commun. Math. Phys. 256 (2005) 305–374

[139] H. Grosse and R. Wulkenhaar, *Self-dual noncommutative ϕ^4-theory in four dimensions is a non-perturbatively solvable and non-trivial quantum field theory*, Commun. Math. Phys. 329 (2014) 1069–1130

[140] H. Grosse and R. Wulkenhaar, *Solvable 4D noncommutative QFT: Phase transitions and quest for eflection positivity*, Preprint, arXiv:1406.7755 [hep-th]

[141] H. Grosse, A. Sako, and R. Wulkenhaar, *The Φ_4^3 and Φ_6^3 matricial QFT models have reflection positive two-point function*, Preprint, arXiv:1612.07584 [math-ph]

[142] H. Grosse, and R. Wulkenhaar, *Self-dual noncommutative ϕ^4-theory in four dimensions is a non-perturbatively solvable and non-trivial quantum field theory*, Commun. Math. Phys. 329 (2014) 1069–1130

[143] V. Guillemin, *Lectures on spectral theory of elliptic operators*, Duke Math. J. 44 (1977) 485–517

[144] R. Gurau, *Invitation to Random Tensors*, in R. Gurau, ed., *Special Issue on Tensor Models, Formalism and Applications*, SIGMA 12 (2016) 094

[145] U. Haagerup, *Connes bicentralizer problem and uniqueness of the injective factor of type III_1*, Acta Math. 158 (1987) 95–148

[146] P. Hajac, M. Khalkhali, B. Rangipour, and Y. Sommerhauser, *Stable anti-Yetter–Drinfeld modules*, C. R. Math. Acad. Sci. Paris 338 (2004) no. 8, 587–590

[147] P. Hajac, M. Khalkhali, B. Rangipour, and Y. Sommerhauser, *Hopf-cyclic homology and cohomology with coefficients*, C. R. Math. Acad. Sci. Paris 338 (2004) 667–672

[148] G. H. Hardy, *Divergent Series*, Clarendon (1949)

[149] S. Henry, *The convolution algebra of an absolutely locally compact topos*, Preprint, arXiv:1701.00113

[150] L. Hesselholt, *On the p-typical curves in Quillen's K-theory*, Acta Math. 177 (1996) 1–53

[151] L. Hesselholt and I. Madsen, *On the K-theory of finite algebras over Witt vectors of perfect fields*, Topology 36 (1997) 29–102

[152] L. Hesselholt and I. Madsen, *On the De Rham–Witt complex in mixed characteristic*, Ann. Sci. Ecole Norm. Sup. (4) 37 (2004) 1–43

[153] L. Hesselholt, *On the topological cyclic homology of the algebraic closure of a local field*, in *An Alpine Anthology of Homotopy Theory: Proceedings of the Second Arolla Conference on Algebraic Topology (Arolla, Switzerland, 2004)*, Contemp. Math. 399, AMS (2006) 133–162

[154] L. Hesselholt, *Topological Hochschild homology and the Hasse-Weil zeta function: An Alpine Bouquet of Algebraic Topology*, Contemp. Math. 708, AMS (2018)

[155] N. Higson, G. Kasparov, and J. Trout, *A Bott periodicity theorem for infinite dimensional Euclidean space*, Advances in Math. 135 (1998) 1–40

[156] N. Higson and J. Roe, *Amenable group actions and the Novikov conjecture*, J. Reine Angew. Math. 519 (2000) 143–153

[157] N. Higson and J. Roe, *Analytic K-homology*, Oxford University Press (2000)

[158] N. Higson and J. Roe, *Mapping surgery to analysis. I. Analytic signatures*, K-Theory 33 (2005) 277–299

[159] N. Higson and J. Roe, *Mapping surgery to analysis. II. Geometric signatures*, K-Theory 33 (2005) 301–324

[160] N. Higson and J. Roe, *Mapping surgery to analysis. III. Exact sequences*, K-Theory 33 (2005) 325–346

[161] N. Higson, *The Mackey analogy and K-theory*, in R. S. Doran, C. C. Moore, and R. J. Zimmer, eds., *Group representations, ergodic theory, and mathematical physics: a tribute to George W. Mackey*, Contemp. Math. 449, AMS (2008) 149–72

[162] N. Higson and J. Roe, *K-homology, assembly and rigidity theorems for relative eta invariants*, Pure Appl. Math. Q. 6(2), Special Issue: In honor of Michael Atiyah and Isadore Singer (2010) 555–601

[163] N. Higson, V. Lafforgue, and G. Skandalis, *Counterexamples to the Baum–Connes conjecture*, Geom. Funct. Anal. 12 (2002) 330–354

[164] N. Higson, *The tangent groupoid and the index theorem*, Quanta of Maths, Clay Math. Proc. 11, AMS (2010) 241–256

[165] M. Hilsum and G. Skandalis, *Morphismes K-orientés d'espaces de feuilles et fonctorialité en théorie de Kasparov (d'après une conjecture d'A. Connes)*, Ann. Sci. Ecole Norm. Sup. (4) 20 (1987) 325–390

[166] M. Hilsum, *Bordism invariance in KK-theory*, Math. Scand. 107 (2010) 73–89

[167] B. Iochum and T. Masson, *Heat asymptotics for nonminimal Laplace type operators and application to noncommutative tori*, J. Geom. Phys. 129 (2018) 1–24

[168] P. Julg, *Travaux de N. Higson et G. Kasparov sur la conjecture de Baum-Connes*, Séminaire Bourbaki, Vol. 1997/98, Astérisque No. 252 (1998) Exp. No. 841, 4, 151–183

[169] P. Julg, *La conjecture de Baum-Connes à coefficients pour le groupe Sp(n, 1)*, C. R. Math. Acad. Sci. Paris 334 (2002) 533–538

[170] P. Julg, *How to prove the Baum–Connes conjecture for the groups Sp(n,1)?*, J. Geom. Phys. 141 (2019) 105–119

[171] B. Julia, *Statistical theory of numbers*, in J. M. Luck, P. Moussa, and M. Waldschmidt, eds., *Number Theory and Physics, Proceedings of the Winter School*, Les Houches, France, March 7–16, 1989, Springer (1990) 276–293

[172] G. Kasparov, *Hilbert C^*-modules: Theorems of Stinespring and Voiculescu*, J. Operator Theory, 4 (1980) 133–150

[173] G. Kasparov, *The operator K-functor and extensions of C^*-algebras*, Izv. Akad. Nauk. SSSR Ser. Mat. 44 (1980) 571–636

[174] G. Kasparov and G. Skandalis, *Groups acting properly on "bolic" spaces and the Novikov conjecture*, Ann. of Math. (2) 158 (2003) 165–206

[175] G. Kasparov and G. Yu, *The Novikov conjecture and geometry of Banach spaces*, Geom. Topol. 16 (2012) 1859–1880

[176] M. Khalkhali, *Basic Noncommutative Geometry*, Second edition, EMS Series of Lectures in Mathematics, European Mathematical Society (EMS) (2013)

[177] M. Khalkhali and A. Sitarz, *Gauss–Bonnet for matrix conformally rescaled Dirac*, J. Math. Phys. 59 (2018) 6

[178] V. Kolokoltsov and V. P. Maslov, *Idempotent Analysis and Its Applications*, Kluwer Academic Publishers (1997)

[179] M. Laca, N. Larsen, and S. Neshveyev, *Phase transition in the Connes–Marcolli GL_2 system*, J. Noncommut. Geom. 1 (2007) 397–430

[180] M. Laca, N. Larsen, and S. Neshveyev, *On Bost–Connes types systems for number fields*, J. Number Theory 129 (2009) 325–338

[181] M. Laca, S. Neshveyev, and M. Trifkovic, *Bost-Connes systems, Hecke algebras, and induction*. J. Noncommut. Geom. 7 (2013) 525–546

[182] M. Laca, N. Larsen, and S. Neshveyev, *Ground states of groupoid C^*-algebras, phase transitions and arithmetic subalgebras for Hecke algebras*, J. Geom. Phys. 136 (2019) 268–283

[183] V. Lafforgue, *K-théorie bivariante pour les algèbres de Banach et conjecture de Baum–Connes*, Invent. Math. 149 (2002) 1–95

[184] V. Lafforgue, *K-théorie bivariante pour les algèbres de Banach, groupoides et conjecture de Baum-Connes. Avec un appendice d'Hervé Oyono-Oyono*, J. Inst. Math. Jussieu 6 (2007) 415–451

[185] V. Lafforgue, *La conjecture de Baum–Connes à coefficients pour les groupes hyperboliques*, J. Noncommut. Geom. 6 (2012) 1–197

[186] G. Landi and P. Martinetti, *On twisting real spectral triples by algebra automorphisms*, Lett. Math. Phys. 106 (2016) 1499–1530

[187] G. Landi and P. Martinetti, *Gauge transformations for twisted spectral triples*, Lett. Math. Phys. 108 (2018) 2589–2626

[188] G. Landi and C. Pagani, *A class of differential quadratic algebras and their symmetries*, J. Noncommut. Geom. 12 (2018) 1469–1501

[189] E. Leichtnam and P. Piazza, *Dirac index classes and the noncommutative spectral flow*, J. Funct. Anal. 200 (2003) 348–400

[190] E. Leichtnam and P. Piazza, *Elliptic operators and higher signatures*, Ann. Inst. Fourier (Grenoble) 54 (2004) 1197–1277

[191] E. Leichtnam and P. Piazza, *Etale groupoids, eta invariants and index theory*, J. Reine Angew. Math. 587 (2005) 169–233

[192] M. Lesch and H. Moscovici, *Modular curvature and Morita equivalence*, Geom. Funct. Anal. 26 (2016) 818–873

[193] M. Lesch, *Divided differences in noncommutative geometry: rearrangement lemma, functional calculus and expansional formula*, J. Noncommut. Geom. 11 (2017) 193–223

[194] M. Lesch and H. Moscovici, *Modular Gaussian curvature*, Springer Surveys (forthcoming)

[195] G. Litvinov, *Tropical Mathematics, Idempotent Analysis, Classical Mechanics and Geometry*, in *Spectral theory and geometric analysis*, Contemp. Math., 535, AMS (2011) 159–186

[196] Y. Liu, *Hypergeometric function and Modular Curvature I, II*, arXiv:1810.09939 and arXiv:1811.07967

[197] J. L. Loday, *Cyclic Homology*, Grundlehren der Mathematischen Wissenschaften, 301, Springer-Verlag (1998)

[198] S. Lord, F. Sukochev, and D. Zanin, *Advances in Dixmier traces and applications*, Springer Surveys (forthcoming)

[199] S. Lord, F. Sukochev and D. Zanin, *Singular Traces: Theory and Applications*, De Gruyter Studies in Mathematics 46, De Gruyter (2013)

[200] D. Maclagan and B. Sturmfels, *Introduction to Tropical Geometry*, Graduate Studies in Mathematics, 161, AMS (2015)

[201] Y. Manin, *Topics in Noncommutative Geometry*, M. B. Porter Lectures, Princeton University Press (1991)

[202] M. Marcolli and G. Tabuada, *Bost–Connes systems, categorification, quantum statistical mechanics, and Weil numbers*, J. Noncommut. Geom. 11 (2017) 1–49

[203] M. Marcolli and Y. Xu, *Quantum statistical mechanics in arithmetic topology*, J. Geom. Phys. 114 (2017) 153–183

[204] R. Meyer, *On a representation of the idele class group related to primes and zeros of L-functions*, Duke Math. J. 127 (2005) 519–595

[205] R. Meyer, *Local and Analytic Cyclic Homology*, EMS Tracts in Mathematics, 3, European Mathematical Society (EMS) (2007)

[206] P. A. Meyer, *Limites médiales, d'après Mokobodzki*, Séminaire de probabilités VII, (Strasbourg, 1971–72), Lecture Notes in Math. 321, Springer (1973) 198–204

[207] G. Mikhalkin, *Enumerative tropical algebraic geometry in $\mathbb{R}^2$*, J. Amer. Math. Soc. 18 (2005) 313–377

[208] J. Milnor and J. Stasheff, *Characteristic Classes*, Annals of Mathematics Studies 76, Princeton University Press; University of Tokyo Press (1974)

[209] I. Mineyev and G. Yu, *The Baum–Connes conjecture for hyperbolic groups*, Inventiones Mathematicae. 149 (2002) 97–122

[210] A. Mishchenko, *C^*-algebras and K-theory*, in *Algebraic Topology, Aarhus 1978 (Proc. Sympos., Univ. Aarhus, Aarhus, 1978)*, Lecture Notes in Math. 763, Springer (1979) 262–74

[211] B. Monthubert, *Groupoids and pseudodifferential calculus on manifolds with corners*, J. Funct. Anal. 199 (2003) 243–286

[212] B. Monthubert, *Contribution of noncommutative geometry to index theory on singular manifolds*, in *Geometry and Topology of Manifolds*, Banach Center Publ. 76, Polish Acad. Sci. Inst. Math. (2007) 221–37

[213] B. Monthubert and V. Nistor, *A topological index theorem for manifolds with corners*, Compos. Math. 148 (2012) 640–668

[214] H. Moriyoshi and P. Piazza, *Eta cocycles, relative pairings and the Godbillon–Vey index theorem*, Geometric and Functional Analysis 22 (2012) 1708–1813

[215] H. Moriyoshi and P. Piazza, *Relative pairings and the Atiyah–Patodi–Singer index formula for the Godbillon–Vey cocycle*, in *Noncommutative Geometry and Global Analysis*, Contemp. Math. 546, AMS (2011) 225–247

[216] H. Moscovici, *Geometric construction of Hopf cyclic characteristic classes*, Adv. Math. 274 (2015) 651–80

[217] H. Moscovici and B. Rangipour, *Hopf algebras and universal Chern classes*, J. Noncommut. Geom. 11 (2017) 71–109

[218] H. Moscovici and B. Rangipour, *Hopf cyclic cohomology and transverse characteristic classes*, Adv. Math. 227 (2011) 654–729

[219] H. Moscovici and B. Rangipour, *Hopf algebras of primitive Lie pseudogroups and Hopf cyclic cohomology*, Adv. Math. 220 (2009) 706–90

[220] R. Nest and B. Tsygan, *Algebraic index theorem*, Comm. Math. Phys. 172 (1995) 223–62

[221] R. Nest and B. Tsygan, *Algebraic index theorem for families*, Adv. Math. 113 (1995) 151–205

[222] R. Nest and B. Tsygan, *Formal versus analytic index theorems*, Internat. Math. Res. Notices (1996) 557–64

[223] R. Nest and B. Tsygan, *Product structures in (cyclic) homology and their applications*, in *Operator Algebras and Quantum Field Theory*, International Press (1997) 416–49

[224] R. Nest and B. Tsygan, *Deformations of symplectic Lie algebroids, deformations of holomorphic symplectic structures, and index theorems*, Asian J. Math. 5 (2001) 599–635

[225] V. Nistor, *Analysis on singular spaces: Lie manifolds and operator algebras*, J. Geom. Phys. 105 (2016) 75–101
[226] N. Nowak and G. Yu, *Large Scale Geometry*, EMS Textbooks in Mathematics, European Mathematical Society (2012)
[227] A. Pais, *Inward Bound: Of Matter and Forces in the Physical World*, Oxford University Press (1986)
[228] M. Pflaum, H. Posthuma, and X. Tang, *The transverse index theorem for proper cocompact actions of Lie groupoids*, J. Differential Geom. 99 (2015) 443–72
[229] M. Pflaum, H. Posthuma, and X. Tang, *The localized longitudinal index theorem for Lie groupoids and the van Est map*, Adv. Math. 270 (2015) 223–62
[230] P. Piazza and T. Schick, *Rho-classes, index theory and Stolz' positive scalar curvature sequence*, J. Topol. 7 (2014) 965–1004
[231] M. Pimsner and D. Voiculescu, *Exact sequences for K-groups and Ext-groups of certain cross-product C^*-algebras*, J. Operator Theory 4 (1980) 93–118
[232] R. Ponge, *Noncommutative residue for Heisenberg manifolds: Applications in CR and contact geometry*, J. Funct. Anal. 252 (2007) 399–463
[233] R. Ponge, *Noncommutative residue invariants for CR and contact manifolds*, J. Reine Angew. Math. 614 (2008) 117–51
[234] H. Ha and R. Ponge, *Laplace-Beltrami operators on noncommutative tori*, J. Geom. Phys. 150 (2020) 103594
[235] E. Prodan and H. Schulz-Baldes, *Bulk and Boundary Invariants for Complex Topological Insulators: From K-Theory to Physics*, Springer (2016)
[236] E. Prodan, *Disordered topological insulators: A non-commutative geometry perspective*, J. Phys. A 44 (2011) 113001
[237] E. Prodan, B. Leung, and J. Bellissard, *The non-commutative n-th Chern number $(n \geq 1)$*, J. Phys. A 46 (2013) 485202
[238] M. Puschnigg, *Asymptotic Cyclic Cohomology*, Lecture Notes in Mathematics 1642, Springer (1996)
[239] M. Puschnigg, *The Kadison–Kaplansky conjecture for word-hyperbolic groups*, Invent. Math. 149 (2002) 153–94
[240] M. Puschnigg, *New holomorphically closed subalgebras of C^*-algebras of hyperbolic groups*, Geom. Funct. Anal. 20 (2010) 243–59
[241] M. A. Rieffel, *Morita equivalence for C^*-algebras and W^*-algebras*, J. Pure Appl. Algebra 5 (1974) 51–96
[242] V. Rivasseau, F. Vignes-Tourneret, and R. Wulkenhaar, *Renormalization of noncommutative phi**4-theory by multi-scale analysis,* Commun. Math. Phys. 262 (2006) 565
[243] J. Roe, *Index Theory, Coarse Geometry, and Topology of Manifolds*, CBMS Regional Conference Series in Mathematics 90, AMS (1996)
[244] J. Roe, *Surgery and C^*-algebras*, in *Surveys on Surgery Theory, Vol. 1*, Ann. of Math. Stud. 145, Princeton University Press (2000) 365–77
[245] J. Roe, *Lectures on Coarse Geometry*, University Lecture Series 31, AMS (2003)
[246] J. Roe, *Hyperbolic groups have finite asymptotic dimension*, Proc. Amer. Math. Soc. 133 (2005) 2489–90
[247] J. Roe, *Positive curvature, partial vanishing theorems and coarse indices*, Proc. Edinb. Math. Soc. 59 (2016) 223–33

[248] J. Rosenberg, *Novikov Conjectures, Index Theorems and Rigidity*, vol. 1, Cambridge University Press (1995)

[249] E. Semenov, F. Sukochev, A. Usachev, and D. Zanin, *Banach Limits and Traces on $L_{1,\infty}$*, Adv. Math. 285 (2015) 568–628

[250] P. H. Siegel, *Witt spaces: A geometric cycle theory for KO-homology at odd primes*, Amer. J. Math. 105 (1983) 1067–1105

[251] I. Simon, *Limited subsets of the free monoid*, in *Proceedings of the 19th Annual Symposium on Computer Science* (1978) 143–50

[252] I. Singer, *Future extensions of index theory and elliptic operators*, in *Prospects in Mathematics*, Ann. Math. Stud. 70 (1971) 171–85

[253] G. Skandalis and J. L. Tu, *The coarse Baum–Connes conjecture and groupoids*, Topology 41 (2002) 807–34

[254] M. Spivak, *Spaces satisfying Poincaré duality*, Topology 6 (1967) 77–101

[255] W. van Suijlekom, *Perturbations and operator trace functions*, J. Funct. Anal. 260 (2011) 2483–96

[256] W. van Suijlekom, *Renormalizability conditions for almost-commutative geometries*, Phys. Lett. B 711 (2012) 434–38

[257] W. van Suijlekom, *Renormalization of the asymptotically expanded Yang–Mills spectral action*, Comm. Math. Phys. 312 (2012) 883–912

[258] W. van Suijlekom, *Noncommutative Geometry and Particle Physics*, Mathematical Physics Studies, Springer (2015)

[259] D. Sullivan, *Hyperbolic Geometry and Homeomorphisms*, Academic Press (1979)

[260] M. Takesaki, *Duality for crossed products and the structure of von Neumann algebras of type III*, Acta Math. 131 (1973) 249–310

[261] M. Takesaki, *Tomita's Theory of Modular Hilbert Algebras and Its Applications*, Lecture Notes in Mathematics 128, Springer (1970)

[262] X. Tang, Y.-J. Yao, and W. Zhang, *Hopf cyclic cohomology and Hodge theory for proper actions*, J. Noncommut. Geom. 7 (2013) 885–905

[263] X. Tang and Y.-J. Yao, *K-theory of equivariant quantization*, J. Funct. Anal. 266 (2014) 478–86

[264] X. Tang, R. Willett, and Y.-J. Yao, *Roe C^*-algebra for groupoids and generalized Lichnerowicz vanishing theorem for foliated manifolds*, Math. Z. 290 (2018) 1309–38

[265] J. L. Tu, *The Baum–Connes conjecture for groupoids*, in C^**-Algebras*, Springer (2000)

[266] A. Valette, *On the Baum–Connes Assembly Map for Discrete Groups, With an Appendix by Dan Kucerovsky*, Advanced Courses in Mathematics, Birkhauser (2003)

[267] A. Valette, *Proper isometric actions on Hilbert spaces: A-(T)-menability and Haagerup property*, in *Handbook of Group Actions, vol. 4*, Advanced Lectures in Mathematics 41, International Press (2018) 625–52

[268] O. Viro, *From the sixteenth Hilbert problem to tropical geometry*, Jpn. J. Math. 3 (2008) 185–214

[269] O. Viro, *On basic concepts of tropical geometry* (Russian), Tr. Mat. Inst. Steklova 273 (2011) 271–303; translation in Proc. Steklov Inst. Math. 273 (2011) 252–82

[270] D. Voiculescu, *A non-commutative Weyl–von Neumann theorem*, Rev. Roumaine Math. Pures Appl. 21 (1976) 97–113

[271] D. Voiculescu, *Commutants mod normed ideals*, Springer Surveys (forthcoming)

[272] A. Weil, *Sur la théorie du corps de classes*, J. Math. Soc. Japan 3 (1951) 1–35

[273] S. Weinberger, Z. Xie, and G. Yu, *Additivity of higher rho invariants and non-rigidity of topological manifolds*

[274] M. Wodzicki, *Noncommutative Residue, Part I. Fundamentals*, in *K-Theory, Arithmetic and Geometry*, Lecture Notes in Mathematics 1289, Springer (1987) 320–99

[275] R. Wulkenhaar, *Quantum field theory on noncommutative spaces*, Springer Surveys (forthcoming)

[276] Z. Xie and G. Yu, *A relative higher index theorem, diffeomorphisms and positive scalar curvature*, Adv. Math. 250 (2014) 35–73

[277] Z. Xie and G. Yu, *Higher rho invariants and the moduli space of positive scalar curvature metrics*, Preprint, arXiv:1608.03661

[278] Z. Xie and G. Yu, *Positive scalar curvature, higher rho invariants and localization algebras*, Adv. Math. 307 (2016) 1046–69

[279] Z. Xie and and G. Yu, *Delocalized eta invariants, algebraicity, and K-theory of group C^*-algebras*, Preprint, arXiv:1805.07617

[280] Z. Xie and G. Yu, *Higher invariants in noncommutative geometry*, Springer Surveys (forthcoming)

[281] G. Yu, *The Novikov conjecture for groups with finite asymptotic dimension*, Ann. Math. (2) 147 (1998) 325–55

[282] G. Yu, *The coarse Baum–Connes conjecture for spaces which admit a uniform embedding into Hilbert space*, Invent. Math. 139 (2000) 201–40

[283] W. Zhang, *Positive scalar curvature on foliations*, Ann. Math. 2 (2017) 1035–68

Alain Connes
Collège de France, IHES
alain@connes.org

2

The Logic of Quantum Mechanics (Revisited)

Klaas Landsman

Dedicated to Miklos Rédei, on the occasion of his 65th birthday

Contents

1 Introduction	85
2 Gelfand Duality Revisited	87
3 Stone Duality and Its Relatives	93
4 Priestley Duality and Esakia Duality	97
5 Intuitionistic Quantum Logic	102
6 Epilogue: From Topos Theory to Quantum Logic	108
References	111

1 Introduction

Any new approach to some topic that has already been studied by serious people in the past comes with the obligation to explain its necessity. Quantum logic is no exception in this regard, especially since it was founded by the greatest mathematician and logician ever to have occupied himself with quantum mechanics, namely, von Neumann; see von Neumann [44] for the fundamental role of projections, and Birkhoff and von Neumann [7] for the subsequent formalization of quantum logic in terms of specific non-boolean lattices (initially taken to be modular lattices, later generalized to orthomodular lattices).

A noteworthy aspect of the approach of Birkhoff and von Neumann, which will also be adopted in our own theory, is its *semantic* nature: unlike traditional twentieth-century logic, which starts from syntax and subsequently moves on to semantics (i.e., model theory), they defined their quantum logic directly through its class of models. Indeed, they conceptually based their model of quantum logic on Boole's models for classical propositional logic, in which (in a physical setting) elementary propositions correspond to (measurable) subsets of phase space M (up to sets of measure zero). Birkhoff and von Neumann first recalled that such sets (or equivalence classes thereof) define a boolean lattice under the obvious partial order $A \leq B$ iff $A \subseteq B$ (which gives rise to the lattice operations $A \vee B = A \cup B$ and $A \wedge B = A \cap B$) and the complementation $A' = A^c = M \backslash A$. In particular, this lattice is distributive and satisfies the law of the excluded middle

$$A \vee A' = \top, \qquad (1.1)$$

where $\top$ (often called 1) is the top element of the lattice (given by M itself).

Using von Neumann's own mathematical formalism for quantum mechanics, in which each physical system is no longer associated with a phase space but with a Hilbert space H, and each elementary proposition is interpreted by a closed linear subspace $L \subseteq H$, Birkhoff and von Neumann observed that the set $\mathcal{L}(H)$ of all such L again forms a lattice under the natural partial ordering (i.e., inclusion), which this time gives rise to the lattice operations $L \vee M = \overline{L + M}$ (i.e., the closed linear span of L and M), and $L \wedge M = L \cap M$ (the same as in the classical case). They observed that this lattice is no longer distributive (unless $\dim(H) = 1$), but, with the obvious (ortho)complementation $L' = L^\perp$ (i.e., the *orthogonal* complement of L in H), it still satisfies the law of the excluded middle.[1]

All this is easy to generalize if we identify the above lattice $\mathcal{L}(H)$ of all closed subspaces of H with the lattice $\mathcal{P}(B(H))$ of all projections on H (here $B(H)$ is the algebra of all bounded operators on H, of which an element e is a projection iff $e^2 = e^* = e$); if $M \subset B(H)$ is a von Neumann algebra, then its subset of projections $\mathcal{P}(M)$ inherits the lattice structure of $\mathcal{P}(B(H)) \cong \mathcal{L}(H)$, so that each von Neumann algebra (*nomen est omen!*) defines a quantum logic in the spirit of Birkhoff and von Neumann [48].

---

[1] Birkhoff and von Neumann noted that if one works with all linear subspaces of H instead of the closed ones (in which case $L \vee M = L + M$), their lattice satisfies a weakened version of distributivity, in that $L \leq N$ implies $L \vee (M \wedge N) = (L \vee M) \wedge N$ for each M (i.e., if distributivity holds merely if $L \leq N$). This is called the *modular law*; it was later shown that their actual lattice of *closed* subspaces satisfies the modular law at least for $M = L^\perp$. Such lattices are called *orthomodular*; orthomodularity is equivalent to the perhaps more appealing condition that the compatibility relation $\overset{c}{\sim}$ on $\mathcal{P}(H)$ is symmetric (i.e., $L \overset{c}{\sim} M$ iff $M \overset{c}{\sim} L$), where we say that $L \overset{c}{\sim} M$ iff $L = (L \wedge M) \vee (L \wedge M^\perp)$, i.e., the associated projections commute.

However, looking at cases like Schrödinger's cat – at least in the naive view that it is neither alive nor dead, which view may be wrong for macroscopic objects [38] but which certainly holds for microscopic ones – and also submitting that distributivity simply cannot be given up if $\wedge$ and $\vee$ are to preserve anything remotely similar to their usual logical meanings "and" and "or," one cannot avoid the impression that despite its novelty and interest, the quantum logic proposed by Birkhoff and von Neumann is

- *too radical* in giving up distributivity (rendering it problematic to interpret the logical operations $\wedge$ and $\vee$ as conjunction and disjunction, respectively);
- *not radical enough* in keeping the law of excluded middle, which is precisely what an "intuition pump" like Schrödinger's cat challenges.

Thus it would be preferable to have a quantum logic with exactly the *opposite* features, that is, one that remains distributive but drops the law of the excluded middle. This suggest the use of *intuitionistic logic* for quantum mechanics, and actually finding appropriate models thereof has been the main outcome of the quantum toposophy program so far.[2]

The aim of this paper is to put the intuitionistic quantum logic discovered through the topos approach in the light of the great (categorical) dualities that on the one hand deserve the name "spatial," and on the other hand are somehow related to logic, namely, Gelfand duality in (commutative) C*-algebra theory, reconsidered in Section 2, and the dualities in lattice theory named after Stone, Birkhoff, Priestely, and Esakia, which will be reviewed in Sections 3 and 4. In Section 5 we show how all of these dualities culminate in our models for intuitionistic quantum logic, which, more or less as an afterthought, are finally derived from topos theory in Section 6.

2 Gelfand Duality Revisited

Our approach to Gelfand duality (as well as to all other topics treated in this paper) will be constructive, which not only means that proofs by *reductio*

---

[2] Although the initial goals of the topos-theoretic approach to quantum mechanics were quite a bit more ambitious, including quantum gravity and the associated development of an entirely new language for theoretical physics; cf. the founding literature on the subject starting with Isham and Butterfield [32] and ending with the review by Döring and Isham [15] – in my view topos theory is best (and more modestly) seen as a tool providing a new approach to quantum logic. See Caspers et al. [8], Heunen, Landsman, and Spitters [28, 29], Heunen et al. [30] Landsman [38], Hekkelman [22], and Rutgers [49] for our side of the program and Wolters [50, 51] for a comparison between the "contravariant" approach of Isham et al. and the "covariant" Nijmegen approach.

ad absurdum, the law of the excluded middle, and the Axiom of Choice are disabled, but also that the use of points is eschewed; instead, one relies on open sets as much as possible.[3] To this end, recall that a *frame* is a complete lattice L that is "infinitely distributive" in that

$$x \wedge \bigvee S = \bigvee \{x \wedge y, y \in S\}, \qquad (2.1)$$

for arbitrary subsets $S \subset L$. Frame homomorphisms by definition preserve finite infima and arbitrary suprema. This defines the category Frm of frames, whose opposite category is called the category Loc of *locales*. Thus a locale is the same thing as a frame, seen however as an object in the opposite category.[4] The motivating example of a frame is the topology $\mathcal{O}(X)$ of a space X, partially ordered by set-theoretic inclusion. Not all frames are topologies though (see also below), and this fact makes the following notation used in constructive mathematics pretty confusing: *any* frame is denoted by $\mathcal{O}(X)$ and the corresponding locale is called X *whether or nor the given frame is a topology, and despite the fact that even if it is, the locale is actually $\mathcal{O}(X)$ rather than the space X. Oh well!*

A simple frame is $2 = \{0, 1\} \equiv \{\bot, \top\}$, with order $0 \leq 1$; this is just the topology $\mathcal{O}(1)$ of a singleton 1. A frame map $p^{-1} \colon \mathcal{O}(X) \to 2$ is the same as a locale map $p \colon 1 \to X$ and defines a *point* of the locale X. We denote the set of points of X by $\mathrm{Pt}(X)$. If $\mathcal{O}(X)$ is the topology of some space X, then each point $x \in X$ corresponds to a map

$$p_x \colon 1 \to X, \quad p_x(1) = x; \qquad (2.2)$$

whose inverse image map $p_x^{-1} \colon \mathcal{O}(X) \to \underline{2}$ is a frame map and hence defines a point in the above sense. Conversely, if X is sober (see below), each point of $\mathcal{O}(X)$ arises in that way. The set $\mathrm{Pt}(X)$ has a natural topology, with opens

$$\mathrm{Pt}(U) = \{p \in \mathrm{Pt}(X) \mid p(1) \in U\}, \qquad (2.3)$$

where $U \in \mathcal{O}(X)$; here $p(1) \in U$ really means $p^{-1}(U) = 1$. This gives a frame map

$$\mathcal{O}(X) \to \mathrm{Pt}(X); \qquad (2.4)$$
$$U \mapsto \mathrm{Pt}(U). \qquad (2.5)$$

A frame $\mathcal{O}(X)$ (or the corresponding locale X) is called *spatial* if this map is an isomorphism. Spatial frames are topologies, but this does not mean that any

---

[3] See Johnstone [34] for motivation and also Mac Lane and Moerdijk [41] for some of what follows.
[4] See Johnstone [33] and Picado and Pultr [47].

topology $\mathcal{O}(X)$ is isomorphic (as a frame) to $\mathcal{O}(\text{Pt}(X))$, since $\text{Pt}(X)$ may not be homomorphic to X. Spaces X for which this *is* the case are called *sober*; more precisely, in that case the map

$$X \to \text{Pt}(X); \tag{2.6}$$

$$x \mapsto p_x, \tag{2.7}$$

is a homomorphism. Thus a sober space X may be reconstructed (up to homomorphism) from its topology $\mathcal{O}(X)$. The category Frm has a full subcategory Spat of spatial frames, likewise the category Top of topological spaces has a full subcategory Sob of sober spaces, and it is well known (cf. [41, Section IX.3, Corollary 4]) that

$$\text{Spat} \simeq \text{Sob}^{\text{op}}, \tag{2.8}$$

that is, the categories Spat and Sob are dual (here C^{op} is the opposite category C): if X is a sober space, then $\mathcal{O}(X)$ is a spatial frame, and if $\mathcal{O}(X)$ is a spatial frame, then $\text{Pt}(X)$ is a sober space (with the obvious choices of maps making these associations functorial).

For later use (in Gelfand duality), we mention that a frame $\mathcal{O}(X)$ with top element $\top$ (which exists because $\mathcal{O}(X)$ is a complete lattice, whence $\top = \bigvee \mathcal{O}(X)$) is called *compact* if every subset $S \subset \mathcal{O}(X)$ with $\bigvee S = \top$ has a finite subset $F \subset S$ with $\bigvee F = \top$. Furthermore, $\mathcal{O}(X)$ is *regular* if each $V \in \mathcal{O}(X)$ satisfies

$$V = \bigvee \{U \in \mathcal{O}(X) \mid U \ll V\}, \tag{2.9}$$

where $U \ll V$ iff there exists W such that $U \wedge W = \bot$ and $V \vee W = \top$.[5] If some frame $\mathcal{O}(X)$ is a topology, then $\mathcal{O}(X)$ is compact and regular iff X is compact and Hausdorff.

Gelfand duality, at last, states, in its simplest form,[6] that one has a duality

$$\text{CCA}_1 \simeq \text{CH}^{\text{op}}, \tag{2.10}$$

---

[5] Note that $U \ll V$ implies $U \leqslant V$, since $U = U \wedge (V \vee W) = (U \wedge V) \vee (U \wedge W) = U \wedge V \leqslant V$.

[6] Less elementary forms of Gelfand duality refer to the nonunital/noncompact case. One version is CCAn $\simeq$ LCHp^{op}, where CCAn is the category of commutative C*-algebras with nondegenerate homomorphisms and LCHp is the category of locally compact Hausdorff spaces and proper continuous maps. This easily follows from unitization, i.e., adding a formal unit to a C*-algebra without one, see, e.g., Landsman (2017), §C.6. Another, due to An Huef, Raeburn, and Williams [3], is CCAm $\simeq$ LCH^{op}, where CCAm is the category of commutative C*-algebras with nondegenerate homomorphisms into the multiplier algebra as arrows and LCH is the category of locally compact Hausdorff spaces and continuous maps. As far as I know, the explicit categorical perspective on Gelfand duality goes back to Negrepontis [43].

where CCA_1 is the category of commutative unital C*-algebras and unital homomorphisms (by which we mean *-homomorphisms), CH is the category of compact Hausdorff spaces and continuous maps, and $\simeq$ denotes equivalence of categories. The idea of the proof is to map a unital C*-algebra A into its Gelfand spectrum $\Sigma(A)$, which consists of all nonzero multiplicative linear functionals $A \to \mathbb{C}$ (or, equivalently, of all pure states on A), equipped with the topology of pointwise convergence (in which $\Sigma(A)$ is compact and Hausdorff); in the opposite direction, a compact Hausdorff space X is sent to the algebra $C(X)$ of continuous functions $X \to \mathbb{C}$ with pointwise operations and the supremum-norm (in which $C(X)$ is a commutative unital C*-algebra). Functorially, any unital homomorphism $\varphi \colon A \to B$ induces a pullback $\varphi^* \colon \Sigma(B) \to \Sigma(A)$, and similarly any continuous map $f \colon X \to Y$ induces a pullback $f^* \colon C(Y) \to C(X)$. In particular, equation (2.10) implies

$$A \cong C(\Sigma(A)), \quad a \mapsto \hat{a}; \qquad (2.11)$$

$$X \cong \Sigma(C(X)), \quad x \mapsto \mathrm{ev}_x, \qquad (2.12)$$

where $\hat{a} \colon \Sigma(A) \to \mathbb{C}$ is the *Gelfand transform* of a, neatly defined by $\hat{a}(\omega) = \omega(a)$, and $\mathrm{ev}_x \colon C(X) \to \mathbb{C}$ is the evaluation map at $x \in X$, that is, $\mathrm{ev}_x(f) = f(x)$.

All (known) proofs of Gelfand duality are nonconstructive, typically relying on either Zorn's Lemma (in realizations of $\Sigma(A)$ through maximal ideals, as in Gelfand's original approach) or on the (equivalent) Hahn–Banach Theorem (in the above definition of $\Sigma(A)$). Constructive versions of Gelfand duality therefore change the statement of the theorem.

In the most radical approach [23, 24] both sides of the duality are changed: instead of C*-algebras one uses so-called *localic* C*-algebras, while compact Hausdorff spaces are replaced by compact regular *locales*. It is enough for our purposes to make the second change but not the first; this slightly less radical approach to Gelfand duality goes back to Banaschewski and Mulvey [4] and was continued by Coquand and Spitters [10].

Constructive Gelfand duality, then, states that CCA_1 is dual to the category of compact regular locales (i.e., equivalent to the category of compact regular frames). Of course, the point is to define the constructive Gelfand spectrum $\mathcal{O}(\Sigma(A))$ directly from A as a frame (or locale), rather than as the topology of the underlying space $\Sigma(A)$.[7] This may be done as follows.[8] A *hereditary*

---

[7] Indeed, in most toposes different from the topos of sets (cf. [41]) the classical Gelfand spectrum does not even exist.

[8] The following construction of $\mathcal{O}(\Sigma(A))$ is taken from Landsman [38, Section C.11], inspired by Akemann and Bice [1]. In the references cited in footnote 2 we used a much more complicated construction, adopted from Coquand and Spitters [10]. I did not redo our computation of

subalgebra of a C*-algebra A is a C*-subalgebra H of A with the property that $a \leqslant b$ for $b \in H^+$ and $a \in A^+$ implies $a \in H^+$.[9] The set of all hereditary subalgebras of A is denoted by $H(A)$. Similarly, the set of all closed left (right) ideals in A is called $L(A)$ ($R(A)$), and the closed two-sided ideals are denoted by $I(A)$. It is easy to show that there are bijective correspondences between hereditary subalgebras H of A, closed left ideals L of A, and closed right ideals R of A, given by

$$L = \{a \in A \mid a^*a \in H^+\}; \tag{2.13}$$

$$R = \{a \in A \mid aa^* \in H^+\}; \tag{2.14}$$

$$H = L \cap L^* = R \cap R^*. \tag{2.15}$$

The set $H(A)$ is a complete lattice in the partial order given by set-theoretic inclusion, with inf and sup of any subset $S \subset H(A)$ given by

$$\bigwedge S = \bigcap S; \tag{2.16}$$

$$\bigvee S = \bigcap \{I \in H(A) \mid I \supseteq J \text{ for all } J \in S\}. \tag{2.17}$$

If A is commutative, with Gelfand spectrum $\Sigma(A)$, then $H(A)$ is a frame, and one has

$$\mathcal{O}(\Sigma(A)) \cong H(A) \tag{2.18}$$

as a frame isomorphism. Moreover, in that case, $L^* = L$, $R^* = R$, and $L = R = H$, so

$$H(A) = I(A) = L(A) = R(A). \tag{2.19}$$

In the usual description, where $\Sigma(A)$ is a space, the map $U \mapsto C_0(U)$ provides an isomorphism (where $U \in \mathcal{O}(\Sigma(A))$, that is, $U \subset \Sigma(A)$ is open), but constructively it is best to simply *define* the constructive Gelfand spectrum $\mathcal{O}(\Sigma(A))$ as $H(A)$. If this is taken as the starting point (and it will), then the connection with the usual theory is as follows:[10]

$\mathcal{O}(\Sigma(A))$ in terms of $H(A)$, but the result should be the same. It should be mentioned, though, that the *proof* of the constructive formulation of Gelfand duality by Coquand and Spitters [10] is in fact constructive, whereas my proof of (2.18) is not.

[9] Here A^+ is the positive cone in A, defined for example as $A^+ = \{a^*a \mid a \in A\}$, and for self-adjoint a and b we say that $a \leqslant b$ iff $b - a \in A^+$ (so that in particular $b \geqslant 0$ iff $b \in A^+$).

[10] A *prime element* $P \in \mathcal{O}(X)$ of some frame $\mathcal{O}(X)$ is an element $P \neq \top$ such that $U \wedge V \leqslant P$ iff $U \leqslant P$ or $V \leqslant P$. For a point $p^{-1}: \mathcal{O}(X) \to \underline{2}$, we write $\ker(p^{-1})$ for $\{U \in \mathcal{O}(X) \mid p^{-1}(U) = 0\}$. For any frame $\mathcal{O}(X)$ (i.e., locale X), there is a bijective correspondence between points $p^{-1}: \mathcal{O}(X) \to \underline{2}$ of X and prime elements $P \in \mathcal{O}(X)$, given by $P = \bigvee \ker(p^{-1})$ and $p^{-1}(U) = 0$ iff $U \leqslant P$. Under this correspondence, the topology on $\text{Pt}(X)$ is given by the *Zariski topology*, whose *closed* sets F_P consist of all $Q \supseteq P$, where P is some prime element of $\mathcal{O}(X)$. The prime elements of $H(A)$, where A is a commutative C*-algebra, are the *prime*

1. The frame $H(A)$ of hereditary subalgebras of a commutative C*-algebra A is spatial, with $\text{Pt}(H(A)) \cong \Sigma(A)$ as topological spaces.
2. The prime elements of $H(A)$ are the maximal ideals of A, so that, equipping the set $\mathcal{M}(A)$ of maximal ideals of A with the Zariski topology, we have $\mathcal{M}(A) \cong \Sigma(A)$.
3. The Gelfand isomorphism (2.11) of the classical theory is replaced by

$$A \cong \text{Frm}(\mathcal{O}(\mathbb{C}), H(A)), \tag{2.20}$$

where we refrain from using the notation $\mathcal{O}(H(A))$ for the frame $H(A)$, as the underlying locale will not occur. In general, $\text{Frm}(\mathcal{O}(Y), \mathcal{O}(X)) = \text{Hom}_{\text{Frm}}(\mathcal{O}(Y), \mathcal{O}(X))$ denotes the set of frame maps $f^{-1}: \mathcal{O}(Y) \to \mathcal{O}(X)$, often written as $\text{Loc}(X, Y)$ or $\text{Hom}_{\text{Loc}}(X, Y)$ or (confusingly) even as $C(X, Y)$, since in the spatial case these are precisely the continuous maps $f: X \to Y$; with inverse image maps f^{-1} as above. Because of this, equations (2.18) and (2.20) recover the classical Gelfand isomorphism (2.11).

For example, if A is finite-dimensional (and still commutative), so that $A \cong \mathbb{C}^n$, we have

$$\mathcal{P}(A) \xrightarrow{\cong} H(A); \tag{2.21}$$

$$e \mapsto eA = \{a \in A \mid ea = a\}. \tag{2.22}$$

Indeed, if $A = \mathbb{C}^n$, so that $a \in A$ is an n-tuple $(a_0, \ldots, a_{n-1})$ with $a_k \in \mathbb{C}$, then each projection $e = (e_0, \ldots, e_{n-1}) \in \mathcal{P}(\mathbb{C}^n)$ is an n-tuple whose only entries are 0 and 1; the pertinent isomorphism $\mathcal{P}(\mathbb{C}^n) \to H(\mathbb{C}^n)$ then maps e to the ideal $e \cdot \mathbb{C}^n$ consisting of all $(a_0, \ldots, a_{n-1}) \in \mathbb{C}^n$ such that $a_k = 0$ if $e_k = 0$ ($k = 0, \ldots, n-1$). Equivalently,

$$\mathcal{P}(\mathbb{C}^n) \cong P(n); \tag{2.23}$$

$$H(\mathbb{C}^n) \cong P(n), \tag{2.24}$$

where the natural number n is seen (à la von Neumann) as the set $\{0, 1, \ldots, n-1\}$, and $P(n)$ is its power set (partially ordered, as always, by inclusion). The (frame) isomorphism (2.23) comes from the bijection $P(n) \to \mathcal{P}(\mathbb{C}^n)$ that maps $s \in P(n)$ to the projection e with $e_k = 1$ iff $k \in s$ (and hence $e_k = 0$ iff

---

ideals in A, i.e., the proper ideals $J \subset A$ such that $J_1 J_2 \subset J$ iff $J_1 \subseteq A$ or $J_2 \subseteq A$, for any ideals J_1, J_2 of A (closed by definition, like J); note that $J_1 J_2 = J_1 \cap J_2$. The topology on $\text{Pt}(X)$ is given by the *Zariski topology*, whose *closed* sets F_P consist of all $Q \supseteq P$, where P is some prime element of $\mathcal{O}(X)$. A proof of the three claims in the main text may be found in Landsman [38, Theorem C.86].

$k \notin s$), while (2.24) is the bijection $P(n) \to H(\mathbb{C}^n)$ that maps $s \subset n$ to the ideal $I_s = \{a \in \mathbb{C}^n \mid a_k = 0 \,\forall\, k \notin s\}$. Similarly,

$$\mathcal{P}(A) \xrightarrow{\cong} \mathcal{O}(\Sigma(A)) = P(\Sigma(A)); \tag{2.25}$$

$$e \mapsto \{\varphi \in \Sigma(A) \mid \varphi(e) = 1\}. \tag{2.26}$$

It is enough to prove this for the special case $A = \mathbb{C}^n$, where $\Sigma(\mathbb{C}^n) \cong n$ under the bijection $n \to \Sigma(\mathbb{C}^n)$ given by $k \mapsto \varphi_k$, where $\varphi_k(a) = a_k$ ($k \in n$), and hence $\mathcal{O}(\Sigma(\mathbb{C}^n)) \cong P(n)$, that is, the discrete topology on n. For one thing, together with (2.23) this reproduces (2.25)–(2.26). Furthermore, we obtain the classical Gelfand isomorphism $A \to C(n)$ as $a \mapsto \hat{a}$ with $\hat{a}(k) = a_k$, as well as the constructive Gelfand isomorphism (2.20) as $a \mapsto \tilde{a}$, with

$$\tilde{a} : \mathcal{O}(\mathbb{C}) \to P(n); \tag{2.27}$$

$$U \mapsto \{k \in n \mid a_k \in U\}. \tag{2.28}$$

3 Stone Duality and Its Relatives

We now turn to Stone duality, once again starting with its classical (i.e., spatial) form. A space X is called *totally disconnected* if it has no other connected subspaces than its points (so any larger subspace $\neq X$ is the union of two proper clopen sets). A *Stone space* is a totally disconnected compact Hausdorff space, and we have a full subcategory St of CH whose objects are Stone spaces. At the other side of the duality we have the category BL of boolean lattices (i.e., distributive orthocomplemented lattices) with homomorphisms of orthocomplemented lattices as arrows.[11] Like the power set $P(X)$ of any set, the poset Clopen(X) of all clopen subsets of some Stone space X (partially ordered by set-theoretic inclusion, so that suprema are unions and infima are intersections) is a boolean lattice. Conversely, a boolean lattice L is isomorphic to Clopen(X) for some Stone space $X = \mathcal{S}(L)$, called the *Stone spectrum* of L, which is uniquely determined by L up to homomorphism (in a manner reviewed below). This gives *Stone duality*:

$$\mathsf{BL} \simeq \mathsf{St}^{\mathrm{op}}. \tag{3.1}$$

The Stone spectrum $\mathcal{S}(L)$ of a boolean lattice L has a canonical realization resembling the set of points of a frame just discussed. This time, we regard

---

[11] An *orthocomplementation* on a lattice L with 0 and 1 is a map $\perp: L \to L$, $x \to x^\perp$, that satisfies $x^{\perp\perp} = x$, $x \leqslant y$ iff $y^\perp \leqslant x^\perp$, $x \wedge x^\perp = 0$, and $x \vee x^\perp = 1$. A lattice (poset) with an orthocomplementation is called *orthocomplemented*. A *homomorphism* of orthocomplemented lattices is a lattice morphism that also preserves the orthocomplementation, as well as 0 or 1.

$2 = \{0, 1\} = P(1)$ as a boolean lattice, and define Pt(L) as the set of all homomorphisms $\varphi: L \to 2$, with topology generated by the basic opens U_x defined in (3.3) below, where $x \in L$, and their set-theoretic complements U_x^c. Then Pt(L) is a Stone space, and

$$L \xrightarrow{\cong} \text{Clopen}(\text{Pt}(L)); \qquad (3.2)$$
$$x \mapsto U_x = \{\varphi \in \text{Pt}(L) \mid \varphi(x) = 1\}, \qquad (3.3)$$

is an isomorphism of boolean lattices. The (contravariant) functorial nature of the Stone spectrum comes out particularly clearly from the above description: given a homomorphism $h: L \to L'$, one immediately obtains a map $h^*: \text{Pt}(L') \to \text{Pt}(L)$ by pullback. Conversely, a continuous map $f: X \to Y$ induces the inverse image map $f^{-1}: \mathcal{O}(Y) \to \mathcal{O}(X)$, as above, which restricts to $f^{-1}: \text{Clopen}(Y) \to \text{Clopen}(X)$. The duality (3.1) then implies

$$L \cong \text{Clopen}(\mathcal{S}(L)); \qquad (3.4)$$
$$X \cong \mathcal{S}(\text{Clopen}(X)). \qquad (3.5)$$

A key example of this construction comes from classical propositional logic. Let $S = \{p_1, p_2, \ldots\}$ be an alphabet of atomic propositions, with associated set wff(S) of well-formed formulas over S according to the rules of classical propositional logic.[12] Because of the recursive definition of wff(S), any map $v: S \to 2$ (where $2 = \{0, 1\}$) has a unique extension $v: \text{wff}(S) \to 2$ (called a *valuation*) subject to the rules (with abuse of notation):

$$v(\bot) = 0; \qquad (3.6)$$
$$v(\neg \alpha) = \neg v(\alpha); \qquad (3.7)$$
$$v(\alpha \wedge \beta) = v(\alpha) \wedge v(\beta); \qquad (3.8)$$
$$v(\alpha \vee \beta) = v(\alpha) \vee (\beta); \qquad (3.9)$$
$$v(\alpha \to \beta) = v(\alpha) \to (\beta), \qquad (3.10)$$

where the expressions on the right-hand side are determined by the usual truth tables.

Let $\mathcal{T}$ be some theory, that is, a subset of wff(S), with associated Lindenbaum algebra

$$L(S, \mathcal{T}) = \text{wff}(S)/\sim_\mathcal{T}, \qquad (3.11)$$

where, for any $\psi, \varphi \in \text{wff}(S)$, we say that $\psi \sim_\mathcal{T} \varphi$ if $\mathcal{T} \vdash \psi \leftrightarrow \varphi$, that is, $\psi \leftrightarrow \varphi$ (which abbreviates $(\psi \to \varphi) \wedge (\varphi \to \psi)$) is provable from $\mathcal{T}$.

---

[12] See, e.g., Givant and Halmos [19] or van Dalen [11].

Unlike wff(S), the set $L(S, \mathcal{T})$ is a boolean lattice in the partial order defined by $[\psi] \leqslant [\varphi]$ whenever $\mathcal{T} \vdash \psi \to \varphi$, and the orthocomplementation defined by $[\psi]' = [\neg \psi]$; suprema and infima are given (with some abuse of notation) by $[\psi] \vee [\varphi] = [\psi \vee \varphi]$ and $[\psi] \wedge [\varphi] = [\psi \wedge \varphi]$, respectively.

Define $\mathrm{Mod}_2(S, \mathcal{T})$ as the set of binary models of $\mathcal{T}$, that is, the set of all valuations $v \colon \mathrm{wff}(S) \to 2$ that satisfy $v(\alpha) = 1$ for each axiom $\alpha \in \mathcal{T}$. Then any $v \in \mathrm{Mod}_2(S, \mathcal{T})$ descends to a homomorphism $v' \colon L(S, \mathcal{T}) \to 2$ of boolean lattices, and vice versa, each such homomorphism v' comes from a unique binary model $v \in \mathrm{Mod}_2(S, \mathcal{T})$. Hence the Stone spectrum of the boolean lattice $L = L(S, \mathcal{T})$ (realized as explained earlier) is just

$$\mathcal{S}(L(S, \mathcal{T})) = \mathrm{Mod}_2(S, \mathcal{T}), \tag{3.12}$$

topologized as explained above (3.2), and the isomorphism (3.2)–(3.3) is neatly given by

$$L(S, \mathcal{T}) \xrightarrow{\cong} \mathrm{Clopen}(\mathrm{Mod}_2(S, \mathcal{T})); \tag{3.13}$$

$$[\psi] \mapsto \{V \in \mathrm{Mod}_2(S, \mathcal{T}) \mid V(\psi) = 1\}. \tag{3.14}$$

Thus Stone duality maps the logical equivalence class $[\psi]$ of some wff $\psi \in \mathrm{wff}(S)$ with respect to the given theory $\mathcal{T}$ *(syntax)* to the set of all models of $\mathcal{T}$ in which ψ is true *(semantics)*. Alas, this equivalence cannot be achieved in intuitionistic quantum logic.

For a point-free or constructive description of Stone duality, we note that the topology $\mathcal{O}(\mathcal{S}(L))$ of the Stone spectrum $\mathcal{S}(L)$ of a boolean lattice L may be given directly as

$$\mathcal{O}(\mathcal{S}(L)) \cong \mathrm{Idl}(L), \tag{3.15}$$

where $\mathrm{Idl}(L)$ is the set of all ideals in L, partially ordered by inclusion.[13] Indeed, for any lattice L, the poset $\mathrm{Idl}(L)$ is a frame, whose points are its prime elements, and it is well known that the set $\mathcal{U}(L)$ of ultrafilters of L (which in a boolean lattice coincide with the prime filters of L), topologized by declaring the sets

$$U'_x = \{F \in \mathcal{U}(L) \mid x \in F\} \quad (x \in L), \tag{3.16}$$

---

[13] An *ideal* in a lattice L is a subset $I \subseteq L$ such that $x, y \in I$ implies $x \vee y \in I$, and $y \leqslant x \in I$ implies $y \in I$. An ideal I is *proper* if $I \neq L$. A *maximal ideal* is an ideal that is maximal in the set of all proper ideals, ordered by inclusion. In a boolean lattice, maximal ideals coincide with *prime ideals*, which are ideals I that do not contain the top element 1, and where $x \wedge y \in I$ implies $x \in I$ or $y \in I$. *Filters* in L are defined dually, i.e., as nonempty subsets $F \subset L$ such that $x, y \in F$ implies $x \wedge y \in F$, and $y \geqslant x \in F$ implies $y \in F$. The (set-theoretic) complement of a maximal ideal is a maximal filter (i.e., an ultrafilter), so that an ideal I in a boolean lattice is maximal (i.e., prime) iff for any $x \in L$ either $x \in I$ or $x' \in I$.

to be a basis of the topology, is a model of the Stone spectrum of L, too: for any $\varphi \in \text{Pt}(L)$, the set $\varphi^{-1}(\{1\})$ is an ultrafilter in L (and $\varphi^{-1}(\{0\})$ is a maximal ideal).[14] Unfortunately, the constructive Stone spectrum (3.15) of a boolean lattice is less useful than the constructive Gelfand spectrum $H(A)$ of a commutative C*-algebra given by (2.18), since the Gelfand isomorphism (2.20) actually involves $H(A)$, whereas the Stone isomorphism (3.4) uses the boolean lattice $\text{Clopen}(\mathcal{S}(L))$ rather than the (non-boolean) frame $\mathcal{O}(\mathcal{S}(L))$.

Comparing (2.10) and (3.1) and noting that by definition St is a full subcategory of CH, there must be a relationship between Gelfand duality and Stone duality, which is subtle:

1. Commutative C*-algebras are not the same things as boolean algebras; this difference will be overcome by looking at projections.[15] The set of all projections in a C*-algebra A is denoted by $\mathcal{P}(A)$, and if A is commutative and has a unit 1_A, then $\mathcal{P}(A)$ is a boolean lattice in the partial order $e \leq f$ iff $ef = e$, with orthocomplementation $e' = 1_A - e$; infima are simply given by $e \wedge f = ef$, and suprema are most easily stated through De Morgan's Law, that is, $e \vee f = (e' \wedge f')'$. Without any further assumptions on A (i.e., beyond commutativity and unitality), we then have

$$\mathcal{P}(A) \cong \text{Clopen}(\Sigma(A)). \tag{3.17}$$

2. One needs conditions on a commutative unital C*-algebra A that make its Gelfand spectrum $\Sigma(A)$ a Stone space. This turns out to be the case iff A has *real rank zero*, written $\text{rr}(A) = 0$,[16] or, equivalently (given that A is commutative), iff A is an approximately finite-dimensional or AF C*-algebra.[17] We then have[18]

$$\Sigma(A) \cong \mathcal{S}(\mathcal{P}(A)); \tag{3.18}$$

$$H(A) \cong \text{Idl}(\mathcal{P}(A)); \tag{3.19}$$

$$A \cong C(\mathcal{S}(\mathcal{P}(A))) \tag{3.20}$$

---

[14] This was even Stone's original description of his spectrum of a boolean lattice!

[15] Just like the case $A = B(H)$, an element $e \in A$ of any C*-algebra A is called a projection if $e^2 = e^* = e$. This implies that projections are positive and in particular self-adjoint elements, and the natural partial order on projections, i.e., $e \leq f$ iff $ef = e$, is a special case of the order defined in footnote 9.

[16] This is the case iff the invertible self-adjoint elements of A are dense in all self-adjoint elements of A.

[17] This means that A is the norm-closure of the union of some (not necessarily countable) directed set of finite-dimensional C*-subalgebras (which in turn are necessarily direct sums of matrix algebras).

[18] Cf. Landsman [38, Theorem C.168] for further details and a proof.

as topological spaces, frames, and C*-algebras, respectively. Conversely, for any boolean lattice L the C*-algebra $C(\mathcal{S}(L))$ is AF (and has real rank zero), and

$$L \cong \mathcal{P}(C(\mathcal{S}(L))). \tag{3.21}$$

The case $A = \mathbb{C}^n$ remains instructive: equation (3.17) reproduces our earlier isomorphisms $\mathcal{P}(\mathbb{C}^n) \cong P(n)$ and $\Sigma(\mathbb{C}^n) \cong n$ with discrete topology, so that $\mathrm{Clopen}(n) = P(n)$. The Stone spectrum $\mathcal{S}(P(n))$ consists of all homomorphisms $\varphi \colon P(n) \to 2$ of boolean lattices; since each subset $s \subset n$ is the supremum of its elements, φ is determined by its values on each $\{k\} \subset n$, where $k \in n$, and if $\varphi(k) = 1$, then the condition that φ be a homomorphism enforces $\varphi(l) = 1 - \varphi(n\setminus\{l\}) = 1 - 1 = 0$ for each $l \neq k$, since $k \in n\setminus\{l\}$ and hence $\varphi(n\setminus\{l\}) = 1$. Therefore, $\mathcal{S}(P(n)) \cong n$ under the map $n \mapsto \mathcal{S}(P(n))$ defined by $k \mapsto \varphi_k$ with $\varphi_k(s) = 1$ iff $k \in s$. This gives (3.18). Since $P(n)$ is finite, we have $I = \downarrow (\bigvee I)$ for any $I \in \mathrm{Idl}(P(n))$, and hence $\mathrm{Idl}(P(n)) \cong P(n)$ under the bijection $P(n) \to \mathrm{Idl}(P(n))$ given by $s \mapsto \downarrow s = \{t \in P(n) \mid t \subset s\}$, which gives (3.19). Equation (3.20) is the classical Gelfand isomorphism $\mathbb{C}^n \cong C(n)$ discussed in Section 2, and finally, for $L = P(n)$ the isomorphism (3.21) follows by unfolding: $\mathcal{S}(L) \cong n$, $C(n) \cong \mathbb{C}^n$, and $\mathcal{P}(\mathbb{C}^n) \cong P(n)$ as above.

4 Priestley Duality and Esakia Duality

Equations (3.17)–(3.20) relate classical *propositional* logic to *commutative* C*-algebras, at least in so far as the semantic side of the former is concerned.[19] Toward *intuitionistic quantum* logic, we need to move to *intuitionistic* logic as well as to *noncommutative* C*-algebras. In support of the first move, we first review the well-known concept of a Heyting lattice (or Heyting algebra), which plays the role of a boolean lattice in intuitionistic propositional logic. A *Heyting lattice* is a lattice L with top $\top$ and bottom $\bot$, equipped with a (necessarily unique) map $\to \colon L \times L \to L$, called (*material*) *implication*, that satisfies

$$a \leqslant (b \to c) \text{ iff } (a \wedge b) \leqslant c. \tag{4.1}$$

---

[19] Commutative von Neumann algebras are a special case of commutative AF-algebras, with the special property that their spectrum is *Stonean*, which adds a measure-theoretic property to the Stone condition. So from a logical perspective von Neumann algebras do not form a particularly natural class of C*-algebras.

A Heyting algebra is automatically distributive. Negation (which in a boolean lattice is orthocomplementation and belongs to the primary structure) is *derived* from → by

$$\neg a \equiv (a \to \bot); \quad (4.2)$$

in classical logic this is a tautology (which may be used to eliminate negation), but in intuitionistic logic it is a definition. A Heyting algebra is *complete* when it is complete as a lattice, in that arbitrary suprema (and hence also arbitrary infima) exist. In that case, condition (2.1) is satisfied, so that a complete Heyting algebra is a frame. Conversely, a frame becomes a complete Heyting algebra if we define the implication arrow → by

$$b \to c = \bigvee \{a \in L \mid a \wedge b \leq c\}. \quad (4.3)$$

However, frames and complete Heyting algebras drift apart as soon as morphisms are concerned, for although in both cases one requires maps to preserve the partial order, maps between Heyting algebras must preserve → rather than $\bigvee$ (which in the case of incomplete Heyting lattices would not even be defined). This defines a category HL of (not necessarily complete) Heyting lattices, for which we would like to find a natural dual category of spaces, analogous to St, with an ensuing generalization of Stone duality.[20]

Let DL be the category of bounded distributive lattices (possessing $\bot \equiv 0$ and $\top \equiv 1$) with bounded lattice homomorphisms (i.e., maps preserving 0, 1, $\vee$ and $\wedge$) as arrows. For each bounded distributive lattice L, the associated poset $\mathcal{I}_p(L)$ of prime ideals in L is both a topological space and a poset. The topology is generated by all sets

$$U_a = \{I \in \mathcal{I}_p(L) \mid a \notin I\} \quad (a \in L), \quad (4.4)$$

and their set-theoretic complements U_x^c, and this makes $\mathcal{I}_p(L)$ a Stone space. The partial order on $\mathcal{I}_p(L)$ is simply given by set-theoretic inclusion.[21] The topology and the order satisfy a compatibility condition called the *Priestley separation axiom*:[22]

---

[20] The following result originates with Esakia [17], which is in Russian (which I could not read). I am indebted to Nick Bezhanishvili for this reference, and also for Morandi [42], from which I learned it. More general duality results for distributive lattices go back to Birkhoff (Jr.) and Priestley (cf. [12] and the special issue of *Studia Logica* 56, nos. 1–2 [1996] dedicated to such results).

[21] The same analysis may be carried out using prime filters: the set-theoretic complement of a prime ideal in L is a prime filter. The appropriate partial order then of course changes direction.

[22] A *down-set* in any poset $(P, \leq)$ is a subset $D \subset P$ such that $x \leq y \in P$ implies $x \in P$. So an ideal in a lattice is a down-set that is closed under finite suprema (a filter is an up-set closed under finite infima).

If $b \not\leqslant a$, there is a clopen down-set $U \subset \mathcal{I}_p(L)$ such that $a \in U$ and $b \notin U$.

A partially ordered Stone space satisfying the Priestley separation axiom is called a *Priestley space*. Such spaces form a category Pr with continuous order-preserving maps as arrows. This category has been invented to yield *Priestley duality*, stating that

$$\mathsf{DL} \simeq \mathsf{Pr}^{\mathrm{op}}; \qquad (4.5)$$

- a bounded distributive lattice L yields a Priestley space $\mathcal{P}r(L) = \mathcal{I}_p(L)$;
- a Priestley space X gives rise to the poset $\mathrm{Clopen}_\downarrow(X)$ of *clopen down-sets* of X (ordered by set-theoretic inclusion), which form a bounded distributive lattice.

Functorially, a bounded lattice homomorphism $\varphi \colon L \to M$ gives rise to a continuous order morphism $\varphi^{-1} \colon \mathcal{I}_p(M) \to \mathcal{I}_p(L)$ (i.e., the inverse image map), and a continuous order morphism $f \colon X \to Y$ similarly induces a pullback $f^{-1} \colon \mathrm{Clopen}_\downarrow(Y) \to \mathrm{Clopen}_\downarrow(X)$. In particular, for any bounded distributive lattice L and any Priestley space X we have

$$L \cong \mathrm{Clopen}_\downarrow(\mathcal{I}_p(L)), \quad a \mapsto U_a; \qquad (4.6)$$

$$X \cong \mathcal{I}_p(\mathrm{Clopen}_\downarrow(X)), \quad x \mapsto \{U \in \mathrm{Clopen}_\downarrow(X) \mid x \notin U\}. \qquad (4.7)$$

Two special cases may clarify this result. We call $0 \neq a \in L$ *join-irreducible* if $a = b \vee c$ implies $a = b$ or $a = c$ (equivalently, $a \leqslant b \vee c$ implies $a \leqslant b$ or $a \leqslant c$). Let $\mathcal{J}(L)$ be the set of join-irreducible elements in L, which is a poset in the partial order inherited from L.

- If L is finite, we have an order isomorphism [12, Lemma 10.8]

$$\mathcal{J}(L) \xrightarrow{\cong} \mathcal{I}_p(L); \qquad (4.8)$$

$$a \mapsto L \setminus \uparrow a, \qquad (4.9)$$

which maps each down-set $\downarrow a \subset \mathcal{J}(L)$, where $a \in \mathcal{J}(L)$, into the (clopen) subset $U_a \subset \mathcal{I}_p(L)$. Consequently, Priestley duality reduces to *Birkhoff duality* between finite distributive lattices L and finite posets P, according to which we have

$$L \cong \mathrm{Down}(\mathcal{J}(L)), \quad a \mapsto (\downarrow a) \cap \mathcal{J}(L); \qquad (4.10)$$

$$P \cong \mathcal{J}(\mathrm{Down}(P)), \quad p \mapsto \downarrow p, \qquad (4.11)$$

where $\mathrm{Down}(P)$ is the lattice of all down-sets in P, partially ordered by set-theoretic inclusion. In this case the *topology* on $\mathcal{I}_p(L)$ is trivial (discrete) and plays no role.

- If L is boolean we recover Stone duality. Thus for boolean lattices the *partial order* on $\mathcal{I}_p(L)$ is trivial and drops out. If L is boolean and finite, $\mathcal{J}(L)$ coincides with the set $\mathcal{A}(L)$ of atoms in L, and Birkhoff duality reduces to $L \cong P(\mathcal{A}(L))$.

Our goal lies in Heyting lattices L. An *Esakia space* is a Priestley space X such that:

For any open set $U \in \mathcal{O}(X)$, the corresponding down-set $\downarrow U$ (defined as the smallest down-set containing U, that is, the intersection of all down-sets containing U) is open, too.[23]

The appropriate arrows $f: X \to Y$ between Esakia spaces X, Y are called *p-morphisms*, which not only preserve order and topology (i.e., are continuous), but in addition satisfy:

If $y \geqslant f(x)$, there is $x' \geqslant x$ such that $f(x') = y$ (for all $x \in X$, $y \in Y$).

If E is the category of Esakia spaces with p-morphisms, the desired duality is given by

$$\mathsf{HL} \simeq \mathsf{E}^{\mathrm{op}}, \tag{4.12}$$

where the pertinent functors are the restrictions of those just stated for Priestley duality. In particular, the Esakia spectrum $\mathcal{E}(L) = \mathcal{I}_p(L)$ of a Heyting lattice L is the same as the associated Priestley spectrum $\mathcal{P}r(L)$ of L (merely seen as a bounded distributive lattice).[24] Unfortunately, toward the applications to logic we are after, Esakia duality has a drawback compared to Stone duality, in that there seems to be no neat intuitionistic analogue of (3.13)–(3.14). Indeed, the Lindenbaum algebra $L(S, \mathcal{T})$ of an intuitionistic propositional theory remains perfectly well defined and duly yields a Heyting algebra, but the realization of its Esakia spectrum $\mathcal{E}(L(S, \mathcal{T}))$ in terms of binary models of $\mathcal{T}$ is meaningless in intuitionistic logic (since the law of the excluded middle, which intuitionistic logic denies, is automatically valid in binary models). Furthermore, Gödel [20] proved that there cannot be a single *finite* Heyting lattice replacing the boolean lattice 2 in providing a complete semantics of intuitionistic propositional logic, and Bezhanishvili et al. [6] extended this no-go result to arbitrary Heyting lattices. One therefore needs some family of finite Heyting lattices to obtain

---

[23] This may equivalently be stated in terms of clopen sets. Esakia spaces are alternatively called *Heyting spaces*, much as Stones spaces are sometimes called *boolean spaces*.

[24] This means that the constructive description (3.15) of the Priestley or Esakia spectrum remains valid, but it seems less useful here since it carries no information about the order (which is defined on the points).

a complete semantics of intuitionistic propositional logic, such as the well-known *Kripke models*.[25] For any (finite) poset P, the set $\mathrm{Up}(P)$ of all up-sets U of P (i.e., $y \geq x \in U$ implies $y \in U$), is a complete Heyting algebra in the partial order defined by inclusion, with $\vee = \cup$, $\wedge = \cap$, and implication

$$U \to V = \{x \in P \mid (\uparrow x) \cap U \subseteq V\}. \tag{4.13}$$

This is actually a special case of (4.3), since the up-sets form a topology on any poset P, called the *Alexandrov topology* (i.e., $\mathrm{Up}(P) = \mathcal{O}(P)$ in this topology), which has the principal up-sets $\uparrow x = \{y \in P \mid y \geq x\}$ as a basis ($x \in P$). In intuitionistic mathematics, elements $x \in P$ are typically interpreted as states of knowledge or information, so that $x \leq y$ means that y carries more knowledge than x (perhaps because y is 'later' than x).

As in the classical case, one has a set $\mathrm{wff}(S)$ of well-formed formulas over some alphabet S, and once again it follows from the recursive definition of $\mathrm{wff}(S)$ that any map $v: S \to \mathrm{Up}(P)$ uniquely extends to a valuation $v: \mathrm{wff}(S) \to \mathrm{Up}(P)$ subject to the rules (3.6)–(3.10), where this time the expressions on the right-hand sides are defined in the Heyting lattice $\mathrm{Up}(P)$ (with $\neg U \equiv U \to \bot$), rather than in the boolean lattice $\mathbf{2}$. We say that $\varphi \in \mathrm{wff}(S)$ is *valid* with respect to v if $v(\varphi) = 1$ (i.e., the top element $1 = P$ of $\mathrm{Up}(P)$). For any $x \in P$ and $\varphi \in \mathrm{wff}(S)$ we write $x \Vdash \varphi$ iff $x \in v(\varphi)$, and say that x *forces* φ. Then obviously $v(\varphi) = 1$ iff $x \Vdash \varphi$ for all $x \in P$, and we have the *forcing rules*:[26]

$$x \Vdash \varphi \text{ and } y \geq x \text{ imply } y \Vdash \varphi; \tag{4.14}$$

$$x \Vdash \bot \text{ for no } x \in P; \tag{4.15}$$

$$x \Vdash \varphi \wedge \psi \text{ iff } x \Vdash \varphi \text{ and } x \Vdash \psi; \tag{4.16}$$

$$x \Vdash \varphi \vee \psi \text{ iff } x \Vdash \varphi \text{ or } x \Vdash \psi; \tag{4.17}$$

$$x \Vdash \varphi \to \psi \text{ iff for all } y \geq x, y \Vdash \varphi \text{ implies } y \Vdash \psi; \tag{4.18}$$

$$x \Vdash \neg \varphi \text{ iff for all } y \geq x, y \Vdash \varphi \text{ is false}. \tag{4.19}$$

For any theory $\mathcal{T} \subset \mathrm{wff}(S)$, the associated Lindenbaum agebra $L(S, \mathcal{T})$ differs from its classical counterpart, since intuitionistic logic has fewer derivation rules than classical logic (in particular, intuitionistic propositional logic lacks the *reductio ad absurdum* (RAA) rule). This difference makes $L(S, \mathcal{T})$ merely a Heyting lattice (rather than a boolean one), and, similarly to the classical case, any valuation $v: \mathrm{wff}(S) \to \mathrm{Up}(P)$ that satisfies $v(\alpha) = 1$ for each $\alpha \in \mathcal{T}$ (i.e., any model of $\mathcal{T}$ in $\mathrm{Up}(P)$, aptly called a *Kripke model*) descends

---

[25] See, e.g., van Dalen [11] and Dummett [16]. Also, Palmgren [45] is a useful summary.
[26] These rules easily follow from the construction of a valuation $v: \mathrm{wff}(S) \to \mathrm{Up}(P)$. Originally (i.e., in the work of Kripke and his followers), equations (4.14)–(4.18), which imply (4.19), were taken to be *axioms* extending a binary "forcing" relation $x \Vdash p$ on $P \times \mathcal{T}$ to $P \times I_{\mathcal{T}}$.

to a Heyting lattice homomorphism $v': L(S, \mathcal{T}) \to \mathrm{Up}(P)$. Conversely, any such homomorphism comes from a valuation. What seems missing here is a realization of the Esakia spectrum of $L(S, \mathcal{T})$ in terms of Kripke models, but we will come close in the next section, at least for intuitionistic "quantum" logics defined by C*-algebras.

5 Intuitionistic Quantum Logic

We now move straight to intuitionistic quantum logic, explaining its origin in topos theory in the next section. The idea is to associate a Heyting lattice $Q(A)$ to any unital C*-algebra A, in contrast with the Birkhoff–von Neumann idea of associating the (orthomodular) projection lattice $\mathcal{P}(A)$ to A (which, unlike our procedure, only makes sense if A has sufficiently many projections, for example if it is a von Neumann algebra). An important role will be played by the poset $\mathcal{C}(A)$ of all unital commutative C*-subalgebras of A, ordered by set-theoretic inclusion,[27] so we will say a few things about this poset first.[28] The poset $\mathcal{C}(A)$ has a bottom element, namely, $\bot = \mathbb{C} \cdot 1_A$, but no top element unless A is commutative, in which case $\top = A$. Similarly, $\mathcal{C}(A)$ has arbitrary infima (i.e., meets), given by intersection, but it only has suprema (i.e., joins) of families of elements that mutually commute. Indeed, it is easy to show that $\mathcal{C}(A)$ is a complete lattice iff A is commutative. In that case, using the Gelfand isomorphism, $\mathcal{C}(A)$ has a purely topological description, as follows.[29] Let $A = C(X)$. Any $C \in \mathcal{C}(A)$ induces an equivalence relation $\sim_C$ on X by

$$x \sim_C y \text{ iff } f(x) = f(y) \; \forall \; f \in C. \tag{5.1}$$

This, in turn, defines a partition $X = \bigsqcup_\lambda K_\lambda$ of X (henceforth called π), whose blocks $K_\lambda \subset X$ are the equivalence classes of $\sim_C$. This partition is *upper semicontinuous*:

- Each block K_λ of the partition π is closed;
- For each block K_λ of π, if $K_\lambda \subseteq U$ for some open $U \in \mathcal{O}(X)$, then there is $V \in \mathcal{O}(X)$ such that $K_\lambda \subseteq V \subseteq U$ and V is a union of blocks of π (in other words, if K is such a block, then $V \cap K = \emptyset$ implies $K = \emptyset$).

---

[27] One may think of this poset as a mathematical home for Bohr's notion of *complementarity*, in that each $C \in \mathcal{C}(A)$ represents some classical or experimental context, which has been decoupled from the others, *except for the inclusion relations, which relate* compatible *experiments* (in general there seem to be no preferred *pairs* of complementary subalgebras $C, C' \in \mathcal{C}(A)$ that jointly generate A, although Bohr typically seems to have had such pairs in mind, e.g., position and momentum). See Landsman [39].

[28] See also Heunen [27], Lindenhovius [40], and Landsman [38].

[29] This description follows from Firby [18], cf. Lindenhovius [16, Chapter 4], or Landsman [38, Section 9.1].

Conversely, any upper semicontinuous partition π of X defines some $C \in \mathcal{C}(C(X))$ by

$$C = \bigcap_{K_\lambda \in \pi} \dot{I}_{K_\lambda}, \qquad (5.2)$$

where $I_K = \{f \in C(X) \mid f(x) = 0 \,\forall\, x \in K\}$ and $\dot{I}_{K_\lambda}$ is its unitization. Therefore, the poset $\mathcal{C}(C(X))$ is anti-isomorphic to the poset $\mathrm{USC}(X)$ of all upper semicontinuous decompositions of X in the ordering $\leqslant$ in which $\pi \leqslant \pi'$ if π is finer than π', and both posets are actually lattices. If X is finite and hence $A \cong \mathbb{C}^n$, then $\mathcal{C}(\mathbb{C}^n)$ is anti-isomorphic to the partition lattice Π_n of n, such that a partition $n = \bigsqcup_\lambda s_\lambda$ (i.e., $\pi = \{s_\lambda\}$, $s_\lambda \subset n$) corresponds to the set of all $(a_1, \ldots, a_n)$ in $\mathbb{C}^n$ for which $a_i = a_j$ whenever $i, j \in s_\lambda$.[30]

We return to the general case (in which A may be noncommutative). Although the poset $\mathcal{C}(A)$ is not itself our intuitionistic quantum logic, we may nonetheless compare it with the projection lattice $\mathcal{P}(A)$ of traditional quantum logic. For $A = B(H)$, the C*-algebra of all bounded operators on some (not necessarily finite-dimensional) Hilbert space H, the projection lattice $\mathcal{P}(A)$ is already a powerful invariant of A in the following sense: if $\dim(H) > 2$, any order isomorphism $\mathsf{N}\colon \mathcal{P}(B(H)) \to \mathcal{P}(B(H))$ preserving orthocomplementation (i.e., $\mathsf{N}(1_H - e) = 1_H - \mathsf{N}(e)$ for each $e \in \mathcal{P}(B(H))$, where 1_H is the unit operator on H) takes the form $\mathsf{N}(e) = ueu^*$, where the operator u is either unitary or anti-unitary, and is uniquely determined by N up to a phase. This is a corollary of Wigner's Theorem in quantum mechanics, cf. Landsman [38, Theorem 5.4]. Similarly, any order isomorphism $\mathsf{B}\colon \mathcal{C}(B(H)) \to \mathcal{C}(B(H))$ takes the form $\mathsf{B}(C) = uCu^*$, and so on.

More generally, if A and B are unital C*-algebras, we define a *weak Jordan isomorphism* of A and B as an invertible map $\mathsf{J}\colon A \to B$ whose restriction to each $C \in \mathcal{C}(A)$, is a unital homomorphism (of commutative C*-algebras) onto its image, and which also satisfies $\mathsf{J}(a + ib) = \mathsf{J}(a) + i\mathsf{J}(b)$ for all self-adjoint $a, b \in A$. *Hamhalter's Theorem* then states that any order isomorphism $\mathsf{B}\colon \mathcal{C}(A) \to \mathcal{C}(B)$ is implemented by a weak Jordan isomorphism $\mathsf{J}: A \to B$ (whose restrictions to all $C \in \mathcal{C}(A)$ define a map $\mathcal{C}(A) \to \mathcal{C}(B)$). If A is isomorphic to neither $\mathbb{C}^2$ nor $M_2(\mathbb{C})$, then J is uniquely determined by B.[31] The proof of this theorem also gives an explicit reconstruction of A from $\mathcal{C}(A)$, though, of course, as a Jordan algebra rather than as a C*-algebra.[32] In order to recover A as a C*-algebra one needs to endow the poset $\mathcal{C}(A)$ with additional

---

[30] The ordering on Π_n has $\pi' \leqslant \pi$ iff π' is *finer* than π, i.e., iff any $s' \in \pi'$ is contained in some $s \in \pi$.
[31] See Hamhalter [21], Lindenhovius [40, Section 4.7], or Landsman [38, Theorem 9.4].
[32] Connes [9] produced a C*-algebra that is not anti-isomorphic to itself. See also Phillips [46].

structure; see, for example, Heunen and Reyes [31] for AW*-algebras and Döring [14] for von Neumann algebras.[33]

We now define our Heyting lattices. If A is finite-dimensional,[34] then $Q(A)$ is given by

$$Q(A) = \{S \colon \mathcal{C}(A) \to \mathcal{P}(A) \mid S(C) \in \mathcal{P}(C),\ S(C) \leq S(D) \text{ if } C \subseteq D\}. \tag{5.3}$$

As stated before, the partial order on $\mathcal{C}(A)$ is here given by set-theoretic inclusion and the one on $\mathcal{P}(A)$ is $e \leq f$ iff $ef = e$. The partial order $\leq$ on $Q(A)$ is defined by $S \leq T$ iff $S(C) \leq T(C)$ for all $C \in \mathcal{C}(A)$, and in this order $Q(A)$ is a Heyting lattice, with operations

$$(S \wedge T)(C) = S(C) \wedge T(C); \tag{5.4}$$

$$(S \vee T)(C) = S(C) \vee T(C); \tag{5.5}$$

$$(S \to T)(C) = \bigwedge_{D \supseteq C}^{\mathcal{P}(C)} S(D)^\perp \vee T(D), \tag{5.6}$$

where the right-hand side of (5.6) is shorthand for

$$\bigwedge_{D \supseteq C}^{\mathcal{P}(C)} S(D)^\perp \vee T(D) \equiv \bigvee \{e \in \mathcal{P}(C) \mid e \leq S(D)^\perp \vee T(D)\ \forall D \supseteq C\}. \tag{5.7}$$

In contrast to traditional quantum logic, both logical connectives $\wedge$ and $\vee$ on $Q(A)$ are physically meaningful, as they only involve *local* conjunctions $S(C) \wedge T(C)$ and disjunctions $S(C) \vee T(C)$, for which $S(C) \in \mathcal{P}(C)$ and $T(C) \in \mathcal{P}(C)$ commute. With similar notation in (5.8)–(5.9) below, the derived operations $\neg$ and $\neg\neg$ are then given by

$$(\neg S)(C) = \bigwedge_{D \supseteq C}^{\mathcal{P}(C)} S(D)^\perp; \tag{5.8}$$

$$(\neg\neg S)(C) = \bigwedge_{D \supseteq C}^{\mathcal{P}(C)} \bigvee_{E \supseteq D}^{\mathcal{P}(D)} S(E). \tag{5.9}$$

A Heyting algebra is boolean iff $\neg\neg S = S$ for each S, and one sees from (5.9) that (at least if $n > 1$) the property $\neg\neg S = S$ only holds iff S is either $\top$ or $\bot$,

---

[33] A similar problem arises if one wants to reconstruct A from its pure state space (seen as a set with a transition probability; cf. [37]) or from its state space (seen as a compact convex set; see [2]).

[34] As shown by Hekkelman [22] and Rutgers [49, Equation (5.3)] also makes sense for AW*-algebras.

so that already for $A = M_2(\mathbb{C})$ the Heyting algebra $Q(A)$ is non-boolean and hence properly intuitionistic.

In the Birkhoff–Neumann approach each projection $e \in \mathcal{P}(A)$ defines an elementary proposition, whereas in ours (where the "classical context" C is crucial) an elementary proposition is a *pair* (C, e), where $e \in \mathcal{P}(C)$; this is supposed to incorporate Bohr's insight that every proposition in quantum theory ought to be accompanied by the (experimental) context in which it is measured. If for each such pair (C, e) we define

$$S_{(C,e)} : \mathcal{C}(A) \to \mathcal{P}(A); \tag{5.10}$$

$$D \mapsto e \ \ (C \subseteq D); \tag{5.11}$$

$$D \mapsto \bot \ \ \text{otherwise,} \tag{5.12}$$

we see that each pair (C, e) injectively defines an element of $Q(A)$. As pointed out by Hermens [25], each element $S \in Q(A)$ is a disjunction over such elementary propositions:

$$S = \bigvee_{C \in \mathcal{C}(A)} S_{(C, S(e))}. \tag{5.13}$$

In the finite-dimensional commutative case $A \cong \mathbb{C}^n$ it is straightforward to compute $Q(A)$, since we already know that $\mathcal{C}(A) \cong \Pi_n$ (i.e., the partition lattice of n, partially ordered by the *opposite* of the usual refinement order), and $\mathcal{P}(A) \cong P(n)$. Hence

$$Q(\mathbb{C}^n) \cong \tilde{Q}(\mathbb{C}^n) = \{\tilde{S} : \Pi_n \to P(n) \mid \pi \leqslant \{\tilde{S}(\pi), n \setminus \{\tilde{S}(\pi)\}, \tilde{S}(\pi) \subseteq \tilde{S}$$
$$(\pi') \text{ if } \pi' \leqslant \pi\}. \tag{5.14}$$

Here the first condition after the bar means that, for any $\pi \in \Pi_n$, any cell $s \in \pi$ must be contained in either $\tilde{S}(\pi) \subset n$ or in its complement, and the second condition simply states that $\tilde{S}$ is (opposite) order preserving. Let us initially ignore the first condition, however, and compute the poset $\tilde{Q}'(\mathbb{C}^n) \subset \tilde{Q}(\mathbb{C}^n)$ defined by

$$\tilde{Q}'(\mathbb{C}^n) = \{\tilde{S} : \Pi_n \to P(n) \mid \tilde{S}(\pi) \subseteq \tilde{S}(\pi') \text{ if } \pi' \leqslant \pi\}. \tag{5.15}$$

For any poset $(X, \leqslant)$, the Hom-set of homomorphisms of posets from X to $P(n)$ is

$$\text{Hom}(X, P(n)) \cong \text{Hom}(X, 2^n) \cong (\text{Hom}(X, 2))^n \cong (\text{Up}(X))^n, \tag{5.16}$$

and so $\tilde{Q}'(\mathbb{C}^n)$ is isomorphic (as poset and even as a Heyting lattice) to

$$\tilde{Q}'(\mathbb{C}^n) \cong (\text{Down}(\Pi_n))^n, \tag{5.17}$$

where we have $\mathrm{Down}(\Pi_n)$ instead of $\mathrm{Up}(\Pi_n)$ in view of the opposite order. Therefore,[35]

$$Q(\mathbb{C}^n) \cong \{(U_1, \ldots, U_n) \in (\mathrm{Down}(\Pi_n))^n \mid \forall_{\pi \in \Pi_n} \forall_{s \in \pi} ((\forall_{k \in s} \pi \in U_k)$$
$$\vee (\forall_{k \in s} \pi \notin U_k))\}. \tag{5.18}$$

Here, like in (5.16), the partial order is given by inclusion, and the condition after the bar may equivalently be stated as $(k \sim_\pi l) \to (\pi \in U_k \leftrightarrow \pi \in U_l)$, where $k \sim_\pi l$ means that k and l lie in the same cell s of the partition π. See also the description (5.29)–(5.30) below.

It is an open question what the Esakia spectrum $\mathcal{E}(Q(A))$ of the Heyting algebra $Q(A)$ is. The closest approximation to the classical case would be to replace the value set $\{0, 1\}$, seen as the discrete topology of a singleton, by the Alexandrov topology of the poset $P = \mathcal{C}(A)$, and hence to replace (3.12) by the set $\mathrm{Mod}_{\mathcal{C}(A)}(Q(A))$ of all Heyting lattice homomorphisms from $Q(A)$ to $\mathrm{Up}(\mathcal{C}(A))$. Indeed,[36] any state ω on A defines a function

$$V_\omega : Q(A) \to \mathrm{Up}(\mathcal{C}(A)); \tag{5.19}$$
$$V_\omega(S) = \{C \in \mathcal{C}(A) \mid \omega(S(C)) = 1\}. \tag{5.20}$$

If we say that $S \in Q(A)$ is *true* in a state ω provided $V_\omega(S) = \mathcal{C}(A)$ (i.e., the top element of the frame $\mathrm{Up}(\mathcal{C}(A))$), and call S *false* if $V_\omega(S) = \emptyset$ (i.e., the bottom element of $\mathrm{Up}(\mathcal{C}(A))$), then $\neg S$ is true iff S is false, and $S \vee T$ is true iff either S or T is true.[37] Consequently, (5.20) simply lists the contexts C in which $S(C)$ is true, and we have $C \Vdash S$ iff $\omega(S(C)) = 1$.

The problem, however, is that the Kochen–Specker Theorem implies that for reasonably noncommutative A (and $A = M_n(\mathbb{C})$ for $n > 1$ is already a case in point) the set of Heyting lattice homomorphisms from $Q(A)$ to $\mathrm{Up}(\mathcal{C}(A))$ is empty.[38] The ensuing disappointment is only limited, since, as already pointed out in the text following (4.12), the poset $\mathcal{C}(A)$ would not be able to do the job on its own in any case. Nonetheless, it would be desirable to map propositions in $Q(A)$ to the (clopen down-) sets of ($\mathcal{C}(A)$-valued Kripke) models (replacing binary models) in which they are true, as in the classical case.

---

[35] I am indebted to Nick Bezhanishvili, Guram Bezhanishvili, David Gabelaia, and Mamuka Jibladze for help with the computation (5.18), in response to an erroneous conjecture in an earlier draft of this chapter.

[36] Note that $V_\omega(S)$ indeed defines an up-set in $\mathcal{C}(A)$, for if $C \subseteq D$ then $S(C) \leq S(D)$, so that $\omega(S(C)) \leq \omega(S(D))$ by positivity of states, and hence $\omega(S(D)) = 1$ whenever $\omega(S(C)) = 1$ (given that $\omega(S(D)) \leq 1$, which is true since $0 \leq \omega(e) \leq 1$ for any projection e).

[37] Since $V_\omega(S) = \mathcal{C}(A)$ iff $S(\mathbb{C} \cdot 1) = 1$, which forces $S(C) = 1$ for all C.

[38] See, e.g., Landsman [38, Section 12.5], based on Heunen, Landsman, and Spitters [28, 29]. I am very grateful to my students Evert-Jan Hekkelman and Quinten Rutgers for reminding me of our own result; their B.Sc. theses [22, 49] contain many interesting results in this direction.

This analysis can be generalized to any unital C*-algebra A.[39] First, we define the set

$$\Sigma_A = \bigsqcup_{C \in \mathcal{C}(A)} \Sigma(C), \qquad (5.21)$$

that is, the disjoint union over all Gelfand spectra $\Sigma(C)$, where $C \in \mathcal{C}(A)$. We then equip Σ_A with the weakest topology making the canonical projection

$$\pi : \Sigma_A \to \mathcal{C}(A); \qquad (5.22)$$
$$\pi(\sigma) = C, \qquad (5.23)$$

where $\sigma \in \Sigma(C) \subset \Sigma_A$, continuous with respect to the Alexandrov topology on $\mathcal{C}(A)$. To be more specific, note that any $\mathcal{U} \subset \Sigma_A$ takes the form

$$\mathcal{U} = \bigsqcup_{C \in \mathcal{C}(A)} \mathcal{U}_C; \qquad (5.24)$$
$$\mathcal{U}_C = \mathcal{U} \cap \Sigma(C). \qquad (5.25)$$

Then $\mathcal{U}$ is open iff the following two conditions are satisfied for each $C \in \mathcal{C}(A)$:

1. $\mathcal{U}_C \in \mathcal{O}(\Sigma(C))$.
2. For all $D \supseteq C$, if $\lambda \in \mathcal{U}_C$ and $\lambda' \in \Sigma(D)$ such that $\lambda'_{|C} = \lambda$, then $\lambda' \in \mathcal{U}_D$.

Being a frame, the topology $\mathcal{O}(\Sigma_A)$ is a Heyting lattice, which generalizes our earlier Heyting lattice $Q(A)$ in (5.3) to arbitrary unital C*-algebras A. To see that $Q(A)$ is indeed a special case of $\mathcal{O}(\Sigma_A)$, where A is taken to be finite-dimensional (e.g., $A = M_n(\mathbb{C})$), we use (2.25)–(2.26), where A is now replaced by $C \in \mathcal{C}(A)$, so that we have an isomorphism (of boolean lattices) $\beta : \mathcal{P}(C) \to \mathcal{O}(\Sigma(C))$. We then obtain a Heyting lattice isomorphism

$$\mathcal{O}(\Sigma_A) \xrightarrow{\cong} Q(A); \qquad (5.26)$$
$$\mathcal{U} \mapsto S_\mathcal{U}; \qquad (5.27)$$
$$S_\mathcal{U}(C) = \beta^{-1}(\mathcal{U}_C). \qquad (5.28)$$

Conversely, each $S \in Q(A)$ defines $\mathcal{U} \in \mathcal{O}(\Sigma(C))$ by (5.24) with $\mathcal{U}_C = \beta(S(C))$.

In our running example $A \cong \mathbb{C}^n$ this leads to the description

$$\Sigma_{\mathbb{C}^n} \cong \bigsqcup_{\pi \in \Pi_n} \pi, \qquad (5.29)$$

---

[39] The discovery that $Q(A)$ is a topology, which is far from obvious even in the special case (5.3), is due to Wolters [50, 51]. See also Heunen [30] and Landsman [38, Section 12.4].

whose elements we denote by pairs (π, s) with $s \in \pi$ (and hence $s \subset n$). The topology $\mathcal{O}(\Sigma_{\mathbb{C}^n})$ is then given by all subsets $\mathcal{U} \subset \Sigma_{\mathbb{C}^n}$ such that if $(\pi, s) \in \mathcal{U}$, then $(\pi', s') \in \mathcal{U}$ whenever $\pi' \leqslant \pi$ (i.e., π' is finer than π) and $s' \subset s$ (where $\subset$ is the same as $\subseteq$). The previous description (5.18) is then recovered through

$$U_k = \{\pi \in \Pi_n \mid \exists_{s \subset n}(((\pi, s) \in \mathcal{U}) \wedge (k \in s))\}. \tag{5.30}$$

The topological condition on $\mathcal{U}$ then precisely gives rise to the condition on the U_k in (5.18).

Also in the general case we may write $\mathcal{U}$ as a disjunction à la (5.13), viz.

$$\mathcal{U} = \bigvee_{C \in \mathcal{C}(A)} \mathcal{U}_C, \tag{5.31}$$

which is even almost trivial, since $\vee = \cup$ in the frame $\mathcal{O}(\Sigma_A)$. The elementary propositions $\mathcal{U}_C \subset \Sigma(C)$ (which in the finite-dimensional case may be identified with projections in C and hence in A, as we have seen) are open subsets of the "classical phase spaces" $\Sigma(C)$, which, in the spirit of Bohr, carry the contextual label C. Consequently, the intuitionistic quantum logic of (unital) C*-algebras would be largely understood at the level of propositional logic, except for the possible functoriality of the map $A \mapsto \mathcal{O}(\Sigma_A)$, that is, of $A \mapsto Q(A)$ for a finite-dimensional C*-algebra A.[40] As soon as this is solved, we would have a new duality between *arbitrary* (unital) C*-algebras and a particular class of Heyting lattices, which is meant to replace Gelfand duality for *commutative* C*-algebras.

What remains is an extension to first-order logic, which we suggest to be the internal logic of the sheaf topos $\mathsf{Sh}(\Sigma_A)$. To understand this suggestion, we now briefly review the topos-theoretic background of the above construction of the Heyting lattice $Q(A)$, which framework (as pointed out before) we now regard as a means rather than as an end.

6 Epilogue: From Topos Theory to Quantum Logic

Let A be a unital C*-algebra, with associated poset $\mathcal{C}(A)$ of all unital commutative C*-subalgebras of A, as before. Regarding $\mathcal{C}(A)$ as a (posetal) category, in which there is a unique arrow $C \to D$ iff $C \subseteq D$ and there are no other arrows, we obtain the topos $\mathsf{T}(A)$ of *covariant* functors $\underline{F} \colon \mathcal{C}(A) \to \mathsf{Sets}$

---

[40] This problem is highly nontrivial (cf. [5, 13]).

from $\mathcal{C}(A)$ into the category Sets of sets, that is,[41]

$$T(A) = [\mathcal{C}(A), \mathsf{Sets}]. \tag{6.1}$$

Since for any poset X we have an equivalence (even an isomorphism) of categories $[X, \mathsf{Sets}] \simeq \mathsf{Sh}(X)$, where X is endowed with the Alexandrov topology,[42] we may alternatively write

$$T(A) \simeq \mathsf{Sh}(\mathcal{C}(A)). \tag{6.2}$$

This category is a *topos*,[43] which makes it a "universe of discourse" in which to do mathematics, replacing set theory.[44] One major difference with set theory is that the logic in most toposes (including $T(A)$) is intuitionistic.[45] Nonetheless, (Dedekind) real numbers and C*-algebras can be defined in toposes (with a natural numbers object), and one even has various constructive versions of Gelfand duality.[46] All we need is the fact that each *commutative* C*-algebra $\underline{A}$ in a topos T has a constructive Gelfand spectrum $\underline{\Sigma}(\underline{A})$, defined as a locale in T, and an associated Gelfand isomorphism à la (2.20), also within T, where $\underline{\Sigma}(\underline{A})$ may be either defined or computed as $\underline{H}(\underline{A})$, the lattice of hereditary C*-subalgebras of $\underline{A}$. Here we underline objects in T, especially the internal C*-algebra $\underline{A}$, in order to distinguish (underlined) constructions internal to T from constructions in Sets, like the given C*-algebra A, on which our reasoning in T ultimately relies. The entire argument hinges on the following C*-algebra $\underline{A}$ in our topos $T(A)$:[47]

$$\underline{A} : \mathcal{C}(A) \to \mathsf{Sets}; \tag{6.3}$$

$$C \mapsto C; \tag{6.4}$$

$$(C \subseteq D) \mapsto (C \hookrightarrow D), \tag{6.5}$$

---

[41] One usually works with presheaves on a given category C, i.e., *contravariant* functors C → Sets. Thus $T(A)$ consists of presheaves on $\mathcal{C}(A)^{\mathrm{op}}$, in which the order on $\mathcal{C}(A)$ is reversed.

[42] This isomorphism maps a functor $\underline{F}: X \to$ Sets to a sheaf $F: \mathcal{O}(X)^{\mathrm{op}} \to$ Sets, by defining the latter on a basis of the Alexandrov topology as $F(\uparrow x) = \underline{F}(x)$ extended to general Alexandrov opens by the sheaf property. Vice versa, a sheaf F on X defines $\underline{F}$ by reading the previous equation from right to left.

[43] A topos is a cartesian closed category (i.e., having a terminal object, binary products, and function spaces) with pullbacks and a subobject classifier. See Mac Lane and Moerdijk [41] or Johnstone [35].

[44] Although in our approach set theory remains the metamathematics in which topos theory is studied.

[45] This implies in particular that all definitions and proofs have to be constructive, in that *reduction ad absurdum*, the law of the excluded middle, and the Axiom of Choice are not enabled.

[46] See Banaschewski and Mulvey [4], Coquand and Spitters [10], and Henry [23, 24].

[47] This C*-algebra was introduced in Heunen, Landsman, and Spitters [28]. The proof that $\underline{A}$ is actually a C*-algebra in $T(A)$ is nontrivial and is somewhat incomplete in the above reference. An improved version may be found in Landsman [38, Section 12.1], relying on details in Banaschewski and Mulvey [4].

where, despite the identical notation, on the left-hand side of (6.4) C is an element of $\mathcal{C}(A)$, whereas on the right-hand side it is the corresponding C*-subalgebra of A seen as a set, and in (6.5) the notation $C \subseteq D$ denotes the unique arrow in $\mathsf{T}(A)$ from C to D, which the functor $\underline{A}$ maps to the inclusion map $\hookrightarrow$ from C into D in the category of sets.

The point is that $\underline{A}$ is a *commutative* C*-algebra in $\mathsf{T}(A)$ under pointwise operations, called the *Bohrification* of A (which may be as noncommutative as one desires). Its constructive Gelfand spectrum $\underline{\Sigma}(\underline{A})$ has been explicitly computed within $\mathsf{T}(A)$,[48] but one of the virtues of toposes of sheaves $\mathsf{Sh}(X)$ (at least for beginners, like the author) is that any locale $\underline{Y}$ in $\mathsf{Sh}(X)$ has a so-called *external* description in set theory,[49] namely, as a locale map $\pi : Y \to X$ (in set theory), or, more precisely, as a frame map

$$\pi^{-1} : \mathcal{O}(X) \to \mathcal{O}(Y). \tag{6.6}$$

Here $\mathcal{O}(Y) = \underline{\mathcal{O}(Y)}(X)$, which is a frame in Sets. Applied to the Gelfand spectrum $\underline{\Sigma}(\underline{A})$ in $\mathsf{T}(A)$, where $X = \mathcal{C}(A)$ in the Alexandrov topology, and $\underline{Y} = \underline{\Sigma}(\underline{A})$, it turns out that $\mathcal{O}(Y)$ is spatial, so that the associated frame $\mathcal{O}(Y)$ is the topology of a genuine space and (6.6) is the inverse image map of a continuous map $\pi : Y \to X$, which is given by (5.22).

This may be seen as a 'derivation' of our intuitionistic quantum logic from topos theory, but of course this derivation is based on certain categorical input that may be hardly more convincing than just postulating the Heyting lattice $\mathcal{O}(\Sigma_A)$ or its finite-dimensional case $Q(A)$. Whichever way one looks at its origin, the advantages of $\mathcal{O}(\Sigma_A)$ or $Q(A)$ over the Birkhoff–von Neumann lattice $\mathcal{P}(A)$ are impressive: the logic is distributive (which is needed in order to interpret the lattice operations $\wedge$ and $\vee$ as "and" and "or," respectively, even in quantum theory), lacks the excluded middle third property (which indeed is highly questionable in quantum theory), and it is spatial in two different senses of the word:

- Seen as a frame, the lattice $\mathcal{O}(\Sigma_A)$ is spatial;
- Seen as a Heyting lattice, $\mathcal{O}(\Sigma_A)$ has an associated Esakia spectrum $\mathcal{E}(\mathcal{O}(\Sigma_A))$.

---

[48] See Heunen, Landsman, and Spitters [28], Wolters [50, 51], and Landsman [38, Section 12.4].
[49] See Joyal and Tierney [36] and Johnstone [35, Section C1.6], summarized in Landsman [38, Section E.4].

References

[1] C. A. Akemann and T. Bice, *Hereditary C*-subalgebra lattices*, arXiv: 1410.0093

[2] E. M. Alfsen and F. W. Shultz, *Geometry of State Spaces of Operator Algebras*, Birkhäuser (2003)

[3] A. An Huef, I. Raeburn, and D. P. Williams, *Functoriality of Rieffel's generalised fixed-point algebras for proper actions*, Proc. Symp. Pure Math. 81 (2010) 9–26

[4] B. Banaschewski and C. J. Mulvey, *A globalisation of the Gelfand duality theorem*, Ann. Pure and Appl. Logic 137 (2006) 62–103

[5] B. van den Berg and C. Heunen, *Noncommutativity as a colimit*, Appl. Categorical Struct. 20 (2012) 393–414. Erratum: 21 (2013) 103–4

[6] G. Bezhanishvili, N. Bezhanishvili, D. Gabelaia, and A. Kurz, *Bitopological duality for distributive lattices and Heyting algebras*, Mathe. Struct. Comput. Sci. 20 (2010) 359–93

[7] G. Birkhoff and J. von Neumann, *The logic of quantum mechanics*, Ann. Math. 37 (1936) 823–43

[8] M. Caspers, C. Heunen, N. P. Landsman, and B. Spitters, *Intuitionistic quantum logic of an n-level system*, Found. Phys. 39 (2009) 731–59

[9] A. Connes, *A factor not anti-isomorphic to itself*, Ann. Math. 101 (1975) 536–54

[10] T. Coquand and B. Spitters, *Constructive Gelfand duality for C*-algebras*, Math. Proc. Cambridge Philos. Soc. 147 (2009) 339–44

[11] D. van Dalen, *Logic and Structure*, 5th ed., Springer (2013)

[12] B. A. Davey and H. A. Priestley, *Introduction to Lattices and Order*, 2nd ed., Cambridge University Press (2002)

[13] A. Döring, *Generalised Gelfand spectra of nonabelian unital C*-algebras*, arXiv:1212.2613

[14] A. Döring, *Two new complete invariants of von Neumann algebras*, arXiv:1411.5558

[15] A. Döring and C. J. Isham, *"What is a thing?": Topos theory in the foundations of physics*, Lecture Notes Phys. 813 (2010) 753–937

[16] M. Dummett, *Elements of Intuitionism*, 2nd ed., Clarendon Press (2000)

[17] L. Esakia, *Heyting Algebras I. Duality Theory* (in Russian), Metsniereba Press (1985)

[18] P. A. Firby, *Lattices and compactifications, II*, Proc. London Math. Soc. 27 (1973) 51–60

[19] S. Givant and P. R. Halmos, *Introduction to Boolean Algebras*, Springer (2009)

[20] K. Gödel, *Zum intuitionistischen Aussagenkalkül 1*, Anzeiger Akad. Wissensch. Wien 69 (1932) 65–66. (*Collected Works* Vol. I, pp. 222–25)

[21] J. Hamhalter, *Isomorphisms of ordered structures of abelian C*-subalgebras of C*-algebras*, J. Math. Anal. Appl. 383 (2011) 391–99

[22] E. J. Hekkelman, *Properties of the lattice $O(\Sigma(A))$ concerning intuitionistic quantum logic*, BSc thesis, Radboud University Nijmegen (2018), www.math.ru.nl/~landsman/EvertJan.pdf

[23] S. Henry, *Localic Metric spaces and the localic Gelfand duality*, arXiv: 1411.0898

[24] S. Henry, *Constructive Gelfand duality for non-unital commutative C*-algebras*, arXiv:1412.2009
[25] R. Hermens, *Philosophy of quantum probability: An empiricist study of its formalism and logic*, PhD thesis, Rijksuniversiteit Groningen (2016)
[26] C. Heunen, *Categorical quantum models and logics*, PhD thesis, Radboud University Nijmegen (2009)
[27] C. Heunen, *Characterizations of categories of commutative C*-algebras*, Commun. Math. Phys. 331 (2014) 215–38
[28] C. Heunen, N. P. Landsman, and B. Spitters, *A topos for algebraic quantum theory*, Commun. Math. Phys. 291 (2009) 63–110
[29] C. Heunen, N. P. Landsman, and B. Spitters, *Bohrification of operator algebras and quantum logic*, Synthese 186 (2012) 719–52
[30] C. Heunen, N. P. Landsman, B. Spitters, and S. Wolters, *The Gelfand spectrum of a noncommutative C*-algebra: A topos-theoretic approach*, J. Aust. Math. Soc. 90 (2012) 32–59
[31] C. Heunen and M. L. Reyes, *Active lattices determine AW*-algebras*, J. Math. Anal. Appl. 416 (2014) 289–313
[32] C. J. Isham and J. Butterfield, *Topos perspective on the Kochen–Specker theorem. I. Quantum states as generalized valuations*, Int. J. Theor. Phys. 37 (1998) 2669–2733
[33] P. T. Johnstone, *Stone Spaces*, Cambridge University Press (1982)
[34] P. T. Johnstone, *The point of pointless topology*, Bull. Am. Math. Soc. New Ser. 8 (1983) 41–53
[35] P. T. Johnstone, *Sketches of an Elephant: A Topos Theory Compendium*, 2 vols., Clarendon Press (2002)
[36] A. Joyal and M. Tierney, *An extension of the Galois theory of Grothendieck*, Mem. Am. Math. Soc. 51 (1984) 1–71
[37] N. P. Landsman, *Mathematical Topics between Classical and Quantum Mechanics*, Springer (1998)
[38] N. P. Landsman, *Foundations of Quantum Theory: From Classical Concepts to Operator Algebras*, Springer Open (2017), www.springer.com/gp/book/9783319517766
[39] N. P. Landsman, *Bohrification*, in J. Faye and H. J. Folse, eds., *Niels Bohr and the Philosophy of Physics: Twenty-First-Century Perspectives*, Bloomsbury (2017) 335–66
[40] B. Lindenhovius, $\mathcal{C}(A)$, PhD thesis, Radboud University Nijmegen (2016), www.math.ru.nl/~landsman/Lindenhovius.pdf
[41] S. Mac Lane and I. Moerdijk, *Sheaves in Geometry and Logic: A First Introduction to Topos Theory*, Springer (1992)
[42] P. J. Morandi, *Dualities in lattice theory* (2005), sierra.nmsu.edu/morandi/notes/Duality.pdf
[43] J. W. Negrepontis, *Applications of the theory of categories to analysis*, PhD thesis, McGill University (1969)
[44] J. von Neumann, *Mathematische Grundlagen der Quantenmechanik*, Springer (1932). English translation: *Mathematical Foundations of Quantum Mechanics*, Princeton University Press (1955)

[45] E. Palmgren, *Semantics of intuitionistic propositional logic*, Lecture Notes, Uppsala University (2009), www2.math.uu.se/~palmgren/tillog/heyting3.pdf
[46] N. C. Phillips, *Continuous-trace C*-algebra s not isomorphic to their opposite algebras*, Int. J. Math. 12 (2001) 263–76
[47] J. Picado and A. Pultr, *Frames and Locales: Topology without Points*, Birkhäuser (2012)
[48] M. Rédei, *Quantum Logic in Algebraic Approach*, Kluwer Academic (1998)
[49] Q. Rutgers, *Intuitionistic quantum logic*, BSc thesis, Radboud University Nijmegen (2018), www.math.ru.nl/~landsman/Quinten.pdf
[50] S. A. M. Wolters, *Quantum toposophy*, PhD thesis, Radboud University Nijmegen (2013), www.math.ru.nl/~landsman/Wolters.pdf
[51] S. A. M. Wolters, *A comparison of two topos-theoretic approaches to quantum theory*, Commun. Math. Phys. 317 (2013) 3–53

Klaas Landsman
Radboud University Nijmegen, Institute for Mathematics,
Astrophysics, and Particle Physics
landsman@math.ru.nl

3
Supergeometry in Mathematics and Physics

Mikhail Kapranov*

Contents

1	Introduction	114
2	Supergeometry as Understood by Mathematicians	116
3	Supergeometry as Understood by Physicists	124
4	Homotopy-Theoretic Underpinnings of Supergeometry	138
	References	149

1 Introduction

In die Traum- und Zaubersphäre
Sind wir, scheint es, eingegangen.

(Goethe, Faust I*)*

Supergeometry is a geometric tool that is supposed to describe supersymmetry ("symmetry between bosons and fermions") in physics. In the impressive arsenal of physico-geometric tools, it occupies a special, somewhat mysterious place.

To an outside observer, it projects a false sense of accessibility, presenting itself as a simple modification of the familiar formalism. However, this is only

---

* My understanding of supergeometry owes a lot to lectures and writings of Y. I. Manin. In particular, the idea of square roots provided by the super formalism, was learned from him a long time ago. The homotopy-theoretic considerations of Section 3 were stimulated by the joint work with N. Ganter [23]. I would also like to thank M. Anel, G. Catren, N. Ganter, N. Gurski, and Y. I. Manin for remarks and suggestions on the preliminary versions of this text. This work was supported by World Premier International Research Center Initiative (WPI Initiative), MEXT, Japan.

an appearance. Like a bewitched place, it is protected not by barriers but by something less tangible and therefore much more powerful. It is not even so clear where this place is localized. A naive attempt to "forge ahead" may encounter little resistance but may also end up missing the real point.

Indeed, the physical origin of supergeometry lies in the comparison of the Bose and Fermi statistics for identical particles. So mathematically, supergeometry zooms in on our mental habits related to questions of commutativity and identity. These habits predate quantum physics and therefore are deeply ingrained, but they do not correspond to the ultimate physical reality. The new way of thinking, offered by supergeometry in mathematics and by supersymmetry in physics, requires changes that cannot be reduced to simple recipes (such as "introducing $\pm$ signs here and there," which is, in the popular mind, the best known feature of supergeometry).

Put differently, if introducing $\pm$ signs can take us so far, then something is happening that we truly do not understand. In addition, one can almost say that mathematicians and physicists mean different things when they speak about supergeometry. This contributes to the enduring feeling of wonder and mystery surrounding this subject.

The present chapter is subdivided into three parts. Following the specifications for the volume, Section 1 presents a very short but self-contained exposition of supergeometry as it is generally understood by mathematicians. Not surprisingly, the presentation revolves around the Koszul sign rule (2.1), which appears first at the level of elements of a ring, and later again at the level of categories. More systematic expositions can be found in the books [19, 43] and in many articles, especially [12].

In Section 2 we discuss the aspects of supergeometry that are used by physicists in relation to supersymmetry. From the mathematical point of view, this amounts to much more than the study of supermanifolds or of the Koszul rule. Some of the basic references are [13, 19, 43, 60]. The entry point for a mathematician here could be found in the idea of taking natural "square roots" of familiar mathematical and physical quantities.

In our presentation, we introduce the abstract concept of a quadratic space (data of Γ-matrices) of which various situations involving spinors form particular cases. This has the advantage of simplifying the general discussion and also of relating the subject to the mathematical theory of intersections of quadrics. In particular, the familiar dichotomy of complete intersections of quadrics versus noncomplete ones has a direct significance for understanding the "constraints" which are usually imposed on superfields.

Finally, Section 3 is devoted to an attempt to uncover deeper roots for the mysterious power of the supergeometric formalism, of which the remarkable consistency of its sign rules is just one manifestation. It seems that the right language to speak about such things is given by homotopy theory. Indeed, this theory provides a systematic modern way to talk about the issues of identity: instead of saying that two things are "the same," we say that they are "homotopic" and specify the homotopy (the precise reason why they should be considered the same). We can also consider homotopies between homotopies and so on. From this point of view, the group $\{\pm 1\}$ of signs hovering over supergeometry, is nothing but π_1^{st}, the first homotopy group of the *sphere spectrum* $\mathbb{S}$. As emphasized by Grothendieck in a more categorical language, $\mathbb{S}$ can be seen as the most fundamental homotopy commutative object. This suggests that "mining the sphere spectrum" beyond the first level should lead to generalizations of supergeometry (and possibly, supersymmetry) involving not just two types of quantities (even and odd) subject to $\pm$ sign rules but, for instance, 24 "sectors" (types of higher level categorical objects) that would account for $\pi_3^{\text{st}} = \mathbb{Z}/24$ and obey some multilevel sign rules that could use higher roots of 1 as well.

2 Supergeometry as Understood by Mathematicians

2.1 Commutative Superalgebras

For a mathematician, supergeometry is the study of supermanifolds and superschemes: objects whose rings of functions are commutative superalgebras.

We fix a field **k** of characteristic $\neq 2$. By an *associative superalgebra* over **k** one means simply a $\mathbb{Z}/2$-graded associative algebra $A = A^{\bar{0}} \oplus A^{\bar{1}}$. Elements of $A^{\bar{0}}$ are called *even* (or *bosonic*), elements of $A^{\bar{1}}$ are called *odd* (or *fermionic*).

An associative superalgebra A is called *commutative* if it satisfies the Koszul sign rule for commutation of homogeneous elements:

$$ab = (-1)^{\deg(a) \cdot \deg(b)} ba. \tag{2.1}$$

The terms *commutative superalgebra* and *supercommutative algebra* are used without distinction.

Often, one considers $\mathbb{Z}$-graded supercommutative algebras $A = \bigoplus_{i \in \mathbb{Z}} A^i$, with the same commutation rule (2.1) imposed on homogeneous elements.

Example 2.1 (a) Any usual commutative algebra A becomes supercommutative, if put in degree $\bar{0}$. The most important example is $A = \mathbf{k}[x_1, \cdots, x_m]$ (the polynomial algebra).

(b) The exterior (Grassmann) algebra $\Lambda[\xi_1, \cdots \xi_n]$ over $\mathbf{k}$, generated by the symbols ξ_i of degree $\bar{1}$, subject only to the relations

$$\xi_i^2 = 0, \quad \xi_i \xi_j = -\xi_j \xi_i, \ i \neq j,$$

is supercommutative. It can be seen as the *free supercommutative algebra* on the ξ_i: the relations imposed are the minimal ones to ensure supercommutativity.

(c) If A, B are the commutative superalgebras, then the tensor product $A \otimes_{\mathbf{k}} B$ with the standard product grading $\deg(a \otimes b) = \deg(a) + \deg(b)$ and multiplication

$$(a_1 \otimes b_1) \cdot (a_2 \otimes b_2) = (-1)^{\deg(b_1) \cdot \deg(a_2)} (a_1 \cdot b_1) \otimes (a_2 \cdot b_2)$$

(a_i, b_i homogeneous), is again a supercommutative algebra. For example,

$$\mathbf{k}[x_1, \cdots, x_m] \otimes \Lambda[\xi_1, \cdots, \xi_n], \quad \deg(x_i) = \bar{0}, \quad \deg(\xi_j) = \bar{1},$$

is a commutative superalgebra. This is a general form of a free commutative superalgebra on a finite set of even and odd generators.

(d) The de Rham algebra $\Omega_X^\bullet$ of differential forms on a C^∞-manifold X is a $\mathbb{Z}$-graded supercommutative algebra over $\mathbb{R}$.

(e) The cohomology algebra $H^\bullet(X, \mathbf{k})$ of any CW-complex X is a $\mathbb{Z}$-graded supercommutative algebra over $\mathbf{k}$.

The importance and appeal of such studies are based on the following heuristic

Principle 2.2 (Principle of naturality of supers) All constructions and features that make commutative algebras special among all algebras can be extended to supercommutative algebras, and make them just as special.

The most important of such features is the relation to geometry: a commutative algebra A can be seen as the algebra of functions on a geometric object $\mathrm{Spec}(A)$ which can be constructed from A and used to build more complicated geometric objects by gluing. Supergeometry (as understood by mathematicians) is the study of similar geometric objects for supercommutative algebras.

2.2 The Symmetric Monoidal Category of Supervector Spaces

The main guiding principle for extending properties of usual commutative algebras to supercommutative ones is also sometimes called the Koszul sign rule. It goes like this:

Principle 2.3 When we move any quantity (vector, tensor, operation) of parity $p \in \mathbb{Z}/2$ past any other quantity of parity q, this move should be accompanied by multiplication with $(-1)^{pq}$.

An instance is given by the first formula in Example 2.1(c). This rule can be formalized as follows.

Let $\text{SVect}_\mathbf{k}$ be the category of *supervector spaces* over $\mathbf{k}$, that is, $\mathbb{Z}/2$-graded vector spaces $V = V^{\bar{0}} \oplus V^{\bar{1}}$. This category has a monoidal structure $\otimes$, the usual graded tensor product. The operation $\otimes$ is associative up to natural isomorphisms, and has a unit object $\mathbf{1} = \mathbf{k}$ (put in degree $\bar{0}$). Define the symmetry isomorphisms

$$R_{V,W} : V \otimes W \longrightarrow W \otimes V, \quad v \otimes w \mapsto (-1)^{\deg(v) \cdot \deg(w)} w \otimes v. \tag{2.2}$$

Proposition 2.4 *The family $R = (R_{V,W})$ makes $(\text{SVect}_\mathbf{k}, \otimes, \mathbf{1}, R)$ into a symmetric monoidal category.*

For background on symmetric monoidal categories we refer to [9, 42]. A basic example of a symmetric monoidal category is given by the usual category of vector spaces $\text{Vect}_\mathbf{k}$, with the usual tensor product, the unit object $\mathbf{1} = \mathbf{k}$, and the symmetry given by $v \otimes w \mapsto w \otimes v$. The meaning of the proposition is that $\text{SVect}_\mathbf{k}$ with symmetry (2.2) satisfies all the same formal properties as the "familiar" category $\text{Vect}_\mathbf{k}$.

If $\dim(V^{\bar{0}}) = m$ and $\dim(V^{\bar{1}}) = n$, we write $\dim(V) = (m|n)$. In particular, we have the standard coordinate superspaces $\mathbf{k}^{m|n}$. We denote by Π the *parity change functor* on $\text{SVect}_\mathbf{k}$ given by multiplication with $\mathbf{k}^{0|1}$ on the left.

It is well known that one can develop linear algebra formalism (tensor, symmetric, exterior powers, etc.) in any symmetric monoidal $\mathbf{k}$-linear abelian category $(\mathcal{V}, \otimes, \mathbf{1}, R)$. This gives a way to define commutativity. That is, a *commutative algebra in* $\mathcal{V}$ is an object $A \in \mathcal{V}$ together with a morphism $\mu_A : A \otimes A \to A$ satisfying associativity and such that the composition

$$A \otimes A \xrightarrow{R_{A,A}} A \otimes A \xrightarrow{\mu_A} A$$

is equal to μ_A. Given two commutative algebras (A, μ_A) and (B, μ_B), the object $A \otimes B$ is again a commutative algebra with respect to $\mu_{A \otimes B}$ given by the composition

$$A \otimes B \otimes A \otimes B \xrightarrow{\text{Id} \otimes R_{B,A} \otimes \text{Id}} A \otimes A \otimes B \otimes B \xrightarrow{\mu_A \otimes \mu_B} A \otimes B.$$

For $\mathcal{V} = \text{SVect}_\mathbf{k}$, this gives Example 2.1(c).

Remark 2.5 (a) Principle 2.2 can be expressed more formally by saying that most constructions involving commutative algebras can be expressed in terms of the symmetry isomorphism of the tensor product and so can be reproduced in any symmetric monoidal category. The category SVect_k is of course a prime example. More generally, the supergeneralizations of Tannakian categories studied by Deligne [11] provide a wider framework for extending the formalism of commutative algebra.

(b) Symmetric monoidal categories can be seen as categorical analogs of commutative algebras: instead of the equality $ab = ba$, we now have canonical isomorphisms $V \otimes W \simeq W \otimes V$. In Section 4.4 we will define categorical analog of *super*commutative algebras.

2.3 Superschemes and Supermanifolds

Given a supercommutative algebra A, the even part $A^{\bar{0}}$ is commutative. At the same time, any element a of the odd part $A^{\bar{1}}$ is nilpotent: $a^2 = 0$.

The commutative algebra $A^{\bar{0}}$ has the associated affine scheme $\text{Spec}(A^{\bar{0}})$. Explicitly, it is a ringed space $\left(\underline{\text{Spec}(A^{\bar{0}})}, \mathcal{O}_{\text{Spec}(A^{\bar{0}})}\right)$, where $\underline{\text{Spec}(A^{\bar{0}})}$ is the set of prime ideals in $A^{\bar{0}}$ with the Zariski topology and $\mathcal{O}_{\text{Spec}(A^{\bar{0}})}$ is a sheaf of local rings on this space obtained by localization of $A^{\bar{0}}$. That is, the value of $\mathcal{O}_{\text{Spec}(A^{\bar{0}})}$ on a "principal open set"

$$U_f = \{\mathfrak{p} \in \underline{\text{Spec}(A^{\bar{0}})} : f \in \mathfrak{p}\}, \quad f \in A^{\bar{0}},$$

is given by $\mathcal{O}_{\text{Spec}(A^{\bar{0}})}(U_f) = A^{\bar{0}}[f^{-1}]$.

Furthermore, $A^{\bar{0}}$ is not only commutative but lies in the center of A as an associative algebra. Therefore, associating to U_f the commutative superalgebra $A[f^{-1}]$, we get a sheaf of commutative superalgebras

$$\mathcal{O}_{\text{Spec}(A)} = \mathcal{O}^{\bar{0}}_{\text{Spec}(A)} \oplus \mathcal{O}^{\bar{1}}_{\text{Spec}(A)}, \quad \mathcal{O}^{\bar{0}}_{\text{Spec}(A)} = \mathcal{O}_{\text{Spec}(A^{\bar{0}})}$$

on the same topological space $\underline{\text{Spec}(A^{\bar{0}})}$ as before. The pair (ringed space)

$$\text{Spec}(A) := \left(\underline{\text{Spec}(A^{\bar{0}})}, \mathcal{O}_{\text{Spec}(A)}\right)$$

is the fundamental geometric object associated to A (see [39, 43]).

The stalks of $\mathcal{O}_X$ are commutative superalgebras which, considered as ordinary associative algebras, are *local rings*. Indeed, we have the following:

Proposition 2.6 *If B is a commutative superalgebra such that $B^{\bar{0}}$ is a local ring with maximal ideal $\mathfrak{m}^{\bar{0}}$, then B itself is a local ring with maximal ideal $\mathfrak{m} = \mathfrak{m}^{\bar{0}} \oplus B^{\bar{1}}$.*

Proof: Odd elements of a supercommutative algebra are nilpotent. This implies at once that $b = b_{\bar{0}} + b_{\bar{1}} \in B$ is invertible, if and only if $b_{\bar{0}}$ is invertible in $B^{\bar{0}}$, that is, $b \notin \mathfrak{m}$. In other words, B is a ring in which noninvertible elements form an ideal, that is, a local ring. ∎

The nilpotency of odd elements mentioned above is the reason why the underlying space of $\mathrm{Spec}(A)$ depends only on the even part of A. Similarly to the usual commutative algebra intuition about spectra of rings with nilpotents, $\mathrm{Spec}(A)$ may be thought of, roughly, as some kind of "infinitesimal neighborhood thickening" of $\mathrm{Spec}(A^{\bar{0}})$.

Working with sheaves of local rings is a fundamental technical feature of Grothendieck's theory of schemes. So we introduce the following:

Definition 2.7 A *superlocally ringed space over $\mathbf{k}$* is a pair $X = (\underline{X}, \mathcal{O}_X)$, where $\underline{X}$ is a topological space and $\mathcal{O}_X = \mathcal{O}_X^{\bar{0}} \oplus \mathcal{O}_X^{\bar{1}}$ is a sheaf of commutative $\mathbf{k}$-superalgebras on $\underline{X}$, with stalks being local rings.

A *morphism* of superlocally ringed spaces $f : (\underline{X}, \mathcal{O}_X) \to (\underline{Y}, \mathcal{O}_Y)$ is a pair $f = (f_\sharp, f^\flat)$, where:

- $f_\sharp : \underline{X} \to \underline{Y}$ is a continuous map of topological spaces.
- $f^\flat : f_\sharp^{-1}(\mathcal{O}_Y) \to \mathcal{O}_X$ is a morphism of sheaves of commutative superalgebras, which, in addition, takes the maximal ideal of each local ring $(f_\sharp^{-1}\mathcal{O}_Y)_x := \mathcal{O}_{Y, f_\sharp(x)}$, $x \in X$ into the maximal ideal of the local ring $\mathcal{O}_{X,x}$.

The resulting category of superlocally ringed spaces over $\mathbf{k}$ will be denoted $\mathcal{SLRS}_\mathbf{k}$.

Proposition 2.8 *For any two commutative superalgebras A, B we have an isomorphism*

$$\mathrm{Hom}_{\mathcal{SA}_k}(A, B) \simeq \mathrm{Hom}_{\mathcal{SLRS}_k}(\mathrm{Spec}(B), \mathrm{Spec}(A)).$$

Definition 2.9 A *superscheme* over $\mathbf{k}$ is a superlocally ringed space over $\mathbf{k}$ locally isomorphic to $\mathrm{Spec}(A)$ for a commutative superalgebra A. The category $\mathcal{SSch}_\mathbf{k}$ of superschemes over $\mathbf{k}$ is defined to be the full subcategory of $\mathcal{SLRS}_\mathbf{k}$ formed by superschemes.

Example 2.10 (a) The *affine superspace* of dimension $(m|n)$ is the superscheme

$$\mathbb{A}^{m|n} = \mathrm{Spec}\big(\mathbf{k}[x_1, \cdots, x_m] \otimes \Lambda[\xi_1, \cdots, \xi_n]\big).$$

(b) The next class of examples is provided by *algebraic supermanifolds* of dimension $(m|n)$. By definition, they are superschemes locally of the form $\mathrm{Spec}\big(A \otimes \Lambda[\xi_1, \cdots, \xi_n]\big)$, where A is the ring of functions on a smooth m-dimensional affine algebraic variety over $\mathbf{k}$.

Each superscheme $X = (\underline{X}, \mathcal{O}_X)$ gives rise to an ordinary scheme $X^{\bar{0}} = (\underline{X}, \mathcal{O}_X^{\bar{0}}/(\mathcal{O}_X^{\bar{1}})^2)$ called the *even part* of X. If X is an algebraic supermanifold of dimension $(m|n)$, then $X^{\bar{0}}$ is a smooth algebraic variety of dimension m.

As with usual schemes, a superscheme X is defined by its *functor of points* on the category of supercommutative algebras

$$A \mapsto X(A) = \mathrm{Hom}_{\mathcal{SS}ch_\mathbf{k}}(\mathrm{Spec}(A), X).$$

Differentiable or analytic supermanifolds are similarly defined as locally ringed spaces $X = (X^0, \mathcal{O}_X)$ where X^0 is an ordinary differentiable or analytic manifold and $\mathcal{O}_X$ is a sheaf of commutative superalgebras locally isomorphic to $\mathcal{O}_X \otimes_\mathbf{k} \Lambda[\xi_1, \cdots, \xi_n]$. Here $\mathcal{O}_X$ is the sheaf of C^∞ or analytic functions on X and $\mathbf{k} = \mathbb{R}$ for differentiable or real analytic manifolds and $\mathbb{C}$ for complex analytic manifolds.

2.4 Lie Supergroups and Superalgebras

Here is an illustration of the Principle of Naturality of Supers: the extension of the formalism of differential geometry to the supercase looks totally straightforward.

Given a $\mathbf{k}$-linear symmetric monoidal category $\mathcal{A}$, one can speak about algebras in $\mathcal{A}$ of any given type, for example Lie algebras. In the case $\mathcal{A} = \mathrm{SVect}$, this gives the familiar concept.

Definition 2.11 A *Lie superalgebra* over $\mathbf{k}$ is a $\mathbf{k}$-supervector space $\mathfrak{g} = \mathfrak{g}^{\bar{0}} \oplus \mathfrak{g}^{\bar{1}}$ equipped with a homogeneous (degree $\bar{0}$) operation $(x, y) \mapsto [x, y]$ satisfying

$$[x, y] = -(-1)^{\deg(x)\cdot\deg(y)}[y, x];$$
$$[x, [y, z]] = [[x, y], z] + (-1)^{\deg(x)\cdot\deg(y)}[y, [x, z]].$$

The following examples can also be given in any symmetric monoidal category. We use the case $\mathcal{V} = \text{SVect}_\mathbf{k}$.

Example 2.12 (a) If R is any associative superalgebra, we can make it into a Lie superalgebra by using the *supercommutator*

$$[x, y] = x \cdot y - (-1)^{\deg(x) \cdot \deg(y)} y \cdot x.$$

(b) If A is an associative superalgebra, then we have the Lie superalgebra $\text{Der}(A) = \text{Der}^{\bar{0}}(A) \oplus \text{Der}^{\bar{1}}(A)$ of *superderivations* of A. Here $\text{Der}^i(A)$ is the space of $\mathbf{k}$-linear maps $D: A \to A$ satisfying the *super-Leibniz rule*

$$D(a \cdot b) = D(a) \cdot b + (-1)^{i \cdot \deg(a)} a \cdot D(b).$$

The bracket is given by the supercommutator as in (a).

Given a supermanifold $X = (\underline{X}, \mathcal{O}_X)$ of dimension $(m|n)$ in the smooth, analytic or algebraic category, superderivations of $\mathcal{O}_X$ form a sheaf T_X of Lie superalgebras on X called the *tangent sheaf*. It is also a sheaf of $\mathcal{O}_X$-modules, locally free of rank $(m|n)$. Its fiber at a point $x \in X^0$ will be denoted $T_x X$ and called the *tangent space to X at x*. It is a supervector space of dimension $(m|n)$.

A *Lie supergroup*, resp. an *algebraic supergroup* over $\mathbf{k}$, is a group object G in the category of smooth supermanifolds, resp. algebraic supermanifolds over $\mathbf{k}$. In particular, it has a unit element $e \in G^0$. The tangent space $T_e G$ is a Lie superalgebra denoted by $\mathfrak{g} = \text{Lie}(G)$. Note that $\mathfrak{g}^{\bar{0}} = \text{Lie}(G^0)$ is the usual Lie algebra corresponding to a Lie or algebraic group. Conversely, given a finite-dimensional Lie superalgebra $\mathfrak{g}$ and a Lie or algebraic group G^0 such that $\text{Lie}(G^0) = \mathfrak{g}^{\bar{0}}$, one can integrate it to a Lie or algebraic supergroup G with $\text{Lie}(G) = \mathfrak{g}$.

We will specially use the case when $\mathfrak{g}$ is *nilpotent*, that is, the iterated commutators $[x_1, [x_2, \cdots, x_n]\ldots]$ vanish for n greater than some fixed N (degree of nilpotency). We assume $\text{char}(\mathbf{k}) = 0$. In this case we can associate to $\mathfrak{g}$ a canonical algebraic group $G = e^{\mathfrak{g}}$ by means of the classical Hausdorff series

$$x \cdot y = x + y + \frac{1}{2}[x, y] + \cdots \tag{2.3}$$

This means the following. Suppose we want to find the set-theoretic group of A-points $e^{\mathfrak{g}}(A) = \text{Hom}(\text{Spec}(A), e^{\mathfrak{g}})$, where A is a supercommutative algebra. As a set, this group is defined to be $(\mathfrak{g} \otimes_\mathbf{k} A)^{\bar{0}}$, which has a structure of an *ordinary* (purely even) nilpotent Lie algebra and therefore can be made into a group by means of the *ordinary* Hausdorff series above. This is the group $e^{\mathfrak{g}}(A)$.

Informally, one says that $e^{\mathfrak{g}}$ "is" the Lie algebra $\mathfrak{g}$ considered as a manifold with the multiplication given by (2.3) but the formal meaning is that the arguments in commutators in (2.3) always belong to some ordinary Lie algebra.

Example 2.13 (The superparticle and its automorphism group) In classical mechanics, a "particle" is thought of as a point, that is, mathematically, as the variety Spec(**k**). It is natural to call the *superparticle* the affine superspace $\mathbb{A}^{0|1} = \text{Spec}(\Lambda[\xi])$. Although it has only one classical (**k**-)point, the superstructure around this point is different. As shown by Witten [63], the study of such a superparticle moving in an (ordinary) Riemannian manifold leads naturally to the de Rham complex $\Omega_M^\bullet$ and to Morse theory.

In particular, we have the algebraic supergroup $\text{Aut}(\mathbb{A}^{0|1})$ formed by "all automorphisms" of $\mathbb{A}^{0|1}$, even as well as odd. For a commutative **k**-superalgebra A the group of A-points of $\text{Aut}(\mathbb{A}^{0|1})$ consists of A-superalgebra automorphisms of $A \otimes_{\mathbf{k}} \Lambda[\xi]$, that is, of transformations

$$\xi \mapsto a\xi + b, \quad a \in A_{\bar{0}}^* \text{ (invertible)}, \quad b \in A_{\bar{1}}.$$

This means that $\text{Aut}(\mathbb{A}^{0|1})$ has dimension $(1|1)$ and its even part is the multiplicative group $\mathbb{G}_m$. The following proposition is due to M. Kontsevich [38]; see [36, Proposition 2.2.2] for a discussion. It explains the appearance of the de Rham differential in Witten's theory.

Proposition 2.14 *Let V be a supervector space over **k**. An (algebraic) action of $\text{Aut}(\mathbb{A}^{0|1})$ on V is the same as the following system of data:*

(1) A $\mathbb{Z}$-grading $V = \bigoplus_{i \in \mathbb{Z}} V^i$ such that $V_{\bar{0}} = \bigoplus V^{2i+1}$ and $V_{\bar{1}} = \bigoplus V^{2i}$.
(2) A differential d on V of degree $+1$ in the above $\mathbb{Z}$-grading such that $d^2 = 0$, that is, making V into a cochain complex.

In other words, the symmetric monoidal category of supervector spaces with an $\text{Aut}(\mathbb{A}^{0|1})$-action is identified with the symmetric monoidal category of cochain complexes.

Here the $\mathbb{Z}$-grading is given by the action of $\mathbb{G}_m \subset \text{Aut}(\mathbb{A}^{0|1})$, and the differential is given by the action of the odd part of the Lie superalgebra $\text{Lie}(\text{Aut}(\mathbb{A}^{0|1}))$ of superderivations of $\Lambda[\xi]$.

2.4.1 Supergeometry and Derived Geometry

It is instructive to compare supergeometry (as it is described in this section) with the formalism of *derived geometry* which seeks to form "nonabelian derived functors" of several familiar algebro-geometric constructions such as forming moduli spaces or intersections of subvarieties (see [7, 41, 42]).

In the simplest setting, objects studied in derived geometry are glued out of local pieces which correspond to commutative differential graded (dg-) algebras $(A^\bullet, d)$, that is, commutative algebra objects in the symmetric monoidal category of cochain complexes. By Proposition 2.14, such pieces can be seen as affine superschemes with action of $\mathrm{Aut}(\mathbb{A}^{0|1})$.

So one can say that, in some approximate sense, derived geometry can be regarded as supergeometry but in an equivariant context, with respect to the action of the supergroup $\mathrm{Aut}(\mathbb{A}^{0|1})$. However, this is a rather simplified point of view for several reasons. Even in the context of dg-algebras, the distinction between negative and positive grading is very important – it corresponds to the distinction between "geometry in the small" (intersections, singularities) and "geometry in the large" (stacks, homotopy type). The precise definitions impose "geometry in the large" in a more global "stacky" way.

2.4.2 Pfaff Systems and Frobenius Theorem

Let X be a supermanifold. A *Pfaff system* in X is a subbundle (i.e., a subsheaf of $\mathcal{O}_X$-modules which is locally a direct summand) $C \subset T_X$. For a Pfaff system C the Lie algebra structure in sections of T_X induces the $\mathcal{O}_X$-linear map known as the *Frobenius pairing*

$$F_C : \Lambda^2_{\mathcal{O}_X} C \longrightarrow T_X/C.$$

A Pfaff system C is called *integrable*, if $F_C = 0$, that is, if C is closed under the bracket of vector fields. In this case we have a superanalog of the *Frobenius theorem*: in the smooth or analytic case C can be locally represented as the relative tangent bundle to a submersion of supermanifolds $X \to Y$.

3 Supergeometry as Understood by Physicists

For a physicist, the really important concept is *supersymmetry*, and supermanifolds per se are of interest only tangentially. For example, they arise as infinite-dimensional spaces of bosonic and fermionic classical fields, over which Feynman integrals are taken. One way (not the only one!) to construct supersymmetric field theories is by using *superspace*, a concept not synonymous with "supermanifold" of mathematicians. In fact, superspace is a supermanifold with a rather special "spinor-conformal" structure.

3.1 The Idea of Nonobservable Square Roots

To understand the idea of supersymmetry, it is useful to include it into the following more general heuristic principle.

Principle 3.1 (**Principle of square roots**) It is useful to represent observable quantities of immediate physical interest (real, positive, bosonic) as bilinear combinations of more fundamental quantities which can be complex, fermionic and not even observable by themselves.

In other words, it is useful to take "square roots" of familiar objects. Let us give several examples.

Example 3.2 (**Wave functions and probability density**) In elementary quantum mechanics, the wave function $\psi(x)$ of a particle (say, electron), is a complex quantity which can not be measured. But the expression

$$P(x) = |\psi(x)|^2 = \overline{\psi}(x) \cdot \psi(x) \geq 0$$

represents the probability density of the electron which is real, nonnegative, and measurable.

Example 3.3 (**Laplace operator on forms**) Let X be a C^∞ Riemannian manifold. The space $\Omega^\bullet(X)$ of all differential forms on X is $\mathbb{Z}$-graded and so can be considered as a supervector space. The Laplace operator on forms is defined as

$$\Delta = d \circ d^* + d^* \circ d = [d, d^*],$$

where d is the exterior derivative and d^* is its adjoint with respect to the Riemannian metric. Thus Δ is nonnegative definite, real (self-adjoint) and bosonic, while d and d^* are fermionic.

Example 3.4 (**Spinors as square roots of vectors**) Let V be a d-dimensional vector space over $\mathbf{k} = \mathbb{R}$ or $\mathbb{C}$, equipped with a nondegenerate quadratic form q. We refer to V as the spacetime, or the Minkowski space. Elements of V (*vectors* in the physical sense of the word) can be represented as bilinear combinations of more fundamental quantities, *spinors*. By definition, spinors are vectors in minimal (spinor) representations of $\mathrm{Spin}(V)$, a double covering of the group $SO(V)$. There is one spinor representation S, if $d = \dim(V)$ is odd, and two representations, S_+ and S_-, if d is even. Further properties of these representations depend on the residues modulo 8 of d and of the signature of q, if $k = \mathbb{R}$ (Bott periodicity). We refer to [3, 10] for a detailed treatment.

The expression of vectors as bilinear combinations of spinors comes from Spin(V)-equivariant maps γ whose nature we indicate in three cases $d = 2, 4, 10$ with Minkowski signature:

$d = 2$: S_+, S_- can be defined over $\mathbb{R}$ and have real dimension 2. They are self-dual: $S_\pm^* = S_\pm$. The gamma maps are $\gamma : \text{Sym}^2(S_\pm) \to V$.

$d = 4$: S_+, S_- are complex and have complex dimension 2. They are hermitian conjugate of each other: $\overline{S}_\pm^* = S_\mp$. The gamma map is $\gamma : S_+ \otimes S_- \to V$.

$d = 10$: S_+, S_- are real, of dimension 16 and self-dual. The gamma maps are $\gamma : \text{Sym}^2(S_\pm) \to V$.

The maps γ can be seen as transposed versions of the systems of Dirac's gamma-matrices. For instance, in the case $d = 10$, we can transpose γ to get a map $\gamma^t : V \to \text{Hom}(S_\pm, S_\pm)$, and the gamma-matrix γ_μ, $\mu = 1, \cdots, 10$, is the operator $\gamma^t(e_\mu) : S_\pm \to S_\pm$ corresponding to the basis vector $e_\mu \in V$.

Needless to say, forming the double covering $\text{Spin}(V) \to SO(V)$ itself can be seen as taking square roots of rotations. The appearance of spinors is a reflection of this procedure at the level of representations of groups.

Example 3.5 (**Weil conjecture over finite fields**) It is tempting to add to this list of "square roots" the following classical example. Let X be a smooth projective curve over a finite field $\mathbb{F}_q$. The étale cohomology group $H^1(X \otimes \overline{\mathbb{F}}_q, \mathbb{Q}_l)$ is acted upon by the Frobenius element Fr, generating $\text{Gal}(\overline{\mathbb{F}}_q/\mathbb{F}_q)$. It was proved by A. Weil (and motivated the more general Weil conjectures) that each eigenvalue λ of Fr is an algebraic integer whose image in each complex embedding satisfies $|\lambda| = \sqrt{q}$, that is, $\lambda \cdot \overline{\lambda} = q$. So the first cohomology, a fermionic structure, gives rise to factorizations $q = \lambda \cdot \overline{\lambda}$.

Furthermore, it was suggested by Y. I. Manin that the motive of a supersingular elliptic curve over $\mathbb{F}_q$ can be seen as a "spinorial square root" of the Tate motive. See [47] for developments in this direction.

Remark 3.6 Example 3.2 (understood more widely, as an instance of the special role played by scalar products and Hilbert spaces in quantum mechanics), can be seen as the source of many appearances of square roots. For instance, a natural L_2-space associated to a manifold M consists of half-forms, that is, of sections of $\omega_M^{\otimes(1/2)}$, a square root of the line bundle of volume forms.

The appearance of the metaplectic double cover $\widetilde{\text{Sp}}(n) \to \text{Sp}(n)$ of the symplectic group can also be traced, via the Maslov index and the WKB approximation, to the same source.

Example 3.7 (Connections and curvature) A gauge (say, electromagnetic) field is represented by a connection ∇ in a principal bundle over the spacetime manifold. The immediately observable quantity (at least for the electromagnetic field which can be measured classically) is the *field strength*, represented by the curvature $F = \nabla^2$. The connection itself, a somewhat more elusive object, can be viewed therefore as a square root of the field strength. The idea that "the fundamental quantities that we observe are curvature data of something" appears, of course, in several areas of physics. Its subtle match with the basic principles of quantum mechanics is a truly remarkable phenomenon.

3.2 A Square Root of d/dt and Theta-Functions

The simplest example of supersymmetry can be obtained by looking at the differential operator (superderivation)

$$Q = \frac{\partial}{\partial \xi} + \xi \frac{\partial}{\partial t} = \begin{pmatrix} 0 & 1 \\ \partial/\partial t & 0 \end{pmatrix}, \quad Q^2 = \frac{\partial}{\partial t} \tag{3.1}$$

in the algebra $\mathcal{O}(\mathbb{A}^{1|1}) = \mathbb{C}[t] \otimes \Lambda[\xi]$. One can replace $\mathbb{C}[t]$ by the algebras of smooth or analytic functions of t. The second equality sign in (3.1) is an instance of *component analysis*: we view $\mathbb{C}[t] \otimes \Lambda[\xi]$ as a free $\mathbb{C}[t]$-module with basis $1, \xi$ and write Q as a 2×2 matrix differential operator in t alone. One verifies immediately that $Q^2 = \partial/\partial t$, so Q gives a square root of the Hamiltonian (time translation).

Already this example is quite nontrivial: it provides a natural explanation of Sato's approach to theta functions [49, 50]. We do it in two stages. First, consider the exponential

$$e^Q = 1 + Q + \frac{Q^2}{2!} + \frac{Q^3}{3!} + \cdots \tag{3.2}$$

as a series of operators acting on complex analytic functions in t, ξ.

Proposition 3.8 e^Q *converges to a local operator, that is, to an endomorphism of the sheaf $\mathcal{O}_{\mathbb{C}^{1|1}}$ of analytic functions on $\mathbb{C}^{1|1}$.*

A standard example of a local operator in this sense is a linear differential operator P on the line. An operator with constant coefficients is just a polynomial in d/dt:

$$P = h(d/dt), \quad h(z) = \sum_{n=0}^{N} a_n z^n.$$

Replacing polynomials by entire analytic functions $h(z) = \sum_{n=0}^{\infty} a_n z^n$, we get expressions which, even when they converge, need not give local operators. For instance, $h(z) = e^z$ gives the shift operator

$$(e^{d/dt} f)(t) = f(t+1). \tag{3.3}$$

However, if $h(z)$ is *subexponential*, that is, if the series $k(z) = \sum n! a_n z^n$ still represents an entire function, then $h(d/dt)$ acts on analytic functions in a local way. Indeed, by the Cauchy formula,

$$\bigl(h(d/dt)f\bigr)(t) = \sum_{n=0}^{\infty} a_n f^{(n)}(t) = \frac{-1}{2\pi i} \oint_{|t'-t|=\varepsilon} f(t') k\left(\frac{1}{t-t'}\right) dt',$$

where ε can be arbitrarily small. So if f is analytic in an open $U \subset \mathbb{C}$, then so is $h(d/dt)f$. For example, $h(z) = \cos\sqrt{z}$ gives a local operator. One can similarly make sense of series $\sum_{n=0}^{\infty} a_n(t)(d/dt)^n$, where $a_n(t)$ are analytic functions in t of subexponential growth in n. These series define local operators on analytic functions on $\mathbb{C}$ known simply as *differential operators of infinite order* [50]. They form a sheaf of rings on $\mathbb{C}$, denoted $\mathcal{D}_\mathbb{C}^\infty$. Similarly for any complex analytic (super-)manifold such as $\mathbb{C}^{1|1}$.

Returning to the situation of Proposition 3.8, we see that Q, being a square root of $\partial/\partial t$, has exponential e^Q which is a local operator: a global section of $\mathcal{D}_{\mathbb{C}^{1|1}}^\infty$ or, after component analysis, of $\mathrm{Mat}_2(\mathcal{D}_\mathbb{C}^\infty)$.

Remark 3.9 Although Q is a superderivation, it is not a derivation in the usual sense and e^Q is not a ring automorphism. In particular, forming e^Q is not an instance of exponentiating a Lie superalgebra to a Lie supergroup.

We now consider the simplest theta-function

$$\theta(t,x) = \sum_{n\in\mathbb{Z}} e^{n^2 t + nx}, \quad \Re(t) < 0, \, x \in \mathbb{C}.$$

Its value at $x = 0$ (the Thetanullwert)

$$\theta(x,0) = \sum_{n\in\mathbb{Z}} q^{n^2}, \quad q = e^t, \, |q| < 1,$$

is a modular form, so there is a relation (modular transformation) relating its values at t and at $1/t$ (in our normalization). The main result of [50] is a characterization of $\theta(t,0)$ by two differential equations of infinite order in t alone (i.e., by local conditions in t). They are then used to deduce the modular transformation because the system of equations is invariant under it.

This can be done as follows. Note that $\theta(t,x)$, as a function of two variables, satisfies the heat equation

$$\frac{\partial \theta}{\partial t} = \frac{\partial^2 \theta}{\partial x^2}.$$

This means that $\partial/\partial x$ acts on θ as a square root of $\partial/\partial t$. On the other hand, θ has periodicity properties

$$\begin{cases} \theta(t, x+2\pi i) = \theta(t,x) \\ \theta(t, x+2t) = e^{-x-t}\theta(t,x). \end{cases}$$

Using (3.3) in the x-variable and the heat equation, we can write this formally as

$$\begin{cases} e^{2\pi i \sqrt{\partial/\partial t}} \theta(t,x) = \theta(t,x) \\ e^{2t\sqrt{\partial/\partial t}} \theta(t,x) = e^{-t-x}\theta(t,x), \end{cases}$$

after which we can specialize to $x=0$. Now, replacing $\sqrt{\partial/\partial t}$ with Q, we get a system of two differential equations of infinite order in t alone, satisfied by $\theta(t,0)$.

3.3 Square Roots of Spacetime Translations

Representing just the Hamiltonian (the operator of energy, or time translation) as a bilinear combination of fermionic operators (Section 3.2, or Example 3.3), is a nonrelativistic procedure. Relativistically, we cannot separate $\partial/\partial t$ from any other constant vector field ∂_v (momentum operator) on the Minkowski space V and so we should represent them all. This leads to the following concept.

Definition 3.10 A *quadratic space* over a field **k** is a datum of finite-dimensional **k**-vector spaces V, B and a surjective linear map $\Gamma: \text{Sym}^2(B) \to V$. We will often view Γ as a V-valued scalar product $(b_1, b_2) \mapsto \Gamma(b_1, b_2) \in V$ on B

A quadratic space Γ gives rise to the Lie superalgebra

$$\mathfrak{t} = \mathfrak{t}_\Gamma, \quad \mathfrak{t}^{\bar{0}} = V, \quad \mathfrak{t}^{\bar{1}} = B, \tag{3.4}$$

with the only nonzero component of the bracket being Γ. We call $\mathfrak{t}$ the *supersymmetry algebra* associated to Γ. The abelian central subalgebra $\mathfrak{t}^{\bar{0}} = V$ is the usual Lie algebra of infinitesimal spacetime translations, and we denote by $T^{\bar{0}} = e^{\mathfrak{t}^{\bar{0}}}$ the corresponding algebraic group (i.e., V considered as an algebraic variety and equipped with vector addition). Since $\mathfrak{t}$ is nilpotent:

$[x, [y, z]] = 0$ for any homogeneous x, y, z, it is easily integrated to an algebraic supergroup $T = e^t$ called the *supersymmetry group* using the truncated Hausdorff series (2.3).

Example 3.11 In physical applications (see [13, 19] for a detailed exposition), V is the Minkowski space (i.e., a vector space over $\mathbb{R}$ or $\mathbb{C}$ with a nondegenerate quadratic form) and B is a direct sum of (possibly several copies of) the spinor bundle(s). Different possible choices of such B are known as different types of *extended supersymmetry*. More precisely,

$N = p$ *supersymmetry (SUSY)* means that $d = \dim(V)$ is odd and $B = S^{\oplus p}$;

$N = (p, q)$ *supersymmetry (SUSY)* means that d is even and $B = S_+^{\oplus p} \oplus S_-^{\oplus q}$.

The map Γ of the quadratic space is constructed out of the gamma-maps γ for spinors. Quadratic spaces obtained in this way will be called *spinorial*. For example,

(a) $d = 10$, $N = (1, 0)$ SUSY: here $B = S_+$ and $\Gamma = \gamma: \text{Sym}^2(S_+) \to V$. This is the most fundamental example in many respects.

(b) $d = 4$, $N = (1, 1)$ SUSY: here $B = S_+ \oplus S_-$ and Γ is the composition

$$\text{Sym}^2(S_+ \oplus S_-) \longrightarrow S_+ \otimes S_- \xrightarrow{\gamma} V.$$

(c) $d = 2$, $N = (1, 1)$ SUSY is often written explicitly using two operators H (energy $= \partial/\partial t$) and P (momentum $= \partial/\partial x$). The space S_+ is spanned by two vectors Q_+, Q_+^* and S_- by Q_-, Q_-^*, and the supersymmetric algebra (i.e., the commutation rule in t) is written as

$$[Q_+, Q_+] = H + P, \quad [Q_-, Q_-] = H - P, \quad [Q_+, Q_-] = 0.$$

For a spinorial quadratic space Γ the group $\text{Spin}(V)$ acts on t and on $T = e^t$. The corresponding semidirect product $\mathcal{P} = \text{Spin}(V) \ltimes T$ is known as the *super-Poincaré group* (corresponding to the type of extended supersymmetry represented by Γ).

3.4 Quadratic Spaces and Intersections of Quadrics

In algebraic geometry, the term *intersection of quadrics* [48, 58] means one of two closely related objects:

(1) A *homogeneous intersection of quadrics*: a subscheme Y in a vector space B, given by homogeneous quadratic equations. That is, the ideal

$I(Y) \subset \mathrm{Sym}^\bullet(B^*)$ is generated by $I_2(Y) \subset \mathrm{Sym}^2(B^*)$, its degree 2 homogeneous part.

(2) A *projective intersection of quadrics*: the projectivization $Z = \mathbb{P}(Y) \subset \mathbb{P}(B)$ of Y as above. In this way, projective intersections of quadrics $Z \subset \mathbb{P}(B)$ are in bijection with those homogeneous intersections of quadrics $Y \subset B$ which are not entirely supported at 0.

Each quadratic space $\Gamma \colon \mathrm{Sym}^2(B) \to V$ gives a homogeneous intersection of quadrics

$$Y_\Gamma = \{b \mid \Gamma(b,b) = 0\} \subset B,$$

and we denote by $Z_\Gamma \subset \mathbb{P}(B)$ its projectivization. Alternatively,

$$Y_\Gamma = \{b \in \mathfrak{t}_\Gamma^{\bar{1}} \mid [b,b] = 0\}$$

is the *scheme of Maurer–Cartan elements* of the supersymmetry algebra $\mathfrak{t}_\Gamma$.

Proposition 3.12 *Each homogeneous intersection of quadrics $Z \subset B$ can be obtained in this way from some quadratic space $\Gamma \colon \mathrm{Sym}^2(B) \to V$, defined uniquely up to an isomorphism.*

Proof: Take $V = I_2(Y)^*$, the space dual to the space of quadratic equations of Y and take for Γ the canonical projection. ∎

So quadratic spaces are simply data encoding intersections of quadrics. Classically, the simplest intersections of quadrics are as follows.

Definition 3.13 (1) A projective intersection of quadrics $Z \subset B$ is called a *complete intersection*, if dim $I_2(Z)$, the number of quadratic equations of Z, is equal to the codimension of Z in $\mathbb{P}(B)$.
(2) A quadratic space $\Gamma \colon \mathrm{Sym}^2(B) \to V$ is called of *complete intersection type*, if Z_Γ is a complete intersection, that is, $\dim(V) = \mathrm{codim}(Z_\Gamma)$.

Example 3.14 (a) For the spinorial quadratic space $\Gamma \colon \mathrm{Sym}^2(S_+) \to V$ of $d = 10$, $n = (1,0)$ SUSY (Example 3.11(a)), X_Γ is the 10-dimensional *space of pure spinors* $\Sigma^{10} \subset \mathbb{P}(S_+) = \mathbb{P}^{15}$. It can be identified with one component of the variety of isotropic 5-planes in V. It is *not* a complete intersection: it has codimension 5 but is given by $d = 10$ equations.
(b) For the quadratic space $\Gamma \colon \mathrm{Sym}^2(S_+ \oplus S_-) \to V$ of $d = 4$, $N = (1,1)$ SUSY (Example 3.11(b)), $X_\Gamma \subset \mathbb{P}(S_+ \oplus S_-) = \mathbb{P}^3$ is the disjoint union of

two skew lines $\mathbb{P}(S_+) \sqcup \mathbb{P}(S_-) \simeq \mathbb{P}^1 \sqcup \mathbb{P}^1$. It is *not* a complete intersection: it has codimension 2 but is given by $d=4$ equations.

(c) The variety $\Sigma^{10} \subset \mathbb{P}^{15}$ is a particular case of the following: a partial flag variety G/P (G reductive algebraic group, $P \subset G$ parabolic), equivariantly embedded into $\mathbb{P}(E)$, where E is an irreducible highest weight representation of G. All such varieties are known to be intersections of quadrics.

So our physical spacetime is really *the space of equations* of an auxiliary intersection (typically, not a complete intersection!) of quadrics.

We will also need families of quadratic spaces parameterized by superschemes.

Definition 3.15 Let X be a superscheme. A *quadratic module* over X is a datum of two locally free sheaves B, V of $\mathcal{O}_X$-modules, both of purely even rank, and of an $\mathcal{O}_X$-linear map $\Gamma \colon \operatorname{Sym}^2_{\mathcal{O}_X}(B) \to V$ with two properties:

(1) Γ is surjective, and therefore the dual map $\Gamma^\vee \colon V^\vee \to \operatorname{Sym}^2_{\mathcal{O}_X}(B^\vee)$ is an embedding of a locally direct summand;
(2) The $\mathcal{O}_X$-algebra $\operatorname{Sym}^\bullet_{\mathcal{O}_X}(B^\vee)/(\Gamma^\vee(V^\vee))$ is flat over $\mathcal{O}_X$.

In other words, a quadratic module gives a family of intersections of quadrics, parameterized by X, and the condition (2) means that this family is flat, in particular, its fibers have the same dimension. This is important for noncomplete intersections.

3.5 Supersymmetry, Superspace, and Constraints

We start with a quick explanation of some physical terms.

3.5.1 Supersymmetry

Supersymmetry is a feature of a field theory (say, a collection of fields φ plus a Lagrangian action $S[\varphi]$) defined, a priori, on the usual (non-super!) Minkowski space V. It means that the action of the usual Poincaré group $SO(V) \ltimes V$ on fields by changes of variables (which leaves any relativistic Lagrangian invariant) is extended, *in some way*, to an action of the super-Poincaré group $\mathcal{P}$ so that $S[\varphi]$ is still invariant. Here $\mathcal{P}$ is constructed out of one of the spinorial quadratic spaces $\Gamma \colon \operatorname{Sym}^2(B) \to V$ (Example 3.11).

Thus the new datum in supersymmetry is the extension of the action of V to an action of $\mathfrak{t}_\Gamma$. This means that we need to represent all the momentum operators $\partial_v, v \in V$, as bilinear combinations of fermionic "supercharges" $D_b, b \in B$ so that we have the commutation relations

$$[D_b, D_{b'}] = \partial_{\Gamma(b,b')}, \quad [D_b, \partial_v] = [\partial_v, \partial_{v''}] = 0.$$

For this to be possible, there should be about equally many bosonic and fermionic fields in the theory. This explains why supersymmetry is sometimes called "symmetry between bosons and fermions."

3.5.2 Superspace

Superspace is a tool to construct supersymmetric theories by replacing the mysterious "in some way" above by a natural construction. More precisely, a superspace is a supermanifold $\mathcal{S}$ extending the spacetime V (so that $V = \mathcal{S}^0$ is its even part) and which admits a natural action of t.

The simplest choice (*flat superspace*) is $\mathcal{S} = T$, the underlying manifold of the supersymmetry group, on which $\mathfrak{t} = \Pi(B) \oplus V$ acts by left-invariant vector fields D_b, ∂_v (see [13, 19]). Any field on $\mathcal{S}$ (referred to as *superfield*) gives an entire multiplet of usual fields on V by *component analysis*, that is by writing $\mathcal{O}_\mathcal{S} = \mathcal{O}_V \otimes \Lambda^\bullet(B^*)$ as a free $\mathcal{O}_V$-module of rank $2^{\dim(B)}$. The Lie algebra t acts naturally on superfields. So working only with such fields, we get supersymmetry seemingly "for free."

This construction can, of course, be done for an arbitrary quadratic space $\Gamma: \mathrm{Sym}^2(B) \to V$. If Γ is spinorial, then the action of t on superfields extends to an action of the super-Poincaré group $\mathcal{P}$.

Remark 3.16 Similarly to Section 3.2, the exponentials e^{D_b} of the supercharges are local operators on analytic superfields, while the shifts e^{∂_v} are not. It would be interesting to understand the consequences of this phenomenon. The situation of Section 3.2 corresponds to the simplest example of a quadratic space: when B is one-dimensional, $V = \mathrm{Sym}^2(B)$ is also one-dimensional, and $\Gamma = \mathrm{Id}$.

3.5.3 The Difficulty

The difficulty with supersymmetry is that it tends to require too many fields (on V) for all of them to make physical sense. The following result [45] is usually intepreted by saying that "supersymmetry in > 11 dimensions is not sensible."

Principle 3.17 (**Nahm's theorem**) *Any supersymmetric theory with $d > 11$ contains fields of spin $\geqslant 2$.*

For the superspace construction this is easy to understand. Already the simplest kind of superfield, a function on $\mathcal{S}$, is a section of $\mathcal{O}_V \otimes \Lambda^\bullet(B^*)$, where B is the direct sum of one or several spinor spaces. As d grows, the decomposition of $\Lambda^\bullet(B^*)$ into Spin(V)-irreducibles quickly begins to contain higher spin representations such as $\mathrm{Sym}^j(V)$, $j \geq 2$.

However, even in the remaining dimensions $d \leq 11$, the superspace construction typically gives too many component fields. To eliminate some of the components, one usually imposes (in a seemingly ad hoc way) some additional restrictions on superfields known as *constraints*. In the next subsection we discuss a conceptual point of view on such constraints.

Supergeometry as understood by physicists is the study of various versions of (not necessarily flat) superspaces. All the examples that have been considered fit into the following concept.

Definition 3.18 An abstract *superspace* is a supermanifold $\mathcal{S}$ (smooth, analytic or algebraic) of dimension $(m|n)$ together with a Pfaff system $C \subset T_\mathcal{S}$ of rank $(0|n)$ satisfying the following properties:

(1) The restriction $C|_{\mathcal{S}^{\bar{0}}}$ coincides with the odd part of $(T_\mathcal{S})|_{\mathcal{S}^{\bar{0}}}$.
(2) Denote $B = \Pi(C)$ and $V = (T_\mathcal{S})/C$. Then, the Frobenius pairing $F_C \colon \Lambda^2 C \to (T_\mathcal{S})/C$, written as an $\mathcal{O}_\mathcal{S}$-linear map $\Gamma \colon \mathrm{Sym}^2_{\mathcal{O}_\mathcal{S}}(B) \to V$, is a quadratic module over $\mathcal{S}$; see Definition 3.15.

Informally, a superspace enhancement $\mathcal{S}$ of an ordinary manifold $\mathcal{S}_{\bar{0}}$ provided a framework for taking square roots of vector fields on $\mathcal{S}_{\bar{0}}$. A choice of such an enhancement can be viewed as a choice of a square root of $\mathcal{S}_{\bar{0}}$ itself, that is, as a way of realizing the spacetime directions as some kind of curvature data, compare Example 3.7.

Example 3.19 (Supercurves, as understood by physicists) For a mathematician, an (algebraic) *supercurve* is an algebraic supermanifold of dimension $1|n$ for some n. For a physicist, a supercurve is a superspace of dimension $(1|n)$, so the Pfaff system C is a necessary part of the structure. See [14, 15, 43]. The geometry of a supercurve of dimension $(1|1)$ is locally modeled on the setting of Section 3.2.

Example 3.20 (Spinorial curved superspaces) Definition 3.18 is quite general. It does not require that the fibers of the quadratic module Γ be spinorial quadratic spaces. However, the intersections of quadrics related to spinorial spaces (such as the space of pure spinors) are *rigid* both as abstract algebraic varieties and as intersections of quadrics. This means that a quadratic module whose one fiber is a spinorial quadratic space has all neighboring fibers spinorial of the same type. Therefore if $\mathcal{S}$ is a superspace (in our sense) with the commutator pairing Γ spinorial at one even point, then we have a similar isomorphic "spinorial structure" in each neighboring tangent space.

This amounts to a differential-geometric structure on S including a conformal structure in the quotient bundle $(T_S)/C$ (in particular, on the ordinary manifold $S^{\bar 0}$) and a "choice of spinors" for this conformal structure (cf. [24, 40]).

Cobordism categories of such curved spinorial superspaces provide a language for an Atiyah-style approach to supersymmetric quantum field theories [55, 56].

3.6 Constraints and Complete Intersection Slices

Various recipes for imposing constraints on superfields can be understood by using the idea of *simple plane slices of complicated intersections of quadrics*.

Let $Z \subset \mathbb{P}(B)$ be an intersection of quadrics. A *plane slice* of Z is a scheme of the form $Z \cap M$, where $M \subset \mathbb{P}(B)$ is a projective subspace. It is an intersection of quadrics in M. The two simplest possibilities are as follows:

(0) $Z \cap M = Z$, that is, M is contained in Z entirely;
(1) $Z \cap M$ is a complete intersection of quadrics in M.

Let $\Gamma: \operatorname{Sym}^2(B) \to V$ be the quadratic space corresponding to Z. A projective subspace M corresponds to a linear subspace $B' \subset B$, and $Z \cap M$ corresponds to the quadratic space $\Gamma': \operatorname{Sym}^2(B') \to V'$, where $V' = \Gamma(\operatorname{Sym}^2(B'))$ and Γ' is the restriction of Γ. The supersymmetry algebra $\mathfrak{t}_{\Gamma'}$ is the Lie sub(super) algebra in $\mathfrak{t}$ generated by $B' \subset \mathfrak{t}_\Gamma^{\bar 1}$. We will call such quadratic spaces *slices of Γ*.

The case (0) above means that $\mathfrak{t}_{\Gamma'} = B'$ is abelian and purely even. In the case (1) (which includes the case (0)) we will say that $\mathfrak{t}_{\Gamma'}$ is a *null-subalgebra*.

Example 3.21 (a) Let $Z = C \sqcup C'$ be the union of two skew lines in $\mathbb{P}^3$ (Example 3.14(b)). Clearly, each of the two lines gives an instance of Case (0) above. Other than that, we have a $\mathbb{P}^1 \times \mathbb{P}^1$ worth of complete intersection slices $Z \cap M$, with M being a chord passing through one point on C and one point on C'.

(b) Let $\Gamma: \operatorname{Sym}^2(S_+) \to V$ be the $d = 10$ $N = (1, 0)$ SUSY quadratic space. Each vector $v \in V, q(v) = 0$, gives a linear operator $\Gamma_v: S_+ \to S_+$ given by the Clifford multiplication by v (transpose of the quadratic space map Γ). If v is a nonzero null vector, that is, $q(v) = 0$, then $\Gamma_v^2 = 0$ and $\operatorname{Ker}(\Gamma_v) = \operatorname{Im}(\Gamma_v)$ is an eight-dimensional subspace in S_+ which we denote L_v. The intersection $\mathbb{P}(L_v) \cap \Sigma^{10}$ is a quadric hypersurface in $\mathbb{P}(L_v)$, and the space of equations of this hypersurface is spanned by v. In other words, $\Gamma(\operatorname{Sym}^2(L_v)) = \mathbb{C} \cdot v$, and

$$\mathfrak{h} = L_v \oplus \mathbb{C} \cdot v \subset S_+ \oplus V = \mathfrak{t}$$

is a (maximal) null subalgebra. The $(1|8)$-dimensional linear subspaces $L_v \oplus \mathbb{C} \cdot v \subset \mathfrak{t} = \mathbb{C}^{10|16}$ are known as *super-null-geodesics* [13, 35, 44, 62].

It seems that the $\mathbb{P}(L_v) \cap \Sigma^{10}$ are precisely the maximal complete intersection slices of Σ^{10}.

Assuming this, we can formulate the constraints on superfields as follows.

3.6.1 Spin 0 Constraints

Spin 0 constraints are imposed on scalar superfields which are functions on $\mathcal{S}$ or, more generally, maps $\Phi \colon \mathcal{S} \to X$ where X is a given target manifold. They have the form

$$D_b \Phi = 0, \quad b \in B',$$

where B' is one fixed subspace of B on which Γ vanishes (case (0) above), maximal with this property. Maps Φ satisfying the constraints are known as *chiral superfields*.

3.6.2 Spin 1 Constraints

Spin 1 constraints are imposed on gauge superfields (connections ∇ in principal bundles on $\mathcal{S}$). They have the form of integrability $(F_\nabla)|_{g \cdot \mathfrak{h}} = 0$, where $\mathfrak{h} = \mathfrak{t}_{\Gamma'}$ runs over all null-subalgebras in $\mathfrak{t}_\Gamma$ and $g \cdot \mathfrak{h}$ is the left translation of $\mathfrak{h}$ in $\mathcal{S}$. In other words,

$$[\nabla_{b_1}, \nabla_{b_2}] = \nabla_{\Gamma'(b_1, b_2)}, \quad [\nabla_b, \nabla_a] = [\nabla_{a_1}, \nabla_{a_2}] = 0, \quad \forall b, b_1, b_2 \in B', \ a, a_1, a_2 \in V'$$

where B' runs over all the maximal complete intersection slices of Y (case 1 above).

Remark 3.22 When imposing constraints on superfields, it is obviously desirable not to end up restricting their dependence in the usual, even directions of spacetime. In case 3.6.2 this is ensured by the fact that the null-subalgebras $\mathfrak{h}$ have $\dim(\mathfrak{h}^{\bar{0}}) = 1$ (they are the usual null-lines). In other words, *all the complete intersection slices of* $Z = \Sigma^{10}$ *are quadric hypersurfaces*: $Z \cap M$ is a hypersurface in M. It would be interesting to study other intersections of quadrics Z with this property.

3.6.3 Lie Algebra Meaning of Complete Intersections

Given a quadratic space $\Gamma \colon \mathrm{Sym}^2(B) \to V$, we can associate to it another $\mathbb{Z}$-graded Lie algebra (i.e., a Lie algebra in $\mathrm{Vect}_{\mathbf{k}}^{\mathbb{Z}}$)

$$\widetilde{\mathfrak{t}}_\Gamma = \mathrm{FL}(B[-1])/(\mathrm{Ker}(\Gamma)).$$

Here $FL(-)$ means the free graded Lie algebra generated by a graded vector space. In our case $B[-1]$ is B put in degree $+1$, so the degree 2 part of $FL(B[-1])$ is $\text{Sym}^2(B)$. The Lie algebra $\tilde{\mathfrak{t}}_\Gamma$ is obtained by quotienting $FL(B[-1])$ by the graded Lie ideal generated by $\text{Ker}(\Gamma) \subset \text{Sym}^2(B)$. Thus

$$\tilde{\mathfrak{t}}_\Gamma^1 = B, \quad \tilde{\mathfrak{t}}_\Gamma^2 = V,$$

but it is not required that V commutes with the generators and therefore $\tilde{\mathfrak{t}}_\Gamma$ can be nontrivial in degrees $\geqslant 3$ (cf. [13, Section 11.3]). We note that $\mathfrak{t}_\Gamma$ can be considered as a $\mathbb{Z}$-graded Lie algebra by lifting the degree $\bar{0}$ part to degree 2 and degree $\bar{1}$ part to degree 1 (this is possible since the degree $\bar{0}$ part lies in the center). With this understanding, we have a surjective homomorphism of graded Lie algebras

$$p: \tilde{\mathfrak{t}}_\Gamma \longrightarrow \mathfrak{t}_\Gamma.$$

Denote by

$$R_\Gamma = \text{Sym}^\bullet(B^*)/I_2(Y_\Gamma)$$

the graded coordinate algebra (commutative in the usual sense) of the homogeneous intersection of quadrics $Y_\Gamma \subset B$. Then, the enveloping algebra $U(\tilde{\mathfrak{t}}_\Gamma)$ is identified with $R_\Gamma^!$, the quadratic dual of the quadratic algebra R_Γ (see [46] for background). In particular, we have a homomorphism of graded algebras

$$\eta: U(\tilde{\mathfrak{t}}_\Gamma) \longrightarrow \text{Ext}^\bullet_{R_\Gamma}(\mathbf{k}, \mathbf{k}).$$

The algebra R_Γ is called *Koszul* if η is an isomorphism. This is the case in all spinorial examples. The role of complete intersections from this point of view is as follows.

Proposition 3.23 *The following are equivalent, and if they are true, then R_Γ is Koszul:*

(i) *Γ is of complete intersection type, that is, Y_Γ is a complete intersection of quadrics.*
(ii) *We have $\tilde{\mathfrak{t}}_\Gamma^i = 0$ for $i \geqslant 3$, that is, the morphism $p: \tilde{\mathfrak{t}}_\Gamma \to \mathfrak{t}_\Gamma$ is an isomorphism. In particular, the condition of commutativity of $\tilde{\mathfrak{t}}_\Gamma^2 = V$ with $\tilde{\mathfrak{t}}_\Gamma^1 = B$ already follows from the defining relations of $\tilde{\mathfrak{t}}_\Gamma$.*

Proof: This is a particular case of the general principle in commutative algebra that locally complete intersections are characterized by the cotangent complex being quasi-isomorphic to a 2-term complex. The special case of intersections of quadrics was studied in [34]. ∎

Furthermore, in many cases (including those related to spinors) the algebra $\widetilde{\mathfrak{t}}_\Gamma$ can be identified with the amalgamated free product of all its null-subalgebras $\widetilde{\mathfrak{t}}_{\Gamma'} = \mathfrak{t}_{\Gamma'}$. This relates the integrability conditions on null-subalgebras with the Koszul duality point of view on constraints for supersymmetric Yang–Mills theories advocated in [44].

4 Homotopy-Theoretic Underpinnings of Supergeometry

4.1 The Skeleton of the Koszul Sign Rule

To understand the nature of the Koszul sign rule 2.3, let us "minimize" the symmetric monoidal category $SVect_{\mathbf{k}}$ by incorporating it.

To account just for the signs, we can disregard all morphisms in $SVect_{\mathbf{k}}$ which are not isomorphisms, as well as all objects which have total dimension >1. Restricted to one-dimensional supervector spaces ($\dim(V^{\bar{0}}) + \dim(V^{\bar{1}}) = 1$) and their isomorphisms, we get a symmetric monoidal category 1-$SVect_{\mathbf{k}}$. Similarly, to capture the $\mathbb{Z}$-graded sign rule, we can restrict to the category 1-$Vect_{\mathbf{k}}^{\mathbb{Z}}$ of $\mathbb{Z}$-graded one-dimensional vector spaces. These categories are examples of the following concept.

Definition 4.1 A *Picard groupoid* is a symmetric monoidal category $(\mathcal{G}, \otimes, \mathbf{1}, R)$ in which all objects are invertible under $\otimes$ and all morphisms are invertible under composition.

A Picard groupoid $\mathcal{G}$ gives rise to two abelian groups:

- The *Picard group* of $\mathcal{G}$, denoted $\text{Pic}(\mathcal{G})$, or $\pi_0(\mathcal{G})$. It is formed by the isomorphism classes of objects, with the operation given by $\otimes$.
- The group $\pi_1(\mathcal{G}) = \text{Aut}_{\mathcal{G}}(\mathbf{1})$ of automorphisms of the unit object. It is canonically identified with the group of automorphisms of any other object.

In our case,

$$1\text{-}SVect_{\mathbf{k}}: \quad \pi_0 = \mathbb{Z}/2, \quad \pi_1 = \mathbf{k}^*;$$
$$1\text{-}Vect_{\mathbf{k}}^{\mathbb{Z}}: \quad \pi_0 = \mathbb{Z}, \quad \pi_1 = \mathbf{k}^*.$$

Here, $\mathbf{k}^*$ is still unnecessarily big: to formulate the sign rule, we need only the subgroup $\{\pm 1\} \subset \mathbf{k}^*$. So we cut these Picard groupoids further.

For this, we replace $\mathbf{k}$ with the ring $\mathbb{Z}$, since $\{\pm 1\} = \mathbb{Z}^*$ is precisely its group of invertible elements. Accordingly, we replace one-dimensional $\mathbf{k}$-vector spaces with free abelian groups of rank 1. This gives Picard groupoids

1-SAb, 1-Ab$^\mathbb{Z}$. Their objects are $\mathbb{Z}/2$- or $\mathbb{Z}$-graded abelian groups which are free of rank 1. As before, the morphisms are isomorphisms, $\otimes$ is the graded tensor product over $\mathbb{Z}$ and the symmetries are given by the Koszul sign rule. The π_i of these Picard groupoids are now as follows:

$$\text{1-SAb:} \quad \pi_0 = \mathbb{Z}/2, \quad \pi_1 = \{\pm 1\} = \mathbb{Z}/2,$$
$$\text{1-Ab}^\mathbb{Z}: \quad \pi_0 = \mathbb{Z}, \quad \pi_1 = \{\pm 1\} = \mathbb{Z}/2.$$

We can call 1-SAb and 1-Ab$^\mathbb{Z}$ the *sign skeleta* of the Koszul sign rule ($\mathbb{Z}/2$-graded and $\mathbb{Z}$-graded versions). They contain all the data needed to write the sign rule but nothing more.

The following simple but remarkable fact can be seen as a mathematical explanation of the Principle of Naturality of Supers 2.2.

Proposition 4.2 *1-Ab$^\mathbb{Z}$ is equivalent to $\mathcal{F}_L$, the free Picard groupoid generated by one formal object (symbol) L.*

By definition, $\mathcal{F}_L$ has, as objects, formal tensor powers $L^{\otimes n}, n \in \mathbb{Z}$. It further has only those morphisms that are needed to write the symmetry isomorphisms

$$R_{L^{\otimes m}, L^{\otimes n}} : L^{\otimes m} \otimes L^{\otimes n} = L^{\otimes(m+n)} \longrightarrow L^{\otimes n} \otimes L^{\otimes m} = L^{\otimes(n+m)}$$

satisfying the axioms of a symmetric monoidal category (as well as composition, tensor products, etc., of such morphisms).
Sketch of proof: L corresponds to the group $\mathbb{Z}$ placed in degree 1, so $L^{\otimes n}$ is $\mathbb{Z}$ in degree n. Furthermore, $R_{L,L} \in \text{Aut}(L^{\otimes 2}) = \text{Aut}(\mathbf{1})$ corresponds to $(-1) \in \mathbb{Z}^*$ (note that $R_{L,L} \circ R_{L,L} = \text{Id}$ by symmetry). The axioms of a symmetric monoidal category give that $R_{L^{\otimes m}, L^{\otimes n}}$ corresponds to $(-1)^{mn}$, so we recover the Koszul rule.

In other words, the category $\text{Vect}_\mathbf{k}^\mathbb{Z}$ which is at the basis of all supergeometry can be obtained as a kind of **k**-*linear envelope of a free Picard groupoid*. More precisely, we have the following construction.

Definition 4.3 Let $\mathcal{G}$ be a Picard groupoid, and $\chi : \pi_1(\mathcal{G}) \to \mathbf{k}^*$ be a homomorphism. By a $(\mathcal{G}, \chi)$-*graded* **k**-*vector space* we will mean a functor $V : \mathcal{G} \to \text{Vect}_\mathbf{k}$, whose value on objects will be denoted $A \mapsto V^A$, satisfying the following condition. For each object A, the action of each $\lambda \in \pi_1(\mathcal{G}) \simeq \text{Aut}(A)$ on A is taken into the multiplication by $\chi(\lambda)$ on V^A. We denote by $\text{Vect}_\mathbf{k}^{(\mathcal{G}, \chi)}$ the category of $(\mathcal{G}, \chi)$-graded **k**-vector spaces.

Since V is a functor, the spaces V^A and $V^{A'}$ for isomorphic objects A, A' are identified, so a $(\mathcal{G}, \chi)$-graded **k**-vector space V can be viewed as a $\pi_0(\mathcal{G})$-graded vector space in the usual sense.

Proposition 4.4 *(a) The category* $\mathrm{Vect}_{\mathbf{k}}^{(\mathcal{G},\chi)}$ *has a structure of a monoidal category with the operation given by*

$$(V \otimes W)^A \;=\; \varinjlim_{\{B \otimes C \to A\}}^{\mathrm{Vect}_{\mathbf{k}}} V^B \otimes_{\mathbf{k}} W^C,$$

the colimit taken over the category formed by pairs of objects $B, C \in \mathcal{G}$ *together with an (iso)morphism* $B \otimes C \to A$. *Furthermore, the symmetry in* $\mathcal{G}$ *makes* $\mathrm{Vect}_{\mathbf{k}}^{(\mathcal{G},\chi)}$ *into a symmetric monoidal category.*

(b) If $\mathcal{G} = \mathcal{F}_L$ *and* $\chi : \pi_1(\mathcal{F}_L) = \mathbb{Z}/2 \to \mathbf{k}^*$ *is the embedding of* $\{\pm 1\}$, *then* $\mathrm{Vect}_{\mathbf{k}}^{(\mathcal{G},\chi)}$ *is identified with the category* $\mathrm{Vect}_{\mathbf{k}}^{\mathbb{Z}}$ *with the symmetry given by the Koszul sign rule.*

4.2 (Higher) Picard Groupoids and Spectra

One of the insights of Grothendieck in his manuscript "Pursuing stacks" (cf. [8], p. 114) was the correspondence between Picard groupoids and a particular class of *spectra* in the sense of homotopy topology. See [16] for a discussion and [23] for a slightly more detailed treatment which we follow here. A systematic account can be found in [32].

The concept of a spectrum arises as a result of stabilizing the homotopy category of pointed topological spaces (say CW-complexes) under the two operations (adjoint functors)

$$\Sigma = \text{reduced suspension}, \quad \Omega = \text{loop space},$$
$$\mathrm{Hom}(\Sigma X, Y) \;=\; \mathrm{Hom}(X, \Omega Y).$$

For example, the spheres S^n satisfy $\Sigma(S^n) = S^{n+1}$. We always have a canonical map (unit of adjunction)

$$\varepsilon_X : X \longrightarrow \Omega \Sigma X.$$

A spectrum Y can be seen as a topological space $\Omega^\infty Y$ together with a sequence of deloopings: spaces $\Omega^{\infty-j} Y$ equipped with compatible homotopy equivalences $\Omega^j(\Omega^{\infty-j} Y) \sim \Omega^\infty Y$. A spectrum Y has homotopy groups $\pi_i(Y), i \in \mathbb{Z}$ defined by

$$\pi_i(Y) \;=\; \pi_{i+j}(\Omega^{\infty-j} Y), \quad j \gg 0.$$

Example 4.5 A topological space X gives the *suspension spectrum* $\Sigma^\infty X$, with $\Omega^\infty \Sigma^\infty X = \varinjlim_n \Omega^n \Sigma^n X$ and $\Omega^{\infty-j} \Sigma^\infty X = \varinjlim_n \Omega^{n-j} \Sigma^n X$ (limits under powers of ϵ). The homotopy groups of $\Sigma^\infty X$ are the *stable homotopy groups* of X:

$$\pi_i(\Sigma^\infty X) = \pi_i^{\text{st}}(X) := \varinjlim_n \pi_{i+n} \Sigma^n X.$$

Spectra form (after inverting homotopy equivalences) a triangulated category SHo known as the *stable homotopy category*. This category has a symmetric monoidal structure (smash product of spectra). Let $m \leqslant n$ be integers ($m = -\infty$ or $n = \infty$ allowed). By a $[m,n]$-*spectrum* we mean a spectrum Y with $\pi_i(Y) = 0$ with $i \notin [m,n]$, and we denote by $\text{SHo}_{[m,n]} \subset \text{SHo}$ the full subcategory of $[m,n]$-spectra. There is a canonical "truncation" functor

$$\tau_{[m,n]} \colon \text{SHo} \longrightarrow \text{SHo}_{[m,n]}.$$

Grothendieck's correspondence can be formulated as follows.

Theorem 4.6 *There is an equivalence of categories*

$$\mathbb{B} \colon \{Picard\ groupoids\}[eq^{-1}] \longrightarrow SHo_{[0,1]},$$

so that $\pi_i(\mathbb{B}(\mathcal{G}c)) = \pi_i(\mathcal{G})$, $i = 0, 1$. *Here* $[eq^{-1}]$ *means that equivalences of Picard groupoids are inverted, similarly to inverting homotopy equivalences in forming SHo.*

A more precise result is proved in [32]. The spectrum $\mathbb{B}\mathcal{G}$ corresponding to a Picard groupoid $\mathcal{G}$ is a version of the classifying space of $\mathcal{G}$. That is, the space $\Omega^\infty \mathbb{B}\mathcal{G} = B\mathcal{G}$ is the usual classifying space of $\mathcal{G}$ as a category, and the deloopings are constructed using the symmetric monoidal structure; see [23, Section 3.1.6] for an explicit construction.

The further point of Grothendieck is that more general $[0,n]$-spectra should have a description in terms of *Picard n-groupoids*, an algebraic concept to be defined, meaning "symmetric monoidal n-categories with all the objects and higher morphisms invertible in all possible senses." Here we can formally allow the case $n = \infty$.

Incredible complexity of the stable homotopy category, well known to topologists, prevents us from hoping for a simple algebraic definition of Picard n-groupoids. Nevertheless, for small values of $n = 2, 3$, this can be accessible and useful. The case $n = 2$ is being treated in the paper [28] building on the theory of symmetric monoidal 2-categories [27, 29].

4.3 The Sphere Spectrum and the Free Picard n-Groupoid

The fundamental role in homotopy theory is played by the *sphere spectrum* $\mathbb{S} = \Sigma^\infty S^0$ defined as the suspension spectrum of the 0-sphere. Its homotopy groups are the *stable homotopy groups of spheres*

$$\pi_i(\mathbb{S}) = \pi_i^{st} := \varinjlim_n \pi_{i+n}(S^n),$$

which vanish for $i < 0$, so $\mathbb{S}$ is a $[0,\infty]$-spectrum. The spectrum $\mathbb{S}$ is the unit object in the symmetric monoidal structure on SHo and for this reason can be considered as a homotopy-theoretic analog of the ring $\mathbb{Z}$ of integers.

This motivates the further installment in Grothendieck's vision of a dictionary between spectra and Picard n-groupoids:

Conjecture 4.7 *The Picard n-groupoid corresponding to $\tau_{[0,n]}\mathbb{S}$, should be identified with $\mathcal{F}_L^{(n)}$ the free Picard n-groupoid on one formal object L.*

The concept of a free Picard n-groupoid presumes that we already have a system of axioms for what a Picard n-groupoid is. If we have such axioms, then $\mathcal{F}_L^{(n)}$ contains as objects formal tensor powers $L^{\otimes n}$ and only those higher morphisms which are needed to write the necessary "higher symmetry isomorphisms."

As before, for large n this seems unattainable directly, but for small $n \leq 2$ this can be made into a theorem. In particular, the case $n = 1$, proved in [32, Proposition 3.1] has enormous significance for supergeometry. Indeed, combining it with Proposition 4.2, we arrive at the following:

Corollary 4.8 *The sign skeleton 1-$SAb^{\mathbb{Z}}$ of the $\mathbb{Z}$-graded Koszul sign rule, is the Picard groupoid corresponding to the $[0,1]$-truncation of $\mathbb{S}$. The groups $\mathbb{Z} = \pi_0(1\text{-}Ab^{\mathbb{Z}})$ and $\mathbb{Z}/2 = \pi_1(1\text{-}Ab^{\mathbb{Z}})$ are the first two stable homotopy groups of spheres.*

In other words, *the entire supermathematics is obtained by unraveling the first two layers of the sphere spectrum.*

In Table 3.1 (which expands, somewhat, a table from the online encyclopedia nLab) we give the values of the π_i^{st} for $i \leq 3$ and indicate mathematical and physical phenomena that these groups govern. We also compare $\mathbb{S}$ with another spectrum, the *algebraic K-theory spectrum* $\mathbb{K}(\mathbf{k})$ of a field $\mathbf{k}$, which has $\pi_i(\mathbb{K}(\mathbf{k})) = K_i(\mathbf{k})$, the Quillen K-groups. These groups are indicated at the bottom row. We notice that the first two groups are π_i of our intermediate Picard groupoid 1-$\text{Vect}_{\mathbf{k}}^{\mathbb{Z}}$ which corresponds to the $[0,1]$-truncation of $\mathbb{K}(\mathbf{k})$ (see [8, Section 4]). A philosophy going back to [4] and to Quillen says

Table 3.1. *The first few π_i^{st} and their significance*

$\pi_0^{st} = \mathbb{Z}$	$\pi_1^{st} = \mathbb{Z}/2$	$\pi_2^{st} = \mathbb{Z}/2$	$\pi_3^{st} = \mathbb{Z}/24$
Cardinality of sets, dimension of vector spaces $K_0(\mathbf{k}) = \mathbb{Z}$	Sign of permutation, determinant, Koszul rule $K_1(\mathbf{k}) = \mathbf{k}^*$	Spin, central extensions of symmetric groups $K_2(\mathbb{R}) \to \pi_1(SL_n(\mathbb{R}))$ $\pi_1(SL_n(\mathbb{R})) = \mathbb{Z}/2$	"String," $\sqrt[24]{1}$ in transformation of $\eta(q)$, $\chi(K3) = 24$ Dilogarithms, etc.

that $\mathbb{S}$ can be heuristically considered as the K-theory spectrum of $\mathbb{F}_1$, the (non-existent) field with one element, the symmetric group S_n being the "limit," as $q \to 1$, of the general linear groups $GL_n(\mathbb{F}_q)$.

The first two columns are self-explanatory. In the third column, the phenomenon of spin(ors) is based on the fundamental group $\pi_1(SO_n) = \pi_1(SL_n(\mathbb{R})) = \mathbb{Z}/2$ which is the same as $\pi_1^{st}(SO_n)$ and is identified with $\pi_1^{st} = \mathbb{Z}/2$ via the map $SO_n \to \Omega^{n+1}(S^{n+1})$ (this is known as J-homomorphism). The existence of central extensions of symmetric and alternating groups S_n and A_n (with center $\mathbb{Z}/2$) and of corresponding projective representations [37, 52] is a related phenomenon: A_n embeds into SO_n, and taking the preimage in Spin(n), we get a $\mathbb{Z}/2$-extension.

One thing is worth noticing. Supergeometry, as understood by mathematicians, tackles only the first two columns of Table 3.1. A similar-sounding concept (supersymmetry), used by physicists, dips into the third column as well: fermions are always wedded to spinors in virtue of the Spin-Statistics Theorem. In fact, there is something in the very structure of the sphere spectrum that seems to relate spin (third column) and statistics (the second column). At the most naive level, this is the coincidence of $\pi_1^{st} = \mathbb{Z}/2$ with $\pi_2^{st} = \mathbb{Z}/2$.

Furthermore, consider the $[1, 2]$-truncated spectrum $\tau_{[1,2]}\mathbb{S}$. Its loop spectrum $\Omega\tau_{[1,2]}\mathbb{S}$ is a $[0, 1]$-spectrum with $\pi_0 = \pi_1^{st} = \mathbb{Z}/2$ and $\pi_1 = \pi_2^{st} = \mathbb{Z}/2$.

Theorem 4.9 (Homotopy-theoretic Spin-Statistics Theorem) *The Picard groupoid corresponding to $\Omega\tau_{[1,2]}\mathbb{S}$ is equivalent to 1-SAb, the sign skeleton of the $\mathbb{Z}/2$-graded Koszul sign rule. In other words, $\Omega\tau_{[1,2]}\mathbb{S}$ is homotopy equivalent to $(\tau_{[0,1]}\mathbb{S})/2$, the reduction of the spectrum $\tau_{[0,1]}\mathbb{S}$ by the element 2 of its π_0.*

So there is not one, but *two ways* in which the same Koszul sign rule appears out of the sphere spectrum, one through statistics, the other one through spin. Note that the "topological proof" of the usual physical Spin-Statistics Theorem, going back to Feynman [18] is based on the intuitive claim that interchanging two particles is "equivalent" to tracing a nontrivial loop in the rotation group, and this claim needs something like 4.9 to be consistent. See [17, Chapter 20] for a more detailed discussion of this claim and the whole issue.

Proof of 4.9: This is an exercise on known facts in homotopy theory. A $[0, 1]$-spectrum Y (or a Picard groupoid) is classified by its π_0, π_1 and the *Postnikov invariant* which is a group homomorphism $k_Y : \pi_0(Y) \to \pi_1(Y)$ satisfying $2k_Y = 0$. Explicitly, k_Y is given by the composition product with the generator $\eta \in \pi_1^{st}$. For $Y = \tau_{[0,1]}\mathbb{S}$ we have therefore that $k_Y : \mathbb{Z} \to \mathbb{Z}/2$ is the surjection, and for $Y = \tau_{[0,1]}\mathbb{S}/2$ we have $k_Y = \mathrm{Id}_{\mathbb{Z}/2}$.

If both $\pi_0 = \pi_1 = \mathbb{Z}/2$ and $k_Y \neq 0$, then $k_Y = \mathrm{Id}$, like for $Y = (\tau_{[0,1]}\mathbb{S})/2$ and so $Y \sim (\tau_{[0,1]}\mathbb{S})/2$. Now, for $Y = \Omega \tau_{[1,2]}\mathbb{S}$ the Postnikov invariant is the map $\pi_1^{st} \to \pi_2^{st}$ given by composition with η. It is known [61] that $\eta^2 \in \pi_2^{st}$ is the generator and so $k_Y \neq 0$. ∎

The fourth column of Table 3.1, headed by $\pi_3^{st} = \mathbb{Z}/24$, is related to various "string-theoretic" mathematics such as the appearance of twenty-fourth roots of 1 in Dedekind's formula for the modular transformation of his η-function [2], the Euler characteristic of a K3 surface being 24, the importance of the central charge modulo 24 in conformal field theory and so on (cf. [30]).

4.4 Toward Higher Supergeometry

Conjecture 4.7 means that the sphere spectrum $\mathbb{S}$ is the ultimate source for meaningful twists of commutativity, that is, for designing truly commutative-like structures, flexible enough to serve as a basis of geometry. The existing supermathematics uses only the first two levels (zeroth and first) of $\mathbb{S}$, with physical applications exploiting the parallelism between the first and the second levels.

This opens up a fantastic possibility of *higher supermathematics* which would use, as its "sign skeleton," the spectrum $\mathbb{S}$ in its entirety or, at least, the truncations $\tau_{[0,n]}\mathbb{S}$ and the free Picard n-groupoids $\mathcal{F}_L^{(n)}$ for as long as we can make sense of them algebraically. Here we sketch the first step in this direction, the formalism for $n = 2$. For convenience, we adopt a genetic approach.

4.4.1 Idea of Supersymmetric Monoidal Categories

Supermathematics begins with replacing commutativity $ab = ba$ with supercommutativity (2.1). Categorical analogs of commutative algebras are symmetric monoidal categories, where we have coherent isomorphisms $V \otimes W \simeq W \otimes V$. So we introduce "categorical minus signs" into these isomorphisms as well.

More precisely, by a **k**-*superlinear category* we mean a module category $\mathcal{V}$ (a category tensored over) the symmetric monoidal category SVect$_\mathbf{k}$. In such a category we have the *parity change functor* Π given by tensoring with the supervector space $\mathbf{k}^{0|1}$. We take Π as the categorical analog of the minus sign. In doing so, we use the identification of $\pi_0(\mathbb{S}/2) = \mathbb{Z}/2$ (the $\mathbb{Z}/2$-grading) with $\pi_1^{\text{st}} = \mathbb{Z}/2$ (the ± 1 signs).

We now consider **k**-superlinear categories $\mathcal{A}$ which are $\mathbb{Z}$- or $\mathbb{Z}/2$-*graded*, that is, split into a categorical direct sum $\mathcal{A} = \boxplus_i \mathcal{A}^i$, where $i \in \mathbb{Z}$ or $\mathbb{Z}/2$. We assume that $\mathcal{A}$ is equipped with graded SVect$_\mathbf{k}$-bilinear bifunctors

$$\otimes = \otimes_{i,j} : \mathcal{A}^i \times \mathcal{A}^j \longrightarrow \mathcal{A}^{i+j}$$

subject to associativity isomorphisms of the usual kind and we want to impose *twisted commutativity isomorphisms*

$$R_{V,W} : V \otimes W \longrightarrow \Pi^{\deg(V) \cdot \deg(W)}(W \otimes V) \tag{4.1}$$

subject to natural axioms, in which, further, various numerical minus signs will be introduced.

4.4.2 Definition of Supersymmetric Monoidal Categories

A possible more precise definition can go as follows. For simplicity consider the $\mathbb{Z}$-graded version. We use Proposition 4.4 as a guideline, and start with $\mathcal{F} = \mathcal{F}_L^{(2)}$, the free Picard 2-groupoid on one object L. It corresponds to the truncation $\tau_{[0,2]}\mathbb{S}$. Thus $\pi_0(\mathcal{F})$, the group of equivalence classes of objects, is identified with $\pi_0^{\text{st}} = \mathbb{Z}$ and will account for the grading.

The category Aut$_\mathcal{F}(\mathbf{1})$ formed by automorphisms of the unit object and 2-morphisms between them, is a usual Picard groupoid which corresponds to $\Omega\tau_{[1,2]}\mathbb{S}$ and so, by the Spin-Statistics Theorem 4.9, it is identified with 1-SAb, the sign skeleton of SVect$_\mathbf{k}$.

We denote by SCat$_\mathbf{k}$ the 2-category of **k**-superlinear categories. It serves as a categorical analog of the category of ordinary vector spaces. We will denote by $\mathcal{V} \boxtimes_{\text{SVect}-\mathbf{k}} \mathcal{W}$ the categorical tensor product [25] of two **k**-superlinear categories $\mathcal{V}$ and $\mathcal{W}$.

The Picard groupoid 1-SVect$_\mathbf{k}$ plays the role of the multiplicative group $\mathbf{k}^*$ for SCat$_\mathbf{k}$: it acts on each object by equivalences. The monoidal functor (embedding) χ: 1-SAb $\to$ 1-SVect$_k$ is therefore an analogue of the homomorphism χ from Definition 4.3.

We now consider the 2-category SVect$_k^{(\mathcal{F}, \chi)}$ formed by all 2-functors $\mathcal{V} \colon \mathcal{F} \to$ SCat$_k$ which take the action of Aut$_\mathcal{F}(\mathbf{1})$ on each object A into the action on $\mathcal{V}^A$ given by χ. As before, a datum of such $\mathcal{V}$ is the same as a datum of a family of superlinear categories $\mathcal{V}^i = \mathcal{V}^{L^{\otimes i}}$, $i \in \mathbb{Z}$, one for each equivalence class of objects of $\mathcal{F}$. Now, the formula

$$(\mathcal{V} \boxtimes \mathcal{W})^A = 2\varinjlim_{\{B \otimes C \to A\}} \mathcal{V}^B \boxtimes_{\text{SVect}_k} \mathcal{W}^C$$

makes SVect$_k^{(\mathcal{F}, \chi)}$ into a symmetric monoidal 2-category. It can be seen as the 2-categorical analog of the category of supervector spaces.

By definition, a *supersymmetric monoidal k-category* is a symmetric monoidal object $\mathcal{A}$ in SVect$_k^{(\mathcal{F}, \chi)}$. An explicit algebraic model for $\mathcal{F}$ was proposed in [5, ex. 5.2]; see also [51, ex. 2.30]. Taking this model for $\mathcal{F}$, we can unravel the data involved in $\mathcal{A}$. This data contains in particular superlinear categories $\mathcal{A}^i$, bifunctors $\otimes_{i,j}$ and isomorphisms $R_{V,W}$ as outlined in Section 4.4.1.

The definition of a $\mathbb{Z}/2$-graded supersymmetric monoidal category is similar, using the Picard 2-groupoid $\mathcal{F}/2$.

If V is an object of a symmetric monoidal category, then there is an action of the symmetric group S_n on $V^{\otimes n}$. If instead $V \in \mathcal{A}^{2m+1}$ is an odd object of a supersymmetric monoidal category $\mathcal{A}$, then $V^{\otimes n} \oplus \Pi(V^{\otimes n})$ has an action of the (spin) central extension of S_n, first discovered by Schur [52].

Example 4.10 (a) The exterior algebra of a superlinear category Similarly to usual linear algebra (Example 2.1), an example of a supersymmetric monoidal category can be extracted from the *categorical version of the exterior power construction* developed in [23]. This construction is based on the *categorical sign character* which is a functor of monoidal categories

$$\text{sgn}_2 \colon S_n \longrightarrow \text{1-SAb}$$

(S_n is the symmetric group considered as a discrete monoidal category). It combines the usual sign character sgn: $S_n \to \mathbb{Z}/2$ at the level of π_0 and the "spin-cocycle" $c \in H^2(S_n, \mathbb{Z}/2)$ at the level of π_1. The exterior power $\Lambda^n \mathcal{V}$ of a superlinear category $\mathcal{V}$ is obtained from the tensor power $\mathcal{V}^{\boxtimes n}$ by considering objects equipped with sgn$_2$-twisted S_n-equivariance structure (see [23] for precise context and details). The analog of the wedge product is given by the functors

$$\wedge_{m,n} \colon \Lambda^m \mathcal{V} \times \Lambda^n \mathcal{V} \longrightarrow \Lambda^{m+n} \mathcal{V}$$

given by partial Π-antisymmetrization, as in [23, Section 4.2].

(b) Superalgebras of types M and Q and the half-tensor product of Sergeev Let **k** be algebraically closed. It is known since C.T.C. Wall [59] that simple finite-dimensional associative superalgebras over **k** are of two types: *type M*, formed by the matrix superalgebras $M_{p|q} = \mathrm{End}(\mathbf{k}^{p|q})$ and *type Q*, formed by the so-called *queer superalgebras* $Q_n \subset M_{n|n}$ (see [33, 37]). The simplest nontrivial queer algebra is the Clifford algebra Cliff_1 on one generator

$$Q_1 = \mathrm{Cliff}_1, \quad \mathrm{Cliff}_n = \mathbf{k}[\xi_1, \cdots \xi_n]/(\xi_i^2 = 1, \ \xi_i\xi_j = -\xi_j\xi_i),$$
$$\deg(\xi_i) = \bar{1}.$$

Their behavior under tensor multiplication is

$$M_{p|q} \otimes M_{m|n} \simeq M_{pm+qn|pn+qm},$$
$$M_{p|q} \otimes Q_n \simeq Q_{(p+q)n}, \quad Q_m \otimes Q_n \simeq M_{mn|mn}. \quad (4.2)$$

This means that the *super-Brauer group* formed by Morita-equivalence classes of these algebras is identified with $\mathbb{Z}/2$, with type M mapping to $\bar{0}$ and type Q mapping to $\bar{1}$. This $\mathbb{Z}/2$ is nothing but π_2^{st}, responsible for spin. See [20, Section 4].

As a consequence, irreducible objects of any semisimple **k**-superlinear category $\mathcal{V}$ also split into two types M and Q, according to their endomorphism algebras being **k** or Q_1. Denoting $\mathcal{V}^{\bar{0}}$ the subcategory formed by direct sum of objects of type M and $\mathcal{V}^{\bar{1}}$ the subcategory formed by sums of objects of type Q, we get an intrinsic $\mathbb{Z}/2$-grading on $\mathcal{V}$. By (4.2), any exact monoidal structure $\otimes$ on $\mathcal{V}$ preserves this grading.

Furthermore, if V, W are irreducible objects of type Q, then $V \otimes W$ is acted upon by $Q_1 \otimes Q_1 \simeq \mathrm{End}(\mathbf{k}^{1|1})$ and so is identified with the direct sum of some object with its shift:

$$V \otimes W \simeq (2^{-1}V \otimes W) \oplus \Pi(2^{-1}V \otimes W),$$
$$2^{-1}V \otimes W := I \otimes_{Q_1 \otimes Q_1} (V \otimes W)$$

where $I \simeq (\mathbf{k}^{1|1})^*$ is an irreducible right module over $Q_1 \otimes Q_1$. We get in this way a new monoidal operation

$$\otimes_{\bar{1},\bar{1}} : \mathcal{V}^{\bar{1}} \times \mathcal{V}^{\bar{1}} \longrightarrow \mathcal{V}^{\bar{0}}, \quad V \otimes_{\bar{1},\bar{1}} W := 2^{-1}V \otimes W.$$

This operation was introduced by A. Sergeev [53] (see also [37, p. 163], for more discussion). If $\otimes$ is symmetric, then $\otimes_{\bar{1},\bar{1}}$ satisfies

$$V \otimes_{\bar{1},\bar{1}} W \simeq \Pi(W \otimes_{\bar{1},\bar{1}} V).$$

Indeed, the interchange of the factors in $V \otimes W$ corresponds to the interchange of ξ_1 and ξ_2 in $\text{Cliff}_2 = Q_1 \otimes Q_1$, and the pullback of I under this interchange is isomorphic to $\Pi(I)$ (which, as a right Cliff_2-module, is not isomorphic to I). The operation $\otimes_{\bar{1},\bar{1}}$ can be used as a source of examples of supersymmetric monoidal structures.

4.4.3 Further Directions

It seems important to investigate the concept of supersymmetric monoidal categories more systematically. Among other things, in such a category $\mathcal{V}$ we can speak about commutative algebra objects A. They would be further twisted generalizations of supercommutative algebras (which appear for $\mathcal{V} = \text{SVect}_\mathbf{k}$, a symmetric, not just a supersymmetric category). At the same time, such objects should be "just as good" as ordinary commutative algebras, in an extension of the Principle of Naturality of Supers 2.2. In particular, we must be able to associate to them some geometric objects $\text{Spec}(A)$ which should be flexible enough to be glued together to form more global structures.

It is also interesting to weaken the concept to allow, for example, for "superbraidings," in which the twisted commutativity isomorphisms (4.1) are retained but the involutivity requirements on them are relaxed.

Supersymmetric and superbraided monoidal categories may be relevant for the recent activity in condensed matter physics studying so-called fermionic phases (see [6, 21, 22, 26]). In particular, various "supercohomology" constructions appearing there seem to be represented by homotopy types and spectra related to truncations of $\mathbb{S}$, the sphere spectrum and $\mathcal{K}(\mathbb{C})$, the K-theory spectrum of $\mathbb{C}$. For instance, the "group supercohomology" of [26] is represented by the spectrum $\tau_{\leq 1} \mathcal{K}(\mathbb{C})/2$, which corresponds to the Picard groupoid of one-dimensional supervector spaces over $\mathbb{C}$.

Another possible approach to higher supergeometry could be to use the truncations and suspensions not of $\mathbb{S}$ itself but of

$$\check{\mathbb{S}} = R\text{Hom}(\mathbb{S}, \mathbb{C}^*),$$

the Pontryagin (Brown–Comenetz) dual of $\mathbb{S}$ considered in [21]. The homotopy groups of $\check{\mathbb{S}}$ are nontrivial only in degrees ≤ 0 with $\pi_0 = \mathbb{C}^*$ dual to $\pi_0^{\text{st}} = \mathbb{Z}$ and other π_i finite. For example, the spectrum corresponding to the Picard groupoid of one-dimensional supervector spaces above can be also described as

$$\tau_{\leq 1} \mathcal{K}(\mathbb{C})/2 = \Sigma \tau_{[-1,0]} \check{\mathbb{S}},.$$

the suspension of the truncation of $\check{\mathbb{S}}$.

As for the next step, one can imagine a "2-supersymmetric monoidal 2-category" to split into a direct sum of 24 sub(2-)categories ("sectors") instead of just two (even and odd objects), with Hom-categories between objects split into even and odd parts and so on, with an appropriate pattern of sign/shift rules for symmetries. The pattern of 24 sectors is of course suggestive of the features of conformal field theory and Chern–Simons theory (the significance of the central charge modulo 24, the 24 classes of framings of a 3-manifold, etc.).

Expressed more generally, study of higher supergeometry can include two complementary directions. One is *categorification*, investigation of deeper and deeper twists for "commutativity," taking the inspiration from the ultimate commutative structure, the sphere spectrum $\mathbb{S}$. The other is *geometrization*, gluing together local objects of algebraic origin, for which some sort of "commutativity" seems to be necessary (why?). Reconciling these two directions can, hopefully, shed more light on the way geometry emerges from quantum behavior.

References

[1] J. F. Adams, *Infinite Loop Spaces*, Princeton University Press (1978)

[2] M. Atiyah, *The logarithm of the Dedekind η-function*, Math. Ann. 278 (1987) 335–80

[3] M. Atiyah, R. Bott, and A. Shapiro, *Clifford modules*, Topology 3 (1964) 3–38

[4] M. Barratt and S. Priddy, *On the homology of non-connected monoids and their associated groups*, Comment. Math. Helv. 47 (1972) 1–14

[5] B. Bartlett, *Quasistrict symmetric monoidal 2-categories via wire diagrams*, Preprint, arXiv:1409.2148

[6] L.Bhardwaj, D. Gaiotto, and A. Kapustin, *State sum constructions of spin-TFT and string net constructions of fermionic phases of matter*, Preprint, arXiv:1605.01640

[7] I. Ciocan-Fontanine and M. Kapranov, *Derived Quot schemes*, Ann. Sci. ENS, 34 (2001) 403–40

[8] P. Deligne, *Le déterminant de la cohomologie*, in *Current Trends in Arithmetical Algebraic Geometry*, Contemp. Math. 67, AMS (1987) 93–177

[9] P. Deligne, *Catégories Tannakiennes*, in *Grothendieck Festschrift*, vol. II, Birkhäuser (1990)

[10] P. Deligne, *Notes on spinors*, in P. Deligne et al., eds., *Quantum Fields and Strings: A Course for Mathematicians*, vol. I, AMS (1999) 99–136

[11] P. Deligne, *Catégories tensorielles*, Moscow Math. J. 2 (2002) 227–48

[12] P. Deligne and J. Morgan, *Notes on supersymmetry (following J. Bernstein)*, in P. Deligne et al., eds., *Quantum Fields and Strings: A Course for Mathematicians*, vol. I, AMS (1999) 41–98

[13] P. Deligne and D. S. Freed, *Supersolutions*, in P. Deligne et al., eds., *Quantum Fields and Strings: A Course for Mathematicians*, vol. I, AMS (1999) 227–356

[14] R. Donagi and E. Witten, *Supermoduli space is not projected*, Preprint, arXiv:1304.7798

[15] R. Donagi and E. Witten, *Super Atiyah classes and obstructions to splitting of supermoduli space*, Preprint, arXiv:1404.6257

[16] V. Drinfeld, *Infinite-dimensional objects in algebra and geometry*, in *The Unity of Mathematics (in Honor of the Nineteeth Birthday of I.M. Gelfand)*, Birkhäuser (2006) 263–304

[17] I. Duck and E. C. G. Sudarshan, *Pauli and the Spin-Statistics Theorem*, World Scientific (1997)

[18] R. P. Feynman and S. Weinberg, *Elementary Particles and the Laws of Physics: The 1986 Dirac Memorial Lectures*, Cambridge University Press (1999)

[19] D. S. Freed, *Five Lectures on Supersymmetry*, AMS (1999)

[20] D. S. Freed, *Anomalies and invertible field theories*, Preprint, ArXiv:1404.7224

[21] D. S. Freed, *Short-range entanglement and invertible field theories*, Preprint, ArXiv:1406.7278

[22] D. S. Freed and M. J. Hopkins, *Reflection positivity and invertible topological phases*, Preprint, ArXiv:1604.06527

[23] N. Ganter and M. Kapranov, *Symmetric and exterior powers of categories*, Transform. Groups 19 (2014) 57–103

[24] S. J. Gates Jr., M. T. Grisaru, M. Rocek, and W. Siegel, *Superspace, or 1001 Lessons in Supersymmetry*, B. Cummings (1983)

[25] J. Greenough, *Monoidal 2-structure of bimodule categories*, Preprint, arXiv:0911.4979

[26] Z.-C. Gu and X.-G. Wen, *Symmetry-protected topological orders for interacting fermions: Fermionic topological nonlinear σ models and a special group supercohomology theory*, Phys. Rev. B 90 (2014) 115141

[27] N. Gurski, N. Johnson, and A. M. Osorno, *K-theory for 2-categories*, Preprint, arXiv:1503.07824

[28] N. Gurski, N. Johnson, and A. M. Osorno, *Realizing stable 2-types via Picard 2-categories*. In preparation.

[29] N. Gurski and A. M. Osorno, *Infinite loop spaces and coherence for symmetric monoidal bicategories*, Adv. Math. 246 (2013) 1–32

[30] M. J. Hopkins, *Algebraic topology and modular forms*, in *Proceedings of the International Congress of Mathematicians*, Vol. I, Higher Ed. Press (2002) 291–317

[31] M. J. Hopkins and I. Singer, *Quadratic functions in geometry, topology and M-theory*, J. Diff. Geom. 70 (2005) 329–452

[32] N. Johnson and A. M. Osorno, *Modeling stable one-types*, Theory Appl. Categories 26 (2012) 520–37

[33] T. Jozefiak, *Semisimple superalgebras*, in: *Algebra: Some Current Trends*, Lecture Notes in Math. 1352, Springer (1988) 96–113

[34] M. M. Kapranov, *On the derived category and K-functor of coherent sheaves on intersections of quadrics*, Math. USSR Izv. 32 (1989) 191–204

[35] M. M. Kapranov and Y. I. Manin, *The twistor transformation and algebraic-geometric constructions of solutions of the equations of field theory*, Russ. Math. Sur. 41 (1986) 33–61

[36] M. Kapranov and E. Vasserot, *Supersymmetry and the formal loop space*, Adv. Math. 227 (2011) 1078–1128

[37] A. Kleshchev, *Linear and Projective Representations of Symmetric Groups*, Cambridge University Press (2005)

[38] M. Kontsevich, *Notes on deformation theory*, Course lecture notes, Berkeley.

[39] D. A. Leites, *Spectra of graded-commutative rings*, Uspekhi. Mat. Nauk 29 (1974) 209–10

[40] J. Lott, *Twistor constraints in supergeometry*, Comm. Math. Phys. 133 (1990) 563–615

[41] J. Lurie, *Derived algebraic geometry II-V*, Preprint, arXiv math/0702299, math/0703204, 0709.3091, 0905.0459

[42] S. Mac Lane, *Categories for a Working Mathematician*, Springer (1971)

[43] Y. I. Manin, *Gauge Fields and Complex Geometry*, Springer (1997)

[44] M. Movshev and A. Schwarz, *On maximally supersymmetric Yang–Mills theories*, Nucl. Phys. B 681 (2004) 324–50

[45] W. Nahm, *Supersymmetries and their representations*, Nucl. Phys. B 135 (1978) 149–66

[46] A. Polishchuk and L. Positselski, *Quadratic Algebras*, AMS (2005)

[47] N. Ramachandran, *Values of zeta functions at $s = 1/2$*, Int. Math. Res. Notices 25 (2005) 1519–41

[48] M. Reid, *The complete intersection of two or more quadrics*, thesis, Cambridge University (1972)

[49] M. Sato, *Pseudo-differential equations and theta functions*, in *Colloque International CNRS sur les équations aux Dérivées Partielles Linéaires (Univ. Paris-Sud, Orsay, 1972)*, Soc. Math. France (1973) 286–91

[50] M. Sato, M. Kashiwara, and T. Kawai, *Linear differential equations of infinite order and theta functions*, Adv. Math. 47 (1983) 300–25

[51] C. Schommer-Pries, *The classification of two-dimensional extended topological field theories*, Preprint, arXiv: 1112.1000

[52] I. Schur, *Über die Darstellung der symmetrischen und der alternierenden Gruppe durch gebrochene lineare Substitutionen*, J. Reine Angew. Math. 139 (1911) 155–250

[53] A. N. Sergeev, *The tensor algebra of the identity representation as a module over the Lie superalgebras $\mathfrak{Gl}(m,n)$ and $Q(n)$*, Math. USSR Sbornik 51 (1985) 419–27

[54] H. X. Sinh, *Gr-catégories*, thèse, Université Paris-7 (1975)

[55] S. Stolz and P. Teichner, *Supersymmetric field theories and generalized cohomology*, in *Mathematical Foundations of Quantum Field Theory and Perturbative String Theory*, Proc. of Symp. in Pure Math. 83, AMS (2011)

[56] Y. Tachikawa, *A pseudo-mathematical pseudo-review on* 4d $N = 2$ *supersymmetric quantum field theories*, IPMU preprint (2014)
[57] B. Toën and G. Vezzosi, *Homotopical algebraic geometry II: geometric stacks and applications*, Mem. Amer. Math. Soc. 193 (2008) 902
[58] A. N. Tyurin, *On intersections of quadrics*, Russ. Math. Surv., 30 (1975) 51–105
[59] C. T. C. Wall, *Graded Brauer groups*, J. reine angew. Math. 213 (1964) 187–99
[60] J. Wess and J. Bagger, *Supersymmetry and Supergravity*, Princeton University Press (1991)
[61] G. W. Whitehead, *Recent Advances in Homotopy Theory*, AMS (2007)
[62] E. Witten, *Twistor-like transform in ten dimensions*, Nucl. Phys. B 266 (1986) 245–64
[63] E. Witten, *Supersymmetry and Morse theory*, J. Differential Geom. 17 (1982) 661–92
[64] H. Whitney, *The mathematics of physical quantities: Part I: mathematical models for measurement*, Amer. Math. Monthly 75 (1968) 115–138

Mikhail Kapranov
Kavli Institute for Physics and Mathematics of the
Universe (WPI), Kashiwa, Japan
mikhail.kapranov@ipmu.jp

PART II

Symplectic Geometry

4
Derived Stacks in Symplectic Geometry

Damien Calaque*

Contents

1	Introduction	155
2	Symplectic Structures in the Derived Setting	161
3	Derived Interpretation of Classical Constructions	179
4	Conclusion	196
	References	199

1 Introduction

Our goal in this chapter is to explain how derived stacks can be useful for ordinary symplectic geometry, with an emphasis on examples coming from classical topological field theories. More precisely, we will use classical Chern–Simons theory and moduli spaces of flat G-bundles and G-local systems as leading examples in our journey.

* First of all I warmly thank Mathieu Anel for the many discussions we had about derived stacks in symplectic geometry (and plenty other topics). I also thank him, as well as Gabriel Catren, for their very constructive and helpful comments on a preliminary version of this chapter. I then owe a debt of gratitude to Tony Pantev, Bertrand Toën, Michel Vaquié, and Gabriele Vezzosi. Finally, numerous discussions with Pavel Safronov have been very useful. The idea of having classical Chern–Simons theory as a common thread came up when I was preparing a colloquium talk *Derived symplectic geometry and classical Chern–Simons theory* (www.perimeterinstitute.ca/fr/videos/derived-symplectic-geometry-and-classical-chern-simons-theory) for a conference on *Deformation quantization of shifted Poisson structures* at Perimeter Institute. This work has been partly supported by the Institut Universitaire de France, by the Agence Nationale de la Recherche (through the project "Structures supérieures en Algèbre et Topologie"), and by the European Research Council (ERC) under the European Union's Horizon 2020 research and innovation programme (Grant Agreement No. 768679).

We will start in the introduction by reviewing various points of view on classical Chern–Simons theory and moduli of flat connections. In the main body of the chapter we will try to convince the reader how derived symplectic geometry (after Pantev–Toën–Vaquié–Vezzosi [31]) somehow reconciles all these different points of view.

1.1 Physics: Classical Chern–Simons Theory

Let M be a closed oriented 3-manifold and let $X = Conn_G(M)$ be the space of all G-connections on M, where $G \subset GL(n,\mathbb{R})$ is a compact simple Lie group.

Remark 1.1 We consider connections on arbitrary differentiable principal G-bundles here. Nevertheless, as the moduli of differentiable G-bundles is discrete, we will deal with connections on the trivial principal G-bundle for simplicity in this introduction.

The Chern–Simons action functional is then

$$CS(A) := \frac{k}{4\pi} \int_M tr(dA \wedge A + \frac{2}{3} A \wedge A \wedge A).$$

Since we will be interested in *classical* (as opposed to *quantum*) Chern–Simons theory then we can safely assume the factor $\frac{k}{4\pi}$ (which is relevant for quantization) is 1.

The space of classical trajectories of our physical system is given by the space $Crit(CS)$ of critical points of the action functional. The Chern–Simons functional has a huge group of symmetries $\mathcal{G}_M = C^\infty(M,G)$ (gauge symmetries), and we will be essentially interested in the "reduced" space of trajectories $X_{red} = Crit(CS)/\mathcal{G}_M$.

If M is now bounded by an oriented closed surface $\Sigma = \partial M$ then we have a restriction map $r : X = Conn_G(M) \to Conn_G(\Sigma) = P$. The space P is the phase space of our system and r encodes boundary/initial conditions (or constraints). The corresponding reduced phase space P_{red} shall in principle be symplectic and the subspace of admissible boundary/initial conditions (i.e., the image of the induced map $X_{red} \to P_{red}$) shall be Lagrangian (note that we would like to say that X_{red} itself is Lagrangian, but $X_{red} \to P_{red}$ might not be injective, i.e., a gauge equivalence class of classical trajectories might not be uniquely determined by its initial conditions).

Remark 1.2 There are infinite-dimensional spaces that are involved here. We will adopt a functorial approach to differential geometry that will allow us to deal with these infinite-dimensional spaces, for example, seeing a space as "its functor of points": $Conn_G(M)$ is for instance the functor that sends a smooth manifold X to smooth families of G-connections on M parameterized by X. We refer to [4, Chapter 1] for an approach using *diffeologies*, and to [32] for an approach which uses Dubuc's C^∞-*rings*. The last one has two advantages: it is very close to the modern presentation of algebraic geometry, and it provides a model of synthetic differential geometry (for which we refer to [4, Chapter 2]).

Remark 1.3 The reduced spaces, even if they happen to be finite-dimensional, may be very singular. We will consider them as derived stacks in order to resolve this issue (see [4, Chapter 9]).

Remark 1.4 The problem that $X_{red} \to P_{red}$ is not necessarily injective is fine in the derived setting.

1.2 Moduli Spaces via Infinite-Dimensional Reduction

Going back to the case when M is without boundary, one observes that $A \in Conn(M, G)$ is a critical point of the action functional if and only if its curvature $F(A) := dA + A \wedge A$ vanishes. In other words, critical points of CS are zeroes of the curvature map $F: X = Conn_G(M) \to \Omega^2(M, \mathfrak{g}) \subset \Omega^1(M, \mathfrak{g})^*$, where the last inclusion is given by $\alpha \mapsto \int_M tr(\alpha \wedge -)$. Hence the "reduced" space of trajectories $X_{red} = Crit(CS)/\mathcal{G}_M$ is the quotient $F^{-1}(0)/\mathcal{G}_M$.

We are tempted to view the curvature map F as a kind of moment map. Up to infinite-dimensional issues, this is actually the case if we go down one dimension and consider the curvature map on the phase space $P = Conn_G(\Sigma)$:

$$F: P = Conn_G(\Sigma) \longrightarrow \Omega^2(\Sigma, \mathfrak{g}) \subset C^\infty(\Sigma, \mathfrak{g})^*$$
$$A \longmapsto dA + A \wedge A.$$

The inclusion $\Omega^2(\Sigma, \mathfrak{g}) \subset C^\infty(\Sigma, \mathfrak{g})^*$ is again given by $\alpha \mapsto \int_\Sigma tr(\alpha \wedge -)$. Note that $P = Conn_G(\Sigma)$ is a (pre-)symplectic affine space with 2-form ω_P given on its tangent $T_A Conn_G(\Sigma) \cong \Omega^1(\Sigma, \mathfrak{g})$ by

$$\omega_{P,A}(B_1, B_2) := \int_\Sigma tr(B_1 \wedge B_2).$$

The action of the gauge group $\mathcal{G}_\Sigma = C^\infty(\Sigma, G)$ on P is Hamiltonian with moment map F, and the reduced phase space $P_{red} = F^{-1}(0)/\mathcal{G}_\Sigma$ is obtained *via* (infinite-dimensional) Hamiltonian reduction.

Remark 1.5 Following Rem 1.2 the space P_{red} exists as an object in diffeologies. As such we can say what a closed 2-form is, but in order to express what nondegeneracy at a point $[A] \in P_{red}$ means we need to have a nice finite-dimensional tangent space $T_{[A]}P_{red}$. One can show that there is an open subset of P_{red} where this is the case, and then at such a point $[A]$ the tangent is expressed as follows:[1]

$$T_{[A]}P_{red} \cong \frac{\ker(dF_A)}{C^\infty(\Sigma,\mathfrak{g})} \cong \frac{\ker\left(d_{dR}:\Omega^1(\Sigma,\mathfrak{g}) \to \Omega^2(\Sigma,\mathfrak{g})\right)}{d_{dR}(C^\infty(\Sigma,\mathfrak{g}))} = H^1(\Sigma,\mathfrak{g}).$$

Finally, the reduced pairing $\omega_{P_{red},[A]}$ is nondegenerate on $H^1(\Sigma,\mathfrak{g})$, by Poincaré duality.

The above is a crucial ingredient in Atiyah–Bott construction of a symplectic structure [4] on the moduli space.

1.3 Local Systems and the Quasi-Hamiltonian Formalism

It is a standard fact that flat G-bundles up to isomorphisms are exactly local systems up to isomorphisms, that is, conjugacy classes of representations of the fundamental group. In the case of a closed oriented surface Σ the fundamental group $\pi_1(\Sigma)$ admits a presentation with $2g$ generators (g being the genus of Σ) $a_1,\ldots,a_g,b_1,\ldots,b_g$ with the sole relation

$$\overrightarrow{\prod_{i=1,\ldots,g}}(a_i,b_i) = a_0 b_0 a_0^{-1} b_0^{-1} \cdots a_g b_g a_g^{-1} b_g^{-1} = 1. \tag{1.1}$$

Therefore the space of G-local systems $Loc_G(\Sigma) = Hom_{Grp}(\pi_1(\Sigma), G)/G$ is of the form $\Phi^{-1}(1)/G$, where $\Phi: G^{2g} \to G$ sends $((a_i, b_i))_{i=1,\ldots,n}$ to the l.h.s. of (1.1). One would very much like to see Φ as a kind of moment map. Indeed, Alekseev–Malkin–Meinreinken [1] have shown that Φ is a so-called Lie group valued moment map: in their language G^{2g} is a quasi-Hamiltonian G-space and its quasi-Hamiltonian reduction (which is genuinely symplectic) is $Loc_G(\Sigma)$. They even interpret G^{2g}/G as G-local systems on $\Sigma - D$, where D is a two-dimensional disk, and the map $\Phi^{-1}(1)/G \to G^{2g}/G$ as the restriction map $Loc_G(\Sigma) \to Loc_G(\Sigma - D)$, viewing a G-local system on Σ as a G-local system on $\Sigma - D$ with trivial holonomy around ∂D.

Observe that G itself is the quotient $C^\infty(S^1, G)/C_*^\infty(S^1, G)$, where $C_*^\infty(S^1, G) = L_*(G)$ denotes based loops (sending 1 to 1). This allows to identify G^{2g} as the moduli of flat connections on $\Sigma - D$ up to

---

[1] Recall that P_{red} is the quotient of the zero locus of F by $\mathcal{G}_\Sigma$.

gauge equivalences fixing a point in the boundary. Finally, the space $FlatConn_G(\Sigma - D)$ of all flat connections on $\Sigma - D$ is equipped with a restriction map

$$\mu: FlatConn_G(\Sigma - D) \to FlatConn_G(S^1) = \Omega^1(S^1, \mathfrak{g}) \subset C^\infty(S^1, \mathfrak{g})^*,$$

which is a moment map for the action of $L(G) = C^\infty(S^1, G)$. Using the fact that $L(G)/L_*(G) = G$, together with the holonomy map $\Omega^1(S^1, \mathfrak{g}) \to G$, Alekseev–Malkin–Meinreinken [1] show that there is a correspondence between Hamiltonian $L(G)$-spaces and quasi-Hamiltonian G-spaces, and that there are symplectomorphisms:

$$P_{red} \xleftarrow{\sim} \mu^{-1}(0)/L(G) \xrightarrow{\sim} \Phi^{-1}(1)/G,$$

where the left arrow sends a flat connection on $\Sigma - D$ that is zero on the boundary to its extension by zero, and the right arrow sends a flat connection to its holonomy representation.

The work of Alekseev–Malkin–Meinreinken provides a finite-dimensional construction of the symplectic structure on the moduli space of flat connections, for which the topological invariance of the reduced phase space is furthermore transparent.

1.4 Deformation-Theoretic Approach

There is a clean way of showing that there is a nondegenerate pairing on the tangent to the moduli space, viewed as a functor of points, of either flat G-bundles or G-local systems. We have that

$$T_{[\mathcal{P}, \nabla]} Flat_G(\Sigma) = H^1_{dR}\big(\Sigma, (\mathcal{P} \times_G \mathfrak{g}, \nabla)\big)$$

is the first de Rham cohomology of Σ with values in the associated flat vector bundle $\mathcal{P} \times_G \mathfrak{g}$. and

$$T_{[\mathcal{L}]} Loc_G(\Sigma) = H^1_{Betti}(\Sigma, \mathcal{L} \times_G \mathfrak{g})$$

is the first singular/Betti cohomology of Σ with values in the associated linear local system $\mathcal{L} \times_G \mathfrak{g}$.

In both cases, using the trace pairing together with the cup-product we get a nondegenerate skew-symmetric form with values in $H^2_{Betti}(\Sigma, \mathbb{R}) \cong H^2_{dR}(\Sigma, \mathbb{R}) \cong \mathbb{R}$. On can easily show that it varies smoothly with respect to $[\mathcal{P}, \nabla]$ or $[\mathcal{L}]$. But proving that it is closed is not easy (and basically reduces to making use of one of the previous approaches).

1.5 Back to Physics: Local Critical Loci

Going along the lines of [20, 23] one can show that the moduli space of flat G-connections on a compact oriented 3-manifold is locally the critical locus of a function on a finite-dimensional differentiable space.[2] In order to see this, one has to observe that near a given flat G-connection ∇ one can put several constraints (like Laplacian eigenvalue constraints) that allows to identify a neighborhood of ∇ as a critical locus of the Chern–Simons functional restricted to a finite-dimensional constrained subspace.

This allows in particular to get rid of infinite-dimensional complications, at the price of loosing the existence of a global action functional. Nevertheless, one can still use this local structure in order to compute interesting invariants.

1.6 Unifying All These Approaches

Our goal in this chapter is to present a context in which the following questions can be addressed in a meaningful way and get natural answers:

(a) how to relate these different approaches, or more precisely, how to put a natural symplectic structure on our favorite moduli so that all these different approaches somehow appear as *different ways of computing* this symplectic structure;

(b) whenever $\partial M = \emptyset$, what does it means for $X_{red} \cong Loc_G(M)$ to be Lagrangian in $P_{red} = *$;[3]

(c) how to get rid of smoothness issues, namely in the introduction, we have "carefully" avoided to be precise about the fact that most statements are true only when one is over some smooth locus in moduli spaces;

(d) how to get rid of the problem that $r: X_{red} \to P_{red}$ may not be an inclusion (for instance, when P_{red} is a point);

(e) to what extent (and how) one can recover an action functional from X_{red};

(f) how to express the fact that Chern–Simons theory is a topological field theory.

---

[2] In [20, 23] this is done for holomorphic Chern–Simons theory. I do not know a reference where this is done for Chern–Simons theory, but it will be easier than its holomorphic analog.

[3] The analogous situation in classical mechanics would be to look at the space of periodic trajectories of a classical mechanical system, which is also a typical example where classical trajectories are not determined by initial/boundary conditions.

1.7 Our Geometric Context

We refer to Mathieu Anel's contribution in the companion volume [4, Chapter 9] for an introduction to the ideas of derived geometry. We will work in the C^∞ setting, using for instance the theory of derived differentiable stacks from [43] (see also [25]), which uses C^∞-rings.

Smoothness issues appearing in question (c) will be resolved by this use of derived differentiable stacks. There is a large class of derived differentiable stacks, the Artin ones, on which all the calculations we want to perform make sense and that is stable both by pullbacks and by groupoid action quotients. Surprisingly enough, the other questions will get very natural answers once one has set up a suitable theory of symplectic structures on Artin stacks.

2 Symplectic Structures in the Derived Setting

Recall that a *symplectic structure* on a smooth manifold M is a 2-form $\omega \in \Omega^2(M)$ that is

- closed for the de Rham differential: $d_{dR}\omega = 0$.
- nondegenerate: the contraction map $\omega^\flat \colon TX \to T^*X, v \mapsto \iota_v\omega$ is a bundle isomorphism.

A submanifold $L \subset M$ is *Lagrangian* if it is

- isotropic: $\omega_{|L}$ vanishes on TL.
- nondegenerate: the induced map $TL \to N^*L$, where N^*L is the conormal bundle of L, is a bundle isomorphism.

2.1 Closed Forms: Structure versus Property

Working in the derived setting, one shall have a notion of differential form that is homotopy invariant in the sense that it does not depend on the explicit presentation of our derived stack. More precisely, Artin stacks have a cotangent *complex*, so that forms will have a cohomological degree. Different presentations of the same stack will lead to quasi-isomorphic cotangent complexes. Therefore if we get a form that is de Rham closed in the naive sense for a given presentation, it might only be de Rham closed *up to a cocycle* for another presentation.[4] This leads to the idea of considering forms that are closed *up to homotopy* rather than in the strict (naive) sense. We will try to implement

---

[4] This is rather usual that algebraic identities are not strictly preserved under quasi-isomorphisms.

this idea and show that it naturally follows from the synthetic and functorial approach to differential forms.

For the first reading, the reader who is only interested in a concrete and intuitive definition of these closed forms up to homotopy can directly go to Section 2.1.5 dealing with examples, and consider that a *closed p-form of degree n* on a derived stack X is a series $(\alpha_p, \alpha_{p+1}, \dots,)$ where

- α_i is an i-form of total degree $p+n$, where the total degree is the sum of the cohomological degree and the form degree.[5]
- $d_{dR}(\alpha_i) = \partial(\alpha_{i+1})$, where d_{dR} is the de Rham differential and ∂ is the *internal* differential (that comes from that we are doing derived geometry), and with the convention that $\partial(\alpha_p) = 0$.

The leading term α_p is called the underlying p-form of degree n.

2.1.1 The de Rham Complex from the Synthetic Point of View

For a genuine smooth manifold X, one can consider the differentiable space $X^{\Delta^k_{inf}}$ of maps from the infinitesimal k-simplex Δ^k_{inf} to X. Recall that the infinitesimal k-simplex Δ^k_{inf} is the space of $k+1$-tuples of points $(x_0, \dots, x_k)$ in $\mathbb{R}^k$ that are pairwise close at order 1; in other words, its C^∞-ring of functions is the quotient of $C^\infty(\mathbb{R}^{k+1})$ by the relations

$$\sum_{i=0}^k x_i = 1 \quad \text{and} \quad (x_i - x_j)^2 = 0 \text{ if } i \neq j.$$

For instance, the infinitesimal 1-simplex Δ^1_{inf} is a so-called fat point, having function C^∞-ring the quotient of $C^\infty(\mathbb{R})$ by $t^2 = 0$ (here t is $x_0 - x_1$), and thus $X^{\Delta^1_{inf}}$ is the tangent space TX to X.

The collection $(\Delta^k_{inf})_{k \geq 0}$ is cosimplicial, hence the collection $(X^{\Delta^k_{inf}})_{k \geq 0}$ forms a simplicial object in differentiable spaces and thus, $C^\infty(X^{\Delta^k_{inf}})_{k \geq 0}$ is a cosimplicial vector space. It is a fact (see [24, Section I.18]) that the usual differential k-forms $\Omega^k(X)$ may be characterized as the joint kernel of the degeneracy maps in $C^\infty(X^{\Delta^k_{inf}})$. Hence the de Rham complex $DR(X)$ of X is identified with the normalized complex[6] of the cosimplicial vector space $C^\infty(X^{\Delta^k_{inf}})_{k \geq 0}$.

---

[5] The form degree will be named *weight*, and the total degree will simply be named *degree*.
[6] Recall that the normalized complex of a cosimplicial vector space $(V^k)_{k \geq 0}$ is constructed as follows: in degree k one has the joint kernel of degeneracies $V^k \to V^{k-1}$, and the differential is the alternating sum of faces $V^k \to V^{k+1}$.

Closed p-forms on X can then be characterized in the following equivalent ways:

(**c1**) as p-cocycles in $DR(X)$.
(**c2**) as morphisms of complexes from $\mathbb{R}[-p]$ to $DR(X)$.[7]
(**c3**) as morphisms of cosimplicial vector spaces from $\mathbb{R}(-p) := K(\mathbb{R}[-p])$ to $C^\infty(X^{\Delta_{inf}^k})_{k \geq 0}$, where K is inverse to the normalized complex functor.

Note that the cosimplicial vector space $\mathbb{R}(-p)$ admits the following very simple description: in cosimplicial degree $k < p$ it is zero, in cosimplicial degree p it is $\mathbb{R}$, and it is freely generated by faces in higher cosimplicial degree.

Observe that (nonnecessarily closed) p-forms can be characterized in similar terms:

(**nc1**) as p-cochains in $DR(X)$.
(**nc2**) as morphisms of graded vector spaces from $\mathbb{R}[-p]$ to $DR(X)^\sharp$, where $(-)^\sharp$ stands for the underlying graded vector space functor.
(**nc3**) as morphisms of p-truncated cosimplicial vector spaces from $\mathbb{R}(-p)_{\leq p}$ to $C^\infty(X^{\Delta_{inf}^k})_{0 \leq k \leq p}$.

We want to generalize the above three definitions for derived differentiable stacks, from the third to the first.

Remark 2.1 The first definition is obviously the simplest and most concrete of the three. Its generalization appears to be rather intuitive and explicit, but proving general results about closed forms on derived stacks with this definition can be rather complicated. The other definitions as well as their generalizations are rather abstract, but they appear to be more convenient for proving general statements about closed forms on derived stacks.

As we will see, the third definition has a straightforward generalization. The generalization of the second one is less straightforward, but we give it for several reasons: it is rather compact, it is the one that nowadays is used in most references (see, e.g., [12]), and it explains the relation between the third and the first definitions.

---

[7] For a cochain complex C and an integer k, its shift by k, denoted $C[k]$, is another cochain complex whose degree m cochains are the degree $m + k$ cochains of C. For instance, $\mathbb{R}[-p]$ is the cochain complex that only consists of a one-dimensional space of degree p cochains.

2.1.2 Closed Forms on Derived Stacks: Third Definition

If X is a derived differentiable stack, then we have a derived stack

$$X^{\Delta_{inf}^k} = \mathbf{Map}(\Delta_{inf}^k, X)$$

of maps from Δ_{inf}^k to X, whose Y-points are maps $\Delta_{inf}^k \times Y \to X$. Recall that derived global functions on a derived stack form a differential graded algebra, and thus, in particular, a cochain complex. Hence $C^\infty(X^{\Delta_{inf}^k})$ is a cochain complex, and $C^\infty(X^{\Delta_{inf}^k})_{k \geqslant 0}$ is a cosimplicial cochain complex (i.e., a cosimplicial object in cochain complexes).

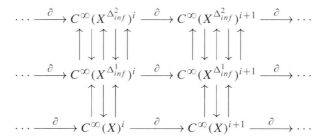

This suggests the following generalization of characterization (**c3**) for closed p-forms:

Definition 2.2 A closed p-form of degree n is a morphism of cosimplicial cochain complexes from $\mathbb{R}(-p)$ to $\left(C^\infty(X^{\Delta_{inf}^k})[n]\right)_{k \geqslant 0}$. One similarly defines a (nonnecessarily closed) p-form of degree n as a morphism of p-truncated cosimplicial cochain complexes from $\mathbb{R}(-p)_{\leqslant p}$ to $\left(C^\infty(X^{\Delta_{inf}^k})[n]\right)_{0 \leqslant k \leqslant p}$.

Note that we have a space[8] $\mathcal{A}^{p,cl}(X, n)$ of such closed p-forms of degree n. For instance, a path between two of these is a homotopy between the corresponding maps of cosimplicial cochain complexes.[9] Similarly, we write $\mathcal{A}^p(X, n)$ for the space of p-forms of degree n. There is an obvious map $\mathcal{A}^{p,cl}(X, n) \to \mathcal{A}^p(X, n)$ which is not necessarily a subspace map. This means that for derived stacks (as opposed to genuine manifolds) being closed is a structure rather than just a property. We will make this more transparent below.

---

[8] Here we mean a space in the sense of homotopy theory, i.e., a homotopy type.
[9] More precisely, we have a class of weak equivalences in the category of cosimplicial cochain complexes: the levelwise quasi-isomorphisms. Localizing at weak equivalences we therefore have an ∞-category of cosimplicial cochain complexes, and thus we have spaces of morphisms. These spaces of morphisms can be computed using an explicit model structure.

2.1.3 Closed Forms and Graded Mixed Complexes: Second Definition

We have seen that, when X is a derived differentiable stack, each $C^\infty(X^{\Delta_{inf}^k})$ is already a cochain complex. We can still consider the normalized complex $DR(X)$, which has the richer structure of a *graded mixed complex*:[10]

- it is a complex, as such it carries a grading that we will refer to as the *cohomological grading*.
- it carries an auxiliary grading, the *weight*, that is reminiscent from the cosimplicial degree: elements in $C^\infty(X^{\Delta_{inf}^k})$ (i.e., k-forms) have weight k.
- its differential splits into a sum $\partial + \epsilon$:

  a) ∂ is the part which comes from the differential of $C^\infty(X^{\Delta_{inf}^k})$, that we call the *internal differential*. It is of zero weight.
  b) ϵ is the part which comes from the alternating sum of the face maps, that we call in general the *mixed differential*, and the *de Rham differential* d_{dR} in this specific example. It has weight one.

If one denotes by $DR(X)_{(p)}$ the weight p part of $DR(X)$, then the graded mixed complex $DR(X)$ can be visualized as follows:[11]

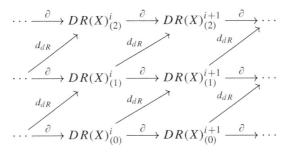

The normalized complex functor N induces an equivalence of categories between cosimplicial complexes and graded mixed complexes sitting in non-negative weight, which induces an equivalence between their ∞-categorical localizations.[12]

Remark 2.3 Note that $N(\mathbb{R}(-p)) = \mathbb{R}[-p](-p)$ is just $\mathbb{R}$ viewed as a graded mixed complex concentrated in weight p and cohomological degree p.

---

[10] The notion of a *mixed complex* (a term coined by Kassel [22]) goes back to Dwyer and Kan's study of Connes's cyclic modules [15], who were calling it a *duchain complex*. *Graded mixed complexes* have been introduced in [31], but we would like to warn the reader that in this chapter we adopt the grading convention from [12].
[11] Observe that $DR(X)_{(0)} = C^\infty(X)$.
[12] Weak equivalences are levelwise quasi-isomorphisms in both cases.

One therefore has the following generalization of characterization **(c2)** for closed p-forms, which is precisely the one appearing in [12] (and most recent references):

Definition 2.4 The space $\mathcal{A}^{p,cl}(X,n)$ of closed p-forms of degree n is the space of maps from $N(\mathbb{R}(-p))$ to $DR(X)[n]$. The space $\mathcal{A}^p(X,n)$ of p-forms of degree n is the space of maps from $N(\mathbb{R}(-p))^\sharp$ to $DR(X)[n]^\sharp$, where $(-)^\sharp$ stands for the underlying graded complex functor (forgetting the mixed differential). The map $\mathcal{A}^{p,cl}(X,n) \to \mathcal{A}^p(X,n)$ comes from applying $(-)^\sharp$.

Remark 2.5 In the nonderived setting (see Section 2.1.1) the graded mixed complex $DR(X)$ is diagonal (in the sense that there is no internal differential and its weight p part is concentrated in cohomological degree p).

Remark 2.6 Graded mixed complexes are very much related to filtered complexes and spectral sequences. For instance, with every bi-complex $(C^{\bullet,\bullet}, d_h, d_v)$ one can associate two graded mixed complexes:

(1) the cohomological grading is the total degree and the weight is the horizontal degree; the internal differential is d_v and the mixed differential is d_h.
(2) the cohomological grading is the total degree and the weight is the vertical degree; the internal differential is d_h and the mixed differential is d_v.

Conversely, one can build two bi-complexes from a graded mixed complex (D, ∂, ϵ) by saying that:

(1) an element of weight p and degree d has bidegree $(p, d-p)$, $d_h = \epsilon$ and $d_v = \partial$.
(2) an element of weight p and degree d has bidegree $(d-p, p)$, $d_h = \partial$, and $d_v = \epsilon$.

For instance, in our example of the de Rham graded mixed complex $DR(X)$, letting $\Omega^p(X) := DR(X)_{(p)}[p]$ we get a bicomplex

$$\begin{array}{ccccccc}
\cdots \xrightarrow{\partial} & \Omega^2(X)^i & \xrightarrow{\partial} & \Omega^2(X)^{i+1} & \xrightarrow{\partial} & \cdots \\
& {\scriptstyle d_{dR}}\uparrow & & {\scriptstyle d_{dR}}\uparrow & & \\
\cdots \xrightarrow{\partial} & \Omega^1(X)^i & \xrightarrow{\partial} & \Omega^1(X)^{i+1} & \xrightarrow{\partial} & \cdots \\
& {\scriptstyle d_{dR}}\uparrow & & {\scriptstyle d_{dR}}\uparrow & & \\
\cdots \xrightarrow{\partial} & C^\infty(X)^i & \xrightarrow{\partial} & C^\infty(X)^{i+1} & \xrightarrow{\partial} & \cdots
\end{array}$$

2.1.4 Closed Forms as Cocycles: First Definition

We would like to have a more concrete definition of (closed) forms, and have an explicit model for the space of morphisms of graded mixed complexes from $N(\mathbb{R}(-p))$ to $DR(X)$ (or any other graded mixed complex). This is a rather standard problem people encounter in homological and homotopical algebra: one needs to replace $N(\mathbb{R}(-p))$ with an equivalent graded mixed complex (a *resolution*) R_p such that mapping out of it is better behaved.[13] Following [12] we have the following explicit nice replacement R_p for $N(\mathbb{R}(-p))$: it is the linear span of $\{x_i, y_i\}_{i \geq p}$ with

- x_is having degree p and weight i;
- y_is having degree $p+1$ and weight $i+1$;
- $\partial(x_i) = y_{i-1}$ and $\epsilon(x_i) = y_i$ (convention: $y_{p-1} = 0$).

In particular, one sees that $\partial(x_p) = 0$ and $\epsilon(x_i) = \partial(x_{i+1})$ for every $i \geq p$.

degree \ weight	p	$p+1$	$p+2$	$\cdots$	i	$i+1$
$p+1$		y_p	y_{p+1}	$\cdots$	y_{i-1}	y_i
p	x_p	x_{p+1}	x_{p+2}	$\cdots$	x_i	x_{i+1}

Remark 2.7 If one accepts that free objects are nice, then one is led to construct R_p in the following way:

- introduce y_p as freely generated by ϵ from x_p: $\epsilon(x_p) = y_p$;
- the new graded mixed complex is no longer equivalent to the original one, hence we introduce x_{p+1} in order to "kill" the new cohomology class corresponding to y_p: $\partial(x_{p+1}) = y_p$;
- iterate the process.

Thus an element in the space $\mathcal{A}^{p,cl}(X, n)$ can be represented by a genuine morphism of graded mixed complexes $R_p \to DR(X)$. It consists in a collection $(\alpha_p, \alpha_{p+1}, \dots)$, where α_i has weight i and degree $p+n$, such that $\partial(\alpha_p) = 0$ and $d_{dR}(\alpha_i) = \partial(\alpha_{i+1})$ for every $i \geq p$. Its underlying p-form is the leading term α_p.

Let $DR(X)_{(\geq p)} := \prod_{i \geq p} DR(X)_{(i)}$ be the completed weight $\geq p$ part of $DR(X)$. We have the following generalization of characterization **(c1)** for closed p-forms:

Definition 2.8 The space $\mathcal{A}^{p,cl}(X, n)$ of closed p-forms of degree n is the space of $(p+n)$-cocycles in $DR(X)_{(\geq p)}$ for the total differential $\partial + d_{dR}$.

---

[13] For the expert reader, R_p will be an explicit cofibrant replacement of $N(\mathbb{R}(-p))$ in the projective model structure on graded mixed complexes.

The space $\mathcal{A}^p(X,n)$ of p-forms of degree n is the space of $(p+n)$-cocycles in $DR(X)_{(p)}$ for the differential ∂. The map $\mathcal{A}^{p,cl}(X,n) \to \mathcal{A}^p(X,n)$ is given by extracting the leading term.

This definition is roughly the one that originally appears in [31, Section 1.2].

2.1.5 Examples

Let us summarize what is known about the de Rham complex:

- if X is a genuine smooth manifold, then $DR(X)$ is the diagonal graded mixed complex associated with the usual de Rham complex $(\Omega^*(X), d_{dR})$.
- for a *geometric* derived stack X, the weight p part $DR(X)_{(p)}$ can be computed in terms of the cotangent complex $DR(X)_{(p)} \cong \Gamma(S^p(\mathbb{L}_X[-1]))$ is the pth symmetric power of a shift of the cotangent complex.[14] In other words, $DR(X)^\sharp \cong \Gamma(S(\mathbb{L}_X[-1](-1)))$ as graded complexes.
- the *de Rham functor* DR satisfies smooth descent: if $X \cong colim_{[n] \in \Delta^{op}} X_n$ is a geometric derived stack that is presented by a nice enough simplicial diagram $X_\bullet$ in stacks then $DR(X) \cong lim_{[n] \in \Delta} DR(X_n)$. Here are examples of nice enough simplicial diagrams of stacks:
  - $X_\bullet$ is the nerve of a Lie groupoid $G_1 \rightrightarrows G_0$: in simplicial degree n,
    $$X_n = \underbrace{G_1 \times_{G_0} \cdots \times_{G_0} G_1}_{n \text{ times}}.$$
    Recall that a Lie groupoid is a groupoid object in manifolds such that the source and target maps $G_1 \to G_0$ are submersions (submersions are also called smooth morphisms in algebraic geometry).
  - $X_\bullet$ is a submersive Segal groupoid in the sense of Toën–Vezzosi [42, Section 1.3] (nerves of Lie groupoids that we just mentionned are examples of these).
  - $X_\bullet$ is a Lie n-groupoid after [47, Definition 1.2], that is, a simplicial manifold for which horn maps $h_{q,k}$ are surjective submersions for $q \geq 1$ and diffeomorphisms for $q \geq n$.

---

[14] The cotangent complex always exists for a derived geometric stack. When it exists, the cotangent complex $\mathbb{L}_X$ of a derived stack X is a far reaching generalization of the cotangent bundle of a manifold, that encodes the infinitesimal structure of the derived stack X. We view it as a $\mathcal{O}_X$-module, where $\mathcal{O}_X$ is the sheaf of functions on X (a sheaf of differential graded commutative algebras). Tensor, symmetric, and skew-symmetric powers of $\mathcal{O}_X$-modules are understood over $\mathcal{O}_X$.

This in particular tells us that, if $X = [G_0/G_1]$ is the quotient stack of a genuine Lie groupoid $G_1 \rightrightarrows G_0$, then

$$DR(X) \cong \lim_{[n] \in \Delta} DR(\underbrace{G_1 \times_{G_0} \cdots \times_{G_0} G_1}_{n \text{ times}})$$

is equivalent to one of the two graded mixed complexes associated with the following first quadrant bi-complex (according to Rem 2.6) from [46, (2)]:

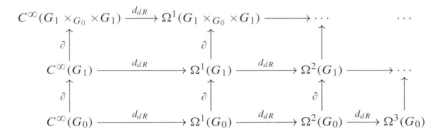

Here the horizontal/mixed differential is the usual de Rham differential d_{dR} and the vertical/internal differential is the alternating sum ∂ of pullbacks of forms along coface maps of the nerve of the groupoid $G_1 \rightrightarrows G_0$. The degree is the total degree while the weight is the degree of forms.

Example 2.9 (Pre-quasi-symplectic groupoids) Let us make explicit what it means to have a closed 2-form ω of degree 1 on a quotient stack $X = [G_0/G_1]$ as above. It is a cocycle of degree 3 in the total complex that is of weight at least 2 (the weight being here the genuine degree of forms). It is thus an element $\omega = \omega_0 + \omega_1 \in \Omega^2(G_1) \oplus \Omega^3(G_0)$ such that $\partial \omega_0 = 0$, $d_{dR}\omega_0 = \partial \omega_1$ and $d_{dR}\omega_1 = 0$. This is exactly the notion of pre-quasi-symplectic groupoid from [46, Definition 2.1] (also called a twisted presymplectic groupoid in [7, Section 2.1]).

If $G_1 \rightrightarrows G_0$ is an action groupoid $G \times M \rightrightarrows M$, with G a compact Lie group, $DR(X)$ can then be described as follows:

- $DR(X) = \left(S^\star(\mathfrak{g}^*) \otimes \Omega^\bullet(M)\right)^G$, with cohomological grading $2 \star + \bullet$ and weight $\star + \bullet$, which we may view as polynomial functions on $\mathfrak{g}$ with values in differential forms on M;
- the internal differential is defined as $\partial(\omega)(x) := \iota_{\vec{x}}(\omega(x))$, where $\vec{x}$ is the fundamental vector field associated with $x \in \mathfrak{g}$;
- the mixed differential is the de Rham differential on M: $\epsilon(\omega)(x) := (d_{dR}\omega)(x)$.

Its total complex is the Cartan model for equivariant cohomology and the weight produces the filtration from [28, Equation (21)], which gives rise to the algebraic counterpart of the Leray spectral sequence.

Remark 2.10 The above is consistent with the fact that $DR(X)^\sharp \cong \Gamma\big(S(\mathbb{L}_X[-1](-1))\big)$. Namely, the cotangent complex of a quotient stack $X = [M/G]$ is the two-term complex of G-equivariant vector bundles

$$\begin{array}{cc} 0 & 1 \\ T_M^* \longrightarrow & \mathfrak{g}^* \times M \end{array}$$

on M, where the differential sends a covector $\alpha \in T_m^* M$ to the linear map $x \mapsto \alpha(\vec{x}_m)$, and $\vec{x}$ is the fundamental vector field associated with $x \in \mathfrak{g}$. When G is compact there is no group cohomology, so that derived global sections over X are just G-invariant global sections over M, and we are done. If G is not compact, then the story is more complicated as it involves nontrivial group cohomology, but we still have a map of graded mixed complexes

$$\big(S^\star(\mathfrak{g}^*) \otimes \Omega^\bullet(M)\big)^G \to DR(X).$$

Example 2.11 (**Closed 2-forms of degree 2 on BG**) Closed 2-forms of degree 2 on $BG = [*/G]$ are exactly G-invariant symmetric bilinear forms, that is, elements in $S^2(\mathfrak{g}^*)^G$.

As we have seen, if $X \cong colim_{[n]\in\Delta^{op}} X_n$ is a geometric stack that is presented by a Lie n-groupoid $X_\bullet$, then we again have that $DR(X) \cong lim_{[n]\in\Delta} DR(X_n)$. In particular, we see that (normalized) multiplicative p-forms on X_q (see, e.g., [27, Definition 2.3]) are exactly p-forms of degree $p-q$ on X. A closed p-form of degree $p-q$ on X would then be a (normalized) element

$$\omega_0 + \cdots + \omega_q \in \Omega^p(X_q) \oplus \cdots \oplus \Omega^{p+q}(X_0)$$

that is closed under the total differential $d_{dR} + \partial$ (where ∂ is the internal differential, and is again given explicitly by the alternating sum of pullbacks of forms along coface maps).

2.2 Nondegeneracy: Shifted Symplectic Structures

2.2.1 Symplectic Linear Algebra in the ∞-Categorical Setting

In this section we follow closely the presentation from [10], where one works within a general ambient stable symmetric monoidal ∞-category

($\mathcal{C}$, $\otimes$, **1**). Here we will restrict our attention to the following two examples of such ∞-categories:

- cochain complexes, with equivalences being quasi-isomorphisms, monoidal product the usual tensor product of cochain complexes, and monoidal unit being $\mathbb{R}$;
- a sheafified verison of the above over a given stack X, denoted $QCoh(X)$. It is a bit larger than the category of complexes of vector bundles over X (which does not necessarily admits fiber products). Its monoidal product is the (derived) tensor product of sheaves of $\mathcal{O}_X$-modules, and its monoidal unit is $\mathcal{O}_X$.

As a matter of notation, we recall that we write ?[1] for the degree shift functor. We define an *n-shifted presymplectic* object as a pair (V, ω) where V is an object of $\mathcal{C}$ and $\omega\colon \wedge^2 V \to \mathbf{1}[n]$ is a morphism in $\mathcal{C}$. We say that it is *n-shifted symplectic* if it is moreover nondegenerate in the following sense: V is dualizable and the adjoint morphism $\omega^\flat \colon V \to V^*[n]$ is an equivalence.[15]

Note that a cochain complex V is dualizable if and only if it is *perfect*, meaning that its cohomology $\oplus_n H^n(V)$ is finite-dimensional. Below we will consider that perfect and dualizable are synonymous.

Example 2.12 When $n = 0$, if $\mathcal{C}$ is the ∞-category of complexes of vector spaces, and if V is concentrated in degree zero, then these notions coincide with the usual notions of (pre-)symplectic vector spaces.

Example 2.13 (Poincaré duality) If M is an n-dimensional oriented compact manifold, then the shifted[16] de Rham complex $DR(X)[1] = (\Omega^{*+1}(M, \mathbb{R}), d_{dR})$ is $(2-n)$-shifted symplectic when equipped with the pairing $\omega(\alpha, \beta) := \int_M \alpha \wedge \beta$.

2.2.2 Shifted Symplectic Structures on Derived Stacks

From here we assume that X is a geometric stack such that the cotangent complex $\mathbb{L}_X$ is dualizable.[17] The cotangent complex thus has a dual $\mathbb{T}_X := \mathbb{L}_X^*$, called the *tangent complex*.

In this case a 2-form of degree n is thus a section ω_0 of $S^2(\mathbb{L}_X[-1])[n+2]$, that is, a map $\mathcal{O}_X \to S^2(\mathbb{L}_X[-1])[n+2]$ in $QCoh(X)$. By

---

[15] Observe that there are two candidates for being the adjoint morphism: but they may only differ by a sign as ω is skew-symmetric. The condition of being an equivalence is not affected by that ambiguity.
[16] Without this shift, the pairing would be symmetric instead of being skew-symmetric. Indeed, recall that $S^2(V[1]) = \wedge^2(V)[2]$.
[17] We will call such a derived stack an *Artin stack*.

duality, this is equivalent to the data of a map $S^2(\mathbb{T}_X[1]) \to \mathcal{O}_X[n+2]$, which in turns is precisely an n-shifted presymplectic structure $\wedge^2 \mathbb{T}_X \to \mathcal{O}_X[n]$ on $\mathbb{T}_X$ in $QCoh(X)$.

Definition 2.14 An n-shifted symplectic structure on X is a closed 2-form ω of degree n such that $(\mathbb{T}_X, \omega_0)$ is an n-shifted symplectic object. In other words, we require that $\omega_0^\flat \colon \mathbb{T}_X \to \mathbb{L}_X[n]$ is an equivalence.

2.2.3 Examples of Shifted Symplectic Structures

Let us first go back to Example 2.9 and check the condition a closed 2-form of degree 1 on a geometric 1-stack $X = [G_0/G_1]$ must satisfy in order to define a 1-shifted symplectic structure. We recall several facts:

- the nondegeneracy condition only concerns the leading term $\omega_0 \in \Omega^2(G_1)$;
- quasi-coherent sheaves on $X = [G_0/G_1]$ are exactly G_1-equivariant sheaves on G_0, and the property of being an equivalence is something that one can check on the underlying sheaf on G_0;[18]
- the underlying sheaf of the tangent complex $\mathbb{T}_X$ is the two-term complex

$$\overset{-1}{\mathcal{L}} \xrightarrow{\rho} \overset{0}{T_{G_0}}$$

of sheaves on G_0, where $\mathcal{L}$ is the Lie algebroid of the groupoid $G_1 \rightrightarrows G_0$, and ρ is the anchor map;[19]
- $e^* T_{G_1} = \mathcal{L} \oplus T_{G_0}$, and ω_0 is compatible with this decomposition ($e \colon G_0 \to G_1$ is the unit map).

---

[18] In other words, the pullback functor $QCoh(X) \to QCoh(G_0)$ is conservative.
[19] Recall that, as a vector bundle on G_0, $\mathcal{L}$ is the restriction to G_0 of the bundle $T_{G_1}^s$ of vectors tangent to the source map $s \colon G_1 \to G_0$. The anchor map is given by the tangent to the target map $t \colon G_1 \to G_0$. This calculation of the underlying sheaf $p^* \mathbb{T}_X$ of $\mathbb{T}_X$, where $p \colon G_0 \to X = [G_0/G_1]$ is the quotient map, actually follows from a smooth descent argument, which we sketch now. Since

$$X = colim_{[n] \in \Delta^{op}} (\underbrace{G_1 \times_{G_0} \cdots \times_{G_0} G_1}_{n \text{ times}})$$

then, denoting $e \colon G_0 \to G_1$ the identity map,

$$p^* \mathbb{T}_X = colim_{[n] \in \Delta^{op}} (\underbrace{e^* T_{G_1} \oplus_{T_{G_0}} \cdots \oplus_{T_{G_0}} e^* T_{G_1}}_{n \text{ times}}).$$

Hence $p^* \mathbb{T}_X$ can be obtained as the normalized complex of the above simplicial diagram of vector bundles on G_0, which can be shown to be the two-term complex $\mathcal{L} \xrightarrow{\rho} T_{G_0}$.

Therefore ω_0^b induces a morphism of two-term complexes of sheaves on X, as follows:

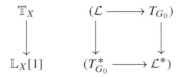

$$\begin{array}{ccc}
\mathbb{T}_X & (\mathcal{L} \longrightarrow T_{G_0}) \\
\downarrow & \downarrow & \downarrow \\
\mathbb{L}_X[1] & (T^*_{G_0} \longrightarrow \mathcal{L}^*)
\end{array}$$

By duality, to check that it is a quasi-isomorphism is equivalent to check that its kernel is acyclic. This amounts to require that at every point $x \in G_0$, the map $\ker(\omega_{0,x}) \cap \mathcal{L}_x \to \ker(\omega_{0,x}) \cap T_x G_0$ is an isomorphism. This is precisely the nondegeneracy condition appearing in the definition of a *quasi-symplectic groupoid* (see [46, Definition 2.5]).

Remark 2.15 Genuine symplectic groupoids from [45] correspond to the situation when the 3-form ω_1 (encoding the closeness of ω_0 up to homotopy) is 0. We provide a nice interpretation of this condition in terms of Lagrangian structures in the next subsection.

Remark 2.16 The above discussion shows in particular that, even if one deals with nonderived stacks, one has to consider cohomologically shifted 2-forms if one wants to grasp any reasonable kind of nondegeneracy property. Indeed, the tangent complex of $[G_0/G_1]$ sits in degree -1 and 0 while the cotangent complex (its dual) sits in degree 0 and 1.

Example 2.17 Let G be a Lie group, with Lie algebra $\mathfrak{g}$, and consider the groupoid $G \times \mathfrak{g}^* \rightrightarrows \mathfrak{g}^*$ of the coadjoint action (of G on $\mathfrak{g}^*$). It is a symplectic groupoid : indeed, $G \times \mathfrak{g}^* \cong T^*G$ carries a canonical symplectic form ω_0. Hence its quotient stack $[\mathfrak{g}^*/G]$ is 1-shifted symplectic. There is an explicit *ad hoc* description of this 1-shifted symplectic structure (see [9, Section 1.2.3] and [8, Example 2.10]), which we now briefly sketch. As we have seen, the tangent complex of $[\mathfrak{g}^*/G]$ is the two-term complex of G-equivariant vector bundles

$$\begin{array}{cc} -1 & 0 \\ \mathfrak{g} \otimes \mathcal{O}_{\mathfrak{g}^*} \longrightarrow T_{\mathfrak{g}^*} \cong \mathfrak{g}^* \otimes \mathcal{O}_{\mathfrak{g}^*} \end{array}$$

on $\mathfrak{g}^*$, and the cotangent complex is

$$\begin{array}{cc} 0 & 1 \\ \mathfrak{g} \otimes \mathcal{O}_{\mathfrak{g}^*} \cong T^*_{\mathfrak{g}^*} \longrightarrow \mathfrak{g}^* \otimes \mathcal{O}_{\mathfrak{g}^*} \end{array}$$

We thus see that the tangent complex and the 1-shifted cotangent complex are canonically identified, *via* the 1-shifted two-form

$$\omega = \sum_i \xi^i d_{dR} x_i \in \left(\mathfrak{g}^* \otimes \Omega^1(\mathfrak{g}^*)\right)^G,$$

where $(x_i)_i$ is a basis of $\mathfrak{g}$ (which defines coordinates on $\mathfrak{g}^*$) and $(\xi^i)_i$ is the dual basis of $\mathfrak{g}^*$. One can easily see that this form is (strictly) d_{dR}-closed.

Example 2.18 If G is a Lie group equipped with an invariant nondegenerate symmetric bilinear form on its Lie algebra, then the groupoid of the conjugation action (of G on itself) is quasi-symplectic, after [46, Proposition 2.8]. Therefore its quotient stack $[G/G]$ is 1-shifted symplectic. We refer to [9, Section 1.2.5] and [37] for more details about this 1-shifted symplectic structure.

Example 2.19 It has also been shown in [31] that, if G is a Lie group equipped with a G-invariant symmetric bilinear form $c \in S^2(\mathfrak{g}^*)^G$ on its Lie algebra, then the induced closed 2-form of degree 2 on the classifying stack $BG = [*/G]$ (see Example 2.11) is nondegenerate if and only if c is nondegenerate (in the usual sense). We will see later that the 1-shifted symplectic structure on $[G/G]$ can be recovered from this 2-shifted symplectic structure on BG.

Example 2.20 One can easily see that a symplectic 2-groupoid $(X_\bullet, \omega)$ in the sense of [27, Definition 2.7] induces a 2-shifted symplectic structure on the geometric stack $X = colim_{[n] \in \Delta^{op}} X_n$. As noticed by the author of [27], their notion of symplectic 2-groupoid is not Morita-invariant.[20]

2.3 Lagrangian Structures: Definition and Examples

2.3.1 Lagrangian Structures in the Linear Setting

Let (V, ω) be a symplectic vector space, and recall that a *Lagrangian* in V is a subspace $L \subset V$ such that

- L is *isotropic*: $\omega_{|L} = 0$;
- L is nondegenerate in the sense that it is maximal.

The isotropy condition is equivalent to the fact that the inclusion $L \subset V$ factors through $L \subset L^\circ \subset V$, where $L^\circ := \{v \in V | \omega(v, L) = 0\}$. Observe that since ω is nondegenerate,[21] we have a canonical identification

$$L^\circ \xrightarrow{\sim} L^\perp := \{\xi \in V^* | \xi_{|L} = 0\}$$

with the conormal $L^\perp$ to L.

---

[20] Indeed, both their notions of closeness and nondegeneracy are too strict from the perspective of shifted symplectic structures.

[21] Meaning that $\omega^\flat : V \to V^*$ is an isomorphism.

The nondegeneracy condition for L is equivalent to any of the following:

- $dim(L) = \frac{1}{2}dim(V)$.
- the inclusion $L \subset L^\circ$ is an equality.
- the map $L \to L^\perp$ given by $\ell \mapsto \omega^\flat(\ell)$ is an isomorphism.[22]

Even though this is rather unusual, the last two characterizations can be reformulated as follows:

- the sequence $0 \to L \to V \to L^* \to 0$, where the map $V \to L^*$ sends v to $\omega^\flat(v)_{|L}$, is exact;
- the sequence $0 \to L \to V^* \to L^* \to 0$, where the map $L \to V^*$ is given by $\ell \mapsto \omega^\flat(\ell)$, is exact.

Note that the two sequences we have written are dual to each other: it is thus clear that one of them is exact if and only if the other is.

This leads us to the following ∞-categorical generalization, which one can guess by observing that a short exact sequence of $0 \to A \to B \to C \to 0$ of vector spaces is the same as a bi-cartesian square

$$\begin{array}{ccc} A & \longrightarrow & B \\ \downarrow & & \downarrow \\ 0 & \longrightarrow & C \end{array}$$

Definition 2.21 Let $(\mathcal{C}, \otimes, \mathbf{1})$ be a stable symmetric monoidal ∞-category,[23] and let (V, ω) be an n-shifted presymplectic object.

a) An isotropic structure on a morphism $L \to V$ is an homotopy between $\omega_{|L}$ and 0.

b) An isotropic structure γ is nondegenerate or Lagrangian if L is perfect and the induced homotopy commuting square

$$\begin{array}{ccc} L & \longrightarrow & V \\ \downarrow & & \downarrow \\ 0 & \longrightarrow & L^*[n] \end{array}$$

is (co)cartesian.[24]

---

[22] Observe that $L^\perp$ is by definition the conormal of $L \subset V$. This equivalent characterization of nondegeneracy corresponds to the one we gave in the beginning of Section 2.

[23] As before, the unaccustomed reader can think about cochain complexes up to quasi-isomorphisms.

[24] First note that we are in a stable ∞-category, so that cartesian squares are cocartesian, and vice versa. The reader who is familiar with triangulated categories can think of such a square

Remark 2.22 The vertical map $V \to L^*[n]$ in the above diagram is adjoint to the composed map $V \otimes L \to V \otimes V \overset{\omega}{\to} \mathbf{1}[n]$. A consequence of the definition is that V is also perfect, and the nondegeneracy condition is equivalent to asking the (shifted) dual homotopy commuting square

$$\begin{array}{ccc} L & \longrightarrow & V^*[n] \\ \downarrow & & \downarrow \\ 0 & \longrightarrow & L^*[n] \end{array}$$

to be (co)cartesian. Furthermore (see [10, Lemma 1.3]), any n-shifted pre-symplectic object that has a Lagrangian is automatically n-shifted symplectic (i.e., ω is nondegenerate if γ is).

Below we give two examples when $\mathcal{C}$ is the ∞-category of complexes of vector spaces.

Example 2.23 Of course, ordinary Lagrangian subspaces in ordinary symplectic vector spaces are examples of Lagrangian structures for 0-shifted symplectic. Indeed, the inclusion map carries a Lagrangian structure (which is just the constant self-homotopy of 0).

Example 2.24 (Relative Poincaré duality) Let N be an oriented $(n+1)$-dimensional compact manifold with oriented boundary $\partial N = M$. Recall from Example 2.13 that $\Omega^{*+1}(M)$ is $(2-n)$-shifted symplectic. We claim that the pullback $\iota^*\Omega^{*+1}(N) \to \Omega^{*+1}(M)$ along the boundary inclusion $\iota : M \hookrightarrow N$ carries a Lagrangian structure. The isotropic structure comes from Stokes formula:

$$\int_M \iota^*(\alpha \wedge \beta) = \int_N d_{dR}(\alpha \wedge \beta).$$

The homotopy γ is given by $\alpha \wedge \beta \mapsto \int_N \alpha \wedge \beta$.

2.3.2 Lagrangian Structures on Derived Stacks

Let $f : L \to X$ be a morphism of derived stacks and assume that X carries an n-shifted presymplectic structure ω. An *isotropic* structure on f (or, abusing language, "on L") is a path γ between $f^*\omega$ and 0 in the space $\mathcal{A}^{2,cl}(L,n)$ of closed 2-forms of degree n on L.

In terms of the cocycle characterization of closed forms, this can be understdood as follows: $\gamma = \gamma_0 + \gamma_1 + \cdots$ with

_{as a distinguished triangle. Readers who are not familiar neither with stable ∞-categories nor with triangulated categories can think of such a square as inducing a long exact sequence in cohomology $\cdots \to H^k(L) \to H^k(V) \to H^{k+n}(L^*) \to H^{k+1}(L) \to \cdots$.}

- γ_i has weight $2+i$ and degree $1+n$;
- $(\partial + d_{dR})(\gamma) = f^*\omega$, meaning that $\partial \gamma_0 = f^*\omega_0$, $\partial(\gamma_1) + d_{dR}(\gamma_0) = f^*\omega_1$, etc.

degree\weight	2	3	4	$\cdots$
$2+n$	$f^*\omega_0$	$f^*\omega_1$	$f^*\omega_2$	$\cdots$
	$\partial \quad d_{dR}$	$\partial \quad d_{dR}$	$\partial \quad d_{dR}$	
$1+n$	γ_0	γ_1	γ_2	$\cdots$

Let us assume that both L and X are Artin stacks and that ω is nondegenerate. An isotropic structure γ on f is *Lagrangian* if the leading term γ_0, which can be viewed as an isotropic structure on the morphism $\mathbb{T}_L \to f^*\mathbb{T}_X$ in the sense of Defn 2.21, is nondegenerate.[25]

Remark 2.25 Having an isotropic structure on $\mathbb{T}_L \to f^*\mathbb{T}_X$ tells us that we have a morphism from $\mathbb{T}_L$ to the homotopy fiber of $f^*\mathbb{L}_X[n] \to \mathbb{L}_L[n]$, which is nothing but $\mathbb{L}_f[n-1]$. The nondegeneracy condition tells us then that this morphism $\mathbb{T}_L \to \mathbb{L}_f[n-1]$ is an equivalence. In the case when $n=0$ and the map $L \to X$ is an inclusion of genuine manifolds, then $\mathbb{T}_L$ is the usual tangent bundle TL and $\mathbb{L}_f[-1]$ is the conormal bundle N^*L. Hence it coincides with the usual notion of a Lagrangian submanifold that we recalled at the beginning of Section 2.

Observe that the notion of a Lagrangian structure on a morphism fully addresses problem (d) from the Introduction.

2.3.3 Examples of Lagrangian Structures

Example 2.26 Of course, any genuine Lagrangian submanifold in a genuine symplectic manifold provides an example of a Lagrangian structure, on the inclusion morphism. Indeed, as in Example 2.23 the Lagrangian structure is the constant self-homotopy of the closed 2-form 0.

Example 2.27 (Symplectic is Lagrangian) Let $*_{(n)}$ be the point equipped with the trivial n-shifted symplectic structure, given by the zero 2-form. Surprisingly enough, it was noticed in [9, Example 2.3] that the space of

---

[25] Observe that one could define nondegenerate isotropic structures on f even without assuming that ω is nondegenerate. These would nevertheless not deserve to be called Lagrangian structures as it would not imply that ω is symplectic (it would only imply that $f^*\omega$ is nondegenerate).

Lagrangian structures on the morphism $X \to *$ is equivalent to the space of $(n-1)$-shifted symplectic structures on X. This answers part of question (b) in the Introduction: a Lagrangian "in" $P_{red} = *$ is a (-1)-shifted symplectic stack.

Example 2.28 If $X \to Y$ is a morphism of Artin stacks, then it has been shown in [10] that the cotangent stack T^*Y is 0-shifted symplectic and that the morphism $T_X^*Y \to T^*Y$ from the conormal stack to the cotangent stack is Lagrangian.

Example 2.29 (Symplectic groupoids) Let $G_1 \rightrightarrows G_0$ be a Lie groupoid together with a quasi-symplectic structure $\omega = \omega_0 + \omega_1$ in the sense of [46, Definition 2.5]. We have seen above in Section 2.2.3 that the stack $[G_0/G_1]$ then carries a 1-shifted symplectic structure. Notice that the pullback of ω along the quotient map $G_0 \to [G_0/G_1]$ is ω_1. Hence the quotient map carries a Lagrangian structure if and only if $\omega_1 = 0$, meaning that $G_1 \rightrightarrows G_0$ actually is a symplectic groupoid.

Remark 2.30 One actually has an equivalence between the following three sets of data:

- a 1-shifted symplectic structure on the quotient $[G_0/G_1]$ and a Lagrangian structure on the quotient map $G_0 \to [G_0/G_1]$.
- a quasi-symplectic structure on $G_1 \rightrightarrows G_0$ is such that $\omega_1 = 0$ (i.e., a multiplicative symplectic structure on G_1).
- a symplectic structure on G_1 such that the submanifold $G_0 \subset G_1$ of units of the groupoid is Lagrangian.

This can been seen in the case of the coadjoint action groupoid $G \times \mathfrak{g}^* \rightrightarrows \mathfrak{g}^*$ from Example 2.17:

- pulling back the 1-shifted two-form $\sum_i \xi^i d_{dR} x_i$ along $\mathfrak{g}^* \to [\mathfrak{g}^*/G]$ amounts to setting $\xi^i = 0$, and thus gives 0.
- the symplectic structure on $G \times \mathfrak{g}^* \cong T^*G$ is multiplicative.
- $\{e\} \times \mathfrak{g}^* \subset G \times \mathfrak{g}^* \cong T^*G$ is Lagrangian.

Example 2.31 (Hamiltonian groupoid actions) Let $(G_1 \rightrightarrows G_0, \omega = \omega_0 + \omega_1)$ be a quasi-symplectic groupoid and let $J: X_0 \to G_0$ be a G_1-space, that is to say a family over G_0 together with a G_1-action. There is a nice description of Lagrangian structures on the map $[J]: [X_0/G_1] \to [G_0/G_1]$. First of all, let $X_\bullet$ be the nerve of the action groupoid $X_1 \rightrightarrows X_0$, where $X_1 = G_1 \times_{G_0} X_0$. Then the map $[J]$ is presented by the obvious morphism of simplicial manifolds $J_\bullet : X_\bullet \to G_\bullet$ (note that $J_0 = J$). Then the pullback $[J]^*\omega$ is $J^*\omega_1 + J_1^*\omega_0$. The condition that $[J]$ is isotropic therefore reads as follows: there exists a 2-form $\eta \in \Omega^2(X_0)$ such that $[J]^*\omega = \partial\eta + d_{dR}\eta$. In other words:

$$J^*\omega_1 = d_{dR}\eta \quad \text{and} \quad J_1^*\omega_0 = \partial\eta.$$

The second condition says that the graph of the action in $G_1 \times X_0 \times X_0$ is isotropic with respect to the 2-form $(\omega_0, \eta, -\eta)$. Hence $[J]$ is isotropic if and only if X_0 is a pre-Hamiltonian G_1-space in the sense of [46, Definition 3.1]. It is an exercise to check that the isotropic structure given by η is Lagrangian if and only if it turns X_0 into a Hamiltonian G_1-space in the sense of [46, Definition 3.5].

Example 2.32 (Moment maps) Hamiltonian spaces for the symplectic groupoid $G \times \mathfrak{g}^* \rightrightarrows \mathfrak{g}^*$ are symplectic G-spaces X together with a map $\mu : X \to \mathfrak{g}^*$ satisfying the moment condition $\omega^\flat(\vec{x}) = d\mu_x$ for every $x \in \mathfrak{g}$. According to the previous Example, we recover the fact that the induced map $[X/G] \to [\mathfrak{g}^*/G]$ carries a Lagrangian structure (see, e.g., [9, Section 2.2.1] or [8, Example 2.14]). Indeed, the moment condition implies that

$$[\mu]^* \sum_i \xi^i d_{dR} x_i = \sum_i \xi^i \mu^* d_{dR} = \sum_i \xi^i d\mu_{x_i} = \sum_i \xi^i d\omega_X^\flat(\vec{x}_i) = \partial\omega_X$$

Example 2.33 (Lie group valued moment maps) Hamiltonian spaces for the conjugation quasi-symplectic groupoid $G \times G \rightrightarrows G$ are precisely the quasi-Hamiltonian G-spaces of Alekseev–Malkin–Meinrenken [1], the map $J : X \to G$ being called a Lie group valued moment map. It again follows from Example 2.31 that the induced map $[X/G] \to [G/G]$ then carries a Lagrangian structure (see, e.g., [9, Section 2.2.2] or [37, Section 2.3]).

Example 2.34 (Symplectic groupoids again) If we are given a symplectic groupoid $(G_1 \rightrightarrows G_0, \omega_0)$ then G_1 naturally becomes a Hamiltonian G_1-space for the action by left multiplication. In particular, we recover the fact that, for symplectic groupoids, the map $G_0 = [G_1/G_1] \to [G_0/G_1]$ carries a Lagrangian structure (see Example 2.29).

3 Derived Interpretation of Classical Constructions

3.1 Weinstein's Symplectic Category

The cotangent space $X = T^*M$ of a differentiable manifold M is symplectic. Let us look at it as the phase space of a classical mechanical system: T^*M

is the space of all possible pairs of position and momentum, and we have a Hamiltonian function $H: X \to \mathbb{R}$ leading to a Hamiltonian vector field $\mathfrak{X}_H$ on X. Inside the *space of histories*, here the space of paths $\gamma: [0, 1] \to X$, one is interested in the solution space $\mathcal{S}$ of the equations of motions

$$\dot{\gamma}(t) = \mathfrak{X}_{H, \gamma(t)}.$$

It happens to be isomorphic to X itself, as a solution is uniquely determined by the initial condition $\gamma(0) \in X$. The evaluation map sending γ to the pair $(\gamma(0), \gamma(1))$ exhibits $\mathcal{S}$ as a Lagrangian submanifold of $X \times \overline{X}$, where $\overline{X}$ stands for X equipped with the opposite symplectic structure.

This is actually a special case of the graph of a symplectomorphism[26] $\varphi: X \to Y$ being a Lagrangian submanifold in $X \times \overline{Y}$. The graph of the composition of two symplectomorphisms actually coincides with the composition of their associated correspondences. Elaborating on this observation, Weinstein suggested in [44] to construct a category with objects being symplectic manifolds and morphisms being Lagrangian correspondences.[27] The main problem being that this is not quite a category, since there are well-known transversality issues when it comes to compose Lagrangian correspondences. These issues are resolved in the derived framework.

Roughly speaking, there is an ∞-category $\mathcal{L}ag_n$ where objects are n-shifted symplectic stacks (n being fixed) and morphisms being Lagrangian correspondences $L \to X \times \overline{Y}$. One can compose Lagrangian correspondences (see [9, Theorem 4.4]): given Lagrangian correspondences $L_1 \to X \times \overline{Y}$ and $L_2 \to Y \times \overline{Z}$, the (derived) fiber product $L_1 \times_Y L_2 \to X \times \overline{Z}$ is again a Lagrangian correspondence:

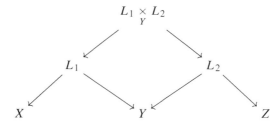

The full construction of this ∞-category has been achieved in [18, Section 11].

---

[26] In the above example, the symplectomorphism is the time 1 flow of the Hamiltonian vector field $\mathfrak{X}_H$.

[27] This idea is at the origin of the definition of a symplectic groupoid, where the graph of the multiplication is required to be Lagrangian.

Derived Stacks in Symplectic Geometry   181

Remark 3.1 Applying the above to the case when $X = Z = *$, and combining it with Example 2.27, one recovers the striking observation from [31] that the (derived) fiber product $L_1 \times_Y L_2$ of two Lagrangian morphisms $L_i \to Y$ is $(n-1)$-shifted symplectic (if we started with an n-shifted symplectic Y).

3.1.1 Examples

Below we give several examples of Lagrangian correspondences and compositions of them.

Example 3.2 (Lagrangian morphisms) Every Lagrangian morphism $L \to X$ can be viewed as a Lagrangian correspondence between X and $*$ (in whichever direction):

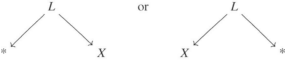

Example 3.3 (Conormal to a graph) The conormal $N^*f \cong X \times_Y T^*Y \to T^*X \times \overline{T^*Y}$ to the graph $X \to X \times Y$ of a morphism $f: X \to Y$ naturally carries a shifted Lagrangian structure. This construction is actually functorial: if $g: Y \to Z$ is another morphism we then get that

$$N^*(g \circ f) \cong N^*f \times_{T^*Y} N^*g$$

as Lagrangian correspondences from T^*X to T^*Z. Indeed, the composition of correspondences

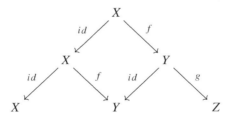

is sent by N^* to the composition of Lagrangian correspondences

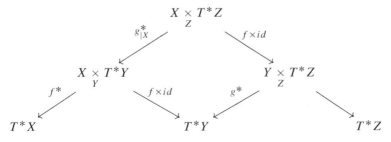

We thus have a functor $N^*\colon d\mathcal{A}rt\mathcal{S}t \to \mathcal{L}ag_0$ of ∞-categories from derived Artin stacks (with their usual morphisms) to 0-shifted symplectic stacks with Lagrangian correspondences.

Example 3.4 (Lagrange multipliers/Constrained critical locus) Following [36], we consider a *variational family*, being the data of a morphism[28] $f\colon P \to X$ and a function $S\colon P \to \mathbb{R}$.

First of all observe that the 1-form $dS\colon P \to T^*P$ carries a Lagrangian structure. This is for instance proven in [10, Section 2.4] (this is actually true for every closed 1-form), but it can also be obtained as a composition of Lagrangian correspondences: indeed, it is the composition of $N^*S \to T^*P \times \overline{T^*\mathbb{R}}$ with the Lagrangian $\mathbb{R} \hookrightarrow T^*\mathbb{R} \cong \mathbb{R}^2$ given by $x \mapsto (x,1)$:

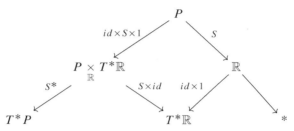

Then consider the *derived constrained critical locus* (or *derived fiber critical locus*, or *derived Lagrange multiplier space*) $\mathbf{Crit}_f(S) := P \times_{T^*P} N^*f$ of the family f. The morphism $\mathbf{Crit}_f(S) \to T^*X$ therefore carries a natural Lagrangian structure:[29]

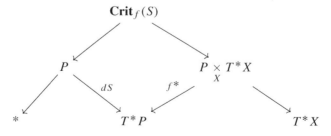

Example 3.5 (Derived critical locus) Notice that when $X = *$ in the above Example then $\mathbf{Crit}_f(S) = \mathbf{Crit}(S)$ is the (absolute) derived critical locus of f and is thus (-1)-shifted symplectic (according to Example 2.27).

---

[28] As we work in the derived setting, and contrary to [36], we do not require that the morphism is a surjective submersion.
[29] Somehow generalizing [36, Proposition 1.1].

Example 3.6 (Hamiltonian bimodules) Clearly, given two quasi-symplectic groupoids $G_1 \rightrightarrows G_0$ and $H_1 \rightrightarrows H_0$, and a Hamiltonian bimodule X (in the sense of [46, Definition 3.13]) for these, then $[X/G_1 \times H_1^{op}]$ provides a Lagrangian correspondence between $[G_0/G_1]$ and $[H_0/H_1]$. When the composition of two such bimodules is well defined (such as in [46, Theorem 3.16] for instance), then the resulting Lagrangian correspondence is the composition of the Lagrangian correspondences associated with these two Hamiltonian bimodules.

A particular case of these compositions is when we have two Hamiltonian G_1-spaces X_0 and Y_0: the derived fiber product $[X_0/G_1] \times_{[G_0/G_1]} [Y_0/G_1] \cong [X_0 \times_{G_0} Y_0/G_1]$ is then 0-shifted symplectic (again, this has to be compared with [46, Theorem 3.21]). Recalling from Example 2.29 that we also have a Lagrangian morphism $G_0 \to [G_0/G_1]$, so that we get the following commuting cube of cartesian squares:

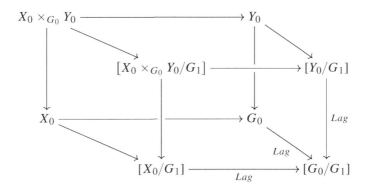

- There are three Lagrangian morphisms to the 1-shifted symplectic stack $[G_0/G_1]$.
- Their three derived intersections give rise to three 0-shifted symplectic stacks.
- $X_0 \times_{G_0} Y_0$ realizes three Lagrangian correspondences (for each pair of these 0-shifted symplectic stacks).

3.1.2 Hamiltonian Reduction

Let us consider a phase space (i.e., a symplectic manifold) X having a Lie group of symmetries G and moments for these symmetries: for each infinitesimal generator $x \in \mathfrak{g}$ of the action there is a Hamiltonian function μ_x, that is, $\omega^\flat(\vec{x}) = d\mu_x$. Collecting all μ_x's we get a map $\mu : X \to \mathfrak{g}^*$, called a *moment map*.

For every weakly regular value $\xi \in \mathfrak{g}^*$ satisfying nice enough hypotheses (see [274, Theorem 1]), the so-called *reduced space* $\mu^{-1}(\mathcal{O}_\xi)/G$ is symplectic, where $\mathcal{O}_\xi$ is the coadjoint orbit of ξ. Observe that these hypotheses actually guarantee that the reduced space $\mu^{-1}(\mathcal{O}_\xi)/G$ is equivalent to the *derived reduced space*

$$[X \times_{\mathfrak{g}^*} \mathcal{O}_\xi/G] \cong [X/G] \times_{[\mathfrak{g}^*/G]} [\mathcal{O}_\xi/G].$$

Now recall from Example 2.32 that the map $[\mu] \colon [X/G] \to [\mathfrak{g}^*/G]$ carries a Lagrangian structure and that the map $[\mathcal{O}_\mu/G] \to [\mathfrak{g}^*/G]$ does as well.[30] Hence we get that the derived reduced space is naturally 0-shifted symplectic (being a "Lagrangian intersection in" a 1-shifted symplectic stack).

As the reader may already have noticed, this is a particular instance of the very general phenomenon that we have already pointed in Example 3.6. We thus again have a commuting cube of cartesian squares as in Example 3.6:

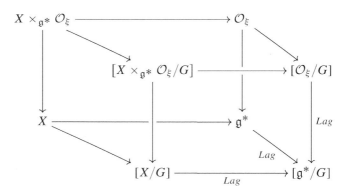

Let us provide below a few more examples of this phenomenon.

Example 3.7 (Hamiltonian spaces for symplectic groupoids are symplectic) Assume we are given a symplectic groupoid $G_1 \rightrightarrows G_0$ and a Hamiltonian G_1-space X. Recall that G_1 itself is a Hamiltonian G_1-space, so that

$$[X/G_1] \times_{[G_0/G_1]} [G_1/G_1] = [X/G_1] \times_{[G_0/G_1]} G_0 \cong X$$

is thus 0-shifted symplectic. The morphism $X \to G_0$ plays the role of a moment map.

Example 3.8 (Quasi-Hamiltonian reduction) Consider the conjugation quasi-symplectic groupoid $G \times G \rightrightarrows G$ and a quasi-Hamiltonian G-space X.

---

[30] Indeed, the inclusion of coadjoint orbit $\mathcal{O}_\mu \hookrightarrow \mathfrak{g}^*$ is a moment map.

Then we get that for any conjugacy class[31] $C \subset G$ the derived fiber product $[X \times_G C/G]$ is 0-shifted symplectic. This gives back the fact that, under suitable assumptions, we have a genuine symplectic manifold $\mu^{-1}(C)/G$ as in [1], where $\mu\colon X \to G$ is the Lie group valued moment map.

3.2 Transgression

A transgression is a map that transfers cohomology classes in a way that changes the cohomological degree, typically when integrating along fibers in differential geometry. This has for instance many manifestations in field theory, where the expression of the action functional in terms of the Lagrangian can sometimes be understood as a transgression procedure.[32]

The so-called *AKSZ formalism* from [2], which allows to present many classical gauge theories as σ-models and make them fit into the BV formalism, is also entirely based on a transgression procedure. In what follows we explain a far reaching generalization of it, called *PTVV formalism*,[33] developed in [31].

3.2.1 Integration Theory on Stacks: Shifted Orientations

In [31] the authors introduce a class of derived stacks that carry a good integration theory of cohomological degree d. Such a derived stack Σ shall be:

(**o1**) such that, for any other stack F, one can extract the $(0, *)$-part of a closed form on the product $\Sigma \times F$. One thus has a morphism $DR(\Sigma \times F) \to C^\infty(\Sigma) \otimes DR(F)$ of graded mixed complexes, where $C^\infty(\Sigma) := \Gamma(\mathcal{O}_\Sigma)$ is the cochain complex of derived global functions on Σ.

(**o2**) such that for any stack F the derived global section functor
$$\Gamma(\Sigma, -) : QCoh(\Sigma \times F) \to QCoh(F)$$
preserves perfect objects.[34]

---

[31] Conjugacy classes are examples of quasi-Hamiltonian G-spaces (see [1, Proposition 3.1]).

[32] We refer to [40] for a systematic approach, and to the survey papers [16, 81] for shorter and less abstract expositions. Observe that these references also deal with variants of shifted presymplectic structures on stacks, but the nondegeneracy condition is almost never satisfied as everything takes place in the realm of *underived* stacks.

[33] The main advantage of the PTVV formalism, compared to the AKSZ one, is that it is model independent. For instance, all notions (e.g., closed forms) and properties (e.g., nondegeneracy) are invariant under appropriate equivalences (e.g., quasi-isomorphisms).

[34] Recall that perfect objects are roughly locally equivalent to bounded complexes of finite-dimensional vector bundles. This condition can thus be understood as saying that the cohomology of finite-dimensional vector bundles on Σ is finite-dimensional.

(o3) equipped with a *d-class* $[\Sigma] : C^\infty(\Sigma) \to \mathbb{R}[-d]$, also called *d-shifted pseudo-orientation*.

(o4) such that for any perfect complex E on Σ, the pairing

$$\Gamma(\Sigma, E) \otimes \Gamma(\Sigma, E^*)[d] \to C^\infty(\Sigma)[d] \to \mathbb{R}$$

is nondegenerate in cohomology.

A pair $(\Sigma, [\Sigma])$ as above is called a *d-oriented* stack. If only (o1) and (o3) are satisfied we call it a *d-pseudo-oriented* stack.

Example 3.9 (The de Rham stack M_{DR}) Let M be a closed compact oriented d-manifold and let $\Sigma = M_{DR}$ be the de Rham stack of M. It is the stack obtained as the quotient of the groupoid $\widehat{M \times M} \rightrightarrows M$, where $\widehat{M \times M}$ is the formal neighborhood of the diagonal in M (i.e., we identify any two infinitesimally close points). A (complex of) vector bundle on M_{DR} is a (complex of) vector bundle on M together with a flat connection, and one has that $C^\infty(M_{DR}) \cong (\Omega^*(M), d_{dR})$ is the de Rham complex of M. We thus have a d-shifted orientation given by $\int_M : (\Omega^*(M), d_{dR}) \to \mathbb{R}[-d]$.

The de Rham stack has first been introduced by Carlos Simpson in the algebro-geometric context [41].

Example 3.10 (The Betti stack M_B) Let M be a closed compact oriented d-manifold and let $\Sigma = M_B$ be the Betti stack of M. It is the constant stack associated with the homotopy type of M. Given a cellular decomposition of M, M_B can be described as the stack with 0-cells as points, 1-cells as isomorphisms, ..., n-cells as n-isomorphisms, etc. One has that $C^\infty(M_B) \cong C^*_{sing}(M, \mathbb{R})$ is the singular cochain complex of M. We have a d-shifted orientation given by evaluating cochains on the fundamental class of M.

Example 3.11 (of a pseudo-orientation) Let M be a closed compact oriented surface and let $\Sigma = B(\wedge^2 T_M)$ be the split first-order degree -1 extension[35] of M by the sheaf $\wedge^2 T_M$: $C^\infty(\Sigma) \cong C^\infty(M) \oplus \Omega^2(M)[-1]$. We have a 1-shifted pseudo-orientation[36] given by $\int_M$.

Remark 3.12 Observe that $\wedge^2 T_M$ gets a nice geometric interpretation as the subspace of $M^{\Delta^2_{inf}}$ consisting of those infinitesimal 2-simplices in M that are either nondegenerate or constant. In other words, $\wedge^2 TM$ is the space of "order one surfaces" in M (like TM is the space of order one curves in M).

---

[35] For a bundle $E \to M$, the split first-order degree -1 extension is also the classifying stack of E, viewed as a bundle of groups on M for the $+$ law. A more standard notation in differential (super)geometry is $E[1]$.

[36] One could imagine a Banach framework in which this example is a 1-shifted orientation.

Given a d-pseudo-oriented derived stack $(\Sigma, [\Sigma])$ and any derived stack F one can compose the morphism of graded mixed complexes $DR(\Sigma \times F) \to C^\infty(\Sigma) \otimes DR(F)$ from (o1) with the d-shifted pseudo-orientation $[\Sigma]$, and get a morphism of graded mixed complexes $DR(\Sigma \times F) \to DR(F)[-d]$, denoted $\int_{[\Sigma]}$. In particular, $\int_{[\Sigma]}$ induces a map $\mathcal{A}^{p,cl}(\Sigma \times F, n) \to \mathcal{A}^{p,cl}(F, n-d)$.

Therefore, if one let F be a derived mapping stack $F := \mathbf{Map}(\Sigma, X)$, then we get a map $\mathcal{A}^{p,cl}(X, n) \to \mathcal{A}^{p,cl}(F, n-d)$: it is given by $\omega \mapsto \int_{[\Sigma]} ev^*\omega$, where $ev : \Sigma \times \mathbf{Map}(\Sigma, X) \to X$ is the evaluation map.

$$\begin{array}{ccc} \Sigma \times \mathbf{Map}(\Sigma, X) \xrightarrow{ev} X & \mathcal{A}^{p,cl}\big(\Sigma \times \mathbf{Map}(\Sigma, X), n\big) \xleftarrow{ev^*} \mathcal{A}^{p,cl}(X, n) \\ \downarrow & \downarrow \int_{[\Sigma]} \\ \mathbf{Map}(\Sigma, X) & \mathcal{A}^{p,cl}\big(\mathbf{Map}(\Sigma, X), n-d\big) \end{array}$$

Example 3.13 Let $\Sigma = S_B^1 \cong B\mathbb{Z}$ and let $X = BG$ be the classifying stack of a Lie group G. Then $\mathbf{Map}(\Sigma, X) \cong [G/G]$. Consider the 2-shifted presymplectic structure ω on BG determined by an invariant symmetric pairing $c \in S^2(\mathfrak{g}^*)^G$. One can show (see [37]) that the transgressed 1-shifted presymplectic form on $[G/G]$ we get is indeed the one coming from the presymplectic groupoid $G \times G \rightrightarrows G$.

Example 3.14 Let $\Sigma = S_{DR}^1$ and let $X = BG$ be the classifying stack of a connected Lie group G. Then $\mathbf{Map}(\Sigma, X) \cong [\Omega^1(S^1, \mathfrak{g})/C^\infty(S^1, G)]$, where the action is given by $g \cdot \alpha = Ad_g\alpha + dgg^{-1}$. One can show (by an explicit calculation) that the transgressed 1-shifted presymplectic form we get is indeed the one pulled-back from $[C^\infty(S^1, \mathfrak{g})^*/C^\infty(S^1, G)]$ along the "pairing+integration map" $\Omega^1(S^1, \mathfrak{g}) \to C^\infty(S^1, \mathfrak{g})^*$.

Example 3.15 Let $\Sigma = M[\wedge^2 T_M]$ be as in Example 3.11 and let $X = BG$ be the classifying stack of a compact and simply connected group G. Consider the 2-shifted presymplectic structure ω on BG determined by an invariant symmetric pairing $c \in S^2(\mathfrak{g}^*)^G$. Then $\int_{[\Sigma]} ev^*\omega$ is a 1-shifted presymplectic structure on $\mathbf{Map}(\Sigma, X) \cong [\Omega^2(M, \mathfrak{g})/C^\infty(M, G)]$.[37] This 1-shifted presymplectic stack is an "infinite-dimensional analog" of $[\mathfrak{g}^*/G]$ and will play a crucial role in the derived interpretation of the infinite-dimensional reduction procedure that we have seen in the introduction.

We now come to a very useful result from [31]:

---

[37] For a more general G, $\mathbf{Map}(\Sigma, X)$ is the derived moduli stack of G-bundles equipped with a basic 2-form.

Theorem 3.16 *If $(\Sigma, [\Sigma])$ is a d-oriented stack, (X, ω) is an n-shifted symplectic stack, and $\mathbf{Map}(\Sigma, X)$ is an Artin stack, then $\int_{[\Sigma]} ev^*\omega$ is nondegenerate. Therefore $\mathbf{Map}(\Sigma, X)$ naturally becomes $(n-d)$-shifted symplectic.*

This is a very nice statement as

- many symplectic structures on moduli spaces can be recovered as particular instances of this result (for $n = d$);
- several examples of so-called perfect obstruction theories (after Behrend–Fantechi [10]) as well (for $n = d - 1$).

3.2.2 De Rham Stack versus Betti Stack

Let X be a d-dimensional manifold. There is a map $X_{DR} \to X_B$ from the de Rham stack to the Betti stack.[38] This tells us in particular that for a derived stack F, we have morphism $\mathbf{Map}(X_B, F) \to \mathbf{Map}(X_{DR}, F)$. It can be shown along the lines of [32] that, whenever $F = BG$, this map is an equivalence.[39]

Example 3.17 Let $F = BG$ and $X = S^1$. On the one hand recall that $S_B^1 \cong B\mathbb{Z}$ and thus $\mathbf{Map}(X_B, F) \cong [G/G]$ (where the action of G on itself is by conjugacy, as usual). On the other hand one can show that $\mathbf{Map}(X_{DR}, F)$ can be described as the quotient stack[40] $[\Omega^1(S^1, \mathfrak{g})/C^\infty(S^1, G)]$, where the action is given by $g \cdot \alpha = Ad_g \alpha + dgg^{-1}$. We therefore get that there is an equivalence between $[\Omega^1(S^1, \mathfrak{g})/C^\infty(S^1, G)]$ and $[G/G]$.

For a more general X this tells us that the derived stack of G-local systems on X is equivalent to the derived stack of flat G-connections on X. Additionally, one can prove that if X is compact and oriented then the d-orientations on X_{DR} and X_B do coincide. This in particular tells us that we have an equivalence of $(2-d)$-shifted symplectic stacks $\mathbf{Map}(X_{DR}, F) \cong \mathbf{Map}(X_B; F)$.

Example 3.18 Going back to the previous example, and assuming now that G is compact and connected, we get an equivalence of 1-shifted symplectic

---

[38] For instance, consider a cover $\pi : \mathfrak{U} = \coprod_i U_i$ by contractible open subsets of X such that all iterated intersections of these open subsets are contractible as well. In other words, the nerve of the map $\pi : \mathfrak{U} \to X$ is made of contractible open subsets of X. Note that X is equivalent to the homotopy colimit (in stacks) of the nerve $N(\pi)$ of the map $\pi : \mathfrak{U} \to X$. Similarly X_{DR} is equivalent to the homotopy colimit of $N(\pi)_{DR}$ (where we have applied the de Rham functor levelwise). Finally, using the fact that we always have a terminal map $(\mathbb{R}^d)_{DR} \to *$ we get a morphism $N(\pi)_{DR} \to \pi_0 N(\pi) = X_B$ (here again we have applied π_0 levelwise).

[39] In [32] this is proven for F being the derived stack of perfect complexes and in the context of derived analytic stacks. The proof carries over in the differentiable context, and for $F = BG$ as well.

[40] Assuming G is connected.

stacks between $[\Omega^1(S^1,\mathfrak{g})/C^\infty(S^1,G)]$ and $[G/G]$. We therefore get a correspondence between Lagrangian morphisms to $[\Omega^1(S^1,\mathfrak{g})/C^\infty(S^1,G)]$ and Lagrangian morphisms to $[G/G]$, providing a very nice interpretation of the correspondence between Hamiltonian $L(G)$-spaces and quasi-Hamiltonian G-spaces from [1].

For a more general d-dimensional compact oriented X, we get that the derived stacks of G-local systems and of flat G-connections on X are equivalent as $(2-d)$-shifted symplectic stacks.

We would now like to calculate the two $(2-d)$-shifted symplectic structures from the above example at an $\mathbb{R}$-point. The calculation actually works for a general mapping stack $\mathbf{Map}(\Sigma, F)$ with a d-oriented stack $(\Sigma, [\Sigma])$ and an n-shifted symplectic stack (F, ω) as in Theorem 3.16. For a $f : \Sigma \to F$ of the mapping stack, the tangent complex at f is

$$\mathbb{T}_f \mathbf{Map}(\Sigma, F) = \Gamma(\Sigma, f^*\mathbb{T}_F).$$

At f, the transgressed $(n-d)$-shifted symplectic form looks as follows:

$$\Gamma(\Sigma, f^*\mathbb{T}_F)^{\otimes 2} \xrightarrow{f^*\omega} C^\infty(\Sigma)[n] \xrightarrow{[\Sigma]} \mathbb{R}[n-d].$$

Example 3.19 Let Σ be either the de Rham or Betti stack of a d-dimensional compact oriented manifold X, and let $F = BG$ for a Lie group G. Then $\Gamma(\Sigma, f^*\mathbb{T}_F) \cong H^*(X, ad(P_f))[1]$, where P_f is the flat G-bundle (or G-local system) on Σ corresponding to the classifying map $f : X_{DR} \to BG$ (or $X_B \to BG$), $ad(P_f) := P_f \times_G \mathfrak{g}$ is the adjoint flat bundle (or local system), and $H^*(X, -)$ means de Rham (or Betti) cohomology with coefficients.

Setting $d = 2$, we thus get a genuine linear symplectic pairing on the degree 0 cohomology of $\Gamma(\Sigma, f^*\mathbb{T}_F)$, which is nothing but $H^1(X, ad(P_f))$. This pairing is precisely the one we have seen in the introduction (in the "deformation-theoretic approach" part).

3.3 Transgression with Boundary

We have seen two systematic ways of constructing new shifted symplectic stacks out of old ones:

- by doing derived intersection of Lagrangian morphisms;
- by transgression.

We would like these two constructions to be compatible with each other. More precisely, we would like to have the following property: if $M \cong M_+ \coprod_N M_-$ is an oriented d-dimensional compact manifold obtained as the gluing of two manifolds sharing a common boundary $N \cong \partial M_+ \cong \partial M_-$, and if F is n-shifted symplectic, then both $\mathbf{Map}(M_B, F)$ and $\mathbf{Map}(M_{DR}, F)$ can be obtained as derived Lagrangian intersections.

More generally we will see that the transgression procedure produces a functor from the ∞-category of cobordisms to the ∞-category of shifted symplectic stacks and Lagrangian correspondences.

3.3.1 Boundary Structures

Let us summarize the structure we need in order to model the situation of a boundary inclusion in our framework. Let $\varphi \colon \Sigma \to \Upsilon$ be a morphism between derived stacks satisfying condition 1 in Section 3.2.1, and assume that Σ is equipped with a d-shifted pseudo-orientation.

Definition 3.20 A *boundary structure* for $(\varphi, [\Sigma])$ is a homotopy $[\Upsilon]$ between $\varphi_*[\Sigma] := [\Sigma] \circ \varphi^*$ and 0 (as morphisms of cochain complexes $C^\infty(\Upsilon) \to \mathbb{R}[-d]$).

Below we list several examples.

Example 3.21 (Pseudo-orientations as boundary structures) Let $\varnothing$ be the initial stack, equipped with its canonical d-shifted orientation. A boundary structure on $\varnothing \to \Sigma$ is exactly the same as a $(d+1)$-shifted pseudo-orientation on Σ. This is very similar to the "Symplectic is Lagrangian" Example 2.27.

Example 3.22 (Boundary inclusions as boundary structures) Let N be a compact $(d+1)$-dimensional manifold with oriented boundary $M = \partial N$. Recall that both M_B and M_{DR} carry a d-shifted orientation. We further have that both maps $M_{DR} \to N_{DR}$ and $M_B \to N_B$, induced by the boundary inclusion $\iota \colon M \hookrightarrow N$, carry boundary structures. In the de Rham stack case this is simply Stokes formula $\int_{\partial M} \iota^* \omega = \int_N d_{dR}\omega$: the homotopy is $\int_N$. In the Betti case, $[M_B]$ is given by the cap product with a fundamental cycle for M, and the homotopy $[N_B]$ is given by the cap product with a compatible fundamental chain for (N, M).

Example 3.23 Let M be a closed oriented surface, and recall from Example 3.11 that $B[\wedge^2 T_M]$ is 1-pseudo-oriented. The projection $\pi \colon \Sigma \to M$ carries an obvious boundary structure, as $\int_M f = 0$ if $f \in C^\infty(M)$. Indeed

- π^* is just the inclusion $C^\infty(M) \hookrightarrow C^\infty(M) \oplus \Omega^2(M)[-1]$;

- the 1-class $[\Sigma]$ is given by $\int_M$ (i.e., it vanishes on functions and sends a 2-form on M to its integral);

- hence $\pi_*[\Sigma] = [\Sigma] \circ \pi^*$ necessarily vanishes (strictly).

Example 3.24 Let M be a closed oriented surface and consider again $\Sigma = B(\wedge^2 T_M)$ with its 1-shifted pseudo-orientation from Example 3.11. We introduce another stack M_∇, being to M_{DR} what connections are to flat connections: M_∇ is the quotient by the groupoid $\mathcal{G} \rightrightarrows M$ generated by the equivalence relation of being close at order 1.[41] We have a morphism $\varphi : \Sigma \to \Upsilon := M_\nabla$ that is roughly given by the map sending a connection to its curvature, which can be described in several ways:

- There is a map that sends every infinitesimal 2-simplex $\Delta^2_{inf} \to M$ to the element in $\mathcal{G}$ given by the sequence of equivalences $x \sim y \sim z \sim x$, where x, y, z are the three vertices of the infinitesimal simplex, which induces a morphism on the quotients $B(\wedge^2 TM) \to M_\nabla$.
- In terms of Lie algebroids, we have a Lie algebroid morphism $Free(\wedge^2 T_M) \to Free(T_M)$ sending $u \wedge v$ to $uv - vu - [u, v]$, leading to a morphism $B(\wedge^2 TM) \to M_\nabla$.

At the level of functions, $C^\infty(\Upsilon)$ is the two-term complex $C^\infty(M) \xrightarrow{d_{dR}} \Omega^1(M)$, $C^\infty(\Sigma)$ is the two-term complex $C^\infty(M) \xrightarrow{0} \Omega^2(M)$, and the morphism looks as follows:

$$C^\infty(\Upsilon) \xrightarrow{\varphi^*} C^\infty(\Sigma) \qquad \text{degree}$$

$$\begin{array}{ccc} C^\infty(M) & \xrightarrow{id} & C^\infty(M) \\ {\scriptstyle d_{dR}} \downarrow & & \downarrow {\scriptstyle 0} \\ \Omega^1(M) & \xrightarrow{d_{dR}} & \Omega^2(M) \end{array} \qquad \begin{array}{c} 0 \\ \\ 1 \end{array}$$

Hence φ carries a boundary structure as the integral of an exact 2-form on a closed manifold M is zero.

The following has been shown in [9, Claim 2.7]:

---

[41] There is yet another description in terms of Lie algebroid on M, using the fact that every Lie algebroid leads to a formal thickening of M (see [11, 17, 30]):
- M itself is associated with the trivial Lie algebroid 0;
- M_{DR} is associated with the Lie algebroid T_M;
- M_∇ is associated with the free Lie algebroid (see [21]) generated by the anchored module $id: T_M \to T_M$;
- $B(\wedge^2 T_M)$ is associated with the free Lie algebroid generated by the anchored module $0: \wedge^2 T_M \to T_M$.

Proposition 3.25 *Given a boundary structure as in Definiton 3.20 above, $\int_{[\Upsilon]} ev^*(-)$ provides a homotopy between the pullback along $\varphi^*\colon \mathbf{Map}(\Upsilon, F) \to \mathbf{Map}(\Sigma, F)$ of $\int_{[\Sigma]} ev^*(-)$ and 0. In particular, if F is equipped with an n-shifted presymplectic structure then φ^* carries an isotropic structure.*

We now provide several incarnations of this very general fact.

Example 3.26 (An infinite-dimensional moment map for the moduli of connections) Applying the above to the boundary structure on $M[\wedge^2 T_M] \to M_\nabla$ from Example 3.24 and to the 2-shifted presymplectic structure on $F = BG$ (G compact and simply connected) arising from $c \in S^2(\mathfrak{g}^*)^G$, we get an isotropic structure on the curvature morphism

$$\mathbf{Conn}_G(M) := \mathbf{Map}(M_\nabla, BG) \to \mathbf{Map}\big(B(\wedge^2 T_M), BG\big)$$
$$= [\Omega^2(M, \mathfrak{g})/C^\infty(M, \mathfrak{g})].$$

If the boundary structure turns out to be nondegenerate in an appropriate sense, defining a *relative d-orientation* (we refer to [9, Definition 2.8] for the details) then, in complete analogy with Theorem 3.16, we have that the isotropic structure on $\varphi^*\colon \mathbf{Map}(\Upsilon, F) \to \mathbf{Map}(\Sigma, F)$ is a Lagrangian morphism.[42]

Example 3.27 (Orientations as relative orientations) Going back to Example 3.21, we have that a shifted relative d-orientation on $\emptyset \to \Sigma$ is exactly the same as a $(d+1)$-shifted orientation on Σ. Therefore if (F, ω) is an n-shifted symplectic stack then we have a Lagrangian stucture on $\mathbf{Map}(\Sigma, F) \to \mathbf{Map}(\emptyset, F) = *$, recovering Theorem 3.16 from its relative analog.

Example 3.28 (Boundary inclusions as relative orientations) The boundary structures from Example 3.22 are always nondegenerate. Therefore, we obtain that whenever G is a compact Lie group, then

- $\mathbf{Loc}_G(M) := \mathbf{Map}(M_B, BG)$ is $(2-d)$-shifted symplectic and the restriction morphism $\mathbf{Loc}_G(N) \to \mathbf{Loc}_G(M)$ is Lagrangian;

- $\mathbf{Flat}_G(M) := \mathbf{Map}(M_{DR}, BG)$ is $(2-d)$-shifted symplectic and the restriction morphism $\mathbf{Flat}_G(N) \to \mathbf{Flat}_G(M)$ is Lagrangian.

---

[42] Whenever both mapping stacks are Artin, as usual.

3.3.2 A Gluing Formula: Transgression as a Topological Field Theory

Just like derived intersections of Lagrangian morphisms (resp. isotropic morphisms) are shifted symplectic (resp. shifted presymplectic), pushouts of relatively oriented morphisms (resp. morphisms equipped with boundary structures) are oriented (resp. pseudo-oriented). We refer to [9, Section 4.2.1] for details.

There is actually an ∞-category of oriented stacks with morphisms being relatively oriented cospans

which works in a way similar (but dual) to our derived/∞-categorical variant of Weinstein's symplectic category from Section 3.1, composition being given by pushout:

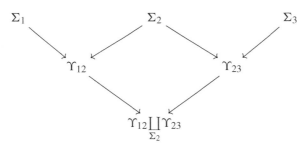

We provide several examples in order to give the reader an intuition of what is going on. We restrict our attention to situations where $\Sigma_1 = \Sigma_3 = \emptyset$: then the composition is a relatively oriented cospan from $\emptyset$ to itself, and thus it is a $(d+1)$-oriented stack.

Example 3.29 (Betti and de Rham gluings) Let $N \cong N_+ \coprod_M N_-$ be an oriented $(d+1)$-dimensional compact manifold obtained as the gluing of two oriented manifolds sharing a common boundary $M \cong \partial N_+ \cong \partial N_-$.

We therefore have d-oriented stacks $\Sigma = M_B$ and relative d-orientations on $\Sigma \to \Upsilon_\pm := (N_\pm)_B$. One can show that we have an equivalence of $(d+1)$-oriented stacks

$$\Upsilon_+ \coprod_\Sigma \Upsilon_- \cong N_B.$$

The same result holds as well for de Rham stacks.

Example 3.30 (Flat connections as connections with zero curvature) Let M be an oriented compact surface. We have a pushout square of derived stacks[43]

$$\begin{array}{ccc} B(\wedge^2 T_M) & \longrightarrow & M_\nabla \\ \downarrow & & \downarrow \\ M & \longrightarrow & M_{DR} \end{array}$$

Recall that

- $B(\wedge^2 T_M)$ is 1-pseudo-oriented (Example 3.11);
- $B(\wedge^2 T_M) \to M_\nabla$ carries a boundary structure (Example 3.24);
- $B(\wedge^2 T_M) \to M$ carries a boundary structure (Example 3.23).

One therefore gets that the pushout M_{DR} is 2-pseudo-oriented. We claim that the 2-shifted pseudo-orientation coincides with the 2-orientation on M_{DR} from Example 3.9.

One can show that the transgression procedure sends compositions of oriented cospans (resp. of cospans with boundary structures) to compositions of Lagrangian (resp. isotropic) correspondences. Restricting it to de Rham or Betti stacks we then get a $3d$-oriented topologicial field theory with values in our derived/∞-variant of Weinstein's symplectic category. This $3d$ TFT can even be shown to be fully extended (see [13]).

We again provide several examples.

Example 3.31 We will apply the transgression procedure to the situation of Example 3.29 with target 2-shifted symplectic stack BG, with G being compact. We thus get that the $(n-d-1)$-shifted symplectic stacks $\mathbf{Loc}_G(N)$ and $\mathbf{Flat}_G(N)$ can be obtained as derived Lagrangian intersections:

$$\begin{array}{ccc} \mathbf{Loc}_G(N) & \longrightarrow & \mathbf{Loc}_G(N_-) \\ \downarrow & & \downarrow{\scriptstyle Lag} \\ \mathbf{Loc}_G(N_+) & \xrightarrow{Lag} & \mathbf{Loc}_G(M) \end{array} \qquad \begin{array}{ccc} \mathbf{Flat}_G(N) & \longrightarrow & \mathbf{Flat}_G(N_-) \\ \downarrow & & \downarrow{\scriptstyle Lag} \\ \mathbf{Flat}_G(N_+) & \xrightarrow{Lag} & \mathbf{Flat}_G(M) \end{array}$$

---

[43] It comes from the following pushout square of Lie algebroids/groupoids over M:

$$\begin{array}{ccc} Free(0:\wedge^2 T_M \to T_M) & \longrightarrow & Free(id:T_M \to T_M) \\ \downarrow & & \downarrow \\ 0 & \longrightarrow & T_M \end{array} \qquad \begin{array}{ccc} (\wedge^2 T_M, +) & \longrightarrow & \mathcal{G} \\ \downarrow & & \downarrow \\ M & \longrightarrow & \widehat{M \times M} \end{array}$$

Example 3.32 Let us specialize the above Example to the case when N is a surface, $M = S^1$ and $N_- = D^2$ is a disk. We get that $\mathbf{Loc}_G(M) \cong [G/G]$ with its 1-shifted symplectic structure and $\mathbf{Loc}_G(N_-) \cong [*/G]$. Hence we get back that the derived stack $\mathbf{Loc}_G(N)$ of G-local systems on the surface N can be obtained as a quasi-Hamiltonian reduction of the derived stack of G-local systems on the same surface with a disk removed. That is, we have an equivalence of 0-shifted symplectic stacks $\mathbf{Loc}_G(N) \cong \mathbf{Loc}_G(N\backslash D^2) \times_{[G/G]} [*/G]$:

$$\begin{array}{ccc} \mathbf{Loc}_G(N) & \longrightarrow & [*/G] \\ \downarrow & & \downarrow {\scriptstyle Lag} \\ \mathbf{Loc}_G(N\backslash D^2) & \xrightarrow[Lag]{} & [G/G] \end{array}$$

If we also further assume that G is connected then we have that $\mathbf{Flat}_G(M) \cong [\Omega^1(S^1, \mathfrak{g})/C^\infty(S^1, G)]$ and $\mathbf{Flat}_G(N_-) \cong [*/G]$. Hence we get back the fact that the derived stack $\mathbf{Flat}_G(N)$ of flat G-bundles on the surface N can be obtained as a kind of infinite-dimensional Hamiltonian reduction of the derived stack of flat G-bundles on the same surface with a disk removed: we have a derived Lagrangian intersection

$$\begin{array}{ccc} \mathbf{Flat}_G(N) & \longrightarrow & [*/G] \\ \downarrow & & \downarrow {\scriptstyle Lag} \\ \mathbf{Flat}_G(N\backslash D^2) & \xrightarrow[Lag]{} & [\Omega^1(S^1, \mathfrak{g})/C^\infty(S^1, G)] \end{array}$$

Example 3.33 We now apply the transgression procedure to the situation of Example 3.30 with target 2-shifted symplectic stack BG again, G being compact and simply connected for the sake of simplicity. We get a 1-shifted presymplectic stack $[\Omega^2(M, \mathfrak{g})/C^\infty(M, G)]$ together with isotropic morphisms to it from $\mathbf{Conn}_G(M)$ and $\mathbf{Bun}_G(M) = [*/C^\infty(M, G)]$. Their derived isotropic intersection is then equivalent, as a 0-shifted presymplectic stack, to the 0-shifted symplectic stack $\mathbf{Flat}_G(M)$:

$$\begin{array}{ccc} \mathbf{Flat}_G(M) & \longrightarrow & [*/C^\infty(M, G)] \\ \downarrow & & \downarrow {\scriptstyle isot.} \\ \mathbf{Conn}_G(M) & \xrightarrow[isot.]{} & [\Omega^2(M, \mathfrak{g})/C^\infty(M, G)] \end{array}$$

4 Conclusion

In this survey of derived symplectic geometry we have shown how one can use derived geometry in order to unify several descriptions of the symplectic structure on the reduced phase space of classical Chern–Simons theory, that is to say the moduli space of flat G-bundles, let us say for a simply connected compact Lie group, on a closed oriented surface M. Let us summarize what we have seen so far:

- The derived moduli space $\mathbf{Flat}_G(M)$ is *defined* as the derived mapping stack $\mathbf{Map}(M_{DR}, BG)$. It naturally gets a 0-shifted symplectic structure by transgression, using the 2-shifted symplectic structure on BG and the 2-shifted orientation on M_{DR}.
- The canonical map $M_{DR} \to M_B$ of 2-shifted oriented stacks induces an equivalence $\mathbf{Loc}_G(M) \xrightarrow{\sim} \mathbf{Flat}_G(M)$ of 0-shifted symplectic stacks, where $\mathbf{Loc}_G(M) := \mathbf{Map}(M_B, BG)$ is the derived stack of G-local systems on M.
- Locally at a point, the 0-shifted symplectic structure is given by a combination of the nondegenerate pairing on $\mathfrak{g}$ with Poincaré duality. This can be seen as a derived extension of the standard fact that $H^1(M, \mathfrak{g})$ is a symplectic vector space.
- The 0-shifted symplectic structure on $\mathbf{Loc}_G(M)$ can be computed through the quasi-Hamiltonian reduction procedure of [1], being understood as a derived Lagrangian intersection. This is derived from a very general compatibility of the transgression procedure with all kinds of "gluings."
- Using a rather surprising gluing data (exhibiting M_{DR} as a kind of pushout of M_∇) we can also compute the 0-shifted symplectic structure on $\mathbf{Flat}_G(M)$ as an infinite-dimensional reduction procedure applied to the derived moduli stack of all G-connections. This reduction procedure is here understood as a derived isotropic intersection.
- Whenever M bounds an oriented $3d$-manifold N, we get a Lagrangian structure on the "boundary condition map" $\mathbf{Flat}_G(N) \to \mathbf{Flat}_G(M)$.
- In the case when N is without boundary (i.e., $M = \emptyset$) we can still make sense of what it means for the derived stack $\mathbf{Flat}_G(N)$ of boundary conditions to be Lagrangian in $* = \mathbf{Flat}_G(\emptyset)$: it means that $\mathbf{Flat}_G(N)$ is (-1)-shifted symplectic.

4.1 What Do We Gain with Shifted Symplectic Structures?

As we have seen, derived geometry provides new tools, at the disposal of symplectic geometers, as well as a rather large framework in which several constructions find a nice interpretation, and do make sense in a wider context.

We would like to point out several differences between ordinary symplectic geometry and its derived counterpart, that we may put in three different boxes:

- First, there are differences coming from the use of Higher Algebra: several properties (e.g., being closed, or isotropic) became structures. This has very interesting and unexpected consequences, such as the derived intersection of two Lagrangians carrying a shifted symplectic structure.
- Then, there are differences that actually come from the very purpose of derived geometry: several constructions do make sense in derived geometry even for pathological situations (including bad quotients, and non-transverse intersections). This allows for instance to deal with symplectic reduction without any hypothesis on the action, or to define Weinstein's symplectic category.
- Finally, there are less expected differences: we discover the existence of a new kind of symplectic structures, that were somehow hidden in ordinary symplectic geometry: the 2-shifted symplectic structure on BG, the 1-shifted symplectic structure on $[\mathfrak{g}^*/G]$, or moment maps being Lagrangian morphisms.

There is one particular instance of a construction that reveals all three aspects that we have just listed: it is the transgression Theorem 3.16 and its generalizations. Indeed, the proof makes a crucial use of structures rather than properties, in many cases the mapping spaces involved become tractable derived stacks (avoiding the use of infinite-dimensional geometry and the need to restrict on some smooth locus), and we get access to negatively shifted symplectic structures just from the 2-shifted symplectic structure on BG.

Let us finally mention a very interesting consequence of being (-1)-shifted symplectic for $\mathbf{Flat}_G(N)$, N being a closed oriented 3-manifold. In [5] it has been proved that (-1)-shifted derived stacks locally look like derived critical loci. In particular this implies that the corresponding *nonderived* Artin stack $Flat_G(N)$ is a d-critical stack in the sense of [19]. Also note that there are sufficient conditions for the existence of a global action functional for (-1)-shifted derived stacks (T. Pantev, private communication). For instance, it is expected that if a (-1)-shifted derived stack carries a Lagrangian foliation $\mathcal{L}$ then there exists a function f on the formal quotient stack $[X/\mathcal{L}]$ and an étale symplectic morphism $X \to \mathbf{Crit}(f)$. This means that somehow, the action functional is hidden in the (-1)-shifted symplectic structures, which may be a more fundamental structure.

4.2 Shifted Poisson Structures

In ordinary symplectic geometry, any symplectic manifold carries a corresponding Poisson structure. A natural question is then: *are there shifted Poisson structures as well?* And the answer is *yes*. Shifted Poisson structures have been introduced and studied in [12, 33], where it is shown that nondegenerate shifted Poisson structures do coincide with shifted symplectic structures. Several observations are in order:

- The theory of shifted Poisson structures is rather technical, involving complications due to the lack of functoriality of the tangent complex (as opposed to the cotangent complex). Simple facts like the bijection between nondegenerate Poisson bivectors and symplectic forms requires a lot of work in the derived setting.
- Many constructions and structures from standard Poisson geometry (such as Poisson–Lie structures, r-matrices, dynamical r-matrices, the Feigin–Odesski algebra, ...) fit very well in the realm of shifted Poisson structures, as it has been shown in Safronov's recent work [39].
- The (-1)-shifted Poisson structures are of particular interest for the classical BV–BRST formalism. In particular, using [12, 33] one can show that the function ring of $\mathbf{Flat}_G(N)$ carries a Poisson bracket of cohomological degree 1: this bracket indeed lies at the heart of the BV–BRST formalism, and is the starting point of perturbative quantization for several classical field theories. More details about the relation between shifted Poisson geometry and the classical and quantum BRST formalism can be found in [38].
- There is an alternative definition of shifted Poisson structures as formal Lagrangian thickenings. The idea being that if X is Poisson then the formal quotient $[X/\mathcal{L}]$ by its symplectic foliation $\mathcal{L}$ is 1-shifted symplectic and the map $X \to [X/\mathcal{L}]$ is Lagrangian. At present it is still a conjecture that shifted Poisson structures are equivalent to formal Lagrangian thickenings.

4.3 Quantization

Having new kinds of (pre)symplectic and Poisson structures at our disposal, we may want to know if there is a notion of quantization for these, especially for examples coming from classical field theory. The theory is still in its infancy, but there has already been some progress:

- Deformation quantization of n-shifted Poisson structures has been discussed in [12], and makes sense for $n \geq -1$. It has been shown to exist whenever

$n > 0$ in [12]. The cases $n = -1$ (BV quantization, or rather BD quantization after [14]) and $n = 0$ (usual case) are the most difficult to deal with (we refer to Pridham's work [34, 35] for partial results).
- Geometric quantization has been initiated in a paper of James Wallbridge [43] (see also the end of [37]).

In [12, 33] there is a notion of compatible pair of an n-shifted presymplectic structure and an n-shifted Poisson structure. It would be interesting to investigate a notion of compatible pair of quantizations (deformation and geometric). Indeed, even in the nonderived situation, comparing the two kinds of quantizations is a relevant question.

References

[1] A. Alekseev, A. Malkin, and E. Meinreinken, *Lie group valued moment maps*, J. Differential Geom. 48 (1998) 445–95

[2] M. Alexandrov, M. Kontsevich, A. Schwarz, and O. Zaboronsky, *The geometry of the master equation and topological quantum field theory*, Int. J. Modern Phys. A 12 (1997) 1405–29

[3] M. Anel and G. Catren, eds., *New Spaces in Mathematics: Formal and Conceptual Reflections*, Cambridge University Press (2021)

[4] M. Atiyah and R. Bott, *The Yang–Mills equations over Riemann surfaces*, Philos. Trans. R. Soc. London, Series A 308 (1983) 523–615

[5] O. Ben-Bassat, C. Brav, V. Bussi, and D. Joyce, *A "Darboux Theorem" for shifted symplectic structures on derived Artin stacks, with applications*, Geom. Topol. 19 (2015) 1287–1359

[6] K. Behrend and B. Fantechi, *The intrinsic normal cone*, Inventiones math. 128 (1997) 45–88

[7] H. Bursztyn, M. Crainic, A. Weinstein, and C. Zhu, *Integration of twisted Dirac brackets*, Duke Math. J. 123 (2004) 549–607

[8] D. Calaque, *Three lectures on derived symplectic geometry and topological field theories*, Indagationes Math. 25 (2014) 926–47

[9] D. Calaque, *Lagrangian structures on mapping stacks and semi-classical TFTs*, Contemporary Math. 643 (2015)

[10] D. Calaque, *Shifted cotangent stacks are shifted symplectic*, Annales de la Faculté des Sciences de Toulouse 28 (2019) no. 1, 67–90

[11] D. Calaque and J. Grivaux, *Formal moduli problems and formal derived stacks*. Preprint, arXiv:1802.09556. To appear in DAGIT.

[12] D. Calaque, T. Pantev, B. Toën, M. Vaquié, and G. Vezzosi, *Shifted Poisson structures and deformation quantization*, J. Topol. 10 (2017) 483–584

[13] D. Calaque, R. Haugseng, and C. Scheimbauer, *The AKSZ construction in derived algebraic geometry as an extended topological quantum field theory*. Manuscript in preparation.

[14] K. Costello and O. Gwilliam, *Factorization Algebras in Quantum Field Theory*, vol. 2, Cambridge University Press (forthcoming)

[15] W. G. Dwyer and D. M. Kan, *Normalizing the cyclic modules of Connes*, Comment. Math. Helv. 60 (1985) 582–600

[16] D. Fiorenza, H. Sati, and U. Screiber, *A higher stacky perspective on Chern–Simons theory*, in *Mathematical Aspects of Quantum Field Theories*, Mathematical Physics Studies (2015) 153–211

[17] R. E. Grady and O. Gwilliam, *Lie algebroids as L_∞-spaces*, J. Inst. Math. Jussieu 19 (2020) 487–535

[18] R. Haugseng, *Iterated spans and classical topological field theories*, Math. Z. 289 (2018) 1427–88

[19] D. Joyce, *A classical model for derived critical loci*, J. Differential Geom. 101 (2015) 289–367

[20] D. Joyce and Y. Song, *A theory of generalized Donaldson–Thomas invariants*, Mem. AMS 1020 (2012) 199 pp.

[21] M. Kapranov, *Free Lie algebroids and the space of paths*, Selecta Math. (N.S.) 13 (2007) 277–319

[22] C. Kassel, *Cyclic homology, comodules and mixed complexes*, J. Alg. 107 (1987) 195–216

[23] Y.-H. Kiem and J. Li, *Categorification of Donaldson–Thomas invariants via perverse sheaves*, Preprint, arXiv:1212.6444v5

[24] A. Kock, *Synthetic Differential Geometry*, 2nd ed., London Math. Soc. Lecture Notes 333 (2006)

[25] J. Lurie, *Derived algebraic geometry V: Structured spaces*, Preprint, www.math.harvard.edu/~lurie/papers/DAG-V.pdf

[26] J. E. Marsden and A. Weinstein, *Reduction of symplectic manifolds with symmetry*, Rep. Math. Phys. 5 (1974) 121–30

[27] R. A. Mehta and X. Tang, *Constant symplectic 2-groupoids*, Lett. Math. Phys. 108 (2018) 1203–23

[28] E. Meinreken, *Equivariant cohomology and the Cartan model*, in *Encyclopedia of Mathematical Physics*, Elsevier (2006)

[29] I. Moerdijk and G. E. Reyes, *Models for Smooth Infinitesimal Analysis*, Springer (1991)

[30] J. Nuiten, *Koszul duality for Lie algebroids*, Advances in Mathematics 354 (2019) 106750

[31] T. Pantev, B. Toën, M. Vaquié, and G. Vezzosi, *Shifted symplectic structures*, Publ. math. l'IHÉS 117 (2013) 271–328

[32] M. Porta, *The derived Riemann–Hilbert correspondence*, Preprint, arXiv:1703.03907

[33] J. P. Pridham, *Shifted symplectic and Poisson structures on derived N-stacks*, J. Topol. 10 (2017) 178–210

[34] J. P. Pridham, *Deformation quantisation for (-1)-shifted symplectic structures and vanishing cycles*, Algebraic Geometry 6 (2019) 747–79

[35] J. P. Pridham, *Deformation quantisation for unshifted symplectic structures on derived Artin stacks*, Selecta Math. 24 (2018) 3027–59

[36] J. Robbin and D. Salamon, *Phase functions and path integrals*, in D. Salamon, ed., *Symplectic Geometry*, LMS Lecture Notes Series 192, Cambridge University Press (1993) 203–26
[37] P. Safronov, *Quasi-Hamiltonian reduction via classical Chern–Simons theory*, Adv. Math. 287 (2016) 733–73
[38] P. Safronov, *Poisson reduction as a coisotropic intersection*, Higher Structures 1 (2017) 87–121
[39] P. Safronov, *Poisson-Lie structures as shifted Poisson structures*, Preprint, arXiv:1706.02623
[40] U. Schreiber, *Differential Cohomology in a Cohesive ∞-Topos*, https://ncatlab.org/schreiber/files/dcct161227.pdf
[41] C. Simpson, *Homotopy over the complex numbers and generalized de Rham cohomology*, in M. Maruyama, ed., *Moduli of Vector Bundles*, Dekker (1996) 229–63
[42] B. Toën and G. Vezzosi, *Homotopical algebraic geometry II: geometric stacks and applications*, Mem. AMS 902 (2008) 224 pp.
[43] J. Wallbridge, *Derived smooth stacks and prequantum categories*, Preprint, arXiv:1610.00441
[44] A. Weinstein, *Symplectic geometry*, Bull. AMS 5 (1981) 1–13
[45] A. Weinstein, *Symplectic groupoids and Poisson manifolds*, Bull. Amer. Math. Soc. (N.S.) 16 (1987) 101–4
[46] P. Xu, *Moment maps and Morita equivalences*, J. Differential Geom. 67 (2004) 289–333
[47] C. Zhu, *Lie n-groupoids ans stacky Lie groupoids*, Preprint, arXiv:math/0609420

Damien Calaque
Université de Montpellier
damien.calaque@umontpellier.fr

5
Higher Prequantum Geometry

Urs Schreiber*

Contents

1	Prequantum Field Theory	202
2	Examples of Prequantum Field Theory	234
References		271

1 Prequantum Field Theory

The geometry that underlies the physics of Hamilton and Lagrange's classical mechanics and classical field theory has long been identified: this is *symplectic geometry* [6] and *variational calculus on jet bundles* [3, 72]. In these theories, configuration spaces of physical systems are differentiable manifolds, possibly infinite-dimensional, and the physical dynamics is all encoded by way of certain globally defined differential forms on these spaces.

But fundamental physics is of course of quantum nature, to which classical physics is but an approximation that applies at nonmicroscopic scales. Of what mathematical nature are systems of quantum physics?

1.1 The Need for Prequantum Geometry

A sensible answer to this question is given by algebraic deformation theory. One considers a deformation of classical physics to quantum physics by

---

* Section 1 is based on joint work with Igor Khavkine. Section 2 is based on joint work with Domenico Fiorenza and Hisham Sati.

deforming a Poisson bracket to the commutator in a noncommutative algebra, or by deforming a classical measure to a quantum BV operator.

However, this tends to work only perturbatively, in the infinitesimal neighborhood of classical physics, expressed in terms of formal (possibly nonconverging) power series in Planck's constant $\hbar$.[1]

There is a genuinely nonperturbative mathematical formalization of quantization, called *geometric quantization* [10, 61, 98, 99]. A key insight is that before genuine quantization even applies, there is to be a *prequantization* step[2] in which the classical geometry is supplemented by *global* coherence data. For actions of global gauge groups, this is also known as cancellation of *classical anomalies* [6, Section 5.A].

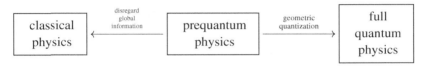

The archetypical example of prequantization is *Dirac charge quantization* [25], [40, Section 5.5], [42]. The classical mechanics of an electron propagating in an electromagnetic field on a spacetime X is all encoded in a differential 2-form on X, called the Faraday tensor F, which encodes the classical Lorentz force that the electromagnetic field exerts on the electron. But this data is insufficient for passing to the quantum theory of the electron: locally, on a coordinate chart U, what the quantum electron really couples to is the "*vector potential*," a differential 1-form A_U on U, such that $dA_U = F|_U$. But globally such a vector potential may not exist. Dirac realized[3] that what it takes to define

---

[1] A detailed introduction to perturbative quantum field theory formulated along the lines discussed here and quantized via formal deformation quantization is in [89]. In particular the relation between Lagrangian densities and symplectic structures on the induced covariant phase of the field theory, which plays some role below, is discussed in more detail there, following [56].

[2] The step of prequantization is that of choosing a complex line bundle whose curvature is the given symplectic form, called the *prequantum line bundle*. Given this, genuine quantization is the further choice of a *polarization* (this is locally a choice of "canonical coordinates" and "canonical momenta") and a choice of a subalgebra of prequantum operators respecting the polarization. Then the Hilbert space of quantum states is the space of polarized sections of the prequantum line bundle and the quantum observables are the chosen prequantum operators acting on this Hilbert space.

[3] Dirac considered this in the special case where spacetime is the complement in four-dimensional Minkowski spacetime of the worldline of a magnetic point charge. The homotopy type of this

the quantized electron globally is, in modern language, a lift of the locally defined vector potentials to a connection on a ($\mathbb{R}/\mathbb{Z}$)-principal bundle over spacetime. The first Chern class of this principal bundle is quantized, and this is identified with the quantization of the magnetic charge whose induced force the electron feels, this is known as *Dirac charge quantization* [25, 42]. This quantization effect, which needs to be present before the quantization of the dynamics of the electron itself even makes sense globally, is an example of *prequantization*.

A variant of this example occupies particle physics these days. As we pass attention from electrons to quarks, these couple to the weak and strong nuclear force, and this coupling is, similarly, locally described by a 1-form A_U, but now with values in a Lie algebra $\mathfrak{su}(n)$, from which the strength of the nuclear force field is encoded by the 2-form $F|_U := dA_U + \frac{1}{2}[A_U \wedge A_U]$. For the consistency of the quantization of quarks, notably for the consistent global definition of *Wilson loop observables*, this local data must be lifted to a connection on a $SU(n)$-principal bundle over spacetime. The second Chern class of this bundle is quantized, and is physically interpreted as the number of *instantons*.[4] In the physics literature instantons are expressed via Chern–Simons 3-forms, mathematically these constitute the prequantization of the 4-form $\text{tr}(F \wedge F)$ to a 2-gerbe with 2-connection, more on this in a moment.[5]

The vacuum which we inhabit is filled with such instantons at a density of the order of one instanton per femtometer in every direction. (The precise quantitative theoretical predictions of this [90] suffer from an infrared regularization ambiguity, but numerical simulations demonstrate the phenomenon [47].) This "instanton sea" that fills spacetime governs the mass of the η'-particle (see [107] or [102]) as well as other nonperturbative chromodynamical phenomena, such as the quark-gluon plasma seen in experiment [96]. It is also at the heart of the standard hypothesis for the mechanism of primordial baryogenesis (see [77] or [48, 74]), the fundamental explanation of a universe filled with matter.

---

space is the 2-sphere and hence in this case principal connections may be exhibited by what in algebraic topology is called a *clutching construction*, and this is what Dirac described. What the physics literature knows as the "Dirac string" in this context is the ray whose complement gives one of the two hemispheres in the clutching construction.

[4] Strictly speaking, the term *instanton* refers to a principal connection that in addition to having nontrivial topological charge, in the sense of Dirac charge quantization [25, 42] also minimizes Euclidean energy. Here we are just concerned with the nontrivial topological charge, which in particular is insensitive to and independent of any "Wick rotation."

[5] Just like a connection on a complex line bundle is a structure that combines local differential 2-form data with topological twists in degree 2, so a higher gerbe with connection pairs the data of local differential forms of higher degrees with that of topological twists (topological charges) of higher degree.

Passing beyond experimentally observed physics, one finds that the qualitative structure of the standard model of particle physics coupled to gravity, namely, the structure of Einstein–Maxwell–Yang–Mills–Dirac–Higgs theory, follows naturally if one assumes that the one-dimensional worldline theories of particles such as electrons and quarks are, at very high energy, accompanied by higher dimensional worldvolume theories of fundamental objects called strings, membranes, and generally p-branes (e.g., [28]). While these are hypothetical as far as experimental physics goes, they are interesting examples of the mathematical formulation of field theory, and hence their study is part of mathematical physics, just as the study of the Ising model or ϕ^4-theory. These p-branes are subject to a higher analog of the Lorentz force, and this is subject to a higher analog of the Dirac charge quantization condition, again a prequantum effect for the worldvolume theory.

For instance the *strong CP-problem* of the standard model of particle physics has several hypothetical solutions; one is the presence of particles called *axions*. The discrete shift symmetry (Peccei–Quinn symmetry) that characterizes these may naturally be explained as the result of $\mathbb{R}/\mathbb{Z}$-brane charge quantization in the hypothetical case that axions are wrapped membranes [100, Section 6].

More generally, p-brane charges are not quantized (in the sense of Dirac charge quantization [25], [40, Section 5.5], [42]) in ordinary integral cohomology, but in generalized cohomology theories. For instance 1-branes (strings) are well known to carry charges whose quantization is in K-theory (see [42]). While the physical existence of fundamental strings remains hypothetical, since the boundaries of strings are particles this does impact on known physics, for instance on the quantization of phase spaces that are not symplectic but just Poisson [71].

Finally, when we pass from fundamental physics to low energy effective physics such as solid state physics, then prequantum effects control topological phases of matter. Indeed, symmetry protected topological phases are described at low energy by higher dimensional Wess–Zumino–Witten models [22], of the same kind as those hypothetical fundamental super p-brane models.

Worldvolume field theory	Prequantum effect
Electron	Dirac charge quantization, magnetic flux quantization
Quark	instantons, baryogenesis
p-Brane	brane charge quantization, axion shift symmetry

These examples show that prequantum geometry is at the heart of the description of fundamental physical reality as well as of effective physical reality (as seen at "low energy"). Therefore, before rushing to discuss the mathematics of quantum geometry proper,[6] it behooves us to first carefully consider the mathematics of prequantum geometry. This is what we do here.

If the prequantization of the Lorentz force potential 1-form A for the electron is a connection on a ($\mathbb{R}/\mathbb{Z}$)-principal bundle, then what is the prequantization of the Chern–Simons 3-form counting instantons, or of the higher Lorentz force potential $(p+1)$-form of a p-brane for higher p?

This question has no answer in traditional differential geometry. It is customary to consider it only after transgressing the $(p+1)$-forms down to 1-forms by splitting spacetime/worldvolume as a product $\Sigma = \Sigma_p \times [0, 1]$ of p-dimensional spatial slices with a time axis, and integrating the $(p+1)$-forms over Σ_p

Global in space	Local in spacetime
1-form A_1	$(p+1)$-form A_{p+1}
$\int_{[0,1]} A_1$	$\overline{\begin{array}{c} A_1 := \int_{\Sigma_p} A_{p+1} \\ \hline \text{fiber integration} \end{array}}$ $\int_{\Sigma_p \times [0,1]} A_{p+1}$

This transgression reduces $(p+1)$-dimensional field theory to one-dimensional field theory, hence to mechanics, on the moduli space of spatial field configurations. That one-dimensional field theory may be subjected to the traditional theory of prequantum mechanics.

But clearly this space/time decomposition is a brutal step for relativistic field theories. It destroys their inherent symmetry and makes their analysis hard. In physics this is called the "noncovariant" description of field theory, referring to covariance under application of diffeomorphisms. We need prequantum geometry for spacetime local field theory where $(p+1)$-forms may be prequantized by regarding them as connections on higher degree analogs of principal bundles. Where an ordinary principal bundle is a smooth manifold, hence a "smooth set," with certain extra structure ([89, Chapter 2]), a higher principal bundle needs to be a smooth homotopy type.

Prequantum bundle	
Global in space	Local in spacetime
Smooth set	smooth homotopy type

---

[6] In genuine quantum theory the observables are promoted to quantum operators acting on the Hilbert space of states given by the polarized sections of the prequantum line bundle.

The generalization of geometry to higher geometry, where sets – which may be thought of as homotopy 0-types – are generalized to homotopy p-types for higher p, had been envisioned in [46] and a precise general framework has eventually been obtained in [67]. This may be specialized to higher differential geometry [80], which is the context we will be using here. The description of prequantum field theory local in spacetime is related to the description of topological quantum field theory local-to-the-point known as "extended" or "multi-tiered" field theory (see [68] or [11]).

	Classical	Prequantum
Global in space	symplectic geometry	prequantum geometry
	classical mechanics	prequantum mechanics
Local in spacetime	jet bundle geometry	higher prequantum geometry
	classical field theory	prequantum field theory

Once we are in a context of higher geometry where higher prequantum bundles exist, several other subtleties fall into place.

Ingredient of variational calculus	New examples available in higher geometry
Spacetime	orbifolds
Field bundle	instanton sectors of gauge fields, integrated BRST complex
Prequantum bundle	global Lagrangians for WZW-type models

A well-kept secret of the traditional formulation of variational calculus on jet bundles is that it does not in fact allow to properly formulate global aspects of local gauge theory. Namely, the only way to make the fields of gauge theory be sections of a traditional field bundle is to fix the instanton number (Chern class) of the gauge field configuration. The gauge fields then are taken to be connections on that fixed bundle. One may easily see [86] that it is impossible to have a description of gauge fields as sections of a field bundle that is both local and respects the gauge principle. However, this *is* possible with a higher field bundle. Indeed, the natural choice of the field bundle for gauge fields has as typical fiber the smooth moduli stack of principal connections. Formulated this way, not only does the space of all field configurations then span all instanton sectors, but it also has the gauge transformations between gauge field configurations built into it.[7]

Moreover, in a context of higher geometry also spacetime itself is allowed to be a smooth homotopy type. This is relevant at least in some hypothetical

---

[7] This may be understood as an aspect of "background independence" of a field theory, in that a "background" choice of instanton sector need not be fixed.

models of fundamental physics, which require spacetime to be an *orbifold*. Mathematically, an orbifold is a special kind of Lie groupoid, which in turn is a special kind of smooth homotopy 1-type.

1.2 The Principle of Extremal Action – Comonadically

Most field theories of relevance in theory and in nature are *local Lagrangian field theories*.[8] This means that their equations of motion are partial differential equations obtained as Euler–Lagrange equations of a local variational principle. This is the modern incarnation of the time-honored *principle of least action* (really, of extremal action).

We review from [57] how this is formalized, from a category-theoretic point of view that will point the way to prequantum field theory in Section 1.4.

The kinematics of a field theory is specified by a smooth manifold Σ of dimension $(p + 1)$ and a smooth fiber bundle E over Σ (typically, but not necessarily, a vector bundle). A *field configuration* is a smooth section of E. If we think of Σ as being spacetime, then typical examples of fields are the electromagnetic field or the field of gravity. But we may also think of Σ as being the worldvolume of a particle (such as the electron in the above examples) or of a higher dimensional "brane" that propagates in a fixed background of such spacetime fields, in which case the fields are the maps that encode a given trajectory.

The dynamics of a field theory is specified by an *equation of motion*, a partial differential equation for such sections. Since differential equations are equations among all the derivatives of such sections, we consider the spaces that these form: the *jet bundle* $J_\Sigma^\infty E$ is the bundle over Σ whose fiber over a point $\sigma \in \Sigma$ is the space of sections of E over the infinitesimal neighborhood $\mathbb{D}_\sigma$ of that point:

$$\left\{ \begin{array}{c} J_\Sigma^\infty E \\ \nearrow \downarrow \\ * \xrightarrow{\sigma} \Sigma \end{array} \right\} \simeq \left\{ \begin{array}{c} E \\ \nearrow \downarrow \\ \mathbb{D}_\sigma \hookrightarrow \Sigma \end{array} \right\}$$

Therefore every section ϕ of E yields a section $j^\infty(\phi)$ of the jet bundle, given by ϕ and all its higher order derivatives.

---

[8] And those that are not tend to be holographic boundary theories of those that are. For instance the 2d chiral WZW model is not itself Lagrangian, but it is the holographic boundary theory of 3d Chern–Simons theory, which is Lagrangian (see, e.g., [44]). Similarly the 6d (2,0)-superconformal field theory is not Lagrangian, but by AdS/CFT duality, at least in a suitable sector, it is the holographic boundary theory of a 7d Chern–Simons theory which is Lagrangian [108].

Higher Prequantum Geometry

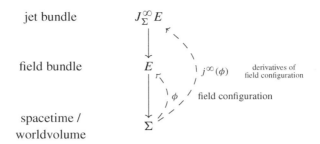

Accordingly, for E, F any two smooth bundles over Σ, then a bundle map

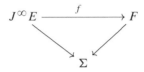

encodes a (nonlinear) differential operator $D_f : \Gamma_\Sigma(E) \longrightarrow \Gamma_\Sigma(F)$ by sending any section ϕ of E to the section $f \circ j^\infty(\phi)$ of F. Under this identification, the composition of differential operators $D_g \circ D_f$ corresponds to the *Kleisli-composite* of f and g, which is

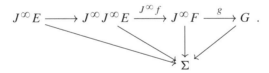

Here the first map is given by reshuffling derivatives and gives the jet bundle construction J_Σ^∞ the structure of a comonad – the *jet comonad*.

Differential operators are so ubiquitous in the present context that it is convenient to leave them notationally implicit and understand *every* morphism of bundles $E \longrightarrow F$ to designate a differential operator $D \colon \Gamma_\Sigma(E) \longrightarrow \Gamma_\Sigma(F)$. This is what we will do from now on. Mathematically this means that we are now in the co-Kleisli category $\mathrm{Kl}(J_\Sigma^\infty)$ of the jet comonad

$$\mathrm{DiffOp}_\Sigma \simeq \mathrm{Kl}(J_\Sigma^\infty).$$

For example the de Rham differential is a differential operator from sections of $\wedge^p T^* \Sigma$ to sections of $\wedge^{p+1} T^* \Sigma$ and hence now appears as a morphism of the form $d_H \colon \wedge^p T^* \Sigma \longrightarrow \wedge^{p+1} T^* \Sigma$. With this notation, a *globally defined local Lagrangian* for fields that are sections of some bundle E over spacetime/worldvolume Σ is simply a morphism of the form

$$L \colon E \longrightarrow \wedge^{p+1} T^* \Sigma.$$

Unwinding what this means, this is a function that at each point of Σ sends the value of field configurations and all their spacetime/worldvolume derivatives at that point to a $(p+1)$-form on Σ at that point. It is this pointwise local (in fact, infinitesimally local) dependence that the term *local* in *local Lagrangian* refers to.

Notice that this means that $\wedge^{p+1} T^*\Sigma$ serves the role of the *moduli space* of horizontal $(p+1)$-forms:

$$\Omega_H^{p+1}(E) = \mathrm{Hom}_{\mathrm{DiffOp}_\Sigma}(E, \wedge^{p+1} T^*\Sigma).$$

Regarding such L for a moment as just a differential form on $J_\Sigma^\infty(E)$, we may apply the de Rham differential to it. One finds that this uniquely decomposes as a sum of the form

$$dL = \mathrm{EL} - d_H(\Theta + d_H(\cdots)), \qquad (1.1)$$

for some Θ and for EL pointwise the pullback of a vertical 1-form on E; such a differential form is called a *source form*: $\mathrm{EL} \in \Omega_S^{p+1,1}(E)$. This particular source form is of paramount importance: the equation

$$\underset{v \in \Gamma(VE)}{\forall} j^\infty(\phi)^* \iota_v \mathrm{EL} = 0$$

on sections $\phi \in \Gamma_\Sigma(E)$ is a partial differential equation, and this is called the *Euler–Lagrange equation of motion* induced by L. Differential equations arising this way from a local Lagrangian are called *variational*.

A little reflection reveals that this is indeed a restatement of the traditional prescription of obtaining the Euler–Lagrange equations by locally varying the integral over the Lagrangian and then applying partial integration to turn all variation of derivatives (i.e., of jets) of fields into variation of the fields themselves. Here we do not consider this under the integral, and hence the boundary terms arising from the would-be partial integration show up as the contribution Θ.

We step back to say this more neatly. In general, a differential equation on sections of a bundle E is what characterizes the kernel of a differential operator. Now such kernels do not in general exist in the Kleisli category DiffOp_Σ of the jet comonad that we have been using, but (as long as it is nonsingular) it does exist in the full Eilenberg–Moore category $\mathrm{EM}(J_\Sigma^\infty)$ of jet-coalgebras. In fact, that category turns out (see [69] or [57]) to be equivalent to the category PDE_Σ whose objects are differential equations on sections of bundles, and whose morphisms are solution-preserving differential operators:

$$\mathrm{PDE}_\Sigma \simeq \mathrm{EM}(J_\Sigma^\infty).$$

Our original category of bundles with differential operators between them sits in PDE_Σ as the full subcategory on the trivial differential equations, those for which every section is a solution. This inclusion extends to (pre-)sheaves via left Kan extension; so we are now in the sheaf topos (see [53]) Sh(PDE$_\Sigma$).

And while source forms such as the Euler–Lagrange form EL are not representable in DiffOp$_\Sigma$, it is still true that for $f: E \longrightarrow F$ any differential operator then the property of source forms is preserved by precomposition with this map, hence we have the induced pullback operation on source forms: $f^*: \Omega_S^{p+1,1}(F) \longrightarrow \Omega_S^{p+1,1}(E)$. This means that source forms do constitute a presheaf on DiffOp$_\Sigma$, hence by left Kan extension an object in the topos over partial differential equations:

$$\mathbf{\Omega}_S^{p+1,1} \in \text{Sh}(\text{PDE}_\Sigma).$$

Therefore now the Yoneda lemma applies to say that $\mathbf{\Omega}_S^{p+1,1}$ is the moduli space for source forms in this context: a source form on E is now just a morphism of the form $E \longrightarrow \mathbf{\Omega}_S^{p+1,1}$. Similarly, the Euler variational derivative is now incarnated as a morphism of moduli spaces of the form $\mathbf{\Omega}_H^{p+1} \xrightarrow{\delta_V} \mathbf{\Omega}_S^{p+1,1}$, and applying the variational differential to a Lagrangian is now incarnated as the composition of the corresponding two modulating morphisms

$$\text{EL} := \delta_V L : E \xrightarrow{L} \mathbf{\Omega}_H^{p+1} \xrightarrow{\delta_V} \mathbf{\Omega}_S^{p+1,1}.$$

Finally, and that is the beauty of it, the Euler–Lagrange differential equation $\mathcal{E}$ induced by the Lagrangian L is now incarnated simply as the kernel of EL:[9]
$\mathcal{E} \xhookrightarrow{\text{ker(EL)}} E$.

In summary, from the perspective of the topos over partial differential equations, the traditional structure of local Lagrangian variational field theory is captured by the following diagram:

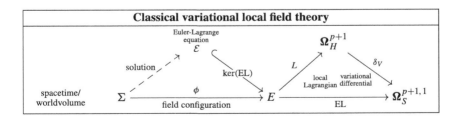

---

[9] That kernel always exists in the topos Sh(PDE$_\Sigma$), but it may not be representable by an actual sub*manifold* of $J_\Sigma^\infty E$ if there are singularities. Without any changes to the general discussion one may replace the underlying category of manifolds by one of "derived manifolds" formally dual to "BV-complexes," where algebras of smooth functions are replaced by higher homotopy-theoretic algebras, for instance by graded algebras equipped with a differential d_{BV}.

So far, all this assumes that there is a globally defined Lagrangian form L in the first place, which is not in fact the case for all field theories of interest. Notably it is in general not the case for field theories of higher WZW type. However, as the above diagram makes manifest, for the purpose of identifying the classical equations of motion, it is only the variational Euler differential $\text{EL} := \delta_V L$ that matters. But if that is so, the variation being a local operation, then we should still call equations of motion $\mathcal{E}$ *locally variational* if there is a cover $\{U_i \to E\}$ and Lagrangians on each patch of the cover $L : U_i \to \Omega_H^{p+1}$, such that there is a globally defined Euler–Lagrange form EL which restricts on each patch U_i to the variational Euler-derivative of L_i. Such *locally variational* classical field theory is discussed in [4, 29].

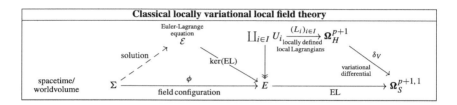

But when going beyond classical field theory, the Euler–Lagrange equations of motion $\mathcal{E}$ are not the end of the story. As one passes to the quantization of a classical field theory, there are further global structures on E and on $\mathcal{E}$ that are relevant. These are the action functional and the Kostant–Souriau prequantization of the covariant phase space. For these one needs to promote a patchwise system of local Lagrangians to a p-gerbe connection. This we turn to now.

1.3 The Global Action Functional – Cohomologically

For a globally defined Lagrangian $(p + 1)$-form L_{p+1} on the jet bundle of a given field bundle, the value of the action functional on a compactly supported field configuration ϕ is simply the integral

$$S(\phi) := \int_\Sigma j^\infty(\phi)^* L_{p+1}$$

of the Lagrangian, evaluated on the field configuration, over the spacetime/worldvolume Σ.

But when Lagrangian forms are only defined patchwise on a cover $\{U_i \to E\}_i$ as in the locally variational field theories mentioned in Section 1.2, then there is no way to globally make invariant sense of the action functional! As soon as sections pass through several patches, then making invariant sense of such an integral requires more data, in particular it requires more than just a compatibility condition of the locally defined Lagrangian forms on double intersections.

The problem of what exactly it takes to define global integrals of locally defined forms has long found a precise answer in mathematics, in the theory of ordinary differential cohomology. This has several equivalent incarnations, the one closest to classical constructions in differential geometry involves Čech cocycles: one first needs to choose on each intersection U_{ij} of two patches U_i and U_j a differential form $(\kappa_p)_{ij}$ of degree p, whose horizontal de Rham differential is the difference between the two Lagrangians restricted to that intersection

$$(L_{p+1})_j - (L_{p+1})_i = d_H(\kappa_p)_{ij} \quad \text{on } U_{ij}.$$

Then further one needs to choose on each triple intersection U_{ijk} a horizontal differential form $(\kappa_{p-1})_{ijk}$ of degree $p-1$ whose horizontal differential is the alternating sum of the relevant three previously defined forms:

$$(\kappa_p)_{jk} - (\kappa_p)_{ik} + (\kappa_p)_{ij} = d_H(\kappa_{p-1})_{ijk} \quad \text{on } U_{ijk}.$$

And so on. Finally on $(p+2)$-fold intersections one needs to choose smooth functions $(\kappa_0)_{i_0 \cdots i_{p+1}}$ whose horizontal differential is the alternating sum of $(p+2)$ of the previously chosen horizontal 1-forms, and, moreover, on $(p+3)$-fold intersections the alternating sum of these functions has to vanish. Such a tuple $(\{U_i\}; \{(L_{p+1})_i\}, \{(\kappa_p)_{ij}\}, \cdots)$ is a horizontal Čech–de Rham cocycle in degree $(p+2)$.

Given such, there is then a way to make sense of global integrals: one chooses a triangulation subordinate to the given cover, then integrates the locally defined Lagrangians $(L_{p+1})_i$ over the $(p+1)$-dimensional cells of the triangulation, integrates the gluing forms $(\kappa_p)_{ij}$ over the p-dimensional faces of these cells, the higher gluing forms $(\kappa_p)_{ijk}$ over the faces of these faces, and so on, and sums all this up. This defines a global action functional, which we may denote by

$$S(\phi) := \int_\Sigma j^\infty(\phi)^*(\{L_i\}, \{(\kappa_p)_{ij}\}, \cdots).$$

This horizontal Čech–de Rham cocycle data is subject to fairly evident coboundary relations (gauge transformations) that themselves are parameterized by systems $(\rho_\bullet)$ of $(p+1) - k$-forms on k-fold intersections:

$$L_i \mapsto L_i + d_H(\rho_p)_i$$
$$(\kappa_p)_{ij} \mapsto (\kappa_p)_{ij} + d_H(\rho_{p-1})_{ij} + (\rho_p)_j - (\rho_p)_i$$
$$\vdots$$

The definition of the global integral as above is preserved by these gauge transformations. This is the point of the construction: if we had only integrated the $(L_{p+1})_i$ over the cells of the triangulation without the contributions of the gluing forms $(\kappa_\bullet)$, then the resulting sum would not be invariant under the operation of shifting the Lagrangians by horizontally exact terms ("total derivatives") $L_i \mapsto L_i + d_H \rho_i$.

It might seem that this solves the problem. But there is one more subtlety: if the action functional takes values in the real numbers, then the functions assigned to $(p+2)$-fold intersections of patches are real valued, and then one may show that there exists a gauge transformation as above that collapses the whole system of forms back to one globally defined Lagrangian form after all. In other words: requiring a globally well-defined $\mathbb{R}$-valued action functional forces the field theory to be globally variational, and hence rules out all locally variational field theories, such as those of higher WZW type.

But there is a simple way to relax the assumptions such that this restrictive conclusion is evaded. Namely, we may pick a discrete subgroup $\Gamma \hookrightarrow \mathbb{R}$ and relax the condition on the functions $(\kappa_0)_{i_0 \cdots i_{p+1}}$ on $(p+2)$-fold intersections to the demand that on $(p+3)$-fold intersections their alternating sum vanishes only modulo Γ. A system of $(p+2) - k$-forms on k-fold intersections with functions regarded modulo Γ this way is called a (horizontal) $\mathbb{R}/\Gamma$-Čech–Deligne cocycle in degree$(p+2)$.

For instance for field theories of WZW-type, as above, we may take Γ to be the discrete group of periods of the closed form ω. Then one may show that a lift of ω to a Čech–Deligne cocycle of local Lagrangians with gluing data always exists. Indeed in general more than one inequivalent lift exists. The choice of these lifts is a choice of prequantization.

However, modding out a discrete subgroup Γ this ways also affects the induced global integral: that integral itself is now only defined modulo the subgroup Γ:

$$S(\phi) = \int_\Sigma j^\infty(\phi)^*(\{(L_{(p+1)})_i\}, \{(\kappa_p)_{ij}\}, \cdots) \quad \in \mathbb{R}/\Gamma.$$

Now, there are not that many discrete subgroups of $\mathbb{R}$. There are the subgroups isomorphic to the integers, and then there are dense subgroups, which make the quotient $\mathbb{R}/\Gamma$ ill behaved. Hence we focus on the subgroup of integers.

The space of group inclusions $i: \mathbb{Z} \hookrightarrow \mathbb{R}$ is parameterized by a nonvanishing real number $2\pi\hbar \in \mathbb{R} - \{0\}$, given by $i: n \mapsto 2\pi\hbar n$. The resulting quotient $\mathbb{R}/_\hbar \mathbb{Z}$ is isomorphic to the circle group $SO(2) \simeq U(1)$, exhibited by the short exact exponential sequence

$$0 \longrightarrow \mathbb{Z} \xhookrightarrow{2\pi\hbar(-)} \mathbb{R} \xrightarrow{\exp(\frac{i}{\hbar}(-))} U(1) \longrightarrow 0 \qquad (1.2)$$

Hence in the case that we take $\Gamma := \mathbb{Z}$, then we get locally variational field theories whose action functional is well defined modulo $2\pi\hbar$. Equivalently the exponentiated action functional is well defined as a function with values in $U(1)$:

$$\exp(\tfrac{i}{\hbar} S(\phi)) \in U(1) \simeq \mathbb{R}/_\hbar \mathbb{Z}.$$

The appearance of Planck's constant $\hbar$ here signifies that requiring a locally variational classical field theory to have a globally well-defined action functional is related to preparing it for quantization. Indeed, if we consider the above discussion for $p = 0$, then the above construction reproduces equivalently Kostant–Souriau's concept of geometric *prequantization*. Accordingly we may think of the Čech–Deligne cocycle data $(\{U_i\}; \{(L_{p+1})_i\}, \{(\kappa_p)_{ij}\}, \cdots)$ for general p as encoding *higher prequantum geometry*.

Coming back to the formulation of variational calculus in terms of diagrammatics in the sheaf topos $\mathrm{Sh}(\mathrm{PDE}_\Sigma)$ as in Section 1.2, what we, therefore, are after is a context in which the moduli object $\mathbf{\Omega}_H^{p+1}$ of globally defined horizontal $(p+1)$-forms may be promoted to an object which we are going to denote $\mathbf{B}_H^{p+1}(\mathbb{R}/_\hbar \mathbb{Z})_{\mathrm{conn}}$ and which modulates horizontal Čech–Deligne cocycles, as above.

Standard facts in homological algebra and sheaf cohomology say that in order to achieve this we are to pass from the category of sheaves on PDE_Σ to the "derived category" over PDE_Σ. We may take this to be the category of chain complexes of sheaves, regarded as a homotopy theory by understanding that a morphism of sheaves of chain complexes that is locally a quasi-isomorphism counts as a weak equivalence. In fact we may pass a bit further. Using the Dold–Kan correspondence to identify chain complexes in nonnegative degree with simplicial abelian groups, hence with group objects in Kan complexes, we think of sheaves of chain complexes as special cases of sheaves of Kan complexes [11]:

$$\mathrm{Sh}(\mathrm{PDE}_\Sigma, \mathrm{ChainCplx}) \xrightarrow{\text{Dold-Kan}} \mathrm{Sh}(\mathrm{PDE}_\Sigma, \mathrm{KanCplx}) \simeq \mathrm{Sh}_\infty(\mathrm{PDE}_\Sigma).$$

In such a homotopy-theoretically enlarged context we find the sheaf of chain complexes that is the $(p+1)$-truncated de Rham complex with the integers included into the 0-forms:

$$\mathbf{B}_H^{p+1}(\mathbb{R}/\hbar\mathbb{Z})_{\mathrm{conn}} := [\mathbb{Z} \stackrel{2\pi\hbar}{\hookrightarrow} \mathbf{\Omega}_H^0 \stackrel{d_H}{\to} \mathbf{\Omega}_H^2 \stackrel{d_H}{\to} \cdots \stackrel{d_H}{\to} \mathbf{\Omega}_H^{p+1}].$$

This chain complex of sheaves is known as the (horizontal) *Deligne complex* in degree $(p+2)$. The horizontal Čech–Deligne cocycles that we saw before are exactly the cocycles in the sheaf hypercohomology with coefficients in the horizontal Deligne complex. Diagrammatically in $\mathrm{Sh}_\infty(\mathrm{PDE}_\Sigma)$ these are simply morphisms $\mathbf{L}: E \to \mathbf{B}^{p+1}(\mathbb{R}/\hbar\mathbb{Z})$ from the field bundle to the Deligne moduli:

$$\{(\{U_i\}, \{(L_{p+1})_i\}, \{(\kappa_p)_{ij}\}, \cdots)\} \simeq \left\{ E \xrightarrow{\mathbf{L}} \mathbf{B}_H^{p+1}(\mathbb{R}/\hbar\mathbb{Z})_{\mathrm{conn}} \right\}.$$

This is such that a smooth homotopy between two maps to the Deligne moduli is equivalently a coboundary of Čech cocycles:

$$\left\{ \begin{array}{c} (\{U_i\}, \{(L_{p+1})_i\}, \{(\kappa_p)_{ij}\}, \cdots) \\ \Big\downarrow {\scriptstyle (\{U_i\}, \{(\rho_p)_i\}, \{(\rho_{p-1})_{ij}\}, \cdots)} \\ (\{U_i\}, \{(L_{p+1})_i + d_h(\rho_p)_i\}, \{(\kappa_p)_{ij} + d_H(\rho_{p-1})_{ij} + (\rho_p)_j - (\rho_p)_i\}, \cdots) \end{array} \right\}$$

$$\simeq \left\{ E \begin{array}{c} \mathbf{L} \\ \Downarrow \\ \mathbf{L}' \end{array} \mathbf{B}_H^{p+1}(\mathbb{R}/\hbar\mathbb{Z}) \right\}$$

Evidently, the diagrammatics serves as a considerable compression of data. In the following all diagrams we display are filled with homotopies as on the right above, even if we do not always make them notationally explicit.

There is an evident morphism $\mathbf{\Omega}_H^{p+1} \longrightarrow \mathbf{B}_H^{p+1}(\mathbb{R}/\hbar\mathbb{Z})_{\mathrm{conn}}$ which includes the globally defined horizontal forms into the horizontal Čech–Deligne cocycles (regarding them as Čech–Deligne cocycles with all the gluing data $(\kappa_\bullet)$ vanishing). This morphism turns out to be the analog of a covering map in traditional differential geometry; it is an atlas of smooth stacks:

Atlas of a smooth manifold	Atlas of a smooth ∞-groupoid
$\coprod_i U_i$ ↓↓ E	Ω_H^{p+1} ↓↓ $\mathbf{B}_H^{p+1}(\mathbb{R}/\hbar\mathbb{Z})_{\mathrm{conn}}$

Via this atlas, the Euler variational differential δ_V on horizontal forms that we have seen in Section 1.2 extends to horizontal Deligne coefficients to induce a curvature map on these coefficients.

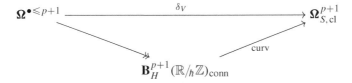

A *prequantization* of a source form EL is a lift through this curvature map, hence a horizontal Čech–Deligne cocycle of locally defined local Lagrangians for EL, equipped with gluing data:

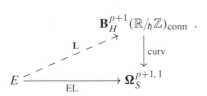

Hence in conclusion we find that in the ∞-topos $\mathrm{Sh}_\infty(\mathrm{PDE}_\Sigma)$ the diagrammatic picture of prequantum local field theory is this:

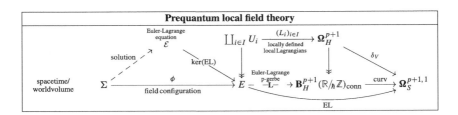

In summary, comparing this to the diagrammatics for variational and locally variational classical field theory which we discussed in Section 1.2, we have the following three levels of description of local Lagrangian field theory:

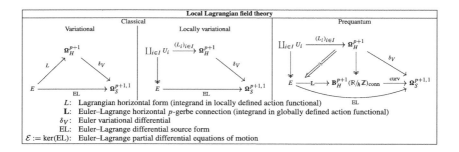

L: Lagrangian horizontal form (integrand in locally defined action functional)
L: Euler–Lagrange horizontal p-gerbe connection (integrand in globally defined action functional)
δ_V: Euler variational differential
EL: Euler–Lagrange differential source form
$\mathcal{E} := \ker(\mathrm{EL})$: Euler–Lagrange partial differential equations of motion

1.4 The Covariant Phase Space – Transgressively

The Euler–Lagrange p-gerbes discussed above are singled out as being exactly the right coherent refinement of locally defined local Lagrangians that may be integrated over a $(p+1)$-dimensional spacetime/worldvolume to produce a *function*, the action functional. In a corresponding manner, there are further refinements of locally defined Lagrangians by differential cocycles that are adapted to integration over submanifolds of Σ_{p+1} of positive codimension. In codimesion k these will yield not functions, but $(p-k)$-gerbes.

We consider this now for codimension 1 and find the covariant phase space of a locally variational field theory equipped with its canonical (pre-)symplectic structure and equipped with a Kostant–Souriau prequantization of that.

First consider the process of transgression in general codimension.

Given a smooth manifold Σ, then the mapping space $[\Sigma, \mathbf{\Omega}^{p+2}]$ into the smooth moduli space of $(p+2)$-forms is the smooth space defined by the property that for any other smooth manifold U, there is a natural identification

$$\left\{ U \longrightarrow [\Sigma, \mathbf{\Omega}^{p+2}] \right\} \simeq \Omega^{p+2}(U \times \Sigma)$$

of smooth maps into the mapping space with smooth $(p+2)$-forms on the product manifold $U \times \Sigma$.

Now suppose that $\Sigma = \Sigma_d$ is an oriented closed smooth manifold of dimension d. Then there is the fiber integration of differential forms on $U \times \Sigma$ over Σ (e.g., [14]), which gives a map

$$\int_{(U \times \Sigma_d)/U} : \Omega^{p+2}(U \times \Sigma_d) \longrightarrow \Omega^{p+2-d}(U).$$

This map is natural in U, meaning that it is compatible with pullback of differential forms along any smooth function $U_1 \to U_2$. This property is precisely what is summarized by saying that the fiber integration map constitutes a morphism in the sheaf topos of the form

$$\int_\Sigma : [\Sigma_d, \mathbf{\Omega}^{p+2}] \longrightarrow \mathbf{\Omega}^{p+2-d} .$$

This provides an elegant means to speak about transgression. Namely, given a differential form $\alpha \in \Omega^{p+2}(X)$ (on any smooth space X) modulated by a morphism $\alpha : X \longrightarrow \mathbf{\Omega}^{p+2}$, then its transgression to the mapping space $[\Sigma, X]$ is simply the form in $\Omega^{p+2-d}([\Sigma, X])$ which is modulated by the composite

$$\int_\Sigma [\Sigma, -] : [\Sigma, X] \xrightarrow{[\Sigma, \alpha]} [\Sigma, \mathbf{\Omega}^{p+2}] \xrightarrow{\int_\Sigma} \mathbf{\Omega}^{p+2-d}$$

of the fiber integration map above with the image of α under the functor $[\Sigma, -]$ that forms mapping spaces out of Σ.

Moreover, this statement has a prequantization [35, Section 2.8]: the fiber integration of curvature forms lifts to a morphism of differential cohomology coefficients

$$\int_\Sigma : [\Sigma, \mathbf{B}^{p+1}(\mathbb{R}/\hbar\mathbb{Z})_{\mathrm{conn}}] \longrightarrow \mathbf{B}^{p+1-d}(\mathbb{R}/\hbar\mathbb{Z})_{\mathrm{conn}}$$

and hence the transgression of a p-gerbe $\nabla : X \longrightarrow \mathbf{B}^{p+1}(\mathbb{R}/\hbar\mathbb{Z})_{\mathrm{conn}}$ (on any smooth space X) to the mapping space $[\Sigma, X]$ is given by the composite $\int_\Sigma \circ [\Sigma, -]$

$$\int_\Sigma [\Sigma, -] : [\Sigma, X] \xrightarrow{[\Sigma, \nabla]} [\Sigma, \mathbf{B}^{p+1}(\mathbb{R}/\hbar\mathbb{Z})_{\mathrm{conn}}] \xrightarrow{\int_\Sigma} \mathbf{B}^{p+1-d}(\mathbb{R}/\hbar\mathbb{Z})_{\mathrm{conn}} .$$

All this works verbatim also in the context of PDEs over Σ. For instance if $L : E \longrightarrow \mathbf{\Omega}_H^{p+1}$ is a local Lagrangian on (the jet bundle of) a field bundle E over Σ_{p+1} as before, then the action functional that it induces, as in Section 1.3, is the transgression to Σ_{p+1}:

$$S : [\Sigma, E]_\Sigma \xrightarrow{[\Sigma, L]_\Sigma} [\Sigma, \mathbf{\Omega}_H^{p+1}]_\Sigma \xrightarrow{\int_\Sigma} \mathbf{\Omega}^0 .$$

But now the point is that we have the analogous construction in higher codimension k, where the Lagrangian does not integrate to a function (a differential 0-form) but to a differential k-form. And all this goes along with passing from globally defined differential forms to Čech–Deligne cocycles.

To apply this for codimension $k = 1$, consider now p-dimensional submanifolds $\Sigma_p \hookrightarrow \Sigma$ of spacetime/worldvolume. We write $N_\Sigma^\infty \Sigma_p$ for the infinitesimal normal neighborhood of Σ_p in Σ. In practice one is often, but not necessarily, interested in Σ_p being a *Cauchy surface*, which means that the induced restriction map

$$[\Sigma_{p+1}, \mathcal{E}] \longrightarrow [N_\Sigma^\infty \Sigma_p, \mathcal{E}]$$

(from field configurations solving the equations of motion on all of Σ to normal jets of solutions on Σ_p) is an equivalence. An element in the solution space $[\Sigma_{p+1}, \mathcal{E}]$ is a *classical state* of the physical system that is being described, a classical trajectory of a field configuration over all of spacetime. Its image in $[\Sigma_{p+1}, \mathcal{E}]$ is the restriction of that field configuration and of all its derivatives to Σ_p.

In many – but not in all – examples of interest, classical trajectories are fixed once their first-order derivatives over a Cauchy surface is known. In these cases the phase space may be identified with the cotangent bundle of the space of field configurations on the Cauchy surface

$$[N_\Sigma^\infty \Sigma_p, \mathcal{E}] \simeq T^*[\Sigma_p, E].$$

The expression on the right is often taken as the definition of *phase spaces*. But since the equivalence with the left-hand side does not hold generally, we will not restrict attention to this simplified case and instead consider the solution space $[\Sigma, \mathcal{E}]_\Sigma$ as the phase space. To emphasize this more general point of view, one sometimes speaks of the *covariant phase space*. Here "covariance" refers to invariance under the action of the diffeomorphism group of Σ, meaning here that no space/time split in the form of a choice of Cauchy surface is made (or necessary) to define the phase space, even if a choice of Cauchy surface is possible and potentially useful for parameterizing phase space in terms of initial value data.

Now it is crucial that the covariant phase space $[\Sigma, \mathcal{E}]_\Sigma$ comes equipped with further geometric structure which remembers that this is not just any old space, but the space of solutions of a locally variational differential equation.

To see how this comes about, let us write $(\boldsymbol{\Omega}_{\mathrm{cl}}^{p+1})_\Sigma$ for the moduli space of all closed $p+1$-forms on PDEs. This is to mean that if E is a bundle over Σ, and regarded as representing the space of solutions of the trivial PDE on sections of E, then morphisms $E \longrightarrow (\boldsymbol{\Omega}^{p+1})_\Sigma$ are equivalent to closed differential $(p+2)$-forms on the jet bundle of E:

$$\left\{ E \longrightarrow (\boldsymbol{\Omega}^{p+1})_\Sigma \right\} \simeq \Omega_{\mathrm{cl}}^{p+1}(J_\Sigma^\infty E).$$

The key now is that there is a natural filtration on these differential forms adapted to spacetime codimension. This is part of a bigrading structure on differential forms on jet bundles known as the *variational bicomplex* [3]. In its low stages it looks as follows:

$$(\Omega^{p+1})_\Sigma \longrightarrow \cdots \longrightarrow \Omega_S^{p+1,1} \oplus \Omega^{p,2} \longrightarrow \Omega_S^{p+1,1}.$$

The lowest item here is what had concerned us in Sections 1.2 and 1.3; it is the moduli of $p+2$-forms which have $p+1$ of their legs along spacetime/worldvolume Σ and whose remaining vertical leg along the space of local field configurations depends only on the field value itself, not on any of its derivatives. This was precisely the correct recipient of the variational curvature, hence the variational differential of horizontal $(p+1)$-forms representing local Lagrangians.

But now that we are moving up in codimension, this coefficient will disappear, as these forms do not contribute when integrating just over p-dimensional hypersurfaces. The correct coefficient for that case is instead clearly $\Omega^{p,2}$, the moduli space of those $(p+2)$-forms on jet bundles which have p of their legs along spacetime/worldvolume, and the remaining two along the space of local field configurations. (There is a more abstract way to derive this filtration from first principles, and which explains why we have restriction to "source forms" (not differentially depending on the jets), indicated by the subscript, only in the bottom row. But for the moment we just take that little subtlety for granted.)

So $\Omega^{p,2}$ is precisely the space of those $(p+2)$-forms on the jet bundle that become (pre-)symplectic 2-forms on the space of field configurations once evaluated on a p-dimensional spatial slice Σ_p of spacetime Σ_{p+1}. We may think of this as a *current* on spacetime with values in 2-forms on fields.

Indeed, there is a *canonical* such *presymplectic current* for every locally variational field theory (see [103] or [56]). To see this, we ask for a lift of the purely horizontal locally defined Lagrangian L_i through the variational bicomplex to a $(p+1)$-form on the jet bundle whose curvature $d(L_i + \Theta_i)$ coincides with the Euler–Lagrange form $\mathrm{EL} = \delta_V L_i$ in vertical degree 1. Such a lift $L_i + \Theta_i$ is known as a *Lepage form* for L_i (e.g., [45, Section 2.1.2]).

Notice that it is precisely the restriction to the shell $\mathcal{E}$ that makes the Euler–Lagrange form EL_i disappear, by construction, so that only the new curvature component Ω_i remains as the curvature of the Lepage form on shell:

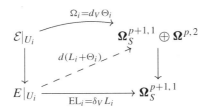

The condition means that the horizontal differential of Θ_i has to cancel against the horizontally exact part that appears when decomposing the differential of L_i as in equation (1.1). Hence, up to horizontal derivatives, this Θ_i is in fact uniquely fixed by L_i:

$$d(L_i + \Theta_i) = (\mathrm{EL}_i - d_H(\Theta_i + d_H(\cdots))) + (d_H\Theta_i + d_V\Theta_i)$$
$$= \mathrm{EL}_i + d_V\Theta_i$$
$$=: \mathrm{EL}_i + \Omega_i.$$

The new curvature component $\Omega \in \Omega^{p,2}(J_\Sigma^\infty E)$ whose restriction to patches is given this way, $\Omega|_{U_i} := d_V\Theta_i$, is known as the *presymplectic current* (see [103] or [56]). Because, by the way we found its existence, this is such that its transgression over a codimension 1 submanifold $\Sigma_p \hookrightarrow \Sigma$ yields a closed 2-form (a "presymplectic 2-form") on the covariant phase space:

$$\omega := \int_{\Sigma_p} [\Sigma_p, \Omega] \in \Omega^2([N_\Sigma^\infty \Sigma_p, \mathcal{E}]).$$

Since Ω is uniquely specified by the local Lagrangians L_i, this gives the covariant phase space canonically the structure of a presymplectic space $([\Sigma, \mathcal{E}], \omega)$. This is the reason why phase spaces in classical mechanics are given by (pre-)symplectic geometry as in [6].

Since Ω is a conserved current, the canonical presymplectic form ω is indeed canonical, it does not depend on the choice of (Cauchy) surface: if $\partial_{\mathrm{in}}\Sigma$ and $\partial_{\mathrm{out}}\Sigma$ are the incoming and outgoing Cauchy surfaces, respectively, in a piece of spacetime Σ, then the corresponding presymplectic forms agree:[10]

$$\omega_{\mathrm{out}} - \omega_{\mathrm{in}} = 0.$$

But by the discussion in Section 1.3, we do not just consider a locally variational classical field theory to start with, but a prequantum field theory. Hence in fact there is more data before transgression than just the new curvature components $d_V\Theta_i$, there is also Čech cocycle coherence data that glues the locally defined Θ_i to a globally consistent differential cocycle.

We write $\mathbf{B}_L^{p+1}(\mathbb{R}/\hbar\mathbb{Z})$ for the moduli space for such coefficients (with the subscript for "Lepage"), so that morphisms $E \longrightarrow \mathbf{B}_L^{p+1}(\mathbb{R}/\hbar\mathbb{Z})$ are equivalent to properly prequantized globally defined Lepage lifts of Euler–Lagrange

---

[10] If the shell $\mathcal{E}$ is taken to be resolved by a derived manifold/BV-complex as in Footnote 9, then any on-shell vanishing condition becomes vanishing up to a d_{BV}-exact term, hence then there is a 2-form ω_{BV} of BV-degeeree -1 such that $\omega_{\mathrm{out}} - \omega_{\mathrm{in}} = d_{\mathrm{BV}}\omega_{\mathrm{BV}}$. (In [18] this appears as BV-BFV axiom (9).) The Poisson bracket induced from this "shifted symplectic form" ω_{BV} is known as the "BV-antibracket" (e.g., [49]).

p-gerbes. In summary then, the refinement of an Euler–Lagrange p-gerbe **L** to a Lepage p-gerbe Θ is given by the following diagram:

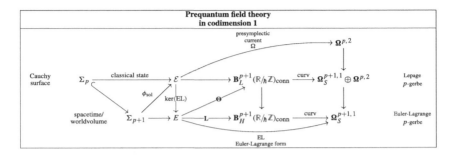

And now higher prequantum geometry bears fruit: since transgression is a natural operation, and since the differential coefficients $\mathbf{B}_H^{p+1}(\mathbb{R}/\hbar\mathbb{Z})_{\mathrm{conn}}$ and $\mathbf{B}_L^{p+1}(\mathbb{R}/\hbar\mathbb{Z})$ precisely yield the coherence data to make the local integrals over the locally defined differential forms L_i and Θ_i be globally well defined, we may now hit this entire diagram with the transgression functor $\int_{\Sigma_p} [N_\Sigma^\infty \Sigma_p, -]$ to obtain this diagram:

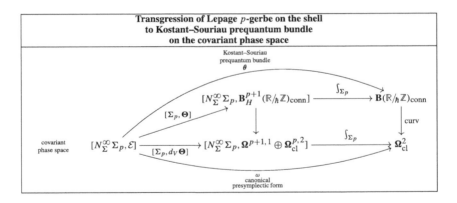

This exhibits the transgression

$$\theta := \int_{\Sigma_p} [N_\Sigma^\infty \Sigma_p, \Theta]$$

of the Lepage p-gerbe Θ as a $(\mathbb{R}/\hbar\mathbb{Z})$-connection whose curvature is the canonical presymplectic form.

But this $([\Sigma, \mathcal{E}]_\Sigma, \theta)$ is just the structure that Souriau originally called and demanded as a prequantization of the (pre)symplectic phase space $([\Sigma, \mathcal{E}]_\Sigma, \omega)$

[61, 98, 99]. Conversely, we see that the Lepage p-gerbe Θ is a "detransgression" of the Kostant–Souriau prequantization of covariant phase space in codimension 1 to a higher prequantization in full codimension. In particular, the higher prequantization constituted by the Lepage p-gerbe constitutes a compatible choice of Kostant–Souriau prequantizations of covariant phase space for *all* choices of codimension 1 hypersurfaces at once. This is a genuine reflection of the fundamental locality of the field theory, even if we look at field configurations globally over all of a (spatial) hypersurface Σ_p.

1.5 The Local Observables – Lie Theoretically

We discuss now how from the previous considerations naturally follow the concepts of local observables of field theories and of the Poisson bracket on them, as well as the concept of conserved currents and the variational Noether theorem relating them to symmetries. At the same time all these concepts are promoted to prequantum local field theory.

In Section 1.4 we have arrived at a perspective of prequantum local field theory where the input datum is a partial differential equation of motion $\mathcal{E}$ on sections of a bundle E over spacetime/worldvolume Σ and equipped with a prequantization exhibited by a factorization of the Euler–Lagrange form $E \xrightarrow{\text{EL}} \Omega_S^{p+1,1}$ and of the presymplectic current form $\mathcal{E} \xrightarrow{\Omega} \Omega^{p,2}$ through higher Čech–Deligne cocycles for an Euler–Lagrange p-gerbe $\mathbf{L}$ and for a Lepage p-gerbe Θ:

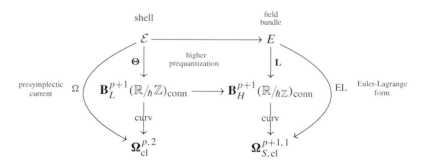

This local data then transgresses to spaces of field configurations over codimension k submanifolds of Σ. Transgressing to codimension 0 yields the globally defined exponentiated action functional

and transgressing to a codimension 1 (Cauchy) surface $\Sigma_p \hookrightarrow \Sigma$ yields the covariant phase space as a prequantized presymplectic manifold

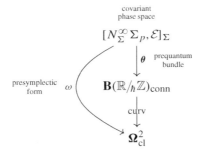

Given any space equipped with a map into some moduli space like this, an automorphism of this structure is a diffeomorphism of the space together with a homotopy that exhibits the preservation of the given map into the moduli space.

We consider now the automorphisms of the prequantized covariant phase space and of the Euler–Lagrange p-gerbe that it arises from via transgression, and find that these recover and make globally well defined the traditional concepts of symmetries and conserved currents, related by the Noether theorem, and of observables equipped with their canonical Poisson bracket.

The correct automorphisms of presymplectic smooth spaces $([\Sigma_p, \mathcal{E}]_\Sigma, \omega) \longrightarrow ([\Sigma_p, \mathcal{E}]_\Sigma, \omega)$ are of course diffeomorphisms $\phi \colon [\Sigma_p, \mathcal{E}]_\Sigma \longrightarrow [\Sigma_p, \mathcal{E}]_\Sigma$ such that the presymplectic form is preserved, $\phi^* \omega = \omega$. In the diagrammatics this means that ϕ fits into a triangle of this form:

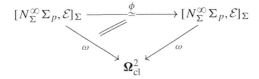

Viewed this way, there is an evident definition of an automorphism of a prequantization $([N_\Sigma^\infty \Sigma_p, \mathcal{E}]_\Sigma, \theta)$ of $([N_\Sigma^\infty \Sigma_p, \mathcal{E}]_\Sigma, \omega)$. This must be a diagram of the following form:

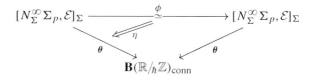

hence a diffeomorphism ϕ together with a homotopy η that relates the modulating morphism of the translated prequantum bundle back to the original prequantum bundle. By the discussion in Section 1.3, such homotopies are equivalently coboundaries between the Čech–Deligne cocycles that correspond to the maps that the homotopy goes between. Here this means that the homotopy in the above diagram is an isomorphism $\eta: \phi^*\theta \xrightarrow{\simeq} \theta$ of circle bundles with connection. These pairs (ϕ, η) are what Souriau called the *quantomorphisms*. Via their canonical action on the space of sections of the prequantum bundle, these become the quantum operators.

To see what this is in local data, consider the special case that θ is a globally defined 1-form and suppose that $\phi = \exp(tv)$ is the flow of a vector field v:

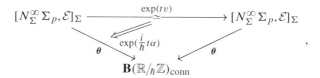

Then the homotopy filling the previous diagram is given by a smooth function $\exp(it\alpha)$ such that

$$\exp(tv)^*\theta - \theta = t\,d\alpha\,.$$

Infinitesimally, for $t \to 0$, this becomes

$$\mathcal{L}_v \theta = d\alpha\,.$$

Using Cartan's formula for the Lie derivative on the left, and the fact that $d\theta = \omega$, by prequantization, this is equivalent to

$$d\underbrace{(\alpha - \iota_v\theta)}_{H} = \iota_v\omega. \tag{1.3}$$

This is the classical formula [6] that says that

$$H := \alpha - \iota_v\theta$$

is a Hamiltonian for the vector field v.

There is an evident smooth group structure on the homotopies as above, and one checks that the induced Lie bracket on Hamiltonians H with Hamiltonian vector fields v is the following:

$$[(v_1, H_1), (v_2, H_2)] = ([v_1, v_2], \iota_{v_2}\iota_{v_1}\omega).$$

Traditionally this is considered only in the special case that ω is symplectic, hence equivalently, in the case that equation (1.3) uniquely associates a Hamiltonian vector field v with any Hamiltonian H. In that case we may identify a pair (v, H) with just H and then the above Lie bracket becomes the *Poisson bracket* on smooth functions induced by ω. Hence the Poisson bracket Lie algebra is secretly the infinitesimal symmetries of the prequantum line bundle θ. This is noteworthy. For instance in the example of the phase space $(T^*\mathbb{R} = \mathbb{R}^2, \omega = dp \wedge dq)$ and writing $q, p \colon \mathbb{R}^2 \to \mathbb{R}$ for the two canonical coordinates (p being called the "canonical momentum"), the Poisson bracket, as above, between these two is

$$\boxed{[q, p] = i\hbar \quad \in \quad i\hbar\mathbb{R} \hookrightarrow \mathfrak{Pois}(\mathbb{R}^2, dp \wedge dq).}$$

This equation is often regarded as the hallmark of quantum theory. In fact, it is a prequantum phenomenon. Notice how the identification of the central term with $i\hbar$ follows here from the first prequantization step back around equation (1.2).

From equation (1.3) it is clear that the Poisson bracket is a Lie extension of the Lie algebra of (Hamiltonian) vector fields by the locally constant Hamiltonians, hence by constant functions in the case that X is connected. The nontrivial Lie integration of this statement is the Kostant–Souriau extension, which says that the quantomorphism group of a connected phase space is a $U(1)$-extension of the diffeological group of Hamiltonian symplectomorphisms.

Hence in summary the situation for observables on the covariant phase space in codimension 1 is as follows:

	Kostant–Souriau extension (connected phase space)		Observables		Flows
Infinitesimally	$i\hbar \mathbb{R}$	$\longrightarrow$	$\mathfrak{Pois}([N_\Sigma^\infty \Sigma_p, \mathcal{E}]_\Sigma, \omega)$ Poisson bracket	$\longrightarrow$	$\mathbf{Vect}(X)$
Finitely	$U(1)$	$\longrightarrow$	**QuantMorph**$([N_\Sigma^\infty \Sigma_p, \mathcal{E}]_\Sigma, \theta)$ quantomorphism group	$\longrightarrow$	**Diff**(X)
Abstractly	$\left\{ \theta \left(\begin{array}{c} [N_\Sigma^\infty \Sigma_p, \mathcal{E}]_\Sigma \\ \text{locally} \\ \text{constant} \\ \text{Hamiltonian} \\ \mathbf{B}(\mathbb{R}/\hbar\mathbb{Z})_{\text{conn}} \end{array} \right) \theta \right\}$	$\longrightarrow$	[diagram: $[N_\Sigma^\infty \Sigma_p, \mathcal{E}]_\Sigma \xrightarrow[\text{Hamiltonian}]{\text{flow}} [N_\Sigma^\infty \Sigma_p, \mathcal{E}]_\Sigma$, θ, $\mathbf{B}(\mathbb{R}/\hbar\mathbb{Z})_{\text{conn}}$, ω, curv, Ω^2_{cl}]	$\longrightarrow$	$\left\{ [N_\Sigma^\infty \Sigma_p, \mathcal{E}] \xrightarrow{\simeq} [N_\Sigma^\infty \Sigma_p, \mathcal{E}]_\Sigma \right\}$

Generally, the symmetries of a p-gerbe connection $\nabla \colon X \to \mathbf{B}^{p+1}(\mathbb{R}/\hbar\mathbb{Z})_{\text{conn}}$ form an extension of the symmetry group of the underlying space by the higher group of flat $(p-1)$-gerbe connections [31]:

	Higher extension		Symmetry of p-gerbe $(p+1)$-connection		Automorphisms of base space
Infinitesimally	$\mathbf{Ch}_{\text{dR,cl}}^{\bullet \leq p}(X)$	$\longrightarrow$	$\mathfrak{sym}_X(F)$ stabilizer L_∞-algebra	$\longrightarrow$	$\mathbf{Vect}(X)$
Finitely	$\mathbf{Ch}_{\text{cl}}^{\bullet \leq p}(X, U(1))$	$\longrightarrow$	**Stab**$_{\mathbf{Aut}(X)}(\nabla)$ stabilizer ∞-group	$\longrightarrow$	**Aut**(X)
Abstractly	$\left\{ \nabla \left(\begin{array}{c} X \\ \simeq \end{array} \right) \nabla \right\}$ $\mathbf{B}^{p+1}(\mathbb{R}/\hbar\mathbb{Z})_{\text{conn}}$	$\longrightarrow$	[diagram: $X \xrightarrow{\text{automorphism}} X$, homotopy stabilization, ∇, ∇, $\mathbf{B}^{p+1}(\mathbb{R}/\hbar\mathbb{Z})_{\text{conn}}$, F, F, curv, Ω^{p+2}_{cl}]	$\longrightarrow$	$\left\{ V \xrightarrow{\simeq} V \right\}$

Specifying this general phenomenon to the Lepage p-gerbes, it gives a Poisson bracket L_∞-algebra on higher currents (local observables) [32] and its higher Lie integration to a higher quantomorphism group constituting a higher Kostant–Souriau extension of the differential automorphisms of the field bundle. This is determined by the (pre)symplectic current $p+2$-form Ω in analogy to how the ordinary Poisson bracket is determined by the

Higher Prequantum Geometry

(pre)symplectic 2-form ω, hence this is a Poisson L_∞-bracket for what has been called "multisymplectic geometry" (see [75]):

	Higher Kostant–Souriau extension		Symmetry of Lepage p-gerbe		Differential automorphisms of dynamical shell
Infinitesimally	$\mathbf{Ch}^{\bullet \leq p}_{\mathrm{dR,cl}}(\mathcal{E})$	$\rightarrow$	$\mathfrak{Pois}(\mathcal{E}, \Omega)$ Poisson bracket L_∞-algebra	$\rightarrow$	$\mathrm{Vect}(\mathcal{E})$
Finitely	$\mathbf{Ch}^{\bullet \leq p}_{\mathrm{cl}}(\mathcal{E}, U(1))$	$\rightarrow$	$\mathbf{Stab}_{\mathbf{Aut}(\mathcal{E})}(\Theta)$ quantomorphism ∞-group	$\rightarrow$	$\mathbf{Aut}(\mathcal{E})$
Abstractly	$\left\{ \begin{array}{c} \mathcal{E} \\ \Theta \xrightarrow{\text{topological Hamiltonian}} \Theta \\ \mathbf{B}^{p+1}_L(\mathbb{R}/\hbar\mathbb{Z})_{\mathrm{conn}} \end{array} \right\}$	$\rightarrow$	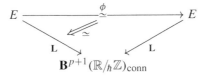	$\rightarrow$	$\left\{ \mathcal{E} \xrightarrow{\text{on-shell symmetry}}_{\simeq} \mathcal{E} \right\}$

So far, this has concerned the covariant phase space with its prequantization via the Lepage p-gerbe. In the same way, there are the higher symmetries of the field space with its prequantization via the Euler–Lagrange p-gerbes $\mathbf{L}$:

$$\begin{array}{ccc} E & \xrightarrow[\simeq]{\phi} & E \\ & \searrow_{\mathbf{L}} \quad \underset{\simeq}{\Downarrow} \quad \swarrow_{\mathbf{L}} & \\ & \mathbf{B}^{p+1}(\mathbb{R}/\hbar\mathbb{Z})_{\mathrm{conn}} & \end{array}$$

To see what these are in components, consider again the special case that $\mathbf{L}$ is given by a globally defined horizontal form, and consider a one-parameter flow of such symmetries

$$\begin{array}{ccc} E & \xrightarrow[\simeq]{\exp(tv)} & E \\ & \searrow_{\mathbf{L}} \quad \underset{\exp(\frac{i}{\hbar}t\Delta)}{\Downarrow} \quad \swarrow_{\mathbf{L}} & \\ & \mathbf{B}^{p+1}_H(\mathbb{R}/\hbar\mathbb{Z})_{\mathrm{conn}} & \end{array}$$

In Čech–Deligne cochain components, this diagram equivalently exhibits the equation

$$\exp(tv)^* L - L = t\, d_H \Delta$$

on differential forms on the jet bundle of E, where v is a vertical vector field. Infinitesimally for $t \to 0$, this becomes

$$\mathcal{L}_v L = d_H \Delta \,.$$

Since L is horizontal while v is vertical, the left-hand side reduces, by equation (1.1), to

$$\iota_v dL = \iota_v (\mathrm{EL} - d_H \Theta) \,.$$

Therefore the infinitesimal symmetry of **L** is equivalent to

$$d_H \underbrace{(\Delta - \iota_v \Theta)}_{J} = \iota_v \mathrm{EL} \,.$$

This says that associated to the symmetry v is a current

$$J := \Delta - \iota_v \Theta$$

that is conserved (horizontally closed) on shell (on the vanishing locus $\mathcal{E}$ of the Euler–Lagrange form EL). This is precisely the statement of the (first variational) Noether theorem. Indeed, in its modern incarnation (see [9, Section 3]), Noether's theorem is understood as stating a Lie algebra extension of the Lie algebra of symmetries by topological currents to the Lie–Dickey bracket on equivalence classes of conserved currents.

Hence the ∞-group extension of symmetries of the Euler–Lagrange p-gerbe promotes Noether's theorem to the statement that higher Noether currents form an L_∞-algebra extension of the infinitesimal symmetries by topological currents:

	Higher topological charge extension		Symmetry of Euler–Lagrange p-gerbe		Differential automorphisms of field bundle
Infinitesimally	$\mathbf{Ch}^{\bullet \leq p}_{\mathrm{dR,cl}}(E)$	$\to$	curr(E, EL) Dickey bracket current L_∞-algebra	$\to$	$\mathrm{Vect}(E)$
Finitely	$\mathbf{Ch}^{\bullet \leq p}_{\mathrm{cl}}(E, U(1))$	$\to$	$\mathrm{Stab}_{\mathrm{Aut}(E)}(\mathbf{L})$ de-transgressed Kac–Moody ∞-group	$\to$	$\mathrm{Aut}(E)$
Abstractly	$\left\{ \begin{array}{c} E \\ \mathbf{L} \swarrow \text{topological current} \searrow \mathbf{L} \\ \mathbf{B}^{p+1}_H (\mathbb{R}/\hbar\mathbb{Z})_{\mathrm{conn}} \end{array} \right\}$	$\to$	$\left\{ \begin{array}{c} E \xrightarrow{\text{variational symmetry}} E \\ \mathbf{L} \searrow \swarrow^{\text{Noether current}} \mathbf{L} \\ \mathrm{EL} \searrow \mathbf{B}^{p+1}_H(\mathbb{R}/\hbar\mathbb{Z})_{\mathrm{conn}} \swarrow \mathrm{EL} \\ \downarrow \text{curv} \\ \Omega^{p+1,1}_{S,\mathrm{cl}} \end{array} \right\}$	$\to$	$\left\{ E \xrightarrow{\text{symmetry}} E \right\}$

In summary, physical local observables arise from symmetries of higher prequantum geometry as follows:

Prequantum geometry	Automorphism up to homotopy	Lie derivative up to differential	Equivalently	Physical quantity
Prequantum shell	$\mathcal{E} \xrightarrow{\exp(tv)} \mathcal{E}$, $\Theta \xrightarrow{\exp(\frac{i}{\hbar}t\alpha)} \Theta$, $\mathbf{B}_L^{p+1}(\mathbb{R}/\hbar\mathbb{Z})$, $\Omega \to \Omega$, curv, $\Omega_{cl}^{p,2}$	$\mathcal{L}_v \Theta = d\alpha$	$\underbrace{d(\alpha - \iota_v\Theta)}_{H} = \iota_v\Omega$	Hamiltonian
Prequantum field bundle	$E \xrightarrow{\exp(tv)} E$, $L \xrightarrow{\exp(\frac{i}{\hbar}t\Delta)} L$, $\mathbf{B}_H^{p+1}(\mathbb{R}/\hbar\mathbb{Z})$, $EL \to EL$, curv, $\Omega_{S,cl}^{p+1,1}$	$\mathcal{L}_v L = d_H \Delta$	$\underbrace{d_H(\Delta - \iota_v\Theta)}_{J} = \iota_v\mathrm{EL}$	conserved current

1.6 The Evolution – Correspondingly

The transgression formula discussed in Section 1.4 generalizes to compact oriented d-manifolds Σ, possibly with boundary $\partial\Sigma \hookrightarrow \Sigma$. Here it becomes transgression *relative* to the boundary transgression.

For curvature forms this is again classical: for $\omega \in \Omega^{p+2}(\Sigma \times U)$ a closed differential form, then $\int_\Sigma \omega \in \Omega^{p+2-d}(U)$ is not in general a closed differential form anymore, but by Stokes's theorem its differential equals the boundary transgression:

$$d_U \int_\Sigma \omega = \int_\Sigma d_U \omega$$

$$= -\int_\Sigma d_\Sigma \omega$$

$$= -\int_{\partial\Sigma} \omega.$$

This computation also shows that a sufficient condition for the bulk transgression of ω to be closed and for the boundary transgression to vanish is that ω be also *horizontally* closed, that is, closed with respect to d_Σ.

Applied to the construction of the canonical presymplectic structure on phase spaces in Section 1.4 this has the important implication that the canonical presymplectic form on phase space is indeed canonical.

Namely, by equation (1.1), the presymplectic current $\Omega \in \Omega^{p,2}(E)$ is horizontally closed on shell, hence is indeed a conserved current:

$$d_H \Omega = d_H d_V \Theta$$
$$= -d_V d_H \Theta$$
$$= -d_V(-d_V L + \mathrm{EL})$$
$$= -d_V \mathrm{EL}.$$

It follows that if Σ is a spacetime/worldvolume with, say, two boundary components $\partial\Sigma = \partial_{\mathrm{in}}\Sigma \sqcup \partial_{\mathrm{out}}\Sigma$, then the presymplectic structures $\omega_{\mathrm{in}} := \int_{\partial_{\mathrm{in}}\Sigma}[\partial_{\mathrm{in}}\Sigma, \omega]$ and $\omega_{\mathrm{out}} := \int_{\partial_{\mathrm{out}}\Sigma}[\partial_{\mathrm{out}}\Sigma, \omega]$ agree on the covariant phase space:[11]

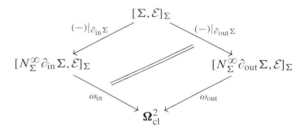

This diagram may be thought of as expressing an *isotropic correspondence* between the two phase spaces, where $[\Sigma, \mathcal{E}]_\Sigma$ is isotropic in the product of the two boundary phase spaces, regarded as equipped with the presymplectic form $\omega_{\mathrm{out}} - \omega_{\mathrm{in}}$. In particular, when both $\partial_{\mathrm{in}}\Sigma$ and $\partial_{\mathrm{out}}\Sigma$ are Cauchy surfaces in Σ, so that the two boundary restriction maps in the above diagram are in fact equivalences, then this is a *Lagrangian correspondence* in the sense of [105, 106].

All this needs to have and does have prequantization: The transgression of a p-gerbe $\nabla : X \to \mathbf{B}^{p+1}(\mathbb{R}/\hbar\mathbb{Z})_{\mathrm{conn}}$ to the bulk of a d-dimensional Σ is no longer quite a $p-d$-gerbe itself, but is a section of the pullback of the $p-d+1$-gerbe that is the transgression to the boundary $\partial\Sigma$. Diagrammatically this means that transgression to maps out of Σ is a homotopy filling a diagram of the following form:

---

[11] If one uses a BV-resolution of the covariant phase space, then they agree up to the BV-differential of a BV (-1)-shifted 2-form, we come back to this in Section 2.1.3.

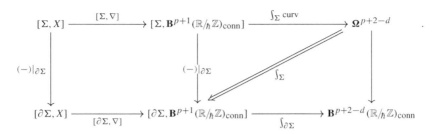

Here the appearance of the differential forms coefficients Ω^{p+2-d} in the top right corner witnesses the fact that the bulk term $\int_\Sigma [\Sigma, \nabla]$ is a trivialization of the pullback of the boundary gerbe $\int_{\partial\Sigma}[\partial\Sigma, \nabla]$ only as a plain gerbe, not necessarily as a gerbe with connection: in general the curvature of the pullback of $\int_{\partial\Sigma}[\partial\Sigma, \nabla]$ will not vanish, but only be exact, as in the above discussion, and the form that it is the de Rham differential of is expressed by the top horizontal morphism in the above diagram.

Hence in the particular case of the transgression of a Lepage p-gerbe to covariant phase space, this formula yields a prequantization of the above Lagrangian correspondence, where now the globally defined action functional

$$\exp(\tfrac{i}{\hbar} S) = \int_\Sigma [\Sigma, \Theta] = \int_\Sigma [\Sigma, \mathbf{L}]$$

exhibits the equivalence between the incoming and outgoing prequantum bundles

$$\theta_{\text{in/out}} = \int_{\partial_{\text{in/out}}\Sigma} [\partial_{\text{in/out}}\Sigma, \theta]$$

on covariant phase space:

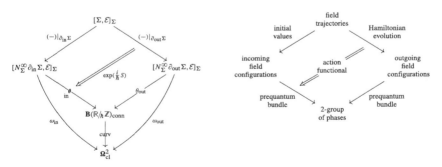

This *prequantized Lagrangian correspondence* hence reflects the prequantum evolution from fields on the incoming piece $\partial_{\text{in}}\Sigma$ of spacetime/worldvolume to the outgoing piece $\partial_{\text{out}}\Sigma$ via trajectories of field configurations along Σ.

2 Examples of Prequantum Field Theory

We now survey classes of examples of prequantum field theory in the sense of Section 1.

2.1 Gauge Fields

Modern physics rests on two fundamental principles. One is the *locality principle*; its mathematical incarnation in terms of differential cocycles on PDEs was the content of Section 1. The other is the *gauge principle*.

In generality, the gauge principle says that given any two field configurations ϕ_1 and ϕ_2 – and everything in nature is some field configuration – then it is physically meaningless to ask whether they are *equal*, instead one has to ask whether they are *equivalent* via a *gauge transformation*

There may be more than one gauge transformation between two field configurations, and hence there may be auto-gauge equivalences that nontrivially reidentify a field configuration with itself. Hence a space of physical field configurations does not really look like a set of points, it looks more like this cartoon:

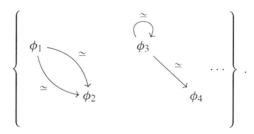

Moreover, if there are two gauge transformations, it is again physically meaningless to ask whether they are equal, instead one has to ask whether they are equivalent via a *gauge-of-gauge* transformation:

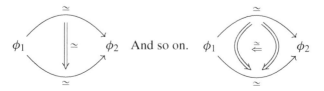

In this generality, the gauge principle of physics is the mathematical principle of *homotopy theory*: in general it is meaningless to say that some objects form a set whose elements are either equal or not, instead one has to consider the *groupoid* which they form, whose morphisms are the equivalences between these objects. Moreover, in general it is meaningless to assume that any two such morphisms are equal or not, rather one has to consider the groupoid which these form, which then in total makes a *2-groupoid*. But in general it is also meaningless to ask whether two equivalences of two equivalences are equal or not, and continuing this way one finds that objects in general form an ∞-*groupoid*, also called a *homotopy type*.

Of particular interest in physics are smooth gauge transformations that arise by integration of *infinitesimal* gauge transformations. An infinitesimal smooth groupoid is a *Lie algebroid* and an infinitesimal smooth ∞-groupoid is an L_∞-*algebroid*. The importance of infinitesimal symmetry transformations in physics, together with the simple fact that they are easier to handle than finite transformations, makes them appear more prominently in the physics literature. In particular, the physics literature is secretly well familiar with smooth ∞-groupoids in their infinitesimal incarnation as L_∞-algebroids: these are equivalently what in physics are called *BRST complexes*. What are called *ghosts* in the BRST complex are the cotangents to the space of equivalences between objects, and what are called higher-order *ghosts-of-ghosts* are cotangents to spaces of higher-order equivalences-of-equivalences. We indicate in a moment how to see this.

While every species of fields in physics is subject to the gauge principle, one speaks specifically of *gauge fields* for those fields which are locally given by differential forms A with values in a Lie algebra (for ordinary gauge fields) or more generally with values in an L_∞-algebroid (for higher gauge fields).

	Infinitesimal		Finite	
	By itself	Acting on fields	By itself	Acting on fields
Symmetry	Lie algebra	Lie algebroid	Lie group	Lie groupoid
Symmetries of symmetries	Lie 2-algebra	Lie 2-algebroid	smooth 2-group	smooth 2-groupoid
Higher-order symmetries	L_∞-algebra	L_∞-algebroid	smooth ∞-group	smooth ∞-groupoid
Physics terminology	FDA (e.g., [20])	BRST complex (e.g., [49])	gauge group	—

We now indicate how such gauge fields and higher gauge fields come about:

- ordinary gauge fields;
- higher gauge fields;
- the BV-BRST complex.

2.1.1 Ordinary Gauge Fields

To start with, consider a plain group G. For the standard applications mentioned in Section 1.1 we would take $G = U(1)$ or $G = \mathrm{SU}(n)$ or products of these, and then the gauge fields we are about to find would be those of electromagnetism and of the nuclear forces, as they appear in the standard model of particle physics.

In order to highlight that we think of G as a *group of symmetries* acting on some (presently unspecified object) $*$, we write

$$G = \left\{ * \xrightarrow{g} * \right\}.$$

In this vein, the product operation $(-) \cdot (-) \colon G \times G \to G$ in the group reflects the result of applying two symmetry operations

$$G \times G \simeq \left\{ \begin{array}{c} * \\ {}^{g_1}\nearrow \quad \searrow{}^{g_2} \\ * \xrightarrow{g_1 \cdot g_2} * \end{array} \right\}.$$

Similarly, the associativity of the group product operation reflects the result of applying three symmetry operations:

$$G \times G \times G \simeq \left\{ \begin{array}{c} \text{(left diagram)} \end{array} = \begin{array}{c} \text{(right diagram)} \end{array} \right\}$$

Here the reader should think of the diagram on the right as a tetrahedron, hence a 3-simplex, that has been cut open only for notational purposes.

Continuing in this way, k-tuples of symmetry transformations serve to label k-simplices whose edges and faces reflect all the possible ways of consecutively applying the corresponding symmetry operations. This forms a *simplicial set*, called the *simplicial nerve* of G, hence a system

$$\mathbf{B}G : k \mapsto G^{\times k}$$

of sets of k-simplices for all k, together with compatible maps between these that restrict $k+1$-simplices to their k-faces (the face maps) and those that regard k-simplices as degenerate $k+1$-simplices (the degeneracy maps). From the above picture, the face maps of $\mathbf{B}G$ in low degree look as follows (where p_i denotes projection onto the ith factor in a cartesian product):

$$\mathbf{B}G := \left[\cdots\cdots G \times G \times G \begin{array}{c} \xrightarrow{(p_1, p_2)} \\ \xrightarrow{-(\mathrm{id},(-)\cdot(-))\to} \\ \xrightarrow{-((-)\cdot(-),\mathrm{id})\to} \\ \xrightarrow{(p_2, p_3)} \end{array} G \times G \begin{array}{c} \xrightarrow{p_1} \\ \xrightarrow{-(-)\cdot(-)\to} \\ \xrightarrow{p_2} \end{array} G \xrightarrow{\quad} * \right].$$

It is useful to remember the smooth structure on these spaces of k-fold symmetry operations by remembering all possible ways of forming smoothly U-parameterized collections of k-fold symmetry operations, for any abstract coordinate chart $U = \mathbb{R}^n$. Now a smoothly U-parameterized collection of k-fold G-symmetries is simply a smooth function from U to $G^{\times k}$, hence equivalently is k smooth functions from U to G. Hence the symmetry group G together with its smooth structure is encoded in the system of assignments

$$\mathbf{B}G : (U, k) \mapsto C^\infty(U, G^{\times k}) = C^\infty(U, G)^{\times k},$$

which is contravariantly functorial in abstract coordinate charts U (with smooth functions between them) and in abstract k-simplices (with cellular maps between them). This is the incarnation of $\mathbf{B}G$ as a *smooth simplicial presheaf*.

Another basic example of a smooth simplicial presheaf is the nerve of an open cover. Let Σ be a smooth manifold and let $\{U_i \hookrightarrow \Sigma\}_{i \in I}$ be a cover of Σ by coordinate charts $U_i \simeq \mathbb{R}^n$. Write $U_{i_0 \cdots i_k} := U_{i_0} \times_X U_{i_1} \times_X \cdots \times_X U_{i_k}$ for the intersection of $(k+1)$ coordinate charts in X. These arrange into a simplicial object like

$$C(\{U_i\}) = \left[\cdots\cdots \coprod_{i_0, i_1, i_2} U_{i_0, i_1, i_2} \rightrightarrows \coprod_{i_0, i_1} U_{i_0, i_1} \rightrightarrows \coprod_{i_0} U_{i_0} \right].$$

A map of simplicial objects

$$C(U_i) \longrightarrow \mathbf{B}G$$

is in degree 1 a collection of smooth G-valued functions $g_{ij} : U_{ij} \longrightarrow G$ and in degree 2 it is the condition that on U_{ijk} these functions satisfy the cocycle condition $g_{ij} \cdot g_{jk} = g_{ik}$. Hence this defines the transition functions for a G-principal bundle on Σ. In physics this may be called the *instanton sector* of a G-gauge field. A G-gauge field itself is a connection on such a G-principal bundle, we come to this in a moment.

We may also think of the manifold Σ itself as a simplicial object, one that does not actually depend on the simplicial degree. Then there is a canonical projection map $C(\{U_i\}) \xrightarrow{\simeq} \Sigma$. When restricted to arbitrarily small open neighborhoods (stalks) of points in Σ, then this projection becomes a *weak homotopy equivalence* of simplicial sets. We are to regard smooth simplicial presheaves which are connected by morphisms that are stalkwise weak homotopy equivalences as equivalent. With this understood, a smooth simplicial presheaf is also called a *higher smooth stack*. Hence a G-principal bundle on Σ is equivalently a morphism of higher smooth stacks of the form

$$\Sigma \longrightarrow \mathbf{B}G.$$

For analyzing smooth symmetries it is useful to focus on infinitesimal symmetries. To that end, consider the (first-order) infinitesimal neighborhood $\mathbb{D}_e(-)$ of the neutral element in the simplicial nerve. Here $\mathbb{D}_e(-)$ is the space around the neutral element that is "so small" that for any smooth function on it which vanishes at e, the square of that function is "so very small" as to actually be equal to zero.

We denote the resulting system of k-fold infinitesimal G-symmetries by $\mathbf{B}\mathfrak{g}$:

$$\mathbf{B}\mathfrak{g} = \left[\cdots\cdots \mathbb{D}_e(G \times G \times G) \begin{array}{c} \xrightarrow{(p_1,p_2)} \\ \xrightarrow{(\mathrm{id},(-)\cdot(-))} \\ \xrightarrow{((-)\cdot(-),\mathrm{id})} \\ \xrightarrow{(p_2,p_3)} \end{array} \mathbb{D}_e(G \times G) \begin{array}{c} \xrightarrow{p_1} \\ \xrightarrow{(-)\cdot(-)} \\ \xrightarrow{p_2} \end{array} \mathbb{D}_e(G) \rightrightarrows * \right].$$

The alternating sum of pullbacks along the simplicial face maps shown above defines a differential d_{CE} on the spaces of functions on these infinitesimal neighborhoods. The corresponding *normalized chain complex* is the differential-graded algebra on those functions which vanish when at least one of their arguments is the neutral element in G. One finds that this is the Chevalley–Eilenberg complex

$$\mathrm{CE}(\mathbf{B}\mathfrak{g}) = \left(\wedge^\bullet \mathfrak{g}^*, d_{\mathrm{CE}} = [-,-]^*\right),$$

which is the Grassmann algebra on the linear dual of the Lie algebra $\mathfrak{g}$ of G equipped with the differential whose component $\wedge^1 \mathfrak{g}^* \to \wedge^2 \mathfrak{g}^*$ is given by the linear dual of the Lie bracket $[-,-]$ and which hence extends to all higher degrees by the graded Leibniz rule.

For example, when we choose $\{t_a\}$ a linear basis for $\mathfrak{g}$, with structure constants of the Lie bracket denoted $[t_a, t_b] = C^c{}_{ab} t_c$, then with a dual basis $\{t^a\}$ of $\mathfrak{g}^*$ we have that

$$d_{\mathrm{CE}} t^a = \tfrac{1}{2} C^a{}_{bc} t^b \wedge t^c.$$

Given any structure constants for a skew bracket like this, then the condition $(d_{\mathrm{CE}})^2 = 0$ is equivalent to the Jacobi identity, hence to the condition that the skew bracket indeed makes a Lie algebra.

Traditionally, the Chevalley–Eilenberg complex is introduced in order to define and to compute Lie algebra cohomology: a d_{CE}-closed element

$$\mu \in \wedge^{p+1} \mathfrak{g}^* \hookrightarrow \mathrm{CE}(\mathbf{B}\mathfrak{g})$$

is equivalently a Lie algebra $(p+1)$-cocycle. This phenomenon will be crucial further below.

Thinking of $\mathrm{CE}(\mathbf{B}\mathfrak{g})$ as the algebra of functions on the infinitesimal neighborhood of the neutral element inside $\mathbf{B}G$ makes it plausible that this is an equivalent incarnation of the Lie algebra of G. This is also easily checked directly: sending finite-dimensional Lie algebras to their Chevalley–Eilenberg algebra constitutes a fully faithful inclusion

$$\mathrm{CE}: \mathrm{LieAlg} \hookrightarrow \mathrm{dgcAlg}^{\mathrm{op}}$$

of the category of Lie algebras into the opposite of the category of differential graded-commutative algebras. This perspective turns out to be useful for computations in gauge theory and in higher gauge theory. Therefore it serves to see how various familiar constructions on Lie algebras look when viewed in terms of their Chevalley–Eilenberg algebras.

Most importantly, for Σ a smooth manifold and $\Omega^\bullet(\Sigma)$ denoting its de Rham dg-algebra of differential forms, then *flat* $\mathfrak{g}$-valued 1-forms on Σ are equivalent to dg-algebra homomorphisms like so:

$$\Omega^\bullet_{\mathrm{flat}}(\Sigma, \mathfrak{g}) := \left\{ A \in \Omega^1(\Sigma) \otimes \mathfrak{g} \mid F_A := d_{\mathrm{dR}} A - \tfrac{1}{2}[A \wedge A] = 0 \right\}$$

$$\simeq \{ \Omega^\bullet(\Sigma) \longleftarrow \mathrm{CE}(\mathbf{B}\mathfrak{g}) \}.$$

To see this, notice that the underlying homomorphism of graded algebras $\Omega^\bullet(\Sigma) \longleftarrow \wedge^\bullet \mathfrak{g}^*$ is equivalently a $\mathfrak{g}$-valued 1-form, and that the respect for the differential forces it to be flat:

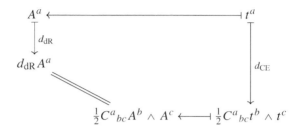

The flat Lie algebra-valued forms play a crucial role in recovering a Lie group from its Lie algebra as the group of finite paths of infinitesimal symmetries. To that end, write $\Delta^1 := [0,1]$ for the abstract interval. Then a $\mathfrak{g}$-valued differential form $A \in \Omega^1_{\text{flat}}(\Delta^1, \mathfrak{g})$ is at each point of Δ^1 an infinitesimal symmetry, hence it encodes the finite symmetry transformation that is given by applying the infinitesimal transformation A_t at each $t \in \Delta^1$ and then "integrating these." This integration is called the *parallel transport* of A and is traditionally denoted by the symbols $P \exp(\int_0^1 A) \in G$. Now of course different paths of infinitesimal transformations may have the same integrated effect. But precisely if A_1 and A_2 have the same integrated effect, then there is a flat $\mathfrak{g}$-valued 1-form on the disk which restricts to A_1 on the upper semicircle and to A_2 on the lower semicircle.

In particular, the composition of two paths of infinitesimal gauge transformations is in general not equal to any given such path with the same integrated effect, but there will always be a flat 1-form $\hat{A}$ on the 2-simplex Δ^2 which interpolates:

Infinitesimal symmetries	Integration	Finite symmetries
A_1 $\hat{A}$ A_2 / $A_{1,2}$	$\mapsto$	$P\exp(\int_0^1 A_1)$, $P\exp(\int_0^1 A_2)$ / $P\exp(\int_0^1 A_1) \cdot P\exp(\int_0^1 A_2)$

In order to remember how the group obtained this way is a Lie group, we simply need to remember how the above composition works in smoothly U-parameterized collections of 1-forms. But a U-parameterized collection of 1-forms on Δ^k is simply a 1-form on $U \times \Delta^k$ which vanishes on vectors tangent to U, hence a vertical 1-form on $U \times \Delta^k$, regarded as a simplex bundle over U.

All this is captured by saying that there is a simplicial smooth presheaf $\exp(\mathfrak{g})$ which assigns to an abstract coordinate chart U and a simplicial degree k the set of flat vertical $\mathfrak{g}$-valued 1-forms on $U \times \Delta^k$:

$$\exp(\mathfrak{g}) := (U, k) \mapsto \Omega^\bullet_{\substack{\text{flat}\\\text{vert}}} (U \times \Delta^k, \mathfrak{g})$$
$$= \left\{ \Omega^\bullet_{\text{vert}}(U \times \Delta^k) \longleftarrow \text{CE}(\mathbf{B}\mathfrak{g}) \right\}$$

By the above discussion, we do not care which of various possible flat 1-forms $\hat{A}$ on 2-simplices are used to exhibit the composition of finite gauge transformations. The technical term for retaining just the information that there is any such 1-form on a 2-simplex at all is to form the *2-coskeleton* $\text{cosk}_2(\exp(\mathfrak{g}))$. And one finds that this indeed recovers the smooth gauge group G, in that there is a weak equivalence of simplicial presheaves:

$$\text{cosk}_3(\exp(\mathfrak{g})) \simeq \mathbf{B}G.$$

So far this produces the gauge group itself from the infinitesimal symmetries. We now discuss how similarly its action on gauge fields is obtained. To that end, consider the *Weil algebra* of $\mathfrak{g}$, which is obtained from the Chevalley-Eilenberg algebra by throwing in another copy of $\mathfrak{g}$, shifted up in degree

$$W(\mathbf{B}\mathfrak{g}) := \left(\wedge^\bullet(\mathfrak{g}^* \oplus \mathfrak{g}^*[1]), d_W = d_{\text{CE}} + \mathbf{d} \right),$$

where $\mathbf{d} \colon \wedge^1 \mathfrak{g}^* \xrightarrow{\simeq} \mathfrak{g}^*[1]$ is the degree shift and we declare d_{CE} and $\mathbf{d}$ to anticommute. So if $\{t^a\}$ is the dual basis of $\mathfrak{g}^*$ from before, write $\{r^a\}$ for the same elements thought of in one degree higher as a basis of $\mathfrak{g}^*[1]$; then

$$d_W \colon t^a \mapsto \tfrac{1}{2} C^a{}_{bc} t^b \wedge t^c + r^a;$$
$$d_W \colon r^a \mapsto C^a{}_{bc} t^b \wedge r^c.$$

A key point of this construction is that dg-algebra homomorphisms out of the Weil algebra into a de Rham algebra are equivalent to unconstrained $\mathfrak{g}$-valued differential forms:

$$\Omega(\Sigma, \mathfrak{g}) := \left\{ A \in \Omega^1(\Sigma) \otimes \mathfrak{g} \right\} \simeq \left\{ \Omega^\bullet(\Sigma) \longleftarrow W(\mathbf{B}\mathfrak{g}) \right\}.$$

This is because now the extra generators r^a pick up the failure of the respect for the d_{CE}-differential, that failure is precisely the curvature F_A:

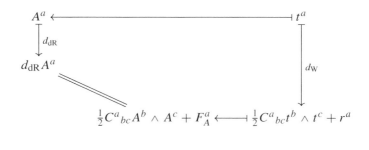

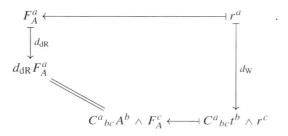

Notice here that once $t^a \mapsto A^a$ is chosen, then the first diagram uniquely specifies that $r^a \mapsto F_A^a$ and then the second diagram is already implied: its commutativity is the *Bianchi idenity* $dF_A = [A \wedge F_A]$ that is satisfied by curvature forms.

Traditionally, the Weil algebra is introduced in order to define and compute invariant polynomials on a Lie algebra. A d_W-closed element in the shifted generators

$$\langle -, -, \cdots \rangle \in \wedge^k \mathfrak{g}^*[1] \hookrightarrow \mathbf{W}(\mathbf{B}\mathfrak{g})$$

is equivalently an invariant polynomial of order k on the Lie algebra $\mathfrak{g}$. Therefore write

$$\mathrm{inv}(\mathbf{B}\mathfrak{g})$$

for the graded commutative algebra of invariant polynomials, thought of as a dg-algebra with vanishing differential.

(For notational convenience we will later often abbreviate, e.g., CE($\mathfrak{g}$) for CE($\mathbf{B}\mathfrak{g}$). This is unambiguous as long as no algebroids with nontrivial bases spaces appear.)

There is a canonical projection map from the Weil algebra to the Chevalley–Eilenberg algebra, given simply by forgetting the shifted generators ($t^a \mapsto t^a$; $r^a \mapsto 0$). And there is the defining inclusion $\mathrm{inv}(\mathbf{B}\mathfrak{g}) \hookrightarrow \mathbf{W}(\mathbf{B}\mathfrak{g})$:

Higher Prequantum Geometry

Cartan had introduced all these dg-algebras as algebraic models of the universal G-principal bundle. We had seen above that homomorphisms $\Omega^\bullet_{\mathrm{vert}}(U \times \Delta^k) \longleftarrow \mathrm{CE}(\mathbf{B}\mathfrak{g})$ constitute the gauge symmetry group G as integration of the paths of infinitesimal symmetries. Here the vertical forms on $U \times \Delta^k$ are themselves part of the sequence of differential forms on the trivial k-simplex bundle over the given coordinate chart U. Hence consider compatible dg-algebra homomorphisms between these two sequences:

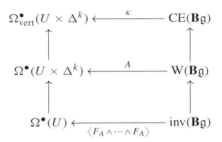

We unwind what this means in components: the middle morphism is an unconstrained Lie algebra-valued form $A \in \Omega^1(U \times \Delta^k, \mathfrak{g})$, hence is a sum

$$A = A_U + A_{\Delta^k}$$

of a 1-form A_U along U and a 1-form A_{Δ^k} along Δ^k. The second summand A_{Δ^k} is the vertical component of A. The commutativity of the top square above says that as a vertical differential form, A_{Δ^k} has to be flat. By the previous discussion this means that A_{Δ^k} encodes a k-tuple of G-gauge transformations. Now we will see how these gauge transformations naturally act on the gauge field A_U.

Consider this for the case $k = 1$, and write t for the canonical coordinate along $\Delta^1 = [0, 1]$. Then A_U is a smooth t-parameterized collection of 1-forms, hence of $\mathfrak{g}$-gauge fields, on U; and $A_{\Delta^1} = \kappa\, dt$ for κ a smooth Lie algebra-valued function, called the *gauge parameter*. Now the equation for the t-component of the total curvature F_A of A says how the gauge parameter together with the mixed curvature component causes infinitesimal transformations of the gauge field A_U as t proceeds:

$$\frac{d}{dt} A_U = d_U \kappa - [\kappa, A] + \iota_{\partial_t} F_A .$$

But now the commutativity of the lower square above demands that the curvature forms evaluated in invariant polynomials have vanishing contraction with ι_{∂_t}. In the case that $\mathfrak{g} = \mathbb{R}$ this means that $\iota_{\partial_t} F_A = 0$, while for nonabelian $\mathfrak{g}$ this is still generically the necessary condition. So for vanishing t-component of the curvature the above equation says that

$$\frac{d}{dt} A_U = d\kappa - [\kappa, A] .$$

This is the traditional formula for infinitesimal gauge transformations κ acting on a gauge field A_U. Integrating this up, κ integrates to a gauge group element $g := P \exp(\int_0^1 \kappa dt)$ by the previous discussion, and this equation becomes the formula for finite gauge transformations (where we abbreviate now $A_t := A_U(t)$):

$$A_1 = g^{-1} A_0 g + g^{-1} d_{\mathrm{dR}} g .$$

This gives the smooth groupoid $\mathbf{B}G_{\mathrm{conn}}$ of $\mathfrak{g}$-gauge fields with G-gauge transformations between them.

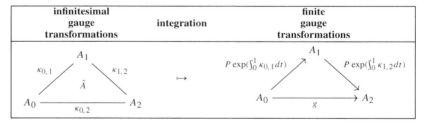

Hence $\mathbf{B}G_{\mathrm{conn}}$ is the smooth groupoid such that for U an abstract coordinate chart, the smoothly U-parameterized collections of its objects are $\mathfrak{g}$-valued differential forms $A \in \Omega(U, \mathfrak{g})$, and whose U-parameterized collections of gauge transformations are G-valued functions g acting by

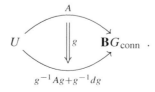

Higher Prequantum Geometry

This dg-algebraic picture of gauge fields with gauge transformations between them now immediately generalizes to higher gauge fields with higher gauge transformations between them. Moreover, this picture allows to produce prequantized higher Chern–Simons-type Lagrangians by Lie integration of transgressive L_∞-cocycles.

2.1.2 Higher Gauge Fields

Ordinary gauge fields are characterized by the property that there are no non-trivial gauge-of-gauge transformations, equivalently that their BRST complexes contain no higher-order ghosts. Mathematically, it is natural to generalize beyond this case to *higher gauge fields*, which do have nontrivial higher gauge transformations. The simplest example is a "2-form field" ("B-field"), generalizing the "vector potential" 1-form A of the electromagnetic field. Whereas such a 1-form has gauge transformations given by 0-forms (functions) κ via

$$A \xrightarrow{\kappa} A' = A + d\kappa \,,$$

a 2-form B has gauge transformations given by 1-forms ρ_1, which themselves then have gauge-of-gauge-transformations given by 0-forms ρ_0:

$$B \xRightarrow[\rho_1' = \rho_1 + d\rho_0]{\rho_1 \;\; \rho_0} B' = B + d\rho_1 = B + d\rho_1'$$

Next a "3-form field" ("C-field") has third-order gauge transformations:

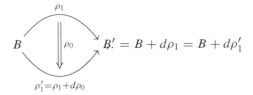

$$\rho_2' = \rho_2 + d\rho_1 = \rho_2 + d\rho_1'$$

Similarly "n-form fields" have order-n gauge-of-gauge transformations and hence have order-n ghost-of-ghosts in their BRST complexes.

Higher gauge fields have not been experimentally observed, to date, as fundamental fields of nature, but they appear by necessity and ubiquitously in higher dimensional supergravity and in the hypothetical physics of strings and p-branes. The higher differential geometry which we develop is to a large extent motivated by making precise and tractable the global structure of higher gauge fields in string and M-theory.

Generally, higher gauge fields are part of mathematical physics just as the Ising model and ϕ^4-theory are, and as such they do serve to illuminate the structure of experimentally verified physics. For instance, the Einstein equations of motion for ordinary (bosonic) general relativity on 11-dimensional spacetimes are equivalent to the full supertorsion constraint in 11-dimensional supergravity with its 3-form higher gauge field [19]. From this point of view one may regard the 3-form higher gauge field in supergravity, together with the gravitino, as auxiliary fields that serve to present Einstein's equations for the graviton in a particularly neat mathematical way.

We now use the above dg-algebraic formulation of ordinary gauge fields in Section 2.1.1 in order to give a quick but accurate idea of the mathematical structure of higher gauge fields.

Above we saw that (finite-dimensional) Lie algebras are equivalently the formal duals of those differential graded-commutative algebras whose underlying graded commutative algebra is freely generated from a (finite-dimensional) vector space over the ground field. From this perspective, there are two evident generalizations to be considered: we may take the underlying vector space to already have contributions in higher degrees itself, and we may pass from vector spaces, being modules over the ground field $\mathbb{R}$, to (finite rank) projective modules over an algebra of smooth functions on a smooth manifold.

Hence we say that an L_∞-*algebroid* (of finite type) is a smooth manifold X equipped with an $\mathbb{N}$-graded vector bundle (degreewise of finite rank), whose smooth sections hence form an $\mathbb{N}$-graded projective $C^\infty(X)$-module $\mathfrak{a}_\bullet$, and equipped with an $\mathbb{R}$-linear differential d_{CE} on the Grassmann algebra of the $C^\infty(X)$-dual $\mathfrak{a}^*$ modules

$$\mathrm{CE}(\mathfrak{a}) := \left(\wedge^\bullet_{C^\infty(X)}(\mathfrak{a}^*),\ d_{\mathrm{CE}(\mathfrak{a})} \right).$$

Accordingly, a homomorphism of L_∞-algebroids we take to be a dg-algebra homomorphism (over $\mathbb{R}$) of their CE-algebras going the other way around. Hence the category of L_∞-algebroids is the full subcategory of the opposite of that of differential graded-commutative algebras over $\mathbb{R}$ on those whose underlying graded-commutative algebra is free on graded locally free projective $C^\infty(X)$-modules:

$$L_\infty \mathrm{Algbd} \hookrightarrow \mathrm{dgcAlg}^{\mathrm{op}}.$$

We say we have a *Lie n-algebroid* when $\mathfrak{a}$ is concentrated in the lowest n-degrees. Here are some important examples of L_∞-algebroids.

When the base space is the point, $X = *$, and $\mathfrak{a}$ is concentrated in degree 0, then we recover **Lie algebras**, as above. Generally, when the base space is the point, then the $\mathbb{N}$-graded module $\mathfrak{a}$ is just an $\mathbb{N}$-graded vector space $\mathfrak{g}$. We write $\mathfrak{a} = \mathbf{B}\mathfrak{g}$ to indicate this, and then $\mathfrak{g}$ is an L_∞**-algebra**. When in addition $\mathfrak{g}$ is concentrated in the lowest n degrees, then these are also called **Lie n-algebras**. With no constraint on the grading but assuming that the differential sends single generators always to sums of wedge products of at most two generators, then we get **dg-Lie algebras**.

The Weil algebra of a Lie algebra $\mathfrak{g}$ hence exhibits a Lie 2-algebra. We may think of this as the Lie 2-algebra inn($\mathfrak{g}$) of inner derivations of $\mathfrak{g}$. By the above discussion, it is suggestive to write $\mathbf{E}\mathfrak{g}$ for this Lie 2-algebra, hence

$$\mathrm{W}(\mathbf{B}\mathfrak{g}) = \mathrm{CE}(\mathbf{B}\mathbf{E}\mathfrak{g})\,.$$

If $\mathfrak{g} = \mathbb{R}[n]$ is concentrated in degree p on the real line (so that the CE-differential is necessarily trivial), then we speak of the *line Lie $(p+1)$-algebra* $\mathbf{B}^p\mathbb{R}$, which as an L_∞-algebroid over the point is to be denoted

$$\mathbf{B}\mathbf{B}^p\mathbb{R} = \mathbf{B}^{p+1}\mathbb{R}\,.$$

All this goes through verbatim, up to additional signs, with all vector spaces generalized to supervector spaces. The Chevalley–Eilenberg algebras of the resulting **super L_∞-algebras** are known in parts of the supergravity literature as **FDA**s [20].

Passing now to L_∞-algebroids over nontrivial base spaces, first of all every smooth manifold X may be regarded as the L_∞-algebroid over X, these are the Lie 0-algebroids. We just write $\mathfrak{a} = X$ when the context is clear.

For the tangent bundle TX over X then the graded algebra of its dual sections is the wedge product algebra of differential forms, $\mathrm{CE}(TX) = \Omega^\bullet(X)$ and hence the de Rham differential makes $\wedge^\bullet \Gamma(T^*X)$ into a dgc-algebra and hence makes TX into a Lie algebroid. This is called the **tangent Lie algebroid** of X. We usually write $\mathfrak{a} = TX$ for the tangent Lie algebroid (trusting that context makes it clear that we do not mean the Lie 0-algebroid over the underlying manifold of the tangent bundle itself). In particular this means that for any other L_∞-algebroid $\mathfrak{a}$ then flat $\mathfrak{a}$-valued differential forms on some smooth manifold Σ are equivalently homomorphisms of L_∞-algebroids like so:

$$\Omega_{\mathrm{flat}}(\Sigma, \mathfrak{a}) \;=\; \{\, T\Sigma \longrightarrow \mathfrak{a}\,\}\,.$$

In particular, ordinary closed differential forms of degree n are equivalently flat $\mathbf{B}^n\mathbb{R}$-valued differential forms:

$$\Omega^n_{cl}(\Sigma) \simeq \{T\Sigma \longrightarrow \mathbf{B}^n\mathbb{R}\}.$$

More generally, for $\mathfrak{a}$ any L_∞-algebroid over some base manifold X, we have its Weil dgc-algebra,

$$W(\mathfrak{a}) := \left(\wedge^\bullet_{C^\infty(X)}(\mathfrak{a}^* \oplus \Gamma(T^*X) \oplus \mathfrak{a}^*[1]), d_W = d_{CE} + \mathbf{d}\right),$$

where $\mathbf{d}$ acts as the degree shift isomorphism in the components $\wedge^1_{C^\infty(X)}\mathfrak{a}^* \longrightarrow \wedge^1_{C^\infty(X)}\mathfrak{a}^*[1]$ and as the de Rham differential in the components $\wedge^k\Gamma(T^*X) \to \wedge^{k+1}\Gamma(T^*X)$. This defines a new L_∞-algebroid that may be called the **tangent L_∞-algebroid** $T\mathfrak{a}$

$$CE(T\mathfrak{a}) := W(\mathfrak{a}).$$

We also write $\mathbf{EB}^p\mathbb{R}$ for the L_∞-algebroid with

$$CE(\mathbf{EB}^p\mathbb{R}) := W(\mathbf{B}^p\mathbb{R}).$$

In direct analogy with the discussion for Lie algebras, we then say that an unconstrained $\mathfrak{a}$-valued differential form A on a manifold Σ is a dg-algebra homomorphism from the Weil algebra of $\mathfrak{a}$ to the de Rham dg-algebra on Σ:

$$\Omega(\Sigma, \mathfrak{a}) := \{\Omega^\bullet(\Sigma) \longleftarrow W(\mathfrak{a})\}.$$

For G a Lie group acting on X by diffeomorphisms, there is the **action Lie algebroid** $X/\mathfrak{g}$ over X with $\mathfrak{a}_0 = \Gamma_X(X \times \mathfrak{g})$ the $\mathfrak{g}$-valued smooth functions over X. Write $\rho: \mathfrak{g} \to \text{Vect}$ for the linearized action. With a choice of basis $\{t_a\}$ for $\mathfrak{g}$ as before and assuming that $X = \mathbb{R}^n$ with canonical coordinates x^i, then ρ has components $\{\rho_a^\mu\}$ and the CE-differential on $\wedge^\bullet_{C^\infty(X)}(\Gamma_X(X \times \mathfrak{g}^*))$ is given on generators by

$$d_{CE}: f \mapsto t^a \rho_a^\mu \partial_\mu f;$$
$$d_{CE}: t^a \mapsto \tfrac{1}{2}C^a{}_{bc}t^b \wedge t^c.$$

In the physics literature this Chevalley–Eilenberg algebra $CE(X/\mathfrak{g})$ is known as the **BRST complex** of X for infinitesimal symmetries $\mathfrak{g}$. If X is thought of as a space of fields, then the t_a are called *ghost fields*.[12]

Given any L_∞-algebroid, it induces further L_∞-algebroids via its extension by higher cocycles. A $p+1$-*cocycle* on an L_∞-algebroid $\mathfrak{a}$ is a closed element

---

[12] More generally the base manifold X may be a derived manifold/BV-complex as in footnote 9. Then $CE(X/\mathfrak{g})$ is known as the "BV-BRST complex."

$$\mu \in (\wedge^{\bullet}_{C^{\infty}(X)} \mathfrak{a}^*)_{p+1} \hookrightarrow \mathrm{CE}(\mathfrak{a}).$$

Notice that now cocycles are *representable* by the higher line L_∞-algebras $\mathbf{B}^{p+1}\mathbb{R}$ from above:

$$\{\mu \in \mathrm{CE}(\mathfrak{a})_{p+1} \mid d_{\mathrm{CE}}\mu = 0\} \simeq \left\{ \mathrm{CE}(\mathfrak{a}) \xleftarrow{\mu^*} \mathrm{CE}(\mathbf{B}^{p+1}\mathbb{R}) \right\}$$
$$= \left\{ \mathfrak{a} \xrightarrow{\mu} \mathbf{B}^{p+1}\mathbb{R} \right\}.$$

It is a traditional fact that $\mathbb{R}$-valued 2-cocycles on a Lie algebra induce central Lie algebra extensions. More generally, higher cocycles μ on an L_∞-algebroid induce L_∞-extensions $\hat{\mathfrak{a}}$, given by the pullback

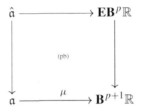

Equivalently this makes $\hat{\mathfrak{a}}$ be the homotopy fiber of μ in the homotopy theory of L_∞-algebras, and induces a long homotopy fiber sequence of the form

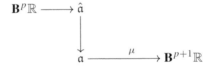

In components this means simply that $\mathrm{CE}(\hat{\mathfrak{a}})$ is obtained from $\mathrm{CE}(\mathfrak{a})$ by adding one generator c in degree p and extending the differential to it by the formula

$$d_{\mathrm{CE}} : c = \mu.$$

This construction has a long tradition in the supergravity literature [20, 37]; we come to the examples considered there in Section 2.4. Iterating this construction, out of every L_∞-algebroid there grows a whole bouquet of further L_∞-algebroids

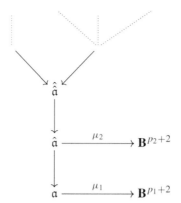

For example for $\mathfrak{g}$ a semisimple Lie algebra with binary invariant polynomial $\langle -,-\rangle$ (the Killing form), then $\mu_3 = \langle -,[-,-]\rangle$ is a 3-cocycle. The L_∞-extension by this cocycle is a Lie 2-algebra called the *string Lie 2-algebra* $\mathfrak{string}_\mathfrak{g}$. If $\{t^a\}$ is a linear basis of $\mathfrak{g}^*$ as before write $k_{ab} := \langle t_a, t_b\rangle$ for the components of the Killing form; the components of the 3-cocycle are $\mu_{abc} = k_{aa'} C^{a'}{}_{bc}$. The CE-algebra of the string Lie 2-algebra then is that of $\mathfrak{g}$ with a generator b added and with CE-differential defined by

$$d_{\mathrm{CE}(\mathfrak{string})}: t^a \mapsto \tfrac{1}{2} C^a{}_{bc} t^b \wedge t^c$$
$$d_{\mathrm{CE}(\mathfrak{string})}: b \mapsto k_{aa'} C^{a'}{}_{cb} t^a \wedge t^b \wedge t^c.$$

Hence a flat $\mathfrak{string}_\mathfrak{g}$-valued differential form on some Σ is a pair consisting of an ordinary flat $\mathfrak{g}$-valued 1-form A and of a 2-form B whose differential has to equal the evaluation of A in the 3-co-cycle:

$$\Omega_{\mathrm{flat}}(\Sigma, \mathfrak{string}_\mathfrak{g})$$
$$\simeq \left\{(A,B) \in \Omega^1(\Sigma,\mathfrak{g}) \times \Omega^2(\Sigma) \mid F_A = 0,\ dB = \langle A \wedge [A \wedge A]\rangle\right\}.$$

Notice that since A is flat, the 3-form $\langle A \wedge [A \wedge A]\rangle$ is its Chern–Simons 3-form. More generally, Chern–Simons forms are such that their differential is the evaluation of the curvature of A in an invariant polynomial.

An invariant polynomial $\langle -\rangle$ on an L_∞-algebroid we may take to be a d_W-closed element in the shifted generators of its Weil algebra $W(\mathfrak{a})$

$$\langle -\rangle \in \wedge^\bullet_{C^\infty(X)}(\mathfrak{a}^*[1]) \hookrightarrow W(\mathfrak{a}).$$

When one requires the invariant polynomial to be binary, that is, in $\wedge^2(\mathfrak{a}^*[1]) \to W(\mathfrak{a})$ and nondegenerate, then it is also called a *shifted symplectic form* and it makes $\mathfrak{a}$ into a "symplectic Lie n-algebroid." For

Higher Prequantum Geometry 251

$n = 0$ these are the symplectic manifolds, for $n = 1$ these are called *Poisson Lie algebroids*, for $n = 2$ they are called *Courant Lie 2-algebroids* [76]. There are also plenty of nonbinary invariant polynomials; we discuss further examples in Section 2.3.

Being d_W-closed, an invariant polynomial on $\mathfrak{a}$ is represented by a dg-homomorphism:

$$W(\mathfrak{a}) \longleftarrow \mathrm{CE}(\mathbf{B}^{p+2}\mathbb{R}) : \langle - \rangle$$

This means that given an invariant polynomial $\langle - \rangle$ for an L_∞-algebroid $\mathfrak{a}$, then it assigns to any $\mathfrak{a}$-valued differential form A a plain closed $(p+2)$-form $\langle F_A \rangle$ made up of the $\mathfrak{a}$-curvature forms, namely, the composite

$$\Omega^\bullet(\Sigma) \xleftarrow{A} W(\mathfrak{a}) \xleftarrow{\langle - \rangle} \mathrm{CE}(\mathbf{B}^{p+2}\mathbb{R}) : \langle F_A \rangle.$$

In other words, A may be regarded as a nonabelian prequantization of $\langle F_A \rangle$.

Therefore we may consider now the ∞-groupoid of $\mathfrak{a}$-connections whose gauge transformations preserve the specified invariant polynomial, such as to guarantee that it remains a globally well-defined differential form. We write $exp(\mathfrak{a})_{conn}$ for the smooth ∞-groupoid of $\mathfrak{a}$-valued connections with such gauge transformations between them. As a smooth simplicial presheaf, it is hence given by the following assignment:

$$exp(\mathfrak{a})_{conn} : (U,k) \mapsto \left\{ \begin{array}{c} \Omega^\bullet_{\mathrm{vert}}(U \times \Delta^k) \longleftarrow \mathrm{CE}(\mathfrak{a}) \\ \uparrow \qquad\qquad \uparrow \\ \Omega^\bullet(U \times \Delta^k) \xleftarrow{A} W(\mathfrak{a}) \\ \uparrow \qquad\qquad \uparrow \\ \Omega^\bullet(U) \xleftarrow{\langle F_A \rangle} \mathrm{inv}(\mathfrak{a}) \end{array} \right\}$$

Here on the right we have, for every U and k, the set of those A on $U \times \Delta^k$ that induce gauge transformations along the Δ^k-direction (that is the commutativity of the top square) such that the given invariant polynomials evaluated on the curvatures are preserved (that is the commutativity of the bottom square).

This $exp(\mathfrak{a})_{conn}$ is the moduli stack of $\mathfrak{a}$-valued connections with gauge transformations and gauge-of-gauge transformations between them that preserve the chosen invariant polynomials [16, 30].

The key example is the moduli stack of $(p+1)$-form gauge fields

$$exp(\mathbf{B}^{p+1}\mathbb{R})_{conn}/\mathbb{Z} \simeq \mathbf{B}(\mathbb{R}/\hbar\mathbb{Z})_{conn}.$$

Generically we write

$$\mathbf{A}_{\text{conn}} := \text{cosk}_{n+1}(\exp(\mathfrak{a})_{\text{conn}})$$

for the n-truncation of a higher smooth stack of $\mathfrak{a}$-valued gauge field connections obtained this way. If $\mathfrak{a} = \mathbf{B}\mathfrak{g}$ then we write $\mathbf{B}G_{\text{conn}}$ for this.

Given such, then an $\mathfrak{a}$-gauge field on Σ (an $\mathbf{A}$-principal connection) is equivalently a map of smooth higher stacks

$$\nabla : \Sigma \longrightarrow \mathbf{A}_{\text{conn}} \,.$$

By the above discussion, a simple map like this subsumes all of the following component data:

1. a choice of open cover $\{U_i \to \Sigma\}$;
2. an $\mathfrak{a}$-valued differential form A_i on each chart U_i;
3. on each intersection U_{ij} of charts a path of infinitesimal gauge symmetries whose integrated finite gauge symmetry g_{ij} takes A_i to A_j;
4. on each triple intersection U_{ijk} of charts a path-of-paths of infinitesimal gauge symmetries whose integrated finite gauge-of-gauge symmetry takes the gauge transformation $g_{ij} \cdot g_{jk}$ to the gauge transformation g_{ik};
5. and so on.

Hence a $\mathfrak{a}$-gauge field is locally $\mathfrak{a}$-valued differential form data which is coherently glued together to a global structure by gauge transformations and higher-order gauge-of-gauge transformations.

Given two globally defined $\mathfrak{a}$-valued gauge fields this way, then a globally defined gauge transformation between them is equivalently a homotopy between maps of smooth higher stacks

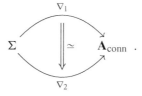

Again, this concisely encodes a system of local data: this is on each chart U_i a path of inifinitesimal gauge symmetries whose integrated gauge transformation transforms the local $\mathfrak{a}$-valued forms into each other, together with various higher-order gauge transformations and compatibilities on higher-order intersections of charts.

Then a gauge-of-gauge transformation is a homotopy of homotopies

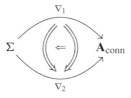

and again this encodes a recipe for how to extract the corresponding local differential form data.

2.1.3 The BV-BRST Complex

The category of partial differential equations that we referred to so far, as in (see [69] or [57]), is modeled on the category of smooth manifolds. Accordingly, it really only contains differential equations that are nonsingular enough such as to guarantee that the shell locus $\mathcal{E} \hookrightarrow J^\infty E$ is itself a smooth manifold. This is not the case for all differential equations of interest. For some pairs of differential operators, their equalizer $E \rightrightarrows F$ does not actually exist in smooth bundles modeled on manifolds.

This is no problem when working in the sheaf topos over PDE_Σ, where all limits do exist as diffeological bundles. However, even though all limits exist here, some do not interact properly with other constructions of interest. For instance intersection products in cohomology will not properly count nontransversal intersections, even if they do exist as diffeological spaces.

To fix this, we may pass to a category of "derived manifolds." In generalization of how an ordinary smooth manifold is the formal dual to its real algebra of smooth functions, via the faithful embedding

$$C^\infty : \mathrm{SmoothMfd} \hookrightarrow \mathrm{CAlg}_\mathbb{R}^{\mathrm{op}},$$

so a derived manifold is the formal dual to a differential graded-commutative algebra in nonpositive degrees, whose underlying graded algebra is of the form $\wedge^\bullet_{C^\infty(X)}(\Gamma(V^*))$ for V a $-\mathbb{N}$-graded smooth vector bundle over X. In the physics literature these dg-algebras are known as *BV-complexes*.

For example, for X a smooth manifold and $S \in C^\infty(X)$ a smooth function on it, then the vanishing locus of S in X is represented by the derived manifold $\ker_d(S)$ that is formally dual to the dg-algebra denoted $C^\infty(\ker_d(S))$ which is spanned over $C^\infty(X)$ by a single generator t of degree -1 and whose differential (linear over $\mathbb{R}$) is defined by

$$d_{\mathrm{BV}} : t \mapsto S.$$

For Σ an ordinary smooth manifold, then morphisms $\Sigma \longrightarrow \ker_d(S)$ are equivalently dg-algebra homomorphisms $C^\infty(\Sigma) \longleftarrow C^\infty(\ker_d(S))$, and these are equivalently algebra homomorphisms $\phi^* : C^\infty(\Sigma) \longleftarrow C^\infty(X)$ such that $\phi^* S = 0$. These, finally, are equivalently smooth functions $\phi : \Sigma \longrightarrow X$ that land everywhere in the 0-locus of S. It is in this way that $\ker_d(S)$ is a resolution of the possibly singular vanishing locus by a complex of nonsingular smooth bundles.

Notice that even if the kernel of S does exist as a smooth submanifold $\ker(S) \hookrightarrow X$ it need not be equivalent to the derived kernel: for instance over $X = \mathbb{R}^1$ with its canonical coordinate function x, then $\ker(x) = \{0\}$ but $\ker_d(x^2) \simeq \mathbb{D}_0^{(1)}$ is the infinitesimal interval around 0.

Given a derived manifold X_d this way, then for each $k \in \mathbb{N}$ the differential k-forms on X_d also inherit the BV-differential, on top of the de Rham differential. We write $\Omega^{k;-s}(X_d)$ to indicate the differential k-forms of BV-degree $-s$. So in particular the 0-forms recover the BV dg-algebra itself $\Omega^{0,-\bullet}(X_d) = C^\infty(X_d)$.

Hence using underived manifolds, the conservation of the presymplectic current, $d_H \Omega = 0$, implies that over a spacetime/worldvolume Σ with two boundary components $\Sigma_{\text{in}} = \partial_{\text{in}} \Sigma$ and $\Sigma_{\text{out}} = \partial_{\text{out}} \Sigma$ the canonical presymplectic forms ω_{in} and ω_{out} agree

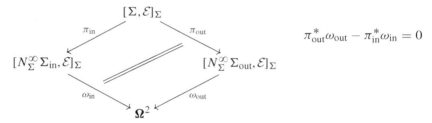

When the covariant phase space is resolved by a derived space $([\Sigma, \mathcal{E}]_\Sigma)_d$, then this equation becomes a homotopy which asserts the existence of a 2-form ω_{BV} of BV-degree -1 which witnesses the invariance of the canonical presymplectic form:

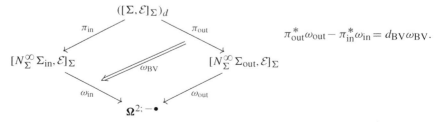

The equation on the right appears in the BV-literature as [18, Equation (9)].

Higher Prequantum Geometry

For the purpose of prequantum field theory, we again wish to de-transgress this phenomenon. Instead of just modeling the covariant phase space by a derived space, we should model the dynamical shell $\mathcal{E} \hookrightarrow J_\Sigma^\infty E$ itself by a derived bundle.

The derived shell $\ker_d(\mathrm{EL})$ is the derived manifold bundle over Σ whose underlying manifold is $J_\Sigma^\infty E$ and whose bundle of "antifields" (in the sense of BV-calculus) is the pullback of $V^*E \otimes \wedge^{p+1}T^*\Sigma$ to the jet bundle (along the projection maps to E).

If ϕ^i are a choice of local vertical coordinates on E (the fields) and ϕ_i^* denotes the corresponding local antifield coordinates with respect to any chosen volume form on Σ, then this BV-differential looks like

$$d_{\mathrm{BV}} = \mathrm{EL}_i \frac{\partial}{\partial \phi_i^*} \quad : \quad \phi_i^* \mapsto \mathrm{EL}_i \,.$$

When regarded as an odd graded vector field, this differential is traditionally denoted by Q.

In such coordinates there is then the following canonical differential form

$$\Omega_{\mathrm{BV}} = d_V \phi_i^* \wedge d_V \phi^i \quad \in \Omega^{p+1,2;-1}(\ker_d(\mathrm{EL}))$$

which, as indicated, is of BV-degree -1 and otherwise is a $(p+3)$-form with horizontal degree $p+1$ vertical degree 2. More abstractly, this form is characterized by the property that

$$\iota_Q \Omega_{\mathrm{BV}} = \mathrm{EL} \quad \in \Omega^{p+1,1;0}(\ker_d(\mathrm{EL})) \,.$$

As before, we write

$$\omega_{\mathrm{BV}} := \int_\Sigma \Omega_{\mathrm{BV}}$$

for the transgression of this form to the covariant phase space. We now claim that there it satisfies the above relation of witnessing the conservation of the presymplectic current up to BV-exact terms.[13] In fact it satisfies the following stronger relation

$$\iota_Q \omega_{\mathrm{BV}} = dS + \pi^* \theta, \tag{2.1}$$

which turns out to be the transgressed and BV-theoretic version of the fundamental variational equation (1.1):

---

[13] This was first pointed out by us informally on the nLab in October 2011, http://ncatlab.org/nlab/revision/diff/phase+space/29.

$$dS = d\int_\Sigma L$$
$$= \int_\Sigma dL$$
$$= \int_\Sigma (\text{EL} - d_H\Theta)$$
$$= \int_\Sigma (\iota_Q \Omega_{\text{BV}} - d_H\Theta)$$
$$= \iota_Q \omega_{\text{BV}} - \pi^*\theta.$$

Equation (2.1) has been postulated as the fundamental compatibility condition for BV-theory on spacetimes Σs with boundary in [18, Equation (7)]. Applying d to both sides of this equation recovers the previous $d_{\text{BV}}\omega_{\text{BV}} = \pi^*\omega$.

Notice that equation (2.1) may be read as saying that the action functional is a Hamiltonian, not for the ordinary presymplectic structure, but for the BV-symplectic structure.

Concept in classical field theory	Local model in in BV-BRST formalism
$[N_\Sigma^\infty \Sigma_p, \mathcal{E}]_\Sigma$ phase space	d_{BV} BV-complex of antifields
$\omega_{\text{in}} = \omega_{\text{out}}$ independence of presymplectic form from choice of Cauchy surface	$\omega_{\text{in}} \xrightarrow{\omega_{\text{BV}}}_{\simeq} \omega_{\text{out}}$ coboundary by BV-bracket
$[N_\Sigma^\infty \Sigma_p, \mathcal{E}]_\Sigma \longrightarrow [\Sigma, \mathbf{B}G_{\text{conn}}]$ smooth groupoid of gauge fields and gauge transformations	d_{BRST} BRST complex of ghost fields
$[\Sigma_p, \mathcal{E}]_\Sigma \longrightarrow [\Sigma, \mathbf{B}^k U(1)_{\text{conn}}]$ higher smooth groupoid of higher gauge fields and higher gauge transformations	d_{BRST} BRST complex of higher-order ghost-of-ghost fields

2.2 Sigma-Model Field Theories

A *sigma-model* is a field theory whose field bundle (as in Section 1) is of the simple form

$$\begin{array}{c} \Sigma \times X \\ \downarrow {\scriptstyle p_1} \\ \Sigma \end{array}$$

for some space X. This means that in this case field configurations, which by definition are sections of the field bundle, are equivalently maps of the form

$$\phi: \Sigma \longrightarrow X.$$

One naturally thinks of such a map as being a Σ-shaped trajectory of a p-dimensional object (a p-brane) on the space X. Hence X is called the *target space* of the model. Specifically, if this models Σ-shaped trajectories of p-dimensional relativistic branes, then X is the *target spacetime*. There are also famous examples of sigma-models where X is a more abstract space, usually some moduli space of certain scalar fields of a field theory that is itself defined on spacetime. Historically, the first sigma-models were of this kind. In fact, in the first examples, X was a linear space. For emphasis that this is not assumed one sometimes speaks of *nonlinear sigma models* for the sigma-models that we consider here. In fact, we consider examples where X is not even a manifold, but a smooth ∞-groupoid, a higher moduli stack.

Given a target space X, then every $(p+1)$-form $A_{p+1} \in \Omega^{p+1}(X)$ on X induces a local Lagrangian for sigma-model field theories with target X: we may simply pull back that form to the jet bundle $J_\Sigma^\infty(\Sigma \times X)$ and project out its horizontal component. Lagrangians that arise this way are known as *topological terms*.

The archetypical example of a sigma-model with topological term is that for describing the electron propagating in a spacetime and subjected to the background forces of gravity and of electromagnetism. In this case $p = 0$ (a point particle, hence a "0-brane"), Σ is the interval $[0,1]$ or the circle S^1, regarded as the abstract *worldline* of an electron. Target space X is a spacetime manifold equipped with a pseudo-Riemannian metric g (modeling the background field of gravity) and with a vector potential 1-form $A \in \Omega^1(X)$ whose differential is the Faraday tensor $F = dA$ (modeling the electromagnetic background field). The local Lagrangian is

$$L = L_{\text{kin}} + \underbrace{q(A_\Sigma)_H}_{L_{\text{int}}} \in \Omega_H^{p+1}(J_\Sigma^\infty(\Sigma \times X)),$$

where L_{kin} is the standard kinetic Lagrangian for (relativistic) point particles, q is some constant, the electric charge of the electron, and $(A_\Sigma)_H$ is the horizontal component of the pullback of A to the jet bundle. The variation of L_{int} yields the Lorentz force that the charged electron experiences.

Now, as in the discussion in Section 1, in general the Faraday tensor F is not globally exact, and hence in general there does not exist a globally defined such 1-form on the jet bundle. But via the sigma-model construction, the prequantization of the worldline field theory of the electron on its jet bundle is naturally induced by a Dirac charge quantization of its background electromagnetic field on target spacetime: given

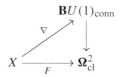

a circle-principal connection on target spacetime for the given field strength Faraday tensor F (hence with local "vector potential" 1-forms $\{A_i\}$ with respect to some cover $\{U_i \to X\}$), then the horizontal projection $(\nabla_\Sigma)_H$ of the pullback of the whole circle-bundle with connection to the jet bundle constitutes a prequantum field theory in the sense of Section 1.3. Similarly, the background electromagnetic field ∇ also serves to prequantize the covariant phase space of the electron, according to Section 1.4. This is related to the familiar statement that in the presence of a magnetic background field the spatial coordinates of the electron no longer Poisson-commute with each other.

This prequantization of sigma-models via $(p + 1)$-form connections on target space works generally: we obtain examples of prequantum field theories of sigma-model type by adding to a globally defined kinetic Lagrangian form a prequantum *topological term* given by the pullback of a $(p + 1)$-form connection on target space. The pullback of that target $(p+1)$-form connection to target space serves to prequantize the entire field theory in all codimensions.

Prequantum sigma-model topological terms			
Background field	$\nabla :$	X	$\longrightarrow \mathbf{B}^{p+1} U(1)_{\text{conn}}$
Prequantum Lagrangian	$(\nabla_\Sigma)_H :$	$\Sigma \times X$	$\longrightarrow \mathbf{B}^{p+1}_H U(1)_{\text{conn}}$
Prequantized phase space	$(\nabla_\Sigma)_L :$	$\mathcal{E}$	$\longrightarrow \mathbf{B}^{p+1}_L U(1)_{\text{conn}}$

While sigma-models with topological terms are just a special class among all variational field theories, in the context of higher differential geometry this class is considerably larger than in traditional differential geometry. Namely, we may regard any of the moduli stacks $\mathbf{A}_{\text{conn}}$ of gauge fields that we discuss in Section 2.1 as target space, that is, we may consider higher stacky field bundles of the form

$$\Sigma \times \mathbf{A}_{\text{conn}} \\ \downarrow \\ \Sigma$$

Everything goes through as before, in particular a field configuration now is a map $\Sigma \longrightarrow \mathbf{A}_{\mathrm{conn}}$ from worldvolume/spacetime Σ to this moduli stack. But by the discussion above in Section 2.1, such maps now are equivalent to gauge fields on Σ. These are, of course, the field configurations of *gauge theories*. Hence, in higher differential geometry, the concepts of sigma-model field theories and of gauge field theories are unified.

In particular, both concepts may mix. Indeed, we find below that higher dimensional WZW-type models generally are "higher gauged," this means that their field configurations are a pair consisting of a map $\phi\colon \Sigma \to X$ to some target spacetime X, together with a ϕ-twisted higher gauge field on Σ.

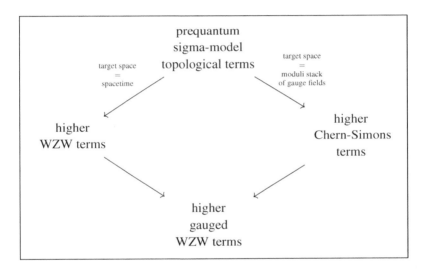

Examples of a (higher) gauged WZW-type sigma model are the Green–Schwarz-type sigma-models of those super p-branes on which other branes may end. This includes the D-branes and the M5-brane. The former are gauged by a 1-form gauge field (the "Chan–Paton gauge field") while the latter is gauged by a 2-form gauge field. We say more about these examples in Section 2.4.

We may construct examples of prequantized topological terms from functoriality of the Lie integration process that already gave the (higher) gauge fields themselves in Section 2.1. There we saw that a $(p+2)$-cocycle on an L_∞-algebroid is a homomorphism of L_∞-algebroids of the form

$$\mu : \mathfrak{a} \longrightarrow \mathbf{B}^{p+2}\mathbb{R}.$$

Moreover, the $\exp(-)$-construction which sends L_∞-algebroids to simplicial presheaves representing universal higher moduli stacks of $\mathfrak{a}$-valued gauge fields is clearly functorial, hence it sends this cocycle to a morphism of simplicial presheaves of the form

$$\exp(\mu) : \exp(\mathfrak{a}) \longrightarrow \mathbf{B}^{p+2}\mathbb{R}.$$

One finds that this descends to the $(p+2)$-coskeleton $\mathbf{A} := \mathrm{cosk}_{p+2}\exp(\mathfrak{a})$ after quotienting out the subgroup $\Gamma \hookrightarrow \mathbb{R}$ of periods of μ [16] (just as in the prequantization of the global action functional in Section 1.3):

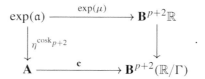

To get a feeling for what the resulting morphism $\mathbf{c}$ is, consider the case that $\mathbf{A} = \mathbf{B}G$ for some group G. There is a geometric realization operation π_∞ which sends smooth ∞-groupoids to plain homotopy types (homotopy types of topological spaces). Under this operation a map $\mathbf{c}$ as above becomes a map c of the form

$$\begin{array}{ccc} \mathbf{B}G & \xrightarrow{\mathbf{c}} & \mathbf{B}^{p+2}(\mathbb{R}/\mathbb{Z}) \\ \downarrow{\eta^{\pi_\infty}} & & \downarrow{\eta^{\pi_\infty}} \\ BG & \xrightarrow{c} & K(\mathbb{Z}, p+3) \end{array},$$

where BG is the traditional classifying space of a (simplicial) topological group G, and where $K(\mathbb{Z}, p+3) = B^{p+3}\mathbb{Z}$ is the Eilenberg–Mac Lane space that classifies integral cohomology in degree $(p+3)$. What BG classifies are G-principal bundles, and hence for each space Σ the map c turns into a *characteristic class* of equivalence classes of G-principal bundles:

$$c_\Sigma : G\mathrm{Bund}(\Sigma)_\sim \longrightarrow H^{p+3}(\Sigma, \mathbb{Z}).$$

Hence c itself is a *universal characteristic class*. Accordingly, $\mathbf{c}$ is a refinement of c that knows about gauge transformations: it sends smooth G-bundles with smooth gauge transformations and gauge-of-gauge transformations between these to integral cocycles and coboundaries and coboundaries-between-coboundaries between these.

Equivalently, we may think of $\mathbf{c}$ as classifying a $(p+1)$-gerbe on the universal moduli stack of G-principal bundles. This is equivalently its homotopy fiber (in direct analogy with the infinitesimal version of this statement in Section 2.1.2) fitting into a long homotopy fiber sequence of the form

$$\begin{array}{ccc} \mathbf{B}^{p+1}(\mathbb{R}/\mathbb{Z}) & \longrightarrow & \mathbf{B}\hat{G} \\ & & \downarrow \\ & \mathbf{B}G & \xrightarrow{\mathbf{c}} \mathbf{B}^{p+2}(\mathbb{R}/\mathbb{Z}) \end{array}$$

Yet another equivalent perspective is that this defines an ∞-group extension $\hat{G}$ of the ∞-group G by the ∞-group $\mathbf{B}^p(\mathbb{R}/\mathbb{Z})$.

So far all this is without connection data; so far these are just higher instanton sectors without any actual gauge fields inhabiting these instanton sectors. We now add connection data to the situation.

Adding connection data to **c** regarded as a higher prequantum bundle on the moduli stack **B**G yields

- Chern–Simons-type prequantum field theory.

Adding instead connection data to **c** regarded as a higher group extension yields

- Wess–Zumino–Witten-type prequantum field theories.

2.3 Chern–Simons-Type Field Theory

For $\mathfrak{g}$ a semisimple Lie algebra with Killing form invariant polynomial $\langle -, - \rangle$, classical 3d Chern–Simons theory [41] has as fields the space of $\mathfrak{g}$-valued differential 1-forms A, and the Lagrangian is the Chern–Simons 3-form

$$L_{\mathrm{CS}}(A) = \mathrm{CS}(A) := \langle A \wedge dA \rangle - \tfrac{1}{3}\langle A \wedge [A \wedge A] \rangle.$$

This Chern–Simons form is characterized by two properties: for vanishing curvature it reduces to the value of the 3-cocycle $\langle -, [-, -] \rangle$ on the connection 1-form A, and its differential is the value of the invariant polynomial $\langle -, - \rangle$ on the curvature 2-form F_A.

There is a slick way to express this in terms of the dg-algebraic description from Section 2.1.2: there is an element cs $\in W(\mathbf{B}\mathfrak{g})$, which in terms of the chosen basis $\{t^a\}$ for $\wedge^1 \mathfrak{g}^*$ is given by

$$\mathrm{cs} : k_{ab}(d_W t^a) \wedge t^b - \tfrac{1}{3} k_{aa'} C^{a'}{}_{bc} t^a \wedge t^b \wedge t^c.$$

Hence equivalently this is a dg-homomorphism of the form

$$W(\mathbf{B}\mathfrak{g}) \xleftarrow{\mathrm{cs}} W(\mathbf{B}^3\mathbb{R})$$

and for $A \in \Omega^1(\Sigma, \mathfrak{g}) = \{\Omega^\bullet(\Sigma) \longleftarrow W(\mathbf{B}\mathfrak{g})\}$ then the Chern–Simons form of A is the composite

$$\Omega^\bullet(\Sigma) \xleftarrow{A} W(\mathbf{B}\mathfrak{g}) \xleftarrow{\mathrm{cs}_3} : \mathrm{CS}(A).$$

Now, the two characterizing properties satisfied by the Chern–Simons equivalently mean in terms of dg-algebra that the map cs makes the following two squares commute:

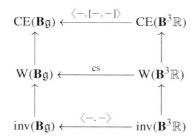

This shows how to prequantize 3d Chern–Simons theory in codimension 3: the vertical sequences appearing here are just the Lie algebraic data to which we apply differential Lie integration, as in Section 2.1, to obtain the moduli stacks of G-connections and of 3-form connections, respectively. Moreover, by the discussion at the end of Section 2.2 and using that $\langle -, [-, -] \rangle$ represents an integral cohomology class on G we get a map

$$(\mathbf{c}_2)_{\mathrm{conn}} := \exp(\mathrm{cs}) : \mathbf{B}G_{\mathrm{conn}} \longrightarrow \mathbf{B}^3(\mathbb{R}/\mathbb{Z})_{\mathrm{conn}}.$$

This is the background 3-connection that induces prequantum Chern–Simons field theory by the general procedure indicated in Section 2.2.

Notice that this map is a refinement of the traditional Chern–Weil homomorphism. This allows for instance to prequantize the Green–Schwarz anomaly cancellation condition for heterotic strings: the higher moduli stack of GS-anomaly free gauge fields is the homotopy fiber product of the prequantum Chern–Simons Lagrangians for the simple groups Spin and SU [79].

This higher Lie-theoretic formulation of prequantum 3-Chern–Simons theory now immediately generalizes to produce higher (and lower) dimensional prequantum L_∞-algebroid Chern–Simons theories.

For $\mathfrak{a}$ any L_∞-algebroid as in Section 2.1.2, we say that a $(p+2)$-cocycle μ on $\mathfrak{a}$ is in transgression with an invariant polynomial $\langle - \rangle$ on $\mathfrak{a}$ if there is an element $cs \in W(\mathfrak{a})$ such that $d_W cs = \langle - \rangle$ and $cs|_{CE} = \mu$. Equivalently this means that cs fits into a diagram of dg-algebras of the form

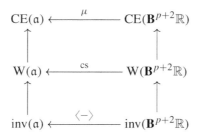

Applying $\exp(-)$ to this induces maps of smooth moduli stacks of the form

$$\mathbf{c}_{\text{conn}} : \mathbf{A}_{\text{conn}} \longrightarrow \mathbf{B}^{p+2}(\mathbb{R}/\Gamma)_{\text{conn}}.$$

This gives a prequantum Chern–Simons-type field theory whose field configurations locally are $\mathfrak{a}$-valued differential forms, and whose Lagrangian is locally the Chern–Simons element cs evaluated on these forms.

For instance if $(\mathfrak{a}, \omega)$ is a symplectic Lie p-algebroid, then we obtain the prequantization of $(p + 1)$-dimensional AKSZ-type field theories [30]. For $p = 1$ this subsumes the topological string A- and B-model. Generally, the prequantum moduli stack of fields for two-dimensional prequantum AKSZ theory is a differential refinement of the symplectic groupoid of a given Poisson manifold [12]. The Poisson manifold canonically defines a boundary condition for the corresponding prequantum 2d Poisson–Chern–Simons theory, and the higher geometric boundary quantization of this 2d prequantum theory reproduces ordinary Kostant–Souriau geometric quantization of the symplectic leaves [71]. This is a nonperturbative improvement of the perturbative algebraic deformation quantization of the Poisson manifold as the boundary of the perturbative 2d AKSZ field theory due to [21].

Generally one expects to find nontopological, nonperturbative p-dimensional quantum field theories arising this way as the higher geometric boundary quantization of $(p + 1)$-dimensional prequantum Chern–Simons-type field theories (see [84] or [81]).

For instance for $(\mathbf{B}^3\mathbb{R}, \omega)$ the line Lie 3-algebra equipped with its canonical binary invariant polynomial, the corresponding prequantum Chern–Simons-type field theory is the 7d abelian cup-product Chern–Simons theory [35]. This has been argued to induce on its boundary the conformal six-dimensional field theory of a self-dual 2-form field (see [108], [51]). This seven-dimensional Chern–Simons theory is one summand in the Chern–Simons term of 11d supergravity compactified on a 4-sphere. The $\text{AdS}_7/\text{CFT}_6$

correspondence predicts that this carries on its boundary the refinement of the self-dual 2-form to a 6d superconformal field theory. There are also nonabelian summands in this 7d Chern–Simons term. For instance for $(\mathbf{B}\mathfrak{string}_{\mathfrak{g}}, \langle -, -, -, -\rangle)$ the string Lie 2-algebra equipped with its canonical degree-4 invariant polynomial, then the resulting prequantum field theory is 7d Chern–Simons field theory on String 2-connection fields [34].

For more exposition of prequantum Chern–Simons-type field theories, see also [36].

2.4 Wess–Zumino–Witten-Type Field Theory

The traditional Wess–Zumino–Witten (WZW) field theory [43, 44] for a semisimple, simply connected compact Lie group G is a 2-dimensional sigma-model with target space G, in the sense of Section 2.2, given by a canonical kinetic term, and with topological term that is locally a potential for the left-invariant 3-form $\langle \theta \wedge [\theta \wedge \theta]\rangle \in \Omega^3(G)_{\mathrm{cl}}$, where θ is the Maurer–Cartan form on G. This means that for $\{U_i \to G\}$ a cover of G by coordinate charts $U_i \simeq \mathbb{R}^n$, then the classical WZW model is the locally variational classical field theory (in the sense discussed in Section 1.2) whose local Lagrangian L_i is (in the notation introduced in Section 2.2) $L_i = (L_{\mathrm{kin}})_i + ((B_i)_\Sigma)_H$ for $B_i \in \Omega^2(U_i)$ a 2-form such that $dB_i = \langle \theta \wedge [\theta \wedge \theta]\rangle|_{U_i}$.

By the discussion in Section 2.2, in order to prequantize this field theory it is sufficient that we construct a $U(1)$-gerbe on G whose curvature 3-form is $\langle \theta \wedge [\theta \wedge \theta]\rangle$. In fact we may ask for a little more: we ask for the gerbe to be *multiplicative* in that it carries 2-group structure that covers the group structure on G, hence that it is given by the 2-group extension classified by the smooth universal class $\mathbf{c} : \mathbf{B}G \longrightarrow \mathbf{B}^3 U(1)$.

An elegant construction of this prequantization, which will set the scene for the general construction of higher WZW models, proceeds by making use of a universal property of the differential coefficients. Namely, one finds that for all $p \in \mathbb{Z}$, then the moduli stack $\mathbf{B}^{p+1}(\mathbb{R}/\mathbb{Z})_{\mathrm{conn}}$ of $(p+1)$-form connections is the homotopy fiber product of $\mathbf{B}^{p+1}(\mathbb{R}/\mathbb{Z})$ with $\Omega_{\mathrm{cl}}^{p+2}$ over $\flat_{\mathrm{dR}}\mathbf{B}^{p+2}\mathbb{R}$.

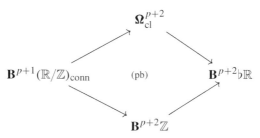

Here "♭" indicates the discrete underlying group, and hence this homotopy pullback says that giving a $(p+1)$-form connection is equivalent to giving an integral $(p+2)$-class and a closed $(p+2)$-form together with a homotopy that identifies the two as cocycles in real cohomology.

In view of this, consider the following classical Lie-theoretic data associated with the semisimple Lie algebra $\mathfrak{g}$:

$\mathfrak{g}$	semisimple Lie algebra
G	its simply connected Lie group
$\theta \in \Omega^1(G, \mathfrak{g})$	Maurer–Cartan form
$\langle -, - \rangle$	Killing metric
$\mu_3 = \langle -, [-, -] \rangle$	Lie algebra 3-cocycle
$k \in H^3(G, \mathbb{Z})$	level
$\mu_3(\theta \wedge \theta \wedge \theta) \xrightarrow[\simeq]{q} k_{\mathbb{R}}$	prequantization condition

Diagrammatically, this data precisely corresponds to a diagram as shown on the left in the following, and hence the universal property of the homotopy pullback uniquely associates a lift ∇_{WZW} as on the right:

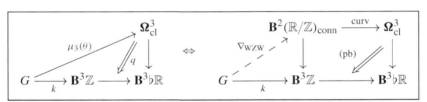

This ∇_{WZW} is the required prequantum topological term for the 2d WZW model. Hence the prequantum 2d WZW sigma-model field theory is the $(p = 2)$-dimensional prequantum field theory with target space the group G and with local prequantum Lagrangian, that is, with Euler–Lagrange gerbe given by

$$\mathbf{L} := \underbrace{\langle \theta_H \wedge \star \theta_H \rangle}_{\mathbf{L}_{\text{kin}}} + \underbrace{(\nabla_{\text{WZW}})_H}_{\mathbf{L}_{\text{WZW}}} \;:\; \Sigma \times G \longrightarrow \mathbf{B}_H^{p+1}(\mathbb{R}/\hbar\mathbb{Z})_{\text{conn}}.$$

This prequantization is a de-transgression of a famous traditional construction. To see this, write $\hat{\Omega}_k G$ for level-k Kac–Moody loop group extension of G. This has an adjoint action by the based path group $P_e G$. Write

$$\text{String}(G) := P_e G /\!/ \hat{\Omega}_k G$$

for the homotopy quotient. This is a differentiable group stack, called the *string 2-group* [8]. It turns out to be the total space of the 2-bundle underlying ∇_{WZW}

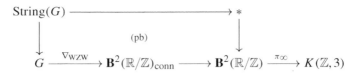

and it is a de-transgression of the Kac–Moody loop group extension $\hat{L}_K G$. Transgressing to fields over the circle gives:

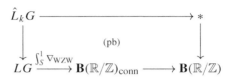

The string 2-group also appears again as the 2-group of Noether symmetries, in the sense of Section 1.5, of the prequantum 2d WZW model. The Noether homotopy fiber sequence for the prequantum 2d WZW model looks as follows [31]:

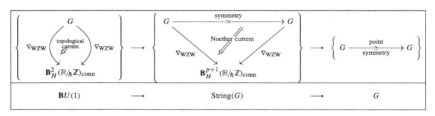

In fact, this extension is classified by the smooth universal characteristic class $\mathbf{c}\colon \mathbf{B}G \longrightarrow \mathbf{B}^3 U(1)$, whose differential refinement gave 3d Chern–Simons theory in Section 2.3.

Given a G-principal bundle $P \to X$, then one may ask for a fiberwise parameterization of ∇_{WZW} over P. If such *definite parameterization* $\nabla\colon P \to \mathbf{B}^2(\mathbb{R}/\mathbb{Z})_{\text{conn}}$ exists, then it defines the prequantum topological term for the *parameterized WZW model* with target space P.

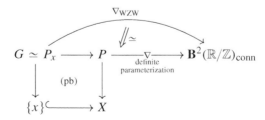

Such a parameterization is equivalent to a lift of the structure group of P through the above extension $\text{String}(G) \longrightarrow G$. Accordingly, the obstruction to parameterizing ∇_{WZW} over P is the universal extension class **c** evaluated on P. Specifically for the case that $G = \text{Spin} \times \text{SU}$, this is the sum of fractional Pontryagin and second Chern class:

$$\tfrac{1}{2} p_1 - c_2 \in H^4(X, \mathbb{Z}).$$

The vanishing of this class is the *Green–Schwarz anomaly cancellation* condition for the 2d field theory describing propagation of the heterotic string on X. This perspective on the Green–Schwarz anomaly via parameterized WZW models had been suggested in [26]. The prequantum field theory we present serves to make this precise and to generalize it to higher dimensional parameterized WZW-type field theories.

Generally, given any L_∞-cocycle $\mu \colon \mathbf{B}\mathfrak{g} \longrightarrow \mathbf{B}^{p+2}\mathbb{R}$ as in Section 2.1.2 with induced smooth ∞-group cocycle $\mathbf{c} \colon \mathbf{B}G \longrightarrow \mathbf{B}^{p+2}(\mathbb{R}/\Gamma)$ as in Section 2.2, then there is a higher analog of the universal construction of the WZW-type topological term ∇_{WZW}.

First of all, the homotopy pullback characterization of $\mathbf{B}^{p+1}(\mathbb{R}/\mathbb{Z})_{\text{conn}}$ refines to one that does not just involve the geometrically discrete coefficients $\mathbf{B}^{p+2}\mathbb{Z}$, but the smooth coefficients $\mathbf{B}^{p+1}(\mathbb{R}/\mathbb{Z})$.

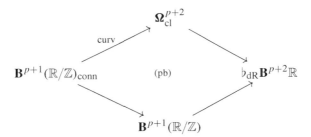

Here $\flat_{\text{dR}}(-)$ denotes the homotopy fiber of the canonical map $\flat(-) \longrightarrow (-)$ embedding the underlying discrete smooth structure of any object into the given smooth object. A key aspect of the theory is that the further homotopy fiber of $\flat_{\text{dR}}(-) \longrightarrow \flat(-)$ has the interpretation of being the Maurer–Cartan form θ on the given smooth ∞-groupoid:

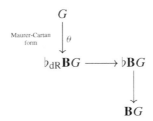

Or rather, one finds that $\flat_{dR}\mathbf{B}G \simeq \mathbf{\Omega}_{\mathrm{flat}}^{1 \leqslant \bullet \leqslant p+2}(-, \mathfrak{g})$ is the coefficient for "hypercohomology" in flat $\mathfrak{g}$-valued differential forms, hence for G a higher smooth group then its Maurer–Cartan form θ is not, in general, a globally defined differential form, but instead a system of locally defined forms with higher coherent gluing data.

But one may universally force θ to become globally defined, so to speak, by pulling it back along the inclusion $\mathbf{\Omega}_{\mathrm{flat}}(-, \mathfrak{g})$ of the globally defined flat $\mathfrak{g}$-valued forms. This defines a differential extension $\tilde{G}$ of G equipped with a globally defined Maurer–Cartan form $\tilde{\theta}$, by the following homotopy pullback diagram

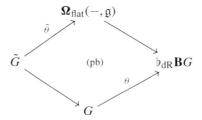

When G is an ordinary Lie group, it so happens that $\flat_{dR}\mathbf{B}G \simeq \mathbf{\Omega}_{\mathrm{flat}}(-, \mathfrak{g})$, and so in this case $\tilde{G} \simeq G$ and $\tilde{\theta} \simeq \theta$, so that nothing new happens.

At the other extreme, when $G = \mathbf{B}^{p+1}(\mathbb{R}/\mathbb{Z})$, then $\theta \simeq$ curv as above, and so in this case one finds that $\tilde{G}$ is $\mathbf{B}^{p+1}(\mathbb{R}/\mathbb{Z})_{\mathrm{conn}}$ and that $\tilde{\theta} \simeq F_{(-)}$ is the map that sends an $(p+1)$-form connection to its globally defined curvature $(p+2)$-form.

More generally, these two extreme cases mix: when G is a $\mathbf{B}^p(\mathbb{R}/Z)$-extension of an ordinary Lie group, then $\tilde{G}$ is a twisted product of G with $\mathbf{B}^p(\mathbb{R}/\mathbb{Z})_{\mathrm{conn}}$, then a single map

$$(\phi, B) \colon \Sigma \longrightarrow \tilde{G}$$

is a pair consisting of an ordinary sigma-model field ϕ together with a ϕ-twisted p-form connection on Σ.

Hence the construction of $\tilde{G}$ is a twisted generalization of the construction of differential coefficients. In particular, given an L_∞-cocycle $\mu \colon \mathbf{B}\mathfrak{g} \longrightarrow \mathbf{B}^{p+2}\mathbb{R}$ Lie-integrating to an ∞-group cocycle $\mathbf{c} \colon \mathbf{B}G \longrightarrow \mathbf{B}^{p+2}(\mathbb{R}/\Gamma)$, then it Lie integrates to a prequantum topological term $\nabla_{\mathrm{WZW}} \colon \tilde{G} \longrightarrow \mathbf{B}^{p+1}(\mathbb{R}/\Gamma)_{\mathrm{conn}}$ via the universal dashed map in the following induced diagram:

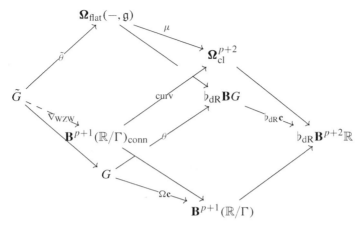

This construction provides a large supply of prequantum WZW-type field theories. Indeed, by the discussion in 2.1.2, from every L_∞-algebroid there emanates a bouquet of L_∞-extensions with L_∞-cocycles on them, hence for every WZW-type sigma model prequantum field theory, we find a whole bouquet of prequantum field theories emanating from it.

Therefore it is interesting to consider the simplest nontrivial L_∞-algebroids and see which bouquets of prequantum field theories they induce. The abelian line Lie algebra $\mathbb{R}$ is arguably the simplest nonvanishing L_∞-algebroid, but it is in fact a little too simple for this purpose. The bouquet it induces is not interesting. But all of the above generalizes essentially verbatim to superalgebra and supergeometry, and in super-Lie-algebra theory we have the "odd lines" $\mathbb{R}^{0|q}$, that is, the superpoints. The bouquet which emanates from these turns out to be remarkably rich [37]; it gives the entire p-brane spectrum of string theory/M-theory.

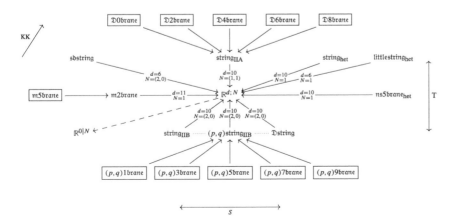

Each entry in this diagram denotes a super L_∞-algebra extension of some super Minkowski spacetime $\mathbb{R}^{d-1,1|N}$ (regarded as the corresponding supersymmetry super Lie algebra), and each arrow denotes a super-L_∞-extension classified by a $p+2$ cocycle for some p. By the above general construction, this cocycle induces a $(p+1)$-dimensional WZW-type sigma-model prequantum field theory with target space a higher extension of super-Minkowski spacetime [37], and the names of the super L_∞-algebras in the above diagram correspond to the traditional names of these super p-branes.

As for the traditional WZW-models, all of this structure naturally generalizes to its parameterized versions: given any higher extended super Minkowski spacetime V equipped with a prequantum topological term $\nabla_{\text{WZW}}\colon V \longrightarrow \mathbf{B}^{p+1}(\mathbb{R}/\Gamma)_{\text{conn}}$ for a super p-brane sigma model, we may ask for globalizations of ∇ over V-manifolds (V-étale stacks) X, hence for topological term ∇ on all of X that is suitably equivalent on each infinitesimal disk $\mathbb{D}^X_x \simeq \mathbb{D}^V_e$ to ∇_{WZW}:

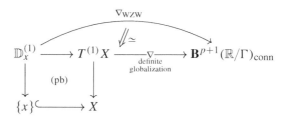

Such globalizations serve as prequantum topological terms for WZW-type sigma-models describing the propagation of super p-branes on V-manifolds X (e.g., [28, Sections 2 and 3]). One finds that such globalizations equip the higher frame bundle of X with a lift of its structure group through a canonical map $\mathbf{Stab}_{\text{GL}(V)}(\nabla) \longrightarrow \text{GL}(V)$ from the homotopy stabilizer group of the WZW term, in direct analogy to the previous examples. Apart from "cancelling the classical anomalies" of making the super p-brane WZW-type sigma-model be globally defined on X, such a lift induces metric structure on X.

Since the homotopy stabilization of ∇ in particular stabilizes its curvature form, there is a reduction of the structure group of the V-manifold in direct analogy to how a globalization of the "associative" 3-form α on $\mathbb{R}^7$ equips a 7-manifold with G_2-structure. For the above super p-brane models the relevant stabilizer is the spin-cover of the Lorentz group, and hence globalizing the prequantum p-brane model over X in particular induces orthogonal structure on X, hence equips X with a field configuration of supergravity.

Given such a globalization of a topological term ∇ over a V-manifold X, it is natural to require it to be infinitesimally integrable. In the present example this comes out to imply that the torsion of the orthogonal structure on X vanishes. This is particularly interesting at the top end of the brane bouquet: for globalization over an 11-dimensional supermanifold, the vanishing of the torsion is equivalent to X satisfying the equations of motion of 11d gravity [19]. The Noether charges of the corresponding WZW-type prequantum field theory are the *supergravity BPS-charges* [78].

Here the relation to G_2-structure is more than an analogy. We may naturally lift the topological term for the M2-brane sigma-model from $\mathbb{R}/\mathbb{Z}$-coefficients to $\mathbb{C}/\mathbb{Z}$-coefficients by adding $\alpha\colon \mathbb{R}^{10,1|32} \to \mathbb{R}^7 \to \mathbf{\Omega}^3_{\mathrm{cl}}$. Then a globalization of the complex linear combination

$$\nabla_{\mathrm{M2}} + i\alpha \colon \mathbb{R}^{10,1|32} \longrightarrow \mathbf{B}^3(\mathbb{C}/\Gamma)_{\mathrm{conn}}$$

over an 11-dimensional supermanifold X equips X with the structure of a G_2-fibration over a 4-dimensional $N = 1$ supergravity spacetime. The volume holonomy of $\nabla_{\mathrm{M2}} + i\alpha$ around supersymmetric 3-cycles are the "M2-instanton contributions." This setup of 11d supergravity Kaluza–Klein-compactified on G_2-manifolds to four spacetime dimensions and with the prequantum M2/M5-brane charges and instantons included – known as *M-theory on G_2-manifolds* [1, 7] – comes at least close to capturing the qualitative structure of experimentally observed fundamental physics:

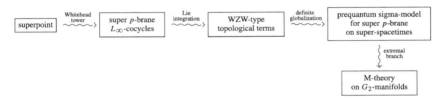

This shows that there is some interesting physics encoded in those prequantum field theories that are canonically induced from a minimum of input data.

References

[1] B. Acharya, *M Theory, G_2-manifolds and four dimensional physics*, Classical Quantum Gravity 19 (2002), http://users.ictp.it/~pub_off/lectures/lns013/Acharya/Acharya_Final.pdf

[2] M. Alexandrov, M. Kontsevich, A. Schwarz, and O. Zaboronsky, *The geometry of the master equation and topological quantum field theory*, Int. J. Modern Phys. A 12 (1997) 1405–29

[3] I. M. Anderson, *The variational bicomplex*, Unpublished manuscript, http://math.uni.lu/~jubin/seminar/bicomplex.pdf

[4] I. M. Anderson and T. Duchamp, *On the existence of global variational principles*, Am. J. Math. 102 (1980) 781–868

[5] M. Ando, A. Blumberg, D. Gepner, M. Hopkins, and C. Rezk, *An ∞-categorical approach to R-line bundles, R-module Thom spectra, and twisted R-homology*, J. Topol. 7 (3) (2013) 869–93

[6] V. I. Arnold, *Mathematical Methods of Classical Mechanics*, 2nd ed., Graduate Texts in Mathematics 60, Springer (1989)

[7] M. Atiyah and E. Witten, *M-theory dynamics on a manifold of G_2-holonomy*, Adv. Theor. Math. Phys 6 (2003) 1–106

[8] J. Baez, A. Crans, U. Schreiber, and D. Stevenson, *From loop groups to 2-groups*, Homol. Homotopy Appl. 9 (2007) 101–35

[9] G. Barnich and M. Henneaux, *Isomorphisms between the Batalin-Vilkovisky antibracket and the Poisson bracket*, J. Math. Phys. 37 (1996) 5273–96

[10] S. Bates and A. Weinstein, *Lectures on the Geometry of Quantization*, American Mathematical Society (1997), www.math.berkeley.edu/~alanw/GofQ.pdf

[11] J. Bergner, *Models for (∞, n)-categories and the cobordism hypothesis*, in H. Sati and U. Schreiber, eds., *Mathematical Foundations of Quantum Field Theory and Perturbative String Theory*, Proceedings of Symposia in Pure Mathematics 83, AMS (2011)

[12] S. Bongers, *Geometric quantization of symplectic and Poisson manifolds*, MSc thesis, Utrecht (2014), http://ncatlab.org/schreiber/show/master+thesis+Bongers

[13] F. Borceux, *Handbook of Categorical Algebra*, 3 vols, Cambridge University Press

[14] R. Bott and L. Tu, *Differential Forms in Algebraic Topology*, Graduate Texts in Mathematics 82, Springer (1982)

[15] K. Brown, *Abstract homotopy theory and generalized sheaf cohomology*, Trans. Amer. Math. Soc. 186 (1973) 419–58, http://ncatlab.org/nlab/files/BrownAbstractHomotopyTheory.pdf

[16] U. Bunke, *Differential cohomology*, Preprint, arXiv:1208.3961

[17] U. Bunke, T. Nikolaus, and M. Völkl, *Differential cohomology as sheaves of spectra*, J. Homotopy Related Struct. October (2014), arxiv:1311.3188

[18] A. Cattaneo, P. Mnev, and N. Reshetikhin, *Classical BV theories on manifolds with boundary*, Commun. Math. Phys. 332 (2014) 535–603

[19] A. Candiello and K. Lechner, *Duality in supergravity theories*, Nucl. Phys. B412 (1994) 479–501

[20] L. Castellani, R. D'Auria, and P. Fré, *Supergravity and Superstrings – A Geometric Perspective*, World Scientific (1991)

[21] A. Cattaneo and G. Felder, *A path integral approach to the Kontsevich quantization formula*, Commun. Math. Phys. 212 (2000) 591–611

[22] X. Chen, Z.-C. Gu, Z.-X. Liu, and X.-G. Wen, *Symmetry protected topological orders and the group cohomology of their symmetry group*, Phys. Rev. B 87 (2013) 155114

[23] L. Corry, *David Hilbert and the Axiomatization of Physics: From Grundlagen der Geometrie to Grundlagen der Physik*, Archimedes: New Studies in the History and Philosophy of Science and Technology 10, Kluwer Academic (2004)

[24] L. Corry, *On the origins of Hilbert's sixth problem: physics and the empiricist approach to axiomatization*, in *Proceedings of the International Congress of Mathematics in Madrid* (2006)

[25] P. A. M. Dirac, *Quantized singularities in the electromagnetic field*, Proc. R. Soc. A 133 (1931) 60–72

[26] J. Distler and E. Sharpe, *Heterotic compactifications with principal bundles for general groups and general levels*, Adv. Theor. Math. Phys. 14 (2010) 335–98

[27] E. Dubuc, *Sur les modèles de la géométrie différentielle synthétique*, Cahiers Topol. Géométrie Différentielle Catégoriques 20 (1979) 231–79, www.numdam.org/item?id=CTGDC_1979__20_3_231_0

[28] M. Duff, *The World in Eleven Dimensions: Supergravity, Supermembranes and M-theory*, IoP (1999)

[29] M. Ferraris, M. Palese, and E. Winterroth, *Local variational problems and conservation laws*, Differential Geometry Appl. 29 (2011) S80–S85

[30] D. Fiorenza, C. L. Rogers, and U. Schreiber, *A higher Chern-Weil derivation of AKSZ sigma-models*, Int. J. Geometric Methods Mod. Phys. 10 (2013)

[31] D. Fiorenza, C. L. Rogers, and U. Schreiber, *Higher $U(1)$-gerbes in geometric prequantization*, Rev. Math. Phys. 28 (2016) 1650012

[32] D. Fiorenza, C. L. Rogers, and U. Schreiber, *L_∞-algebras of local observables from higher prequantum bundles*, Homol. Homotopy Appl. 16 (2014) 107–42

[33] H. Sati and U. Schreiber, *Lie n-algebras of BPS charges*, J. High Energ. Phys. 87 (2017)

[34] D. Fiorenza, H. Sati, and U. Schreiber, *Multiple M5-branes, String 2-connections, and 7d nonabelian Chern–Simons theory*, Adv. Theoret. Math. Phys. 18 (2014) 229–321

[35] D. Fiorenza, H. Sati, and U. Schreiber, *Higher extended cup-product Chern–Simons theories*, J. Geometry Phys. 74 (2013) 130–63

[36] D. Fiorenza, H. Sati, and U. Schreiber, *A higher stacky perspective on Chern–Simons theory*, in D. Calaque et al., eds., *Mathematical Aspects of Quantum Field Theories*, Springer (2014)

[37] D. Fiorenza, H. Sati, and U. Schreiber, *Super Lie n-algebra extensions, higher WZW models and super p-branes with tensor multiplet fields*, Int. J. Geometric Methods Mod. Phys. 12 (2015) 1550018

[38] D. Fiorenza, H. Sati, and U. Schreiber, *The WZW term of the M5-brane and differential cohomotopy*, J. Math. Phys. 56 (2015) 102301

[39] D. Fiorenza, U. Schreiber, and J. Stasheff, *Čech-cocycles for differential characteristic classes*, Adv. Theoretical Math. Phys. 16 (2012)

[40] T. Frankel, *The Geometry of Physics – An Introduction*, 2nd ed., Cambridge University Press (2012)

[41] D. Freed, *Classical Chern–Simons theory Part I*, Adv. Math. 113 (1995) 237–303; *Classical Chern–Simons theory, part II*, Houston J. Math. 28 (2002) 293–310; *Remarks on Chern–Simons theory*, Bull. AMS *(New Ser.)* 46 (2009)

[42] D. Freed, *Dirac charge quantization and generalized differential cohomology surveys*, in *Differential Geometry*, International Press (2000) 129–94

[43] K. Gawedzki, *Topological actions in two-dimensional quantum field theories*, in *Nonperturbative Quantum Field Theory*, NATO Adv. Sci. Inst. Ser. B Phys. 185, Plenum (1988)

[44] K. Gawedzki, *Conformal field theory: A case study*, in Y. Nutku, C. Saclioglu, and T. Turgut, eds., *Frontier in Physics*, Perseus (2000)

[45] G. Giachetta, L. Mangiarotti, and G. Sardanashvily, *Advanced Classical Field Theory*, World Scientific (2009)

[46] A. Grothendieck, *Pursuing stacks* (1983), https://thescrivener.github.io/PursuingStacks/

[47] F. Gruber, *Topology in dynamical Lattice QCD simulations*, PhD thesis (2013), http://epub.uni-regensburg.de/27631/

[48] G. 't Hooft, *Symmetry breaking through Bell–Jackiw anomalies*, Phys. Rev. Lett. 37 (1976), www.staff.science.uu.nl/~hooft101/gthpub/symm_br_bell_jackiw.pdf

[49] M. Henneaux and C. Teitelboim, *Quantization of Gauge Systems*, Princeton University Press (1994)

[50] D. Hilbert, *Mathematical Problems*, Bull. Amer. Math. Soc. (1902) 437–79

[51] M. Hopkins and I. Singer, *Quadratic functions in geometry, topology, and M-theory*, J. Differential Geom. 70 (2005) 329–452

[52] P. Iglesias-Zemmour, *Diffeology*, Mathematical Surveys and Monographs, AMS (2013), www.umpa.ens-lyon.fr/~iglesias/Site/The+Book.html

[53] P. Johnstone, *Sketches of an Elephant: A Topos Theory Compendium*, Oxford Logic Guides, 43, 44, Clarendon Press (2002)

[54] A. Joyal, *Notes on logoi* (2008), www.math.uchicago.edu/~may/IMA/JOYAL/Joyal.pdf

[55] C. Kapulkin, P. LeFanu Lumsdaine, and V. Voevodsky, *The simplicial model of univalent foundations*, Preprint, arXiv:1211.2851

[56] I. Khavkine, *Covariant phase space, constraints, gauge and the Peierls formula*, Int. J. Mod. Phys. A 29 (2014) 1430009

[57] I. Khavkine and U. Schreiber, *Synthetic geometry of differential equations: I. Jets and comonad structure*, Preprint, arXiv:1701.06238

[58] A. Kock, *Formal manifolds and synthetic theory of jet bundles*, Cahiers Topol. Géométrie Différentielle Catégoriques 21 (1980), www.numdam.org/item?id=CTGDC_1980__21_3_227_0

[59] A. Kock, *Synthetic Differential Geometry*, LMS Lecture Notes 333, Cambridge University Press (1981, 2006), http://home.imf.au.dk/kock/sdg99.pdf

[60] A. Kock, *Synthetic Geometry of Manifolds*, Cambridge Tracts in Mathematics 180, Cambridge University Press (2010), http://home.imf.au.dk/kock/SGM-final.pdf

[61] B. Kostant, *On the definition of quantization*, in *Géométrie Symplectique et Physique Mathématique*, Colloques Intern. 237, CNRS (1975) 187–210

[62] W. Lawvere, *An elementary theory of the category of sets*, Proc. Nat. Acad. Sci. U.S.A. 52 (1965) 1506–11. Reprinted in *Reprints in Theory and Applications of Categories*, No. 11 (2005) 1–35

[63] W. Lawvere, *Categorical dynamics*, Lecture in Chicago (1967), www.mat.uc.pt/~ct2011/abstracts/lawvere_w.pdf

[64] W. Lawvere, *Toposes of laws of motion*, Talk in Montreal (1997), www.acsu.buffalo.edu/~wlawvere/ToposMotion.pdf

[65] W. Lawvere, *Outline of synthetic differential geometry*, Lecture in Buffalo (1998), ncatlab.org/nlab/files/LawvereSDGOutline.pdf

[66] D. Licata and M. Shulman, *Adjoint logic with a 2-category of modes*, Logical Foundations of Computer Science (2016), http://dlicata.web.wesleyan.edu/pubs/ls15adjoint/ls15adjoint.pdf

[67] J. Lurie, *Higher Topos Theory*, Annals of Mathematics Studies 170, Princeton University Press (2009)

[68] J. Lurie, *On the classification of topological field theories*, Curr. Develop. Math. 2008 (2009) 129–280

[69] M. Marvan, *A note on the category of partial differential equations*, in *Differential Geometry and Its Applications*, Proceedings of the Conference in Brno, August 24–30 (1986), http://ncatlab.org/nlab/files/MarvanJetComonad.pdf

[70] M. Marvan, *On zero-curvature representations of partial differential equations* (1993), http://citeseerx.ist.psu.edu/viewdoc/summary?doi=10.1.1.45.5631

[71] J. Nuiten, *Cohomological quantization of local prequantum boundary field theory*, MSc thesis, Utrecht (2013), http://ncatlab.org/schreiber/show/master+thesis+Nuiten

[72] P. J. Olver, *Applications of Lie groups to differential equations*, Graduate Texts in Mathematics 107, Springer (1993)

[73] E. Rijke, M. Shulman, and B. Spitters, *Modalities in homotopy type theory*, Preprint, arXiv:1706.07526

[74] A. Riotto and M. Trodden, *Recent progress in baryogenesis*, Ann. Rev. Nucl. Part. Sci. 49 (1999) 35–75

[75] C. Rogers, L_∞-*algebras from multisymplectic geometry*, Lett. Math. Phys. 100 (2012) 29–50

[76] D. Roytenberg, *On the structure of graded symplectic supermanifolds and Courant algebroids*, in T. Voronov, ed., *Quantization, Poisson Brackets and Beyond*, Contemporary Mathematics 315, AMS (2002)

[77] A. Sakharov, *Violation of CP invariance, C asymmetry, and baryon asymmetry of the universe*, J. Exp. Theoretical Phys. 5 (1967) 24–27

[78] H. Sati and U. Schreiber, *Lie n-algebras of BPS charges*, Preprint, arXiv:1507.08692

[79] H. Sati, U. Schreiber, and J. Stasheff, *Twisted differential string- and fivebrane structures*, Commun. Math. Phys. 315 (2012) 169–213

[80] U. Schreiber, *Differential cohomology in a cohesive topos*, http://ncatlab.org/schreiber/show/differential+cohomology+in+a+cohesive+topos

[81] U. Schreiber, *Quantization via linear homotopy-types*, Preprint, arXiv:1402.7041

[82] U. Schreiber, *Differential generalized cohomology in cohesive homotopy-type theory*, Talk at IHP trimester on Semantics of Proofs and Certified Mathematics, Workshop 1: Formalization of Mathematics, Institut Henri Poincaré, Paris, May 5–9 (2014), http://ncatlab.org/schreiber/show/IHP14

[83] U. Schreiber, *Differential cohomology is cohesive homotopy theory*, Talk at Higher Structures along the Lower Rhine, June (2014), http://ncatlab.org/schreiber/show/Differential+cohomology+is+Cohesive+homotopy+theory

[84] U. Schreiber, *What, and for what is higher geometric quantization?*, http://ncatlab.org/schreiber/show/What,+and+for+what+is+Higher+geometric+quantization

[85] U. Schreiber, *Differential cohesion and idelic structure*, Talk at CUNY Workshop on Differential Cohomologies, New York (2014), http://ncatlab.org/nlab/show/differential+cohesion+and+idelic+structure

[86] U. Schreiber, *Higher field bundles for gauge fields*, Talk at Operator and Geometric Analysis on Quantum Theory, Levico Terme (Trento), Italy, September 15–19 (2014), http://ncatlab.org/schreiber/show/Higher+field+bundles+for+gauge+fields

[87] U. Schreiber, *Modern physics formalized in modal homotopy type theory*, Talk at the workshop Applying Homotopy Type Theory to Physics, Bristol, April 7–8 (2015), http://ncatlab.org/schreiber/show/Modern+Physics+formalized+in+Modal+Homotopy+Type+Theory

[88] U. Schreiber, *Some thoughts on the future of modal homotopy type theory*, Talk at German Mathematical Society Meeting, Hamburg (2015), http://ncatlab.org/schreiber/show/Some+thoughts+on+the+future+of+modal+homotopy+type+theory

[89] U. Schreiber, *Mathematical quantum field theory*, Lecture notes, Hamburg (2018), https://ncatlab.org/nlab/show/A+first+idea+of+quantum+field+theory

[90] T. Schaefer and E. Shuryak, *Instantons in QCD*, Rev. Mod. Phys. 70 (1998) 323–426

[91] M. Shulman, *Univalence for inverse diagrams and homotopy canonicity*, in Math. Struct. Comput. Sci. (forthcoming)

[92] M. Shulman, *The univalence axiom for elegant Reedy presheaves*, Homol. Homotopy Appl. (forthcoming)

[93] M. Shulman, *Model of type theory in an $(\infty,1)$-topos*, http://ncatlab.org/homotopytypetheory/revision/model+of+type+theory+in+an+(infinity,1)-topos/3

[94] M. Shulman, *Univalence for inverse EI diagrams*, Preprint, arXiv:1508.02410

[95] M. Shulman, *Brouwer's fixed-point theorem in real-cohesive homotopy type theory*, Preprint, arXiv:1509.07584

[96] E. Shuryak, *Nonperturbative QCD and quark-gluon plasma*, (2001), http://users.ictp.it/~pub_off/lectures/lns010/Shuryak/Shuryak.pdf

[97] J. Simons and D. Sullivan, *Axiomatic characterization of ordinary differential cohomology*, J. Topol. (2008) 45–56

[98] J.-M. Souriau, *Structure des systèmes dynamiques*, Dunod (1970)

[99] J.-M. Souriau, *Modèle de particule à spin dans le champ électromagnétique et gravitationnel*, Ann. Phys. théorique 20 (1974) 315–64, www.numdam.org/item?id=AIHPA_1974__20_4_315_0

[100] P. Svrcek and E. Witten, *Axions in string theory*, Preprint, arXiv:hep-th/0605206

[101] Univalent Foundations Project, *Homotopy Type Theory: Univalent Foundations of Mathematics*, Institute for Advanced Study, Princeton University (2013), http://homotopytypetheory.org/book/
[102] G. Veneziano, *U(1) without instantons,* Nucl. Phys. B 159 (1979) 213–24
[103] G. J. Zuckerman, *Action principles and global geometry*, in S. T. Yau, ed., *Mathematical Aspects of String Theory*, World Scientific (1987) 259–84, http://ncatlab.org/nlab/files/ZuckermanVariation.pdf
[104] S. Weinberg, *The Quantum Theory of Fields*, 3 vols., Cambridge University Press (2005)
[105] A. Weinstein, *Symplectic manifolds and their lagrangian submanifolds*, Adv. Math. 6 (1971)
[106] A. Weinstein, *Lectures on Symplectic Manifolds*, CBMS Regional Conf. Series in Math. 29, AMS (1983)
[107] E. Witten, *Current algebra theorems for the $U(1)$ "Goldstone boson,"* Nucl. Phys. B 156 (1979) 269
[108] E. Witten, *Five-brane effective action in M-Theory*, J. Geom. Phys. 22 (1997)

Urs Schreiber
Institute of Mathematics CAS, Czech Republic
schreiber@math.cas.cz

PART III

Spacetime

6
Struggles with the Continuum

John C. Baez*

Contents

1 Newtonian Gravity	283
2 Quantum Mechanics of Charged Particles	286
3 Classical Electrodynamics of Point Particles	289
4 Quantum Field Theory	298
5 General Relativity	312
6 Conclusions	321
References	321

Is spacetime really a continuum, with points being described – at least locally – by lists of real numbers? Or is this description, though immensely successful so far, just an approximation that breaks down at short distances? Rather than trying to answer this hard question, let us look back at the struggles with the continuum that mathematicians and physicists have had so far.

The worries go back at least to Zeno. Among other things, he argued that an arrow can never reach its target:

> *That which is in locomotion must arrive at the half-way stage before it arrives at the goal.*
>
> *(Aristotle [4])*

and Achilles can never catch up with a tortoise:

> *In a race, the quickest runner can never overtake the slowest, since the pursuer must first reach the point whence the pursued started, so that the slower must always hold a lead.*
>
> *(Aristotle [5])*

---

* I thank Flip Tancredo for his Feynman diagram package and Emory Bunn and Greg Weeks for helpful comments.

These paradoxes can now be dismissed using the theory of convergent sequences: a sum of infinitely many terms can still converge to a finite answer. But this theory is far from trivial. It became fully rigorous long after the rise of Newtonian physics. At first, the practical tools of calculus seemed to require infinitesimals, which seemed logically suspect. Thanks to the work of Dedekind, Cauchy, Weierstrass, Cantor, and others, a beautiful formalism was developed to handle the concepts of infinity, real numbers, and limits in a precise axiomatic manner.

However, the logical problems are not gone. Gödel's theorems hang like a dark cloud over the axioms of set theory, assuring us that any consistent theory as strong as Peano arithmetic, or stronger, will leave some questions unsettled. For example: how many real numbers are there? The continuum hypothesis proposes a conservative answer, but since this is independent of the usual axioms of set theory, the question remains open: there could be *vastly* more real numbers than most people think. Worse, the superficially plausible axiom of choice – which amounts to saying that the product of any collection of nonempty sets is nonempty – has scary consequences, like the existence of nonmeasurable subsets of the real line. This in turn leads to results like that of Banach and Tarski: one can partition a ball of unit radius into six disjoint subsets, and by rigid motions reassemble these subsets into *two* disjoint balls of unit radius. (One can even do the job with five, but no fewer [90].)

Most mathematicians and physicists are inured to these logical problems. Few of us bother to learn about attempts to tackle them head-on, such as

- nonstandard analysis and synthetic differential geometry, which let us work consistently with infinitesimals [52, 53, 60, 73];
- constructivism, in which one must "construct" a mathematical object to prove that it exists [8];
- finitism (which avoids completed infinities altogether) [95];
- ultrafinitism, which even denies the existence of very large numbers [10].

This sort of foundational work proceeds slowly, and is deeply unfashionable. One reason is that it rarely seems to impact "real life." For example, it seems that no question about the *experimental consequences* of physical theories has an answer that depends on whether or not we assume the continuum hypothesis or the axiom of choice.

But even if we take a hard-headed practical attitude and leave logic to the logicians, our struggles with the continuum are far from over. In fact, the infinitely divisible nature of the real line – the existence of arbitrarily small real numbers – is a serious challenge to physics.

One of the main goals of physics is to find theories that systematically generate predictions – for example, predictions of the future state of a system given knowledge of its present state. However, even setting aside the question of whether these predictions are correct, there is a problem. For many of the most widely used physical theories, *we have been unable to rigorously prove that they give predictions in all circumstances*. Integrals may diverge, differential equations may fail to have solutions, and so on. Underlying all these difficulties is a struggle with the continuum nature of spacetime itself.

One might hope that a radical approach to the foundations of mathematics – such as those listed above – would allow us to sidestep these problems. However, I know of no significant progress along these lines. Some of the ideas of constructivism have been embraced by topos theory, which also provides a foundation for calculus with infinitesimals [52, 53]. Topos theory and especially "∞-topos theory" are becoming important in mathematical physics [81]. They shed new light on gauge theory, string theory, and other topics. But as far as I know, they have not yet been used to solve, or get around, the problems we discuss here.

Many physicists believe that a successful theory of quantum gravity will dramatically change our concept of spacetime and shed new light on the problems that plague our existing theories. I, too, am inclined to believe this. However, at present, no theory of quantum gravity has made a single experimentally verified prediction. Moreover, these theories are all beset with their own internal problems. Thus, in this survey, we limit ourselves to theories that apparently *have* been successful in making predictions, and examine the "cracks" in these theories: the problems that arise from assuming spacetime is a continuum.

Let us look at some examples.

1 Newtonian Gravity

In its simplest form, Newtonian gravity describes ideal point particles attracting each other with a force inversely proportional to the square of their distance. It is one of the early triumphs of modern physics. But what happens when these particles collide? Apparently the force between them becomes infinite. What does Newtonian gravity predict then?

Of course real planets are not points: when two planets come too close together, this idealization breaks down. Yet if we wish to study Newtonian gravity *as a mathematical theory*, we should consider this case. Part of working with a continuum is successfully dealing with such issues.

In fact, there is a well-defined "best way" to continue the motion of two point masses through a collision. Their velocity becomes infinite at the moment of collision but is finite before and after. The total energy, momentum, and angular momentum are unchanged by this event. So, a 2-body collision is not a serious problem. But what about a simultaneous collision of three or more bodies? This seems more difficult.

Worse than that, Xia proved in 1992 that with five or more particles, there are solutions where particles shoot off to infinity in a finite amount of time [79, 94]. This sounds crazy at first, but it works like this: a pair of heavy particles orbit each other, another pair of heavy particles orbit each other, and these pairs toss a lighter particle back and forth. Each time they do this, the two pairs move further apart from each other, while the two particles within each pair get closer together. Each time they toss the lighter particle back and forth, the two pairs move away from each other faster. As the time t approaches a certain value T_0, the speed of these pairs approaches infinity, so they shoot off to infinity in opposite directions in a finite amount of time, and the lighter particle bounces back and forth an infinite number of times.

Of course this is not possible in the real world, but Newtonian physics has no "speed limit," and we are idealizing the particles as points. So, if two or more of them get arbitrarily close to each other, the potential energy they liberate can give some particles enough kinetic energy to zip off to infinity in a finite amount of time! After that time, the solution is undefined.

You can think of this as a modern reincarnation of Zeno's paradox. Suppose you take a coin and put it heads up. Flip it over after half a second, and then flip it over after 1/4 of a second, and so on. After one second, which side will be up? There is no well-defined answer. That may not bother us, since this is a contrived scenario that seems physically impossible. It is a bit more bothersome that Newtonian gravity does not tell us what happens to our particles when $t = T_0$.

One might argue that collisions and these more exotic "noncollision singularities" occur with probability zero, because they require finely tuned initial conditions. If so, perhaps we can safely ignore them.

This is a nice fallback position. But to a mathematician, this argument demands proof. A bit more precisely, we would like to prove that the set of initial conditions for which two or more particles come arbitrarily close to each other within a finite time has "measure zero." This would mean that "almost all" solutions are well defined for all times, in a very precise sense.

In 1977, Saari proved that this is true for four or fewer particles [76]. However, to the best of my knowledge, the problem remains open for five or more particles. Thanks to previous work by Saari, we know that the set of initial conditions that lead to collisions has measure zero, regardless of the number of particles [77, 78]. So, the main remaining problem is to prove that for five or more particles, noncollision singularities occur with probability zero. For four particles, nobody knows if such singularities can occur at all. For three or fewer we know they do not.

It is remarkable that even Newtonian gravity, often considered a prime example of determinism in physics, has not been proved to make definite predictions, not even "almost always." In 1840, Laplace [57] wrote,

> *We ought to regard the present state of the universe as the effect of its antecedent state and as the cause of the state that is to follow. An intelligence knowing all the forces acting in nature at a given instant, as well as the momentary positions of all things in the universe, would be able to comprehend in one single formula the motions of the largest bodies as well as the lightest atoms in the world, provided that its intellect were sufficiently powerful to subject all data to analysis; to it nothing would be uncertain, the future as well as the past would be present to its eyes. The perfection that the human mind has been able to give to astronomy affords but a feeble outline of such an intelligence.*

However, this dream has not yet been realized for Newtonian gravity.

I expect that noncollision singularities *will* be proved to occur with probability zero. If so, the remaining question would be why it takes so much work to prove this, and thus prove that Newtonian gravity makes definite predictions in almost all cases. Is this a weakness in the theory, or just the way things go? Clearly it has something to do with three idealizations:

- point particles whose distance can be arbitrarily small;
- potential energies that can be arbitrarily large and negative;
- velocities that can be arbitrarily large.

These are connected: as the distance between point particles approaches zero, their potential energy approaches $-\infty$, and conservation of energy dictates that some velocities approach $+\infty$.

Does the situation improve when we go to more sophisticated theories? For example, does the "speed limit" imposed by special relativity help the situation? Or might quantum mechanics help, since it describes particles as "probability clouds," and puts limits on how accurately we can simultaneously know both their position and momentum?

We begin with quantum mechanics, which indeed does help.

2 Quantum Mechanics of Charged Particles

Few people spend much time thinking about "quantum celestial mechanics" – that is, quantum particles obeying Schrödinger's equation, that attract each other gravitationally, obeying an inverse square force law. But Newtonian gravity is a lot like the electrostatic force between charged particles. The main difference is a minus sign, which makes like masses attract, while like charges repel. In chemistry, people spend a lot of time thinking about charged particles obeying Schrödinger's equation, attracting or repelling each other electrostatically. This approximation neglects magnetic fields, spin, and indeed anything related to the finiteness of the speed of light, but it is good enough to explain quite a bit about atoms and molecules.

In this approximation, a collection of charged particles is described by a wave function ψ, which is a complex-valued function of all the particles' positions and also of time. The basic idea is that ψ obeys Schrödinger's equation

$$\frac{d\psi}{dt} = -iH\psi$$

where H is an operator called the Hamiltonian, and I am working in units where $\hbar = 1$.

Does this equation succeed in predicting ψ at a later time given ψ at time zero? To answer this, we must first decide what kind of function ψ should be, what concept of derivative applies to such functions, and so on. These issues were worked out by von Neumann and others starting in the late 1920s. It required a lot of new mathematics. Skimming the surface, we can say this.

At any time, we want ψ to lie in the Hilbert space consisting of square-integrable functions of all the particles' positions. We can then formally solve Schrödinger's equation as

$$\psi(t) = \exp(-itH)\psi(0)$$

where $\psi(t)$ is the solution at time t. But for this to really work, we need H to be a self-adjoint operator on the chosen Hilbert space. The correct definition of "self-adjoint" is a bit subtler than what most physicists learn in a first course on quantum mechanics. In particular, an operator can be superficially self-adjoint – the actual term for this is *symmetric* – but not truly self-adjoint.

In 1951, based on earlier work of Rellich, Kato proved that H is indeed self-adjoint for a collection of nonrelativistic quantum particles interacting via inverse square forces [47, 70]. So, this simple model of chemistry works fine. We can also conclude that "celestial quantum mechanics" would dodge the nasty problems involving collisions or noncollision singularities that we saw in Newtonian gravity. The reason, simply put, is the uncertainty principle.

In the classical case, bad things happen because the energy is not bounded below. A pair of classical particles attracting each other with an inverse square force law can have arbitrarily large *negative* energy, simply by being very close to each other. Noncollision singularities exploit this fact. Since energy is conserved, if you have a way to make some particles get an arbitrarily large *negative* energy, you can balance the books by letting others get an arbitrarily large *positive* energy and shoot to infinity in a finite amount of time!

When we switch to quantum mechanics, the energy of any collection of particles becomes bounded below. The reason is that to make the potential energy of two particles large and negative, they must be very close. Thus, their difference in position must be very small. In particular, this difference must be accurately known! Thus, by the uncertainty principle, their difference in momentum must be very poorly known: at least one of its components must have a large standard deviation. This in turn means that the expected value of the kinetic energy must be large.

This must all be made quantitative, to prove that as particles get close, the uncertainty principle provides enough positive kinetic energy to counterbalance the negative potential energy. The Kato–Lax–Milgram–Nelson theorem [69, Section X.2], a refinement of the original Kato–Rellich theorem, is the key to understanding this issue. The Hamiltonian H for a collection of particles interacting by inverse square forces can be written as $K + V$, where K is an operator for the kinetic energy and V is an operator for the potential energy. With some clever work one can prove that for any $\epsilon > 0$, there exists $c > 0$ such that if ψ is a smooth normalized wave function that vanishes at infinity and at points where particles collide, then

$$|\langle \psi, V\psi \rangle| \leq \epsilon \langle \psi, K\psi \rangle + c.$$

Remember that $\langle \psi, V\psi \rangle$ is the expected value of the potential energy, while $\langle \psi, K\psi \rangle$ is the expected value of the kinetic energy. Thus, this inequality is a precise way of saying how kinetic energy triumphs over potential energy.

By taking $\epsilon = 1$, it follows that the Hamiltonian is bounded below on such states ψ:

$$\langle \psi, H\psi \rangle \geq -c.$$

But the fact that the inequality holds even for smaller values of ϵ is the key to showing H is "essentially self-adjoint." This means that while H is not self-adjoint when defined only on smooth wave functions that vanish outside a bounded set and at points where particles collide, it has a unique self-adjoint extension to some larger domain. Thus, we can unambiguously take this extension to be the true Hamiltonian for this problem.

To fully appreciate this, one needs to see what could have gone wrong. Suppose space had an extra dimension. In three-dimensional space, Newtonian gravity obeys an inverse square force law because the area of a sphere is proportional to its radius squared. In four-dimensional space, the force obeys an inverse *cube* law:

$$F = -\frac{Gm_1 m_2}{r^3}.$$

Using a cube instead of a square here makes the force stronger at short distances, with dramatic effects. For example, even for the classical 2-body problem, the equations of motion no longer "almost always" have a well-defined solution for all times. For an open set of initial conditions, the particles spiral into each other in a finite amount of time!

The quantum version of this theory is also problematic. The uncertainty principle is not enough to save the day. The inequalities above no longer hold: kinetic energy does not triumph over potential energy. The Hamiltonian is no longer essentially self-adjoint on the space of wave functions I described. It has, in fact, *infinitely many* self-adjoint extensions! Each one describes *different physics*: namely, a different choice of what happens when particles collide [32, 36]. Moreover, when G exceeds a certain critical value, the energy is no longer bounded below.

The same problems afflict quantum particles interacting by the electrostatic force in 4d space, as long as some of the particles have opposite charges. So, chemistry would be quite problematic in a world with four dimensions of space.

With more dimensions of space, the situation becomes even worse. This is part of a general pattern in mathematical physics: our struggles with the continuum tend to become worse in higher dimensions. String theory and M-theory may provide exceptions.

3 Classical Electrodynamics of Point Particles

Now let us consider special relativity. Special relativity prohibits instantaneous action at a distance. Thus, most physicists believe that special relativity requires that forces be carried by fields, with disturbances in these fields propagating no faster than the speed of light. The argument for this is not watertight, but we seem to actually *see* charged particles transmitting forces via a field, the electromagnetic field – that is, light. So, most work on relativistic interactions brings in fields.

Classically, charged point particles interacting with the electromagnetic field are described by two sets of equations: Maxwell's equations and the Lorentz force law. The first are a set of differential equations involving

- the electric field $\vec{E}$ and magnetic field $\vec{B}$ (bundled together into the electromagnetic field F);
- the electric charge density ρ and current density $\vec{j}$ (bundled into another field called the "four-current" J).

By themselves, these equations are not enough to completely determine the future given initial conditions. In fact, you can choose ρ and $\vec{j}$ freely, subject to the conservation law

$$\frac{\partial \rho}{\partial t} + \nabla \cdot \vec{j} = 0.$$

For any such choice, there exists a solution of Maxwell's equations for $t \geqslant 0$ given initial values for $\vec{E}$ and $\vec{B}$ that obey these equations at $t = 0$.

Thus, to determine the future given initial conditions, we also need equations that say what ρ and $\vec{j}$ will do. For a collection of charged point particles, they are determined by the curves in spacetime traced out by these particles. The Lorentz force law says that the force on a particle of charge e is

$$\vec{F} = e(\vec{E} + \vec{v} \times \vec{B}),$$

where $\vec{v}$ is the particle's velocity and $\vec{E}$ and $\vec{B}$ are evaluated at the particle's location. From this law we can compute the particle's acceleration if we know its mass.

The trouble starts when we try to combine Maxwell's equations and the Lorentz force law in a consistent way, with the goal being to predict the future behavior of the $\vec{E}$ and $\vec{B}$ fields, together with particles' positions and velocities, given all these quantities at $t = 0$. Attempts to do this began in the late 1800s. The drama continues today, with no definitive resolution! Good accounts have been written by Feynman [27], Pais [227], Janssen and Mecklenburg [45], and Rohrlich [74]. Here we can only skim the surface.

The first sign of a difficulty is this: the charge density and current associated to a charged particle are singular, vanishing off the curve it traces out in spacetime but "infinite" on this curve. For example, a charged particle at rest at the origin has

$$\rho(t,\vec{x}) = e\delta(\vec{x}), \qquad \vec{j}(t,\vec{x}) = \vec{0}$$

where δ is the Dirac delta and e is the particle's charge. This in turn forces the electric field to be singular at the origin. The simplest solution of Maxwell's equations consistent with this choice of ρ and $\vec{j}$ is

$$\vec{E}(t,\vec{x}) = \frac{e\hat{r}}{4\pi\epsilon_0 r^2}, \qquad \vec{B}(t,\vec{x}) = 0$$

where $\hat{r}$ is a unit vector pointing away from the origin and ϵ_0 is a constant called the permittivity of free space.

In short, the electric field is "infinite," or undefined, at the particle's location. So, it is unclear how to define the "self-force" exerted by the particle's own electric field on itself. The formula for the electric field produced by a static point charge is really just our old friend, the inverse square law. Since we had previously ignored the force of a particle on itself, we might try to continue this tactic now. However, other problems intrude.

In relativistic electrodynamics, the electric field has energy density equal to

$$\frac{\epsilon_0}{2}|\vec{E}|^2.$$

Thus, the total energy of the electric field of a point charge at rest is proportional to

$$\frac{\epsilon_0}{2}\int_{\mathbb{R}^3}|\vec{E}|^2 d^3x = \frac{e^2}{8\pi\epsilon_0}\int_0^\infty \frac{1}{r^4} r^2 dr.$$

But this integral diverges near $r = 0$, so the electric field of a charged particle has an infinite energy!

How, if at all, does this cause trouble when we try to unify Maxwell's equations and the Lorentz force law? It helps to step back in history. In 1902, the physicist Abraham assumed that instead of a point, an electron is a sphere of radius R with charge evenly distributed on its surface [1]. Then the energy of its electric field becomes finite, namely,

$$E = \frac{e^2}{8\pi\epsilon_0}\int_R^\infty \frac{1}{r^4} r^2 dr = \frac{1}{2}\frac{e^2}{4\pi\epsilon_0 R},$$

where e is the electron's charge.

Abraham also computed the extra momentum that a *moving* electron of this sort acquires due to its electromagnetic field. He got it wrong because he did not understand Lorentz transformations. In 1904 Lorentz did the calculation right [59]. Using the relationship between velocity, momentum, and mass, we can derive from his result a formula for the "electromagnetic mass" of the electron:

$$m = \frac{2}{3} \frac{e^2}{4\pi \epsilon_0 R c^2}$$

where c is the speed of light. We can think of this as the extra mass an electron acquires by carrying an electromagnetic field along with it.

Putting the last two equations together, these physicists obtained a remarkable result:

$$E = \frac{3}{4} mc^2.$$

Then, in 1905, a fellow named Einstein came along and made it clear that the only reasonable relation between energy and mass is

$$E = mc^2.$$

What had gone wrong?

In 1906, Poincaré figured out the problem [67]. It is not a computational mistake, nor a failure to properly take special relativity into account. The problem is that like charges repel, so if the electron were a sphere of charge it would explode without something to hold it together. And that something – whatever it is – might have energy. But their calculation ignored that extra energy.

In short, the picture of the electron as a tiny sphere of charge, with nothing holding it together, is incomplete. And the calculation showing $E = \frac{3}{4}mc^2$, together with special relativity saying $E = mc^2$, shows that this incomplete picture is actually inconsistent. At the time, some physicists hoped that *all* the mass of the electron could be accounted for by the electromagnetic field. Their hopes were killed by this discrepancy.

Nonetheless it is interesting to take the energy E computed above, set it equal to $m_e c^2$ where m_e is the electron's observed mass, and solve for the radius R. The answer is

$$R = \frac{1}{8\pi \epsilon_0} \frac{e^2}{m_e c^2} \approx 1.4 \times 10^{-15} \text{ meters.}$$

In the early 1900s, this would have been a remarkably tiny distance: 0.00003 times the Bohr radius of a hydrogen atom. By now we know this is roughly the radius of a proton. We know the electron is not a sphere of this size. So at present it makes more sense to treat the calculations so far as a prelude to some kind of limiting process where we take $R \to 0$. These calculations teach us two lessons.

First, the electromagnetic field energy approaches $+\infty$ as we let $R \to 0$, so it is challenging to take this limit and get a well-behaved physical theory. One approach is to give a charged particle its own "bare mass" m_{bare} in addition to the mass m_{elec} arising from electromagnetic field energy, in a way that depends on R. Then as we take the $R \to 0$ limit we can let $m_{\text{bare}} \to -\infty$ in such a way that $m_{\text{bare}} + m_{\text{elec}}$ approaches a chosen limit m, the physical mass of the point particle. This is an example of "renormalization."

Second, it is wise to include conservation of energy-momentum as a requirement in addition to Maxwell's equations and the Lorentz force law. Here is a more sophisticated way to phrase Poincaré's realization. From the electromagnetic field one can compute a "stress-energy tensor" T, which describes the flow of energy and momentum through spacetime. You can compute the total energy and momentum of the electromagnetic field by integrating T over the hypersurface $t = 0$. The resulting 4-vector will transform correctly under Lorentz transformations if the stress-energy tensor has vanishing divergence:

$$\partial^\mu T_{\mu\nu} = 0,$$

where as usual we sum over repeated indices. This equation says that energy and momentum are locally conserved. However, this equation *fails to hold* for a spherical shell of charge with no extra forces holding it together. The reason is that in absence of extra forces, it violates conservation of momentum for charges to feel an electromagnetic force yet not accelerate.

So far we have only discussed the simplest situation: a single charged particle at rest, or moving at a constant velocity. To go further, we can try to compute the acceleration of a small charged sphere in an arbitrary electromagnetic field. Then, by taking the limit as the radius r of the sphere goes to zero, perhaps we can obtain the law of motion for a charged point particle.

In fact this whole program is fraught with difficulties, but physicists boldly go where mathematicians fear to tread, and in a rough way this program was carried out already by Abraham [2] in 1905. His treatment of special relativistic effects was wrong, but these were easily corrected; the real difficulties lie elsewhere. In 1938 his calculations were carried out much more carefully – though still not rigorously – by Dirac [21]. The resulting law of motion is thus called the *Abraham–Lorentz–Dirac force law*.

There are three key ways in which this law differs from our earlier naive statement of the Lorentz force law:

- We must decompose the electromagnetic field in two parts, the "external" electromagnetic field F_{ext} and the field produced by the particle:

$$F = F_{\text{ext}} + F_{\text{ret}}.$$

  Here F_{ext} is a solution of Maxwell's equations with $J = 0$, while F_{ret} is computed by convolving the particle's 4-current J with a function called the *retarded Green's function*. This breaks the time-reversal symmetry of the formalism, ensuring that radiation emitted by the particle moves outward as time goes into the future, not the past. We then decree that the particle only feels a Lorentz force due to F_{ext}, not F_{ret}. This avoids the problem that F_{ret} becomes infinite along the particle's path as $r \to 0$.
- Maxwell's equations say that an accelerating charged particle emits radiation, which carries energy-momentum. Conservation of energy-momentum implies that there is a compensating force on the charged particle. This is called the *radiation reaction*. So, in addition to the Lorentz force, there is a radiation reaction force.
- As we take the limit $r \to 0$, we must adjust the particle's bare mass m_{bare} in such a way that its physical mass $m = m_{\text{bare}} + m_{\text{elec}}$ is held constant. This involves letting $m_{\text{bare}} \to -\infty$ as $m_{\text{elec}} \to +\infty$.

It is easiest to describe the Abraham–Lorentz–Dirac force law using standard relativistic notation. So, we switch to units where $c = 4\pi\epsilon_0 = 1$, let x^μ denote the spacetime coordinates of a point particle, and use a dot to denote the derivative with respect to proper time. Then the Abraham–Lorentz–Dirac force law says

$$m\ddot{x}^\mu = eF^{\mu\nu}_{\text{ext}} \dot{x}_\nu - \frac{2}{3}e^2 \ddot{x}^\alpha \ddot{x}_\alpha \dot{x}^\mu + \frac{2}{3}e^2 \dddot{x}^\mu.$$

The first term at right is the Lorentz force, which looks more elegant in this new notation. The second term acts to reduce the particle's velocity at a rate proportional to its velocity (as one would expect from friction), but also proportional to the squared magnitude of its acceleration. This is the radiation reaction.

The last term, called the *Schott term*, is the most shocking. Unlike all familiar forces in classical mechanics, this involves the *third* derivative of the particle's position! This seems to shatter our original hope of predicting the electromagnetic field and the particle's position and velocity given their initial values. Now it seems we need to specify the particle's initial position, velocity, *and acceleration*.

Furthermore, unlike Maxwell's equations and the original Lorentz force law, the Abraham–Lorentz–Dirac force law is not symmetric under time reversal. If we take a solution, and replace t with $-t$, the result is not a solution. Like the force of friction, radiation reaction acts to make a particle lose energy as it moves into the future, not the past. The reason is that our assumptions have explicitly broken time reversal symmetry: the splitting $F = F_{ext} + F_{ret}$ says that a charged accelerating particle radiates into the future, creating the field F_{ret}, and is affected only by the remaining electromagnetic field F_{ext}.

Worse, the Abraham–Lorentz–Dirac force law has counterintuitive solutions. Suppose for example that $F_{ext} = 0$. Besides the expected solutions where the particle's velocity is constant, there are solutions for which the particle accelerates indefinitely, approaching the speed of light! These are called *runaway solutions*. In a runaway solution, the acceleration as measured in the frame of reference of the particle grows exponentially with the passage of proper time.

So, the notion that special relativity might help us avoid the pathologies of Newtonian point particles interacting gravitationally – solutions where particles shoot to infinity in finite time – is cruelly mocked by the Abraham–Lorentz–Dirac force law. Particles cannot move faster than light, but even a *single* particle can extract an arbitrary amount of energy-momentum from the electromagnetic field in its immediate vicinity and use this to propel itself forward at speeds approaching that of light.

The energy stored in the field near the particle is sometimes called *Schott energy*. The Schott term describes how this energy can be converted into kinetic energy for the particle. The details have been nicely explained by Grøn [38].

Even worse, suppose we generalize the framework to include more than one particle. Arguments for the Abraham–Lorentz–Dirac force law can be generalized to this case, and the result is simply that each particle obeys this law with an external field F_{ext} that includes the fields produced by all the other particles. But a problem appears when we use this law to compute the motion of two particles of opposite charge. To simplify the calculation, suppose they are located symmetrically with respect to the origin, with equal and opposite velocities and accelerations. Suppose the external field felt by each particle is solely the field created by the other particle. Since the particles have opposite charges, they should attract each other. However, one can prove they will never collide. In fact, if at any time they are moving toward each other, they will later *turn around and move away from each other at ever-increasing speed!*

This fact was discovered by Eliezer [26] in 1943. It is so counterintuitive that several proofs were required before physicists believed it. A self-contained proof and review of the literature can be found in Parrott's book [62], along with a discussion of the runaway solutions mentioned earlier.

None of these strange phenomena have ever been seen experimentally. Faced with this problem, physicists have naturally looked for ways out. First, why not simply *cross out* the Schott term in the Abraham–Lorentz–Dirac force? Unfortunately the resulting simplified equation

$$m\ddot{x}^\mu = eF_{\text{ext}}^{\mu\nu} \dot{x}_\nu - \frac{2}{3}e^2 \ddot{x}^\alpha \ddot{x}_\alpha \dot{x}^\mu$$

has only trivial solutions. The reason is that with the particle's path parameterized by proper time, the vector $\dot{x}^\mu$ has constant length, so the vector $\ddot{x}^\mu$ is orthogonal to $\dot{x}^\mu$. So is the vector $F_{\text{ext}}^{\mu\nu} \dot{x}_\nu$, because F_{ext} is an antisymmetric tensor. So, the last term must be zero, which implies $\ddot{x} = 0$, which in turn implies that all three terms must vanish.

Another possibility is that some assumption made in deriving the Abraham–Lorentz–Dirac force law is incorrect. Of course the theory is *physically* incorrect, in that it ignores quantum mechanics, but that is not the issue. The issue here is one of mathematical physics, of trying to formulate a well-behaved classical theory that describes charged point particles interacting with the electromagnetic field. If we can prove this is impossible, we will have learned something. But perhaps there is a loophole. The original arguments for the Abraham–Lorentz–Dirac force law are by no means mathematically rigorous. They involve a delicate limiting procedure, and approximations that were believed, but not proved, to become perfectly accurate in the $r \to 0$ limit. Could these arguments conceal a mistake?

Calculations involving a spherical shell of charge have been improved by a series of authors, and are nicely summarized by Rohrlich [74, 75]. In all these calculations, nonlinear powers of the acceleration and its time derivatives are neglected, and one hopes this is acceptable in the $r \to 0$ limit.

Dirac [21], struggling with renormalization in quantum field theory, took a different tack. Instead of considering a sphere of charge, he treated the electron as a point from the very start. However, he studied the flow of energy-momentum across the surface of a tube of radius r centered on the electron's path. By computing this flow in the limit $r \to 0$, and using conservation of energy-momentum, he attempted to derive the force on the electron. He did not obtain a unique result, but the simplest choice gives the Abraham–Lorentz–Dirac equation. More complicated choices typically involve nonlinear powers of the acceleration and its time derivatives.

Since this work, many authors have tried to simplify Dirac's rather complicated calculations and clarify his assumptions. Parrott's book is a good guide to much of this work [62]. But more recently, Kijowski and coauthors have made impressive progress in a series of papers that solve many of the problems we have seen [34, 48, 49, 50, 51].

Kijowski's key idea is to impose conditions on precisely how the electromagnetic field is allowed to behave near the path traced out by a charged point particle. He decomposes the field into a "regular" part and a "singular" part:

$$F = F_{\text{reg}} + F_{\text{sing}}.$$

Here F_{reg} is smooth everywhere, while F_{sing} is singular near the particle's path, but only in a carefully prescribed way. Roughly, at each moment, in the particle's instantaneous rest frame, the singular part of its electric field consists of the familiar term proportional to $1/r^2$, together with a term proportional to $1/r^3$ which depends on the particle's acceleration. No other singularities are allowed.

On the one hand, this eliminates the ambiguities mentioned earlier: in the end, there are no "nonlinear powers of the acceleration and its time derivatives" in Kijowski's force law. On the other hand, this avoids breaking time reversal symmetry, as the earlier splitting $F = F_{\text{ext}} + F_{\text{ret}}$ did.

Next, Kijowski defines the energy-momentum of a point particle to be $m\dot{x}$, where m is its physical mass. He defines the energy-momentum of the electromagnetic field to be just that due to F_{reg}, not F_{sing}. This amounts to eliminating the infinite "electromagnetic mass" of the charged particle. He then shows that Maxwell's equations and conservation of total energy-momentum imply an equation of motion for the particle.

This equation is very simple:

$$m\ddot{x}^\mu = e F_{\text{reg}}^{\mu\nu} \dot{x}_\nu.$$

It is just the Lorentz force law. Since the troubling Schott term is gone, this is a second-order differential equation. Thus, we can hope to predict the future behavior of the electromagnetic field, together with the particle's position and velocity, given all these quantities at $t = 0$.

And indeed this is true! In 1998, together with Gittel and Zeidler, Kijowski proved that initial data of this sort, obeying the careful restrictions on allowed singularities of the electromagnetic field, determine a unique solution of Maxwell's equations and the Lorentz force law, at least for a short amount of time [34]. Even better, all this remains true for any number of particles.

There are some obvious questions to ask about this new approach. In the Abraham–Lorentz–Dirac force law, the acceleration was an independent variable that needed to be specified at $t = 0$ along with position and momentum. This problem disappears in Kijowski's approach. But how? I mentioned that the singular part of the electromagnetic field, F_{sing}, depends on the particle's acceleration. But more is true: the particle's acceleration is completely determined by F_{sing}. So, the particle's acceleration is not an independent variable because it is *encoded into the electromagnetic field*.

Another question is: where did the radiation reaction go? The answer is: we can see it if we go back and decompose the electromagnetic field as $F_{\text{ext}} + F_{\text{ret}}$ as we had before. If we take the law

$$m\ddot{x}^\mu = eF^{\mu\nu}_{\text{reg}}\dot{x}_\nu$$

and rewrite it in terms of F_{ext}, we recover the original Abraham–Lorentz–Dirac law, including the radiation reaction term and Schott term.

Unfortunately, this means that "pathological" solutions where particles extract arbitrary amounts of energy from the electromagnetic field are still possible. A related problem is that apparently nobody has yet proved solutions exist for all time. Perhaps a singularity worse than the allowed kind could develop in a finite amount of time – for example, when particles collide.

Thus, classical point particles interacting with the electromagnetic field still present serious challenges to the physicist and mathematician. When you have an infinitely small charged particle right next to its own infinitely strong electromagnetic field, trouble can break out very easily!

Finally, I should also mention attempts, working within the framework of special relativity, to get rid of fields and have particles interact with each other directly. For example, in 1903 Schwarzschild [82] introduced a framework in which charged particles exert an electromagnetic force on each other, with no mention of fields. In this setup, forces are transmitted not instantaneously but at the speed of light: the force on one particle at spacetime point x depends on the motion of some other particle at spacetime point y only if the vector $x - y$ is lightlike. Later Fokker and Tetrode [29, 88] derived this force law from a principle of least action. In 1949, Feynman and Wheeler checked that this formalism gives results compatible with the usual approach to electromagnetism using fields, except for several points:

- Each particle exerts forces only on *other* particles, so we avoid the thorny issue of how a point particle responds to the electromagnetic field produced by itself.
- There are no electromagnetic fields not produced by particles: for example, the theory does not describe the motion of a charged particle in an "external electromagnetic field".

- The principle of least action guarantees that "if A affects B then B affects A." So, if a particle at x exerts a force on a particle at a point y in its future lightcone, the particle at y exerts a force on the particle at x in its past lightcone. This raises the issue of "reverse causality", which Feynman and Wheeler address.

Besides the reverse causality issue, perhaps one reason this approach has not been more pursued is that it does not admit a Hamiltonian formulation in terms of particle positions and momenta. Indeed, there are a number of "no-go theorems" for relativistic multiparticle Hamiltonians [20, 58], saying that these can only describe noninteracting particles. So, most work that takes *both* quantum mechanics and special relativity into account uses fields. Indeed, in quantum electrodynamics, even the charged point particles are replaced by fields.

4 Quantum Field Theory

When we study charged particles interacting electromagnetically in a way that takes both quantum mechanics and special relativity into account, we are led to quantum field theory. The ensuing problems are vastly more complicated than in any of the physical theories discussed so far. They are also more consequential, since at present quantum field theory is our best description of all known forces except gravity. As a result, many of the best minds in 20th-century mathematics and physics have joined the fray, and it is impossible here to give more than a quick summary of the situation. This is especially true because the final outcome of the struggle is not yet known.

It is ironic that quantum field theory originally emerged as a *solution* to a problem involving the continuum nature of spacetime, now called the *ultraviolet catastrophe*. In classical electromagnetism, a box with mirrored walls containing only radiation acts like a collection of harmonic oscillators, one for each vibrational mode of the electromagnetic field. If we assume waves can have arbitrarily small wavelengths, there are infinitely many of these oscillators. In classical thermodynamics, a collection of harmonic oscillators in thermal equilibrium will share the available energy equally: this result is called the *equipartition theorem*.

Taken together, these principles lead to a dilemma worthy of Zeno. The energy in the box must be divided into an infinite number of equal parts. If the energy in each part is nonzero, the total energy in the box must be infinite. If it is zero, there can be no energy in the box.

For the logician, there is an easy way out: perhaps a box of electromagnetic radiation can only be in thermal equilibrium if it contains no energy at all! But this solution cannot satisfy the physicist, since it does not match what is actually observed. In reality, any nonnegative amount of energy is allowed in thermal equilibrium.

Experiment also rules out another cheap solution: simply forbidding, by fiat, waves with wavelength shorter than some fixed length. This makes the infinities go away. However, we find that for any nonzero temperature, most of the radiation in a mirrored box will have very short wavelength. This is not what is observed.

The right way out of the dilemma was to change our concept of the harmonic oscillator. Planck did this in 1900, almost without noticing it [66]. Classically, a harmonic oscillator can have any nonnegative amount of energy. Planck instead treated the energy

> *not as a continuous, infinitely divisible quantity, but as a discrete quantity composed of an integral number of finite equal parts.*

In modern notation, the allowed energies of a quantum harmonic oscillator are integer multiples of $\hbar\omega$, where ω is the oscillator's frequency and $\hbar$ is a new constant of nature, named after Planck. When energy can only take such discrete values, the equipartition theorem no longer applies. Instead, the principles of thermodynamics imply that there is a well-defined thermal equilibrium in which vibrational modes with shorter and shorter wavelengths, and thus higher and higher energies, hold less and less of the available energy. The results agree with experiments when the constant $\hbar$ is given the right value.

The full import of what Planck had done became clear only later, starting with Einstein's 1905 paper on the photoelectric effect [24]. Here he proposed that the discrete energy steps actually arise because light comes in particles, now called *photons*, with a photon of frequency ω carrying energy $\hbar\omega$. It was even later that Ehrenfest emphasized the role of the equipartition theorem in the original dilemma, and called this dilemma the ultraviolet catastrophe. As usual, the actual history is more complicated than the textbook summaries [55].

The theory of the "free" quantum electromagnetic field – that is, photons not interacting with charged particles – is now well understood. It is a bit tricky to deal with an infinite collection of quantum harmonic oscillators, but since each evolves independently from all the rest, the issues are manageable. Many advances in analysis were required to tackle these issues in a rigorous way, but they were erected on a sturdy scaffolding of algebra. The reason is that the quantum harmonic oscillator is exactly solvable in terms of well-understood functions, and so is the free quantum electromagnetic field. By the 1930s, physicists knew precise formulas for the answers to more or less any problem involving the free quantum electromagnetic field. The challenge to mathematicians was then to find a coherent mathematical framework that takes us to these answers starting from clear assumptions. This challenge was taken up and completely met by the mid-1960s [7].

However, for physicists, the free quantum electromagnetic field is just the starting-point, since this field obeys a quantum version of Maxwell's equations where the charge density and current density vanish. Far more interesting is *quantum electrodynamics*, or QED, where we also include fields describing charged particles – for example, electrons and their antiparticles, positrons – and try to impose a quantum version of the full-fledged Maxwell's equations. Nobody has found a fully rigorous formulation of QED, nor has anyone proved such a thing cannot be found.

QED is part of a more complicated quantum field theory, the Standard Model, which describes the electromagnetic, weak and strong forces, quarks and leptons, and the Higgs boson. It is widely regarded as our best theory of elementary particles. Unfortunately, nobody has found a rigorous formulation of this theory either, despite decades of hard work by many smart physicists and mathematicians.

To spur progress, the Clay Mathematics Institute has offered a million-dollar prize for anyone who can prove a widely believed claim about a class of quantum field theories called *pure Yang–Mills theories* [18]. A good example is the fragment of the Standard Model that describes only the strong force – or in other words, only gluons. Unlike photons in QED, gluons interact with each other. To win the prize, one must prove that the theory describing them is mathematically consistent and that it describes a world where the lightest particle is a "glueball": a blob made of gluons, with mass strictly greater than zero. This theory is considerably simpler than the Standard Model. However, it is already very challenging.

This is not the only million-dollar prize that the Clay Mathematics Institute is offering for struggles with the continuum. They are also offering one for a proof of global existence of solutions to the Navier–Stokes equations for fluid flow [17]. However, their quantum field theory challenge is the only one for which the problem statement is not completely precise. The Navier–Stokes equations are a collection of partial differential equations for the velocity and pressure of a fluid. We know how to precisely phrase the question of whether these equations have a well-defined solution for all time given smooth initial data. Describing a quantum field theory is a trickier business!

To be sure, there are a number of axiomatic frameworks for quantum field theory [39, 86]. We can prove physically interesting theorems from these axioms, and also rigorously construct some quantum field theories obeying these axioms [7, 33, 72]. The easiest are the free theories, which describe noninteracting particles. There are also examples of rigorously constructed quantum field theories that describe interacting particles in fewer than four spacetime dimensions. However, no quantum field theory that describes interacting particles in four-dimensional spacetime has been proved to obey the usual axioms. Thus, much of the wisdom of physicists concerning quantum field theory has not been fully transformed into rigorous mathematics.

Worse, the question of whether a particular quantum field theory studied by physicists obeys the usual axioms is not completely precise – at least, not yet. The problem is that going from the physicists' formulation to a mathematical structure that might or might not obey the axioms involves some choices.

This is not a cause for despair; it simply means that there is much work left to be done. In practice, quantum field theory is marvelously good for calculating answers to many physics questions. The answers involve approximations. These approximations seem to work very well: that is, they match experiments. Unfortunately we do not fully understand, in a mathematically rigorous way, what these approximations are supposed to be approximating.

How could this be? I will try to sketch some of the key issues in the case of quantum electrodynamics. The history of QED has been nicely told by Schweber [84], so I will focus on concepts rather than the history, and hope that experts forgive me for cutting corners and trying to get across the basic ideas at the expense of many technical details. The nonexpert is encouraged to fill in the gaps with the help of some textbooks, for example those of Zee [96], Peskin and Schroeder [65], and Itzykson and Zuber [44], or, for a more mathematical view, Ticciati [89].

QED involves just one dimensionless parameter, the fine structure constant:

$$\alpha = \frac{1}{4\pi\epsilon_0} \frac{e^2}{\hbar c} \approx \frac{1}{137.036}.$$

We can think of $\alpha^{1/2}$ as a dimensionless version of the electron charge. It says how strongly electrons and photons interact.

Nobody knows why the fine structure constant has the value it does. In computations, we are free to treat it as an adjustable parameter. If we set it to zero, quantum electrodynamics reduces to a free theory, where photons and electrons do not interact with each other. A standard strategy in QED is to take advantage of the fact that the fine structure constant is small and expand answers to physical questions as power series in $\alpha^{1/2}$. This is called *perturbation theory*, and it allows us to exploit our knowledge of free theories.

One of the main questions we try to answer in QED is this: if we start with some particles with specified energy-momenta in the distant past, what is the probability that they will turn into certain other particles with certain other energy-momenta in the distant future? As usual, we compute this probability by first computing a complex amplitude and then taking the square of its absolute value. The amplitude, in turn, is computed as a power series in $\alpha^{1/2}$.

The term of order $\alpha^{n/2}$ in this power series is a sum over Feynman diagrams with n vertices. For example, suppose we are computing the amplitude for two electrons with some specified energy-momenta to interact and become two electrons with some other energy-momenta. One Feynman diagram appearing in the answer is this:

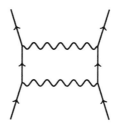

Here the electrons exhange a single photon. Since this diagram has two vertices, it contributes a term of order α. The electrons could also exchange two photons:

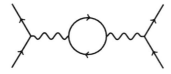

giving a term of α^2. A more interesting term of order α^2 is this:

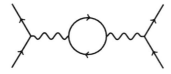

Here the electrons exchange a photon that splits into an electron-positron pair and then recombines. There are infinitely many diagrams with two electrons coming in and two going out. However, there are only finitely many with n vertices. Each of these contributes a term proportional to $\alpha^{n/2}$ to the amplitude.

In general, the external edges of these diagrams correspond to the experimentally observed particles coming in and going out. The internal edges correspond to "virtual particles": that is, particles that are not directly seen, but appear in intermediate steps of a process.

Each of these diagrams is actually a notation for an integral. There are systematic rules for writing down the integral starting from the Feynman diagram [44, 65]. To do this, we first label each edge of the Feynman diagram with an energy-momentum, a variable $p \in \mathbb{R}^4$. The integrand, which we shall not describe here, is a function of all these energy-momenta. In carrying out the integral, the energy-momenta of the external edges are held fixed, since these correspond to the experimentally observed particles coming in and going out. We integrate over the energy-momenta of the internal edges, which correspond to virtual particles, while requiring that energy-momentum is conserved at each vertex.

However, there is a problem: the integral typically diverges! Whenever a Feynman diagram contains a loop, the energy-momenta of the virtual particles in this loop can be arbitrarily large. Thus, we are integrating over an infinite region. In principle the integral could still converge if the integrand goes to zero fast enough. However, we rarely have such luck.

What does this mean, physically? It means that if we allow virtual particles with arbitrarily large energy-momenta in intermediate steps of a process, there are "too many ways for this process to occur," so the amplitude for this process diverges.

Ultimately, the continuum nature of spacetime is to blame. In quantum mechanics, particles with large momenta are the same as waves with short wavelengths. Allowing light with arbitrarily short wavelengths created the ultraviolet catastrophe in classical electromagnetism. Quantum electromagnetism averted that catastrophe – but the problem returns in a different form as soon as we study the interaction of photons and charged particles.

Luckily, there is a strategy for tackling this problem. The integrals for Feynman diagrams become well defined if we impose a "cutoff," integrating only over energy-momenta p in some bounded region, say a ball of some large radius Λ. In quantum theory, a particle with momentum of magnitude greater than Λ is the same as a wave with wavelength less than $\hbar/\Lambda$. Thus, imposing the cutoff amounts to ignoring waves of short wavelength – and for the same reason, ignoring waves of high frequency. We obtain well-defined answers to physical questions when we do this. Unfortunately the answers depend on Λ, and if we let $\Lambda \to \infty$, they diverge.

However, this is not the correct limiting procedure. Indeed, among the quantities that we can compute using Feynman diagrams are the charge and mass of the electron! Its charge can be computed using diagrams in which an electron emits or absorbs a photon:

Similarly, its mass can be computed using a sum over Feynman diagrams where one electron comes in and one goes out.

The interesting thing is this: to do these calculations, we must start by assuming some charge and mass for the electron – but the charge and mass we *get out* of these calculations do not equal the masses and charges we *put in*!

The reason is that virtual particles affect the observed charge and mass of a particle. Heuristically, at least, we should think of an electron as surrounded by a cloud of virtual particles. These contribute to its mass and "shield" its electric field, reducing its observed charge. It takes some work to translate between this heuristic story and actual Feynman diagram calculations, but it can be done.

Thus, there are two different concepts of mass and charge for the electron. The numbers we put into the QED calculations are called the "bare" charge and mass, e_{bare} and m_{bare}. Poetically speaking, these are the charge and mass we would see if we could strip the electron of its virtual particle cloud and see it in its naked splendor. The numbers we get out of the QED calculations are called the "renormalized" charge and mass, e_{ren} and m_{ren}. These are computed by doing a sum over Feynman diagrams. So, they take virtual particles into account. These are the charge and mass of the electron clothed in its cloud of virtual particles. It is these quantities, not the bare quantities, that should agree with experiment.

Thus, the correct limiting procedure in QED calculations is a bit subtle. For any value of Λ and any choice of e_{bare} and m_{bare}, we compute e_{ren} and m_{ren}. The necessary integrals all converge, thanks to the cutoff. We choose e_{bare} and m_{bare} so that e_{ren} and m_{ren} agree with the experimentally observed charge and mass of the electron. The bare charge and mass chosen this way depend on Λ, so call them $e_{\text{bare}}(\Lambda)$ and $m_{\text{bare}}(\Lambda)$.

Next, suppose we want to compute the answer to some other physics problem using QED. We do the calculation with a cutoff Λ, using $e_{\text{bare}}(\Lambda)$

and $m_{\text{bare}}(\Lambda)$ as the bare charge and mass in our calculation. Then we take the limit $\Lambda \to \infty$.

In short, rather than simply fixing the bare charge and mass and letting $\Lambda \to \infty$, we cleverly adjust the bare charge and mass as we take this limit. This procedure is called *renormalization*, and it has a complex and fascinating history [11]. There are many technically different ways to carry out renormalization, and our account so far neglects many important issues. Let us mention three of the simplest.

First, besides the classes of Feynman diagrams already mentioned, we must also consider those where one photon goes in and one photon goes out, such as this:

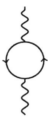

These affect properties of the photon, such as its mass. Since we want the photon to be massless in QED, we have to adjust parameters as we take $\Lambda \to \infty$ to make sure we obtain this result. We must also consider Feynman diagrams where nothing comes in and nothing comes out – so-called *vacuum bubbles* – and make these behave correctly as well.

Second, the procedure just described, where we impose a "cutoff" and integrate over energy-momenta p lying in a ball of radius Λ, is not invariant under Lorentz transformations. Indeed, any theory featuring a smallest time or smallest distance violates the principles of special relativity: thanks to time dilation and Lorentz contraction, different observers will disagree about times and distances. We could accept that Lorentz invariance is broken by the cutoff and hope that it is restored in the $\Lambda \to \infty$ limit, but physicists prefer to maintain symmetry at every step of the calculation. This requires some new ideas: for example, replacing Minkowski spacetime with four-dimensional Euclidean space. In four-dimensional Euclidean space, Lorentz transformations are replaced by rotations, and a ball of radius Λ is a rotation-invariant concept. To do their Feynman integrals in Euclidean space, physicists often let time take imaginary values. They do their calculations in this context and then transfer the results back to Minkowski spacetime at the end. Luckily, there are theorems justifying this procedure [33, 39, 86].

Third, besides infinities that arise from waves with arbitrarily short wavelengths, there are infinities that arise from waves with arbitrarily *long* wavelengths. The former are called *ultraviolet divergences*. The latter are called *infrared divergences*, and they afflict theories with massless particles, like the photon. For example, in QED the collision of two electrons will emit an infinite number of photons with very long wavelengths and low energies, called *soft photons*. In practice this is not so bad, since any experiment can only detect photons with energies above some nonzero value. However, infrared divergences are conceptually important. It seems that in QED any electron is inextricably accompanied by a cloud of soft photons [12]. This is distinct from the "virtual particle cloud" that I mentioned before: these are real particles, emited by the electron whenever it accelerates.

Now let us summarize what we do and do not know about perturbation theory in QED. On the bright side, thanks to the efforts of many brilliant physicists and mathematicians, QED has been proved to be "perturbatively renormalizable" [30, 80]. This means that we can indeed carry out the procedure roughly sketched above, obtaining answers to physical questions as power series in $\alpha^{1/2}$. On the dark side, we do not know if these power series converge. In fact, it is widely believed that they diverge! This puts us in a curious situation.

A good example is the magnetic dipole moment of the electron. An electron, being a charged particle with spin, has a magnetic field. A classical computation says that its magnetic dipole moment is

$$\vec{\mu} = -\frac{e}{2m_e}\vec{S},$$

where $\vec{S}$ is its spin angular momentum. Quantum effects correct this computation, giving

$$\vec{\mu} = -g\frac{e}{2m_e}\vec{S}$$

for some constant g called the *gyromagnetic ratio*. This constant can be computed using QED as a sum over Feynman diagrams in which an electron exchanges a single photon with a massive charged particle:

The answer is a power series in $\alpha^{1/2}$, but since all these diagrams have an even number of vertices, it only contains integral powers of α. The lowest-order term gives simply $g = 2$. In 1948, Schwinger [83] computed the next term and found a small correction to this simple result:

$$g = 2 + \frac{\alpha}{\pi} \approx 2.00232.$$

By now have people have computed g up to order α^5. This requires computing over 13,000 integrals, one for each Feynman diagram of the above form with up to 10 vertices [3]. The answer agrees very well with experiment: in fact, if we also take other Standard Model effects into account we get agreement to roughly one part in 10^{12}. This is the most accurate prediction in all of science!

However, if we continue adding up terms in this power series, there is no guarantee that the answer converges. Indeed, in 1952 Dyson [22] gave a heuristic argument that makes physicists expect that the series *diverges*, along with most other power series in QED.

The argument goes as follows. If these power series converged for small positive α, they would have a nonzero radius of convergence, so they would also converge for small negative α. Thus, QED would make sense for small negative values of α, which correspond to *imaginary* values of the electron's charge. If the electron had an imaginary charge, electrons would attract each other electrostatically, since the usual repulsive force between them is proportional to e^2. Thus, if the power series converged, we would have a theory like QED for electrons that attract rather than repel each other.

However, there is a good reason to believe that QED cannot make sense for electrons that attract. The reason is that it describes a world where the vacuum is unstable. That is, there would be states with arbitrarily large negative energy containing many electrons and positrons. Thus, we expect that the vacuum could spontaneously turn into electrons and positrons together with photons (to conserve energy). Of course, this is not a rigorous proof that the power series in QED diverge: just an argument that it would be strange if they did not.

To see why electrons that attract could have arbitrarily large negative energy, consider a state ψ with a large number N of such electrons inside a ball of radius R. We require that these electrons have small momenta, so that nonrelativistic quantum mechanics gives a good approximation to the situation. Since its momentum is small, the kinetic energy of each electron is a small fraction of its rest energy $m_e c^2$. If we let $\langle \psi, E\psi \rangle$ be the expected value of the total rest energy and kinetic energy of all the electrons, it follows that $\langle \psi, E\psi \rangle$ is approximately proportional to N.

The Pauli exclusion principle puts a limit on how many electrons with momentum below some bound can fit inside a ball of radius R. This number is asymptotically proportional to the volume of the ball. Thus, we can assume N is approximately proportional to R^3. It follows that $\langle \psi, E\psi \rangle$ is approximately proportional to R^3.

There is also the negative potential energy to consider. Let V be the operator for potential energy. Since we have N electrons attracted by a $1/r$ potential, and each pair contributes to the potential energy, we see that $\langle \psi, V\psi \rangle$ is approximately proportional to $-N^2 R^{-1}$, or $-R^5$. Since R^5 grows faster than R^3, we can make the expected energy $\langle \psi, (E+V)\psi \rangle$ arbitrarily large and negative as $N, R \to \infty$.

Note the interesting contrast between this result and some previous ones we have seen. In Newtonian mechanics, the energy of particles attracting each other with a $1/r$ potential is unbounded below. In quantum mechanics, thanks to the uncertainty principle, the energy is bounded below for any fixed number of particles. However, quantum field theory allows for the creation of particles, and this changes everything! Dyson's disaster arises because the vacuum can turn into a state with large numbers of electrons and positrons. This disaster only occurs in an imaginary world where α is negative – but it may be enough to prevent the power series in QED from having a nonzero radius of convergence.

We are left with a puzzle: how can perturbative QED work so well in practice, if the power series in QED diverge?

Much is known about this puzzle. There is an extensive theory of *Borel summation*, which allows one to extract well-defined answers from certain divergent power series. For example, consider a particle of mass m on a line in a potential

$$V(x) = x^2 + \beta x^4.$$

When $\beta \geqslant 0$ this potential is bounded below, but when $\beta < 0$ it is not: classically, it describes a particle that can shoot to infinity in a finite time. Let $H = K + V$ be the quantum Hamiltonian for this particle, where

$$K = -\frac{\hbar^2}{2m} \frac{\partial^2}{\partial x^2}$$

is the usual operator for the kinetic energy and V is the operator for potential energy. When $\beta \geqslant 0$, the Hamiltonian H is essentially self-adjoint on the set of smooth wave functions that vanish outside a bounded set. Moreover, in this case H has a "ground state": a state ψ whose expected energy $\langle \psi, H\psi \rangle$ is as low as possible. Call this expected energy $E(\beta)$. One can show that $E(\beta)$ depends smoothly on β for $\beta \geq 0$, and one can write down a Taylor series for $E(\beta)$.

On the other hand, when $\beta < 0$, the Hamiltonian H is *not* essentially self-adjoint on the set of smooth wave functions that vanish outside a bounded interval. This means that the quantum mechanics of a particle in this potential is ill-behaved when $\beta < 0$. Heuristically speaking, the problem is that such a particle could tunnel through the barrier given by the local maxima of $V(x)$ and shoot off to infinity in a finite time.

This situation is similar to Dyson's disaster, since we have a theory that is well behaved for $\beta \geq 0$ and ill behaved for $\beta < 0$. As before, the bad behavior seems to arise from our ability to convert an infinite amount of potential energy into other forms of energy. However, in this simpler situation one can *prove* that the Taylor series for $E(\beta)$ does not converge. Simon [85] did this around 1969. Moreover, one can prove that Borel summation, applied to this Taylor series, gives the correct value of $E(\beta)$ for $\beta \geq 0$ [37]. The same is known to be true for certain quantum field theories [72]. Analyzing these examples, one can see why summing the first few terms of a power series can give a good approximation to the correct answer even though the series diverges. The terms in the series get smaller and smaller for a while, but eventually they become huge.

Unfortunately, nobody has been able to carry out this kind of analysis for quantum electrodynamics. In fact, the current conventional wisdom is that this theory is inconsistent, due to problems at very short distance scales. In our discussion so far, we summed over Feynman diagrams with $\leq n$ vertices to get the first n terms of power series for answers to physical questions. However, one can also sum over all diagrams with $\leq n$ loops: that is, graphs with genus $\leq n$. This more sophisticated approach to renormalization, which sums over infinitely many diagrams, may dig a bit deeper into the problems faced by quantum field theories.

If we use this alternate approach for QED we find something surprising. Recall that in renormalization we impose a momentum cutoff Λ, essentially ignoring waves of wavelength less than $\hbar/\Lambda$, and use this to work out a relation between the the electron's bare charge $e_{\text{bare}}(\Lambda)$ and its renormalized charge e_{ren}. We try to choose $e_{\text{bare}}(\Lambda)$ that makes e_{ren} equal to the electron's experimentally observed charge e. If we sum over Feynman diagrams with $\leq n$ vertices this is always possible. But if we sum over Feynman diagrams with at most one loop, it ceases to be possible when Λ reaches a certain very large value, namely,

$$\Lambda = \exp\left(\frac{3\pi}{2\alpha} + \frac{5}{6}\right) m_e c \approx e^{647} m_e c.$$

According to this one-loop calculation, the electron's bare charge becomes *infinite* at this point! This value of Λ is known as a *Landau pole*, since it was first noticed in about 1954 by Landau and his colleagues [56].

What is the meaning of the Landau pole? I said that poetically speaking, the bare charge of the electron is the charge we would see if we could strip off the electron's virtual particle cloud. A somewhat more precise statement is that $e_{\text{bare}}(\Lambda)$ is the charge we would see if we collided two electrons head-on with a momentum on the order of Λ. In this collision, there is a good chance that the electrons would come within a distance of $\hbar/\Lambda$ from each other. The larger Λ is, the smaller this distance is, and the more we penetrate past the effects of the virtual particle cloud, whose polarization "shields" the electron's charge. Thus, the larger Λ is, the larger $e_{\text{bare}}(\Lambda)$ becomes. So far, all this makes good sense: physicists have done experiments to actually measure this effect. The problem is that $e_{\text{bare}}(\Lambda)$ becomes infinite when Λ reaches a certain huge value.

Of course, summing only over diagrams with ≤ 1 loops is not definitive. Physicists have repeated the calculation summing over diagrams with ≤ 2 loops, and again found a Landau pole. But again, this is not definitive. Nobody knows what will happen as we consider diagrams with more and more loops. Moreover, the distance $\hbar/\Lambda$ corresponding to the Landau pole is absurdly small! For the one-loop calculation quoted above, this distance is about

$$e^{-647} \frac{\hbar}{m_e c} \approx 6 \times 10^{-294} \text{ meters.}$$

This is hundreds of orders of magnitude smaller than the length scales physicists have explored so far. Currently the Large Hadron Collider can probe energies up to about 10 TeV, and thus distances down to about 2×10^{-20} meters, or about 0.00002 times the radius of a proton. Quantum field theory seems to be holding up very well so far, but no reasonable physicist would be willing to extrapolate this success down to 6×10^{-294} meters, and few seem upset at problems that manifest themselves only at such a short distance scale.

Indeed, attitudes on renormalization have changed significantly since 1948, when Feynman, Schwinger, and Tomonoga developed it for QED. At first it seemed a bit like a trick. Later, as the success of renormalization became ever more thoroughly confirmed, it became accepted. However, some of the most thoughtful physicists remained worried. In 1975, Dirac said,

Most physicists are very satisfied with the situation. They say: "Quantum electrodynamics is a good theory and we do not have to worry about it any more." I must say that I am very dissatisfied with the situation, because this so-called "good theory" does involve neglecting infinities which appear in its equations, neglecting them in an arbitrary way. This is just not sensible mathematics. Sensible mathematics involves neglecting a quantity when it is small—not neglecting it just because it is infinitely great and you do not want it!

As late as 1985, Feynman wrote,

The shell game that we play [...] is technically called "renormalization." But no matter how clever the word, it is still what I would call a dippy process! Having to resort to such hocus-pocus has prevented us from proving that the theory of quantum electrodynamics is mathematically self-consistent. It's surprising that the theory still hasn't been proved self-consistent one way or the other by now; I suspect that renormalization is not mathematically legitimate.

By now renormalization is thoroughly accepted among physicists. The key move was a change of attitude emphasized by Wilson in the 1970s [93]. Instead of treating quantum field theory as the correct description of physics at arbitrarily large energy-momenta, we can assume it is only an approximation. For renormalizable theories, one can argue that even if quantum field theory is inaccurate at large energy-momenta, the corrections become negligible at smaller, experimentally accessible energy-momenta. If so, instead of seeking to take the $\Lambda \to \infty$ limit, we can use renormalization to relate bare quantities at some large but finite value of Λ to experimentally observed quantities.

From this practical-minded viewpoint, the possibility of a Landau pole in QED is less important than the behavior of the Standard Model. Physicists believe that the Standard Model would suffer from Landau pole at momenta low enough to cause serious problems if the Higgs boson were considerably more massive than it actually is. Thus, they were relieved when the Higgs was discovered at the Large Hadron Collider with a mass of about $125 \text{ GeV}/c^2$. However, the Standard Model may still suffer from a Landau pole at high momenta, as well as an instability of the vacuum [46].

Regardless of practicalities, for the *mathematical* physicist, the question of whether or not QED and the Standard Model can be made into well-defined mathematical structures that obey the axioms of quantum field theory remain open problems of great significance. Most physicists believe that this can be done for pure Yang–Mills theories, but actually proving this is the first step toward winning $1,000,000 from the Clay Mathematics Institute.

5 General Relativity

Combining electromagnetism with relativity and quantum mechanics led to QED, and we have seen the immense struggles with the continuum this caused. Combining gravity with relativity led Einstein to *general relativity*.

In general relativity, infinities coming from the continuum nature of spacetime are deeply connected to its most dramatic successful predictions: black holes and the big bang. In this theory, the density of the Universe approaches infinity as we go back in time toward the big bang, and the density of a star approaches infinity as it collapses to form a black hole. Thus we might say that instead of struggling against infinities, general relativity *accepts* them and has learned to live with them.

General relativity does not take quantum mechanics into account, so the story is not yet over. Many physicists hope that quantum gravity will eventually save physics from its struggles with the continuum. Simple dimensional analysis suggests that quantum gravity effects may become important at length scales near the "Planck length":

$$\ell_p = \sqrt{\frac{\hbar G}{c^3}} \approx 1.6 \times 10^{-35} \text{ meters.}$$

Unfortunately, this is too small for direct experiments at present. The hope that something new happens around this length scale has motivated a profusion of new ideas on spacetime: too many to survey here. Instead, I shall focus on the humbler issue of how singularities arise in general relativity – and why they might not rob this theory of its predictive power.

General relativity says that spacetime is a four-dimensional Lorentzian manifold. Thus, it can be covered by patches equipped with coordinates, so that in each patch we can describe points by lists of four numbers. Any curve $\gamma(s)$ going through a point then has a tangent vector v whose components are $v^\mu = d\gamma^\mu(s)/ds$. Furthermore, given two tangent vectors v, w at the same point we can take their inner product

$$g(v, w) = g_{\mu\nu} v^\mu w^\nu$$

where as usual we sum over repeated indices, and $g_{\mu\nu}$ is a 4×4 matrix called the metric, depending smoothly on the point. We require that at any point we can find some coordinate system where this matrix takes the usual Minkowski form:

$$g = \begin{pmatrix} -1 & 0 & 0 & 0 \\ 0 & 1 & 0 & 0 \\ 0 & 0 & 1 & 0 \\ 0 & 0 & 0 & 1 \end{pmatrix}.$$

However, as soon as we move away from our chosen point, the form of the matrix g in these particular coordinates may change.

General relativity says how the metric is affected by matter. It does this in a single equation, Einstein's equation, which relates the "curvature" of the metric at any point to the flow of energy and momentum through that point.

To work with the concept of curvature, Einstein had to learn differential geometry from his mathematician friend Marcel Grossman. One of the great delights of general relativity is how much more can be rigorously proved about this theory than quantum field theory, where even the basic formalism remains problematic. The price to pay is a lot of differential geometry. Instead of explaining all this, I will take some shortcuts and focus on providing intuition. It helps to reformulate Einstein's equation in terms of the motion of particles. For more details, and a list of resources for further study, see [6].

To understand Einstein's equation, let us see what it says about a small round ball of test particles that are initially all at rest relative to each other. The scenario here requires a bit of explanation. First, because spacetime is curved, it only looks like Minkowski spacetime – the world of special relativity – in the limit of a very small region. The concepts of "round" and "at rest relative to each other" only make sense in this limit. Thus, the forthcoming statement of Einstein's equation is precise only in this limit. Of course, taking this limit relies on the fact that spacetime is a continuum.

Second, a *test particle* is a classical point particle with so little mass that while it is affected by gravity, its effects on the geometry of spacetime are negligible. We assume our test particles are affected only by gravity, no other forces. In general relativity this means that they move along timelike geodesics. Roughly speaking, these are paths that go slower than light and bend as little as possible. We can make this precise without much work.

For a path in *space* to be a geodesic means that if we slightly vary any small portion of it, it can only become longer. However, a path $\gamma(s)$ in *spacetime* traced out by a particle moving slower than light must be "timelike," meaning that its tangent vector $v = \gamma'(s)$ satisfies $g(v,v) < 0$. We define the proper time along such a path from $s = s_0$ to $s = s_1$ to be

$$\int_{s_0}^{s_1} \sqrt{-g(\gamma'(s), \gamma'(s))}\, ds.$$

This is the time ticked out by a clock moving along that path. A timelike path is a geodesic if the proper time can only *decrease* when we slightly vary any small portion of it. Particle physicists prefer the opposite sign convention for the metric, and then we do not need the minus sign under the square root. But the fact remains the same: timelike geodesics locally maximize the proper time.

Actual particles are not test particles. First, the concept of test particle does not take quantum theory into account. Second, all known particles are affected by forces other than gravity. Third, any actual particle affects the geometry of the spacetime it inhabits. Test particles are just a mathematical trick for studying the geometry of spacetime. Still, a sufficiently light particle that is affected very little by forces other than gravity should be well approximated by a test particle, though rigorously proving this is difficult [25]. For example, an artificial satellite moving through the Solar System behaves like a test particle if we ignore the solar wind, the radiation pressure of the Sun, and so on.

If we start with a small round ball consisting of many test particles that are initially all at rest relative to each other, to first order in time it will not change shape or size. However, to second order in time it can expand or shrink, due to the curvature of spacetime. It may also be stretched or squashed, becoming an ellipsoid. This should not be too surprising, because any linear transformation applied to a ball gives an ellipsoid.

Let $V(t)$ be the volume of the ball after a time t has elapsed, where time is measured by a clock attached to the particle at the center of the ball. Then in units where $c = G = 1$, Einstein's equation says

$$\left.\frac{\ddot{V}}{V}\right|_{t=0} = -4\pi \begin{pmatrix} \text{flow of } t\text{-momentum in the t direction} + \\ \text{flow of } x\text{-momentum in the x direction} + \\ \text{flow of } y\text{-momentum in the y direction} + \\ \text{flow of } z\text{-momentum in the z direction} \end{pmatrix}.$$

These flows here are measured at the center of the ball at time zero, and the coordinates used here take advantage of the fact that to first order, at any one point, spacetime looks like Minkowski spacetime.

The flows in Einstein's equation are the diagonal components of a 4×4 matrix T called the *stress-energy tensor*. The components $T_{\alpha\beta}$ of this matrix say how much momentum in the α direction is flowing in the β direction through a given point of spacetime. Here α and β range from 0 to 3, corresponding to the $t, x, y,$ and z coordinates. For example, T_{00} is the flow of t-momentum in the t-direction. This is just the energy density, usually denoted ρ. The flow of x-momentum in the x-direction is the pressure in the

x direction, denoted P_x, and similarly for y and z. The reader may be more familiar with direction-independent pressures, but it is easy to manufacture a situation where the pressure depends on the direction: just squeeze a book between one's hands.

Thus, Einstein's equation says

$$\left.\frac{\ddot{V}}{V}\right|_{t=0} = -4\pi(\rho + P_x + P_y + P_z).$$

It follows that positive energy density and positive pressure both curve spacetime in a way that makes a freely falling ball of point particles tend to shrink. Since $E = mc^2$ and we are working in units where $c = 1$, ordinary mass density counts as a form of energy density. Thus a massive object will make a swarm of freely falling particles at rest around it start to shrink. In short, *gravity attracts*.

Already from this, gravity seems dangerously inclined to create singularities. Suppose that instead of test particles we start with a stationary cloud of "dust": a fluid of particles having nonzero energy density but no pressure, moving under the influence of gravity alone. The dust particles will still follow geodesics, but they will affect the geometry of spacetime. Their energy density will make the ball start to shrink. As it does, the energy density ρ will increase, so the ball will tend to shrink ever faster, approaching infinite density in a finite amount of time. This in turn makes the curvature of spacetime become infinite in a finite amount of time. The result is a *singularity*.

In reality, matter is affected by forces other than gravity. Repulsive forces may prevent gravitational collapse. However, this repulsion creates pressure, and Einstein's equation says that pressure also creates gravitational attraction! In some circumstances this can overwhelm whatever repulsive forces are present. Then the matter collapses, leading to a singularity – at least according to general relativity.

When a star more than 8 times the mass of our Sun runs out of fuel, its core suddenly collapses. The surface is thrown off explosively in an event called a supernova. Most of the energy – the equivalent of thousands of Earth masses – is released in a 10-second burst of neutrinos, formed as a byproduct when protons and electrons combine to form neutrons. If the star's mass is below 20 times that of our the Sun, its core crushes down to a large ball of neutrons with a crust of iron and other elements: a neutron star.

However, this ball is unstable if its mass exceeds the Tolman–Oppenheimer–Volkoff limit, somewhere between 1.5 and 3 times that of our Sun. Above this limit, gravity overwhelms the repulsive forces that hold up the neutron star. And indeed, no neutron stars heavier than 3 solar masses have been observed. Thus, for very heavy stars, the endpoint of collapse is not a neutron star, but something else: a *black hole*, an object that bends spacetime so much even light cannot escape.

If general relativity is correct, a black hole contains a singularity. Many physicists expect that general relativity breaks down inside a black hole, perhaps because of quantum effects that become important at strong gravitational fields. The singularity is considered a strong hint that this breakdown occurs. If so, the singularity may be a purely theoretical entity, not a real-world phenomenon. Nonetheless, everything we have observed about black holes matches what general relativity predicts. Thus, unlike all the other theories we have discussed, general relativity predicts infinities that are connected to striking phenomena that are *actually observed*.

The Tolman–Oppenheimer–Volkoff limit is not precisely known, because it depends on properties of nuclear matter that are not well understood [9]. However, there are theorems that say singularities *must* occur in general relativity under certain conditions.

One of the first was proved by Raychauduri [68] and Komar [54] in the mid-1950s. It applies only to "dust," and indeed it is a precise version of our verbal argument above. It introduced the *Raychauduri equation*, which is the geometrical way of thinking about spacetime curvature as affecting the motion of a small ball of test particles. It shows that under suitable conditions, the energy density must approach infinity in a finite amount of time along the path traced out by a dust particle.

The first required condition is that the flow of dust be initally converging, not expanding. The second condition, not mentioned in our verbal argument, is that the dust be "irrotational," not swirling around. The third condition is that the dust particles be affected only by gravity, so that they move along geodesics. Due to the last two conditions, the Raychauduri–Komar theorem does not apply to collapsing stars.

The more modern singularity theorems eliminate these conditions. But they do so at a price: they require a more subtle concept of singularity. There are various possible ways to define this concept. They are all a bit tricky, because a singularity is not a point or region in spacetime.

For our present purposes, we shall define a singularity to be an *incomplete timelike or null geodesic*. As already explained, a timelike geodesic is the kind of path traced out by a test particle moving slower than light. Similarly, a null geodesic is the kind of path traced out by a test particle moving at the speed of light. We say a geodesic is *incomplete* if it ceases to be well defined after a finite amount of time. For example, general relativity says a test particle falling into a black hole follows an incomplete geodesic. In a rough-and-ready way, people say the particle "hits the singularity." But the singularity is not a place in spacetime. What we really mean is that the particle's path becomes undefined after a finite amount of time.

We need to be a bit careful about the role of "time" here. For test particles moving slower than light this is easy, since we can parameterize a timelike geodesic by proper time. However, the tangent vector $v = \gamma'(s)$ of a null geodesic has $g(v, v) = 0$, so a particle moving along a null geodesic does not experience any passage of proper time. Still, any geodesic, even a null one, has a family of preferred parameterizations. These differ only by reparameterizations of the form $s \mapsto as + b$. By "time" we really mean the variable s in any of these preferred parameterizations. Thus, if our spacetime is some Lorentzian manifold M, we say a geodesic $\gamma \colon [s_0, s_1] \to M$ is incomplete if, parameterized in one of these preferred ways, it cannot be extended to a strictly longer interval.

The first modern singularity theorem was proved by Penrose [63] in 1965. It says that if space is infinite in extent, and light becomes trapped inside some bounded region, and no exotic matter is present to save the day, either a singularity or something even more bizarre must occur. This theorem applies to collapsing stars. When a star of sufficient mass collapses, general relativity says that its gravity becomes so strong that light becomes trapped inside some bounded region. We can then use Penrose's theorem to analyze the possibilities.

Shortly thereafter Hawking proved a second singularity theorem, which applies to the big bang [40]. It says that if space is finite in extent, and no exotic matter is present, generically either a singularity or something even more bizarre must occur. The singularity here could be either a big bang in the past, a big crunch in the future, both – or possibly something else. Hawking also proved a version of his theorem that applies to certain Lorentzian manifolds where space is infinite in extent, as seems to be the case in our Universe. This version requires extra conditions.

There are some undefined phrases in this summary of the Penrose–Hawking singularity theorems, most notably these:

- exotic matter;
- singularity;
- something even more bizarre.

So, let me say a bit about each.

These theorems precisely specify what is meant by *exotic matter*. All known forms of matter obey the *dominant energy condition*, which says that

$$|P_x|, |P_y|, |P_z| \leqslant \rho$$

at all points and in all locally Minkowskian coordinates. Exotic matter is anything that violates this condition. For a detailed discussion of this and other energy conditions, see the survey by Curiel [19].

The Penrose–Hawking singularity theorems also say what counts as "something even more bizarre." An example would be a closed timelike curve. A particle following such a path would move slower than light yet eventually reach the same point where it started: and not just the same point in space, but the same point in *spacetime*. If you could do this perhaps you could wait, see if it rains tomorrow, and then go back in time and decide whether to buy an umbrella today. There are certainly solutions of Einstein's equation with closed timelike curves. The first interesting one was found by Einstein's friend Gödel in 1949, as part of an attempt to probe the nature of time [35]. However, closed timelike curves are generally considered less plausible than singularities.

In the Penrose–Hawking singularity theorems, "something even more bizarre" means that spacetime is not "globally hyperbolic." To understand this, we need to think about when we can predict the future or past given initial data. When studying field equations like Maxwell's theory of electromagnetism or Einstein's theory of gravity, physicists like to specify initial data on space at a given moment of time. However, in general relativity there is considerable freedom in how we choose a slice of spacetime and call it "space." What should we require? For starters, we want a three-dimensional submanifold S of spacetime that is *spacelike*: every vector v tangent to S should have $g(v, v) > 0$. However, we also want any timelike or null curve to hit S exactly once. A spacelike surface with this property is called a *Cauchy surface*, and a Lorentzian manifold containing a Cauchy surface is said to be *globally hyperbolic*. Global hyperbolicity excludes closed timelike curves, but also other bizarre behavior. In a globally hyperbolic spacetime, we can predict the future or past given initial data on any Cauchy surface.

By now the original singularity theorems have been greatly generalized and clarified. Hawking and Penrose [43] gave a unified treatment of both theorems in 1970. The textbook by Hawking and Ellis [42] gave the first really systematic introduction to this subject, and Wald's text gives a shorter, more modern treatment [91]. Hawking's 1994 lectures [41] give a beautiful overview of the key ideas, and a paper by Garfinkle and Senovilla [41] reviews the subject and its history up to 2015.

If we accept that general relativity really predicts the existence of singularities in physically realistic situations, the next step is to ask whether they rob general relativity of its predictive power. The *cosmic censorship hypothesis*, proposed by Penrose in 1969, claims they do not [64].

To formulate such a conjecture, we must first think about what behaviors we consider acceptable. Consider first a black hole formed by the collapse of a star. According to general relativity, matter can fall into this black hole and "hit the singularity" in a finite amount of proper time, but nothing can come out of the singularity. The time-reversed version of a black hole, called a *white hole*, is often considered more disturbing. White holes have never been seen, but they are mathematically valid solutions of Einstein's equation. In a white hole, matter can come *out* of the singularity, but nothing can fall *in*. Naively, this seems to imply that the future is unpredictable given knowledge of the past. Of course, the same logic applied to black holes would say the past is unpredictable given knowledge of the future.

If white holes are disturbing, perhaps the big bang should be more so. In the usual solutions of general relativity describing big bang cosmologies, *all matter in the universe* comes out of a singularity! More precisely, if one follows any timelike geodesic back into the past, it becomes undefined after a finite amount of proper time. Naively, this may seem a massive violation of predictability: in this scenario, the whole universe "sprang out of nothing" about 14 billion years ago.

However, in all three examples so far – astrophysical black holes, their time-reversed versions and the big bang – spacetime is globally hyperbolic. Thus, we can specify data on a Cauchy surface and solve Einstein's equation to predict the future (and past) development of the metric throughout all of spacetime. How is this compatible with the naive intuition that a singularity causes a failure of predictability?

For any globally hyperbolic spacetime M, one can find a smoothly varying family of Cauchy surfaces S_t ($t \in \mathbb{R}$) such that each point of M lies on exactly one of these surfaces. This amounts to a way of chopping spacetime into "slices of space" for various choices of the "time" parameter t. For an astrophysical black hole, the singularity is in the future of all these surfaces. That is, a timelike or null geodesic that hits the singularity must go through all the surfaces S_t before it becomes undefined. Similarly, for a white hole or the big bang, the singularity is in the past of all these surfaces. In either case, the singularity cannot interfere with our predictions of what occurs in spacetime. For more on this topic, try Earman's delightful book *Bangs, Crunches, Whimpers and Shrieks: Singularities and Acausalities in Relativistic Spacetimes* [23].

A more challenging example is posed by the Kerr–Newman solution of Einstein's equation coupled to the vacuum Maxwell's equations [91]. When

$$e^2 + (J/m)^2 < m^2,$$

this solution describes a rotating charged black hole with mass m, charge e and angular momentum J in units where $c = G = 1$. In 1968, Carter [15] noted that the Kerr–Newman solution acts like a particle with gyromagnetic ratio $g = 2$, surprisingly close to that of an electron. However, an electron has

$$e^2 + (J/m)^2 \gg m^2.$$

The Kerr–Newman solution still has $g = 2$ in this case, but also some disturbing pathological features. It has closed timelike curves accessible from the outside world! It also has a "naked singularity." Roughly speaking, this is a singularity that can be seen by arbitrarily faraway observers in a spacetime whose geometry asymptotically approaches that of Minkowski spacetime. A spacetime with a naked singularity cannot be globally hyperbolic [42].

The cosmic censorship hypothesis comes in a number of forms. The original version due to Penrose is now called "weak cosmic censorship" [92]. It asserts that in a spacetime whose geometry asymptotically approaches that of Minkowski spacetime, gravitational collapse cannot produce a naked singularity.

In 1991, Preskill and Thorne made a bet against Hawking in which they claimed that weak cosmic censorship was false. Hawking conceded this bet in 1997 when a counterexample was found. This features finely tuned infalling matter poised right on the brink of forming a black hole. It *almost* creates a region from which light cannot escape – but not quite. Instead, it creates a naked singularity!

Given the delicate nature of this construction, Hawking did not give up. Instead he made a second bet, which says that weak cosmic censorship holds "generically" – that is, for an open dense set of initial conditions. In 1999, Christodoulou proved that for spherically symmetric solutions of Einstein's equation coupled to a massless scalar field, weak cosmic censorship holds generically [13]. While spherical symmetry is a very restrictive assumption, this result is a good example of how, with plenty of work, we can make progress in rigorously settling the questions raised by general relativity.

Indeed, Christodoulou has been a leader in this area. For example, the vacuum Einstein equations have solutions describing gravitational waves, much as the vacuum Maxwell's equations have solutions describing electromagnetic waves. However, gravitational waves can actually form black holes when they collide. This raises the question of the stability of Minkowski spacetime. Must sufficiently small perturbations of the Minkowski metric go away in the form of gravitational radiation, or can tiny wrinkles in the fabric of spacetime somehow amplify themselves and cause trouble – perhaps even a singularity? In 1993, together with Klainerman, Christodoulou proved that Minkowski spacetime is indeed stable [16]. Their proof fills a 514-page book.

In 2008, Christodoulou completed an even longer rigorous study of the formation of black holes [14]. This can be seen as a vastly more detailed look at questions which Penrose's original singularity theorem addressed in a general, preliminary way. Nonetheless, there is much left to be done to understand the behavior of singularities in general relativity [71].

6 Conclusions

We have seen that in every major theory of physics, challenging mathematical questions arise from the assumption that spacetime is a continuum. The continuum threatens us with infinities. Do these infinities threaten our ability to extract predictions from these theories – or even our ability to formulate these theories in a precise way? We can answer these questions, but only with hard work. Is this a sign that we are somehow on the wrong track? Is the continuum as we understand it only an approximation to some deeper model of spacetime? Only time will tell. Nature is providing us with plenty of clues, but it will take patience to read them correctly.

References

[1] M. Abraham, *Prinzipien der Dynamik des Elektrons*, Phys. Z. 4 (1902) 57–62, https://de.wikisource.org/wiki/Prinzipien_der_Dynamik_des_Elektrons_(1902)

[2] M. Abraham, *Theorie der Elektrizität: Elektromagnetische Theorie der Strahlung*, Teubner (1905), https://archive.org/details/theoriederelekt04fpgoog

[3] T. Aoyama, M. Hayakawa, T. Kinoshita, and M. Nio, *Tenth-order QED contribution to the electron $g-2$ and an improved value of the fine structure constant*, Phys. Rev. Lett. 109 (2012) 111807

[4] Aristotle, *Physics* VI:9, 239b10, http://classics.mit.edu/Aristotle/physics.6.vi.html #761

[5] Aristotle, *Physics* VI:9, 239b15, http://classics.mit.edu/Aristotle/physics.6.vi.html #764

[6] J. C. Baez and E. F. Bunn, *The meaning of Einstein's equation*, Amer. J. Phys. 73 (2005) 644–52

[7] J. C. Baez, I. E. Segal, and Z. Zhou, *Introduction to Algebraic and Constructive Quantum Field Theory*, Princeton University Press (1992)

[8] E. Bishop, *Foundations of Constructive Analysis*, McGraw-Hill (1967)

[9] I. Bombaci, *The maximum mass of a neutron star*, Astron. Astrophys. 305 (1996) 871–77

[10] A. Boucher, *Arithmetic without the successor axiom* (2006), http://citeseerx.ist.psu.edu/viewdoc/summary?doi=10.1.1.85.3071

[11] L. M. Brown, ed., *Renormalization: From Lorentz to Landau (and Beyond)*, Springer (2012)

[12] D. Buchholz, *Gauss' law and the infraparticle problem*, Phys. Lett. B 174 (1986) 331–34

[13] D. Christodoulou, *The instability of naked singularities in the gravitational collapse of a scalar field*, Ann. Math. 149 (1999) 183–217

[14] D. Christodoulou, *The Formation of Black Holes in General Relativity*, European Mathematical Society (2009)

[15] B. Carter, *Global structure of the Kerr family of gravitational fields*, Phys. Rev. 174 (1968) 1559–71

[16] D. Christodoulou and S. Klainerman, *The Global Nonlinear Stability of the Minkowski Space*, Princeton University Press (1993)

[17] Clay Mathematics Institute, *Navier–Stokes equation*, www.claymath.org/millennium-problems/navier–stokes-equation

[18] Clay Mathematics Institute, *Yang–Mills and mass gap*, www.claymath.org/millennium-problems/yang–mills-and-mass-gap

[19] E. Curiel, *A primer on energy conditions*, Preprint, arXiv:1405.0403

[20] D. G. Currie, T. F. Jordan, and E. C. G. Sudarshan, *Relativistic invariance and Hamiltonian theories of interacting particles*, Rev. Mod. Phys. 35 (1963) 350–75. Erratum, 1032

[21] P. A. M. Dirac, *Classical theory of radiating electrons*, Proc. Roy. Soc. London A 167 (1938) 148–69

[22] F. J. Dyson, *Divergence of perturbation theory in quantum electrodynamics*, Phys. Rev. B 85 (1952) 631–32

[23] J. Earman, *Bangs, Crunches, Whimpers and Shrieks: Singularities and Acausalities in Relativistic Spacetimes*, Oxford University Press, (1993), www.pitt.edu/~jearman/Earman_1995BangsCrunches.pdf

[24] A. Einstein, *Über einen die Erzeugung und Verwandlung des Lichtes betreffenden heuristischen Gesichtspunkt*, Ann. Phys. 17 (1905) 132–48. Available in translation at https://en.wikisource.org/wiki/On_a_Heuristic_Point_of_View_about_the_Creation_and_Conversion_of_Light

[25] A. Einstein, L. Infeld, and B. Hoffmann, *The gravitational equations and the problem of motion*, Ann. Math. 39 (1938) 65–100

[26] C. J. Eliezer, *The hydrogen atom and the classical theory of radiation*, Proc. Camb. Philos. Soc. 39 (1943) 173–80

[27] R. P. Feyman, R. B. Leighton, and M. Sands, *The Feynman Lectures on Physics*, vol. II, Addison–Wesley (1963), www.feynmanlectures.caltech.edu/II_28.html.
[28] R. P. Feynman and J. A. Wheeler, *Classical electrodynamics in terms of direct interparticle action*, Rev. Mod. Phys. 21 (1949) 425–33
[29] A. D. Fokker, *Ein invarianter Variationssatz für die Bewegung mehrerer elektrischer Massenteilchen*, Z. Phys. 58 (1929) 386–93
[30] J. S. Feldman, T. R. Hurd, L. Rosen, and J. D. Wright, *QED: A Proof of Renormalizability*, Lecture Notes in Physics 312, Springer (1988)
[31] D. Garfinkle and J. M. M. Senovilla, *The 1965 Penrose singularity theorem*, Class. Quant. Grav. 32 (2015) 124008
[32] D. M. Gitman, I. V. Tyutin, and B. L. Voronov, *Self-adjoint extensions and spectral analysis in the Calogero problem*, J. Phys. A 43 (2010) 145205
[33] J. Glimm and A. Jaffe, *Quantum Physics: A Functional Integral Point of View*, Springer (1987)
[34] H.-P. Gittel, J. Kijowski, and E. Zeidler, *The relativistic dynamics of the combined particle-field system in renormalized classical electrodynamics*, Comm. Math. Phys. 198 (1998) 711–36, www.cft.edu.pl/~kijowski/Odbitki-prac/GKZ.pdf
[35] K. Gödel, *An example of a new type of cosmological solution of Einstein's field equations of gravitation*, Rev. Mod. Phys. 21 (1949) 447–50
[36] S. Gopalkrishnan, *Self-adjointness and the renormalization of singular potentials*, BA thesis, Amherst College (2006), www.amherst.edu/media/view/10264/original/gopalakrishnan06.pdf
[37] S. Graffi, V. Grecchi, and B. Simon, *Borel summability: application to the anharmonic oscillator*, Phys. Lett. B32 (1970) 631–34
[38] Ø. Grøn, *The significance of the Schott energy for energy-momentum conservation of a radiating charge obeying the Lorentz–Abraham–Dirac equation*, Amer. J. Phys. 79 (2011) 115–22
[39] R. Haag, *Local Quantum Physics: Fields, Particles, Algebras*, Springer (1996)
[40] S. W. Hawking, *Singularities in the universe*, Phys. Rev. Lett. 17 (1966) 444–45
[41] S. W. Hawking and R. Penrose, *The Nature of Space and Time*, Princeton University Press (2010). Hawking's lectures are also available as arXiv:hep-th/9409195
[42] S. W. Hawking and G. F. R. Ellis, *The Large Scale Structure of Space-Time*, Cambridge Univeristy Press (1973)
[43] S. W. Hawking and R. Penrose, *The singularities of gravitational collapse and cosmology*, Proc. R. Soc. London, Ser. A 314 (1970) 529–48
[44] C. Itzykson and J.-B. Zuber, *Quantum Field Theory*, Dover (2006)
[45] M. Janssen and M. Mecklenburg, *From classical to relativistic mechanics: Electromagnetic models of the electron*, in V. F. Hendricks et al., eds., *Interactions: Mathematics, Physics and Philosophy, 1860–1930*, Springer (1990) 65–134, http://philsci-archive.pitt.edu/1990/
[46] F. Jegerlehner, M. Kalmykov, and B. A. Kniehl, *Self-consistence of the Standard Model via the renormalization group analysis*, in L. Fiala, M. Lokajicek, and

N. Tumova, eds., *Proceedings of the 16th International Workshop (ACAT2014)*, IOP Publishing (2015) 012074

[47] T. Kato, *Fundamental properties of Hamiltonian operators of Schrödinger type*, Trans. Amer. Math. Soc. 70 (1951) 195–211

[48] J. Kijowski, *Electrodynamics of moving particles*, Gen. Rel. Grav. 26 (1994) 167–201, www.cft.edu.pl/~kijowski/Odbitki-prac/GRG-NEW.pdf

[49] J. Kijowski, *On electrodynamical self-interaction*, Acta Phys. Polonica A 85 (1994) 771–87, www.cft.edu.pl/ ~kijowski/Odbitki-prac/BIRULA.pdf

[50] J. Kijowski and P. Podleś, *Born renormalization in classical Maxwell electrodynamics*, J. Geom. Phys. 48 (2003) 369–84

[51] J. Kijowski and P. Podleś, *A geometric analysis of the Maxwell field in a vicinity of a multipole particle and new special functions*, J. Geom. Phys. 59 (2009) 693–709

[52] A. Kock, *Synthetic Differential Geometry*, Cambridge University. Press (2006), http://home.imf.au.dk/kock/sdg99.pdf.

[53] A. Kock, *Synthetic Differential Geometry of Manifolds*, Cambridge University Press (2010), http://home.imf.au.dk/kock/SGM-final.pdf

[54] A. Komar, *Necessity of singularities in the solution of the field equations of general relativity*, Phys. Rev. 104 (1956) 544–46

[55] H. Kragh, *Quantum Generations: A History of Physics in the Twentieth Century*, Princeton University Press (1999)

[56] L. D. Landau, A. A. Abrikosov, and I. M. Khalatnikov, *An asymptotic expression for the photon Green function in quantum electrodynamics*, Dokl. Akad. Nauk SSSR 95 (1954) 1177–82. Reprinted in *Collected Papers of L. D. Landau*, D. Ter Haar, ed., Pergamon (1965)

[57] P. Laplace, *Essai Philosophique sur les Probabilités*, Paris Bachelier (1840). Reprinted as *A Philosophical Essay on Probabilities*, trans. W. Truscott and F. L. Emory, Dover (1951)

[58] H. Leutwyler, *A no-interaction theorem in classical relativistic Hamiltonian particle mechanics*, Nuovo Cimento 37 (1965) 556–67

[59] M. A. Lorentz, *Electromagnetic phenomena in a system moving with any velocity smaller than that of light*, Proc. R. Netherlands Acad. Arts Sci. 6 (1904) 809–31, https://en.wikisource.org/wiki/Electromagnetic_phenomena

[60] P. A. Loeb and M. P. H. Wolff, *Nonstandard Analysis for the Working Mathematician*, Springer (2015)

[61] A. Pais, *Electromagnetic mass: The first century*, in *Subtle Is the Lord: The Science and the Life of Albert Einstein*, Oxford University Press (1982) chapter 7

[62] S. Parrott, *Relativistic Electrodynamics and Differential Geometry*, Springer (1987)

[63] R. Penrose, *Gravitational collapse and space-time singularities*, Phys. Rev. Lett. 14 (1965) 57–59

[64] R. Penrose, *Gravitational collapse: The role of general relativity*, Riv. Nuovo Cimento 1 (1969) 252–76

[65] M. E. Peskin and D. V. Schroeder, *An Introduction to Quantum Field Theory*, Westview (1995)

[66] M. Planck, *Ueber das Gesetz der Energieverteilung im Normalspectrum*, Ann. Phys. 309 (1901) 553–63, http://www.physik.uni-augsburg.de/annalen/history/historic-papers/1901_309_553-563.pdf
[67] H. Poincaré, *Sur la dynamique de l'electron, Rendiconti del Circolo Matematico di Palermo* 21 (1906) 129–75. Reprinted in *Œuvres de Henri Poincaré*, vol. 9, Gauthiers-Villars (1954) 494–550
[68] A. K. Raychaudhuri, *Relativistic cosmology I*, Phys. Rev. 98 (1955) 1123–26.
[69] M. Reed and B. Simon, *Methods of Modern Mathematical Physics, vol. 2: Fourier Analysis, Self-Adjointness*, Academic Press (1975)
[70] F. Rellich, *Störungstheorie der Spektralzerlegung, II*, Math. Ann. 116 (1939) 555–70
[71] A. Rendall, *Theorems on existence and global dynamics for the Einstein equations*, Living Rev. Rel. 8 (2002)
[72] V. Rivasseau, *From Perturbative to Constructive Renormalization*, Princeton University Press (1991).
[73] A. Robinson, *Non-standard Analysis*, Princeton University Press (1996)
[74] F. Rohrlich, *The dynamics of a charged sphere and an electron*, Am. J. Phys. 65 (1997) 1051–56
[75] F. Rohrlich, *Classical self-force*, Phys. Rev. D 60 (1999) 084017
[76] D. G. Saari, *A global existence theorem for the four-body problem of Newtonian mechanics*, J. Diff. Eq. 26 (1977) 80–111, http://www.sciencedirect.com/science/article/pii/0022039677901000
[77] D. G. Saari, *Improbability of collisions in Newtonian gravitational systems*, Trans. Amer. Math. Soc. 162 (1971) 267–71
[78] D. G. Saari, *Improbability of collisions in Newtonian gravitational systems, II*, Trans. Amer. Math. Soc. 181 (1973) 351–68
[79] D. G. Saari and Z. Xia, *Off to infinity in finite time*, Notices Amer. Math. Soc. 42 (1995) 538–46, www.ams.org/notices/199505/saari-2.pdf
[80] G. Scharf, *Finite Quantum Electrodynamics: The Causal Approach*, Springer (1995)
[81] U. Schreiber, *Differential cohomology in a cohesive ∞-topos*, Preprint, arXiv:1310.7930
[82] K. Schwarzschild, *Zur Elektrodynamik. II. Die elementare elektrodynamische Kraft*, Göttinger Nachrichten 128 (1903) 132–41, https://eudml.org/doc/58547
[83] J. Schwinger, *On quantum-electrodynamics and the magnetic moment of the electron*, Phys. Rev. 73 (1948) 416–17
[84] S. Schweber, *QED and the Men Who Made It: Dyson, Feynman, Schwinger, and Tomonaga*, Princeton Unversity Press (1994)
[85] B. Simon, *Coupling constant analyticity for the anharmonic oscillator*, Ann. Phys. 58 (1970) 76–136
[86] R. F. Streater and A. S. Wightman, *PCT, Spin and Statistics, and All That*, Benjamin Cummings (1964)
[87] D. M. A. Stuart, *Geodesics and the Einstein nonlinear wave system*, J. Math. Pure Appl. 83 (2004) 541–87, www.sciencedirect.com/science/article/pii/S0021782403001004

[88] H. Tetrode, *Über den Wirkungszusammenhang der Welt. Eine Erweiterung der klassischen Dynamik*, Z. Phys. 10 (1922) 317–28. Also available in translation at www.projects.science.uu.nl/igg/seevinck/Translation_Tetrode.pdf
[89] R. Ticciati, *Quantum Field Theory for Mathematicians*, Cambridge University Press (2008)
[90] S. Wagon, *The Banach–Tarski Paradox*, Cambridge University Press (1986)
[91] R. Wald, *General Relativity*, University of Chicago Press (1984)
[92] R. Wald, *Gravitational collapse and cosmic censorship*, in B. R. Iyer and B. Bhawal, eds., *Black Holes, Gravitational Radiation and the Universe*, Springer Netherlands (1999) 69–86
[93] K. G. Wilson and J. Kogut, *The renormalization group and the ϵ expansion*, Phys. Rep. C12 (1974) 75–200
[94] Z. Xia, *The existence of non-collision singularities in Newtonian systems*, Ann. Math. 135 (1992) 411–68
[95] F. Ye, *Strict Finitism and the Logic of Mathematical Applications*, Springer (2011)
[96] A. Zee, *Quantum Field Theory in a Nutshell*, Princeton University Press (2010)

John C. Baez
Department of Mathematics, University of California,
Riverside and Centre for Quantum Technologies,
National University of Singapore
baez@math.ucr.edu

7

Twistor Theory: A Geometric Perspective for Describing the Physical World

Roger Penrose*

Contents

1	Early Motivations	327
2	The Emergence of Twistor Theory	335
3	Fields, Quantization, and Curved Spacetime	346
4	Palatial Twistor Theory	358
	References	369

1 Early Motivations

1.1 Geometrical Background: Two Roles for a Riemann Sphere

The basic geometrical proposal underlying twistor theory effectively came together in early December 1963, when I was on a nine-month appointment at the University of Texas in Austin [46]. Various motivational notions had been troubling me for several years previously, concerning what I had felt to be a need for a novel approach to foundational physics, in which concepts from both quantum mechanics and relativity theory had significant roles to play. These were interrelated via the theme of complex analysis and complex-number geometry, areas of mathematics that had impressed me deeply from around 1950, during my time as an undergraduate in mathematics at University College, London. These ideas had then featured strongly in my mind in the early 1960s. The thought I had in late 1963 was the initial stage of the proposal

---

*I am deeply grateful for financial assistance through a personal endowment from J. P. Moussouris.

that, a little later, I indeed referred to as "twistor theory," owing to a key role that the twisted configuration of interlocking circles shown in Figure 7.1 (a stereographically projected family of the Clifford parallels on a 3-sphere) had played for me. The reader might well ask what such an intriguing configuration might have to do with a basic theory of physics. We shall see later that this configuration represents the angular momentum of a massless particle with spin, but in order to explain this, it is necessary first to outline some of the various ideas that had been troubling me earlier. I shall come to the specific role of the configuration of Figure 7.1 in Sections 2.1, 2.4, and 2.5, particularly at the end of that section.

One of my main motivations had arisen from my feeling that there was a need for a formalism that was geared to that specific dimensionality of spacetime structure that we directly perceive around us. This line of thinking was very unlike that of various other ideas for an underlying physics of the world that later became popular, for example, string theory [12]. I had earlier become convinced that what was needed would be a formalism that should be very specific to the number of space and time dimensions, namely, 3 and

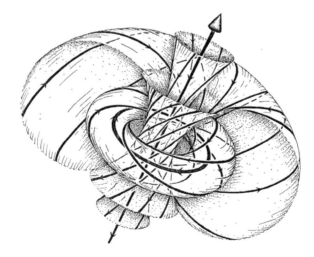

Figure 7.1 A picture representing a nonnull twistor: stereographic projection – to a Euclidean 3-space E – of Clifford parallels on a 3-sphere. The tangent directions to the circles point in the direction (projected into E) of the rays of a Robinson congruence. By continually reassembling itself, the entire configuration travels with the speed of light, as E evolves in time, in the direction of the large arrow at the top right. The configuration represents the angular momentum structure of a massless particle with spin.

1, respectively, that macroscopically present themselves to us, and I took the view that this should be central to the scheme. This indeed goes very much in opposition to the role of spacetime dimensionality underlying many of the current trends, most particularly string theory, where extra space dimensions (and even an extra time dimension, in the case of "F-theory") are regarded as essential ingredients of these various theories [12], taken to be serious proposals for the overall spacetime geometry of the physical world that we inhabit. It also contrasts with the very natural and commendable desire, in pure mathematics, for formalisms that can be applied, generally, to any spatial dimensionality whatever, but the aims of theoretical physics are very different from those of pure mathematics, even though much of theoretical physics depends vitally on the latter.

Another of my basic motivations had been for a formalism that was essentially *complex* in the sense that it would be able to take advantage of what I had regarded, ever since my days as a mathematics undergraduate, as the "magic" of complex analysis and holomorphic (i.e., complex analytic) geometry. I had learned that the complex number system has not only a profoundly deep power and elegance, but that it had also found a basic realization in its underlying role in the formalism of quantum theory. I later began to study quantum mechanics in a serious way, and was particularly impressed by the superb course of lectures given by Paul Dirac, when I was a graduate student (in algebraic geometry), and subsequently a Research Fellow, at St John's College Cambridge. I became fascinated by the quantum description of spin, and how the complex numbers of quantum mechanics were directly related to the three-dimensionality of physical space, via the 2-sphere of spatial directions being appropriately identified as a Riemann (or Bloch) sphere of the ratios of pairs of complex numbers (quantum amplitudes) where, in the case of a massive particle of spin ½ such as an electron (see Figure 7.2), we can think of these as being the complex components of a 2-spinor. Moreover, I had realized that in the relativistic context, there was another role for the Riemann sphere, this time as the celestial sphere that an astronaut in space would observe. The transformation of this celestial sphere to that of a second astronaut, moving at a relativistic speed while passing nearby the first would be one that preserves the complex structure of the Riemann sphere (i.e., conformal without reflection). The special (i.e., nonreflective) Lorentz group is thus seen to be identical with these holomorphic transformations of this Riemann sphere (Möbius transformations). Again this was clear from the 2-spinor formalism, this time in the relativistic context (see [37]).

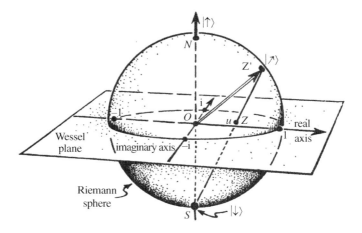

Figure 7.2 The Riemann sphere (here in its role as a Bloch sphere) projects stereographically from its south pole S to the complex (Wessel) plane, whose unit circle coincides with the equator of the sphere. A general spin state $|\nearrow\rangle = w|\uparrow\rangle + z|\downarrow\rangle$, of a spin-½ massive particle is represented by the point Z on the Wessel plane denoting the complex number $u = \frac{z}{w}$, which is the stereographic image of Z' on the sphere (so S, Z, and Z' are collinear). The spin direction $\nearrow$ is then OZ, where O is the sphere's centre.

1.2 The 2-Spinor Formalism

This dual role for the Riemann sphere, one fundamentally to do with quantum mechanics in the case of three spatial dimensions, and the other fundamentally to do with macroscopic relativity, in (3+1)-dimensional spacetime, struck me as being no accident, but something that linked together these two great revolutions of 20th-century physics – of the small and of the large – via the magic of complex numbers. I felt that this might represent a definite clue to a deep unifying relation between the two. Both could be seen as a feature of the 2-spinor calculus, as introduced by Élie Cartan [7] and Bartel van der Waerden [59], and which I had learned how to use from Dirac (see [9]), in an unexpected deviation from his normal Cambridge course on quantum mechanics.

I liked to think of a 2-spinor (often referred to by physicists as a "Weyl spinor") in a very geometrical way, and I realized that, up to an overall sign, a nonzero 2-spinor can be represented as a future-pointing null vector (a vector pointing along the future null cone), referred to as the "flagpole," together with a "flag plane" direction through that flagpole [36, 52]. The flag plane would be a null half-plane bounded by the flagpole. This flag geometry can be thought of in the following way. Imagine the Riemann sphere $\mathcal{S}$ of

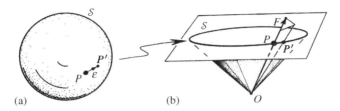

Figure 7.3 (a) The space of null directions at some spacetime point O is represented as a Riemann 2-sphere S. The flagpole direction of a 2-spinor is represented, on S, as the point P. Infinitesimally near to P is P', where the direction $\overrightarrow{PP'}$ provides the 2-spinor's flag plane. (b) In spacetime terms, the 2-spinor's flagpole is shown as the null 4-vector $\overrightarrow{OF}$, where we realize S as a particular 3-plane intersection of the future null cone of O (all this taken in O's tangent 4-space), so that P lies on the line OF. The 2-spinor's flag plane is now seen as the null half-2-plane extending away from the line OF in the direction of P'.

null (i.e., lightlike) directions at some point O in spacetime (see Figure 7.3). We are thinking of the geometry in the tangent 4-space of the point O. The flagpole direction is represented by some point P on a sphere of cross section of the future null cone of O, which we identify with S, and we choose a point P' on S infinitesimally separated from P. The straight line extended out from P in the direction of P', when joined to O, defines the required flag half-plane. We note that as the point P' rotates about P, the flag plane rotates about the flagpole. The spinor itself is defined only up to sign by this geometry, but we must take note that if P' rotates continuously around P through 2π, the spinor becomes replaced by its *negative*. To reach the original 2-spinor by this procedure, the rotation of the flag plane would have to be through 4π.

I had found that 2-spinor methods were surprisingly valuable in giving us insights into the formalism of *general relativity* that were different from those that the standard Lorentzian tensor framework readily provides. Most immediately striking was the very simple-looking 2-spinor expression for Weyl's conformal curvature [38] (see also [64]). Whereas the usual Weyl-tensor quantity C_{abcd}, has a somewhat complicated collection of symmetry and trace-free conditions, the corresponding 2-spinor is simply a *totally symmetric* complex 2-spinor quantity Ψ_{ABCD}.

Some comments concerning the 2-spinor index notation being used here are appropriate. Capital italic Latin index letters $A, B, C, \ldots$ refer to the (2-complex dimensional) spin space if they are upper indices, and to the *dual* of this space if lower ones; *primed* such letters $A', B', C', \ldots$ refer to the complex-conjugate spin space. The tensor product of the spin space with its complex conjugate is identified with the complexified tangent space to the spacetime, at each of its points, here the *real* tangent vectors arise as the *hermitian* members of this tensor product. In general, I shall take these as *abstract* indices, in the sense described in my book with Wolfgang Rindler, *Spinors and Space-Time*, vol. 1 [52], so that no coordinate system is implied, either for the spacetime or to define a basis for the spin-space. This is notationally very handy, because the spacetime indices $a, b, c, \ldots$ can then be thought of as "shorthand" for the spinor index pairs:

$$a = AA', \quad b = BB', \quad c = CC', \quad \ldots$$

The spin-space (and hence also its dual and complex conjugate) has a symplectic structure defined by the skew-symmetric quantities

$$\varepsilon_{AB}, \quad \varepsilon^{AB}, \quad \varepsilon_{A'B'}, \quad \varepsilon^{A'B'},$$

these being used for lowering or raising indices (where we must be a little careful about signs and index orderings):

$$\kappa_B = \kappa^A \varepsilon_{AB}, \quad \kappa^A = \kappa_B \varepsilon^{AB}, \quad \eta_{B'} = \eta^{A'} \varepsilon_{A'B'}, \quad \eta^{A'} = \eta_{B'} \varepsilon^{A'B'}$$

so that in terms of components,

$$\kappa_1 = \kappa^0, \quad \kappa_0 = -\kappa^1, \quad \eta_{1'} = \eta^{0'}, \quad \eta_{0'} = -\eta^{1'},$$

where the component form of each of the epsilons is

$$\begin{pmatrix} 0 & 1 \\ -1 & 0 \end{pmatrix}.$$

The metric tensor, in abstract-index form is

$$g_{ab} = \varepsilon_{AB} \varepsilon_{A'B'}$$

and the abstract-index form of the Weyl conformal curvature tensor for spacetime is

$$C_{abcd} = \Psi_{ABCD} \varepsilon_{A'B'} \varepsilon_{C'D'} + \varepsilon_{AB} \varepsilon_{CD} \tilde{\Psi}_{A'B'C'D'}.$$

Here, I have allowed for the case of a complex metric g_{ab}, both Ψ_{ABCD} and $\tilde{\Psi}_{A'B'C'D'}$ being totally symmetric, where Ψ_{ABCD} describes the anti-self-dual (left-handed) Weyl curvature and $\tilde{\Psi}_{A'B'C'D'}$, the self-dual (right-handed) part. In the case of a real Lorentzian spacetime metric ($\bar{\varepsilon}_{AB} = \varepsilon_{AB}$), $\tilde{\Psi}_{A'B'C'D'}$ is the complex conjugate of Ψ_{ABCD}:

$$\tilde{\Psi}_{A'B'C'D'} = \bar{\Psi}_{A'B'C'D'} (= \overline{\Psi_{ABCD}}),$$

but it will be important for what follows that we consider the complex case also, as we shall be concerned with self-dual (complex vacuum) spacetimes, for which $\Psi_{ABCD} = 0$ and anti-self-dual ones, for which $\tilde{\Psi}_{A'B'C'D'} = 0$, later (these complex fields being regarded as wave functions).

1.3 Zero Rest-Mass Fields

We find that in the case of a (real Lorentzian) vacuum metric (with or without cosmological constant), the Bianchi identities become

$$\nabla^{AA'} \Psi_{ABCD} = 0$$

which may be compared with the Maxwell's equations in charge-free spacetime

$$\nabla^{AA'} \varphi_{AB} = 0,$$

where φ_{AB} relates to a (possibly complex) Maxwell field tensor F_{ab} in the same way as Ψ_{ABCD} relates to C_{abcd}, namely,

$$F_{ab} = \varphi_{AB}\varepsilon_{A'B'} + \varepsilon_{AB}\tilde{\varphi}_{A'B'},$$

where φ_{AB} describes the anti-self-dual (left-handed) part of the field and $\tilde{\varphi}_{A'B'}$ the self-dual (right-handed) part. For a real Maxwell field, they are complex conjugates of each other:

$$\tilde{\varphi}_{A'B'} = \overline{\varphi}_{A'B'}.$$

I had become interested in the issue of finding solutions of the general equation

$$\nabla^{AA'} \varphi_{ABC...E} = 0$$

in (conformally) flat spacetime, $\varphi_{ABC...E}$ being symmetric in its n spinor indices, the equation being the (conformally invariant) free-field equation for a massless field of spin $n/2$ [8, 9, 39]. This equation (together with the wave equation in suitably conformally invariant form, which includes an $R/6$ term, R being the scalar curvature) had a particular importance for me, and I believed it to have a rather basic status in relativistic physics. For I had come to the view that nature might have a "massless" structure at its roots, mass itself being a secondary phenomenon. In around 1961 (see [45]) I had found a formula for obtaining the solution of this field equation from general data freely specified on a null initial hypersurface. I had formed the view that this formula had a certain kinship with the Cauchy integral formula for obtaining the value of a holomorphic function at some point of the complex plane in terms of the function's values along a closed contour surrounding that point. I had felt that, in some sense, this massless field equation might be akin to the Cauchy–Riemann equations. It had to be some unusual "complex" way of looking at Minkowski space, I had surmised, in which the massless field equations were simply a statement of *holomorphicity* – but in what sense could this possibly be true?

There was one remaining feature that I felt sure must be represented, as part of this mysterious "complex" way of looking at spacetime. This arose from a discussion that I had had with Engelbert Schücking when I shared an office with him in the spring of 1961 at Syracuse University in New York State. Engelbert had persuaded me of the key importance to quantum field theory of the splitting of field amplitudes into positive and negative frequency parts. I was not happy with the standard procedure of first resolving these amplitudes into Fourier components and then selecting the positive ones, as not only did this strike me as too "top-heavy," but also the Fourier analysis is not conformally invariant – and I had come to believe that this conformal invariance, being a feature of massless fields, was important (again, something that had been stressed to me by Engelbert).

I had become aware that for complex functions defined on a line (thought of as the timeline) we may understand their splitting into positive- and negative-frequency parts in the following way. We view this timeline as being the equator of real numbers in a *Riemann sphere* which, as before, is the complex plane compactified by the single point labeled by "∞," but where the sphere is now being oriented somewhat differently from that of Figure 7.2, with the real numbers now featuring as the equator (increasing as we proceed in an anticlockwise sense in the horizontal plane), rather than the unit circle. Functions defined on this equatorial circle which extend holomorphically into the southern hemisphere (with usual conventions) are the functions of positive frequency, and those which extend holomorphically into the northern hemisphere are those of negative frequency. An arbitrary complex function defined on this circle can be split into a function extending globally into the southern hemisphere and one globally into the northern hemisphere – uniquely except for an ambiguity with regard to the constant part – and this provides us with the required positive/negative frequency split, without any resort to Fourier analysis. I wanted to extend this picture into something more global, with regard to spacetime, and I had in mind that my sought-for "complex" way of looking at Minkowski space should exhibit something strongly analogous to this division into two halves, where the boundary between the two could be interpreted in "real" terms, in some direct way. This had set the stage for the emergence of twistor theory.

2 The Emergence of Twistor Theory

2.1 Robinson Congruences

A colleague of mine, Ivor Robinson, who had taken up a position at what later became the University of Texas at Dallas, had been working on finding global nonsingular *null* solutions of Maxwell's free-field equations in Minkowski space $\mathbb{M}$, where "null" in this context means that the invariants of the field tensor F_{ab} vanish, that is, $F_{ab}F^{ab} = 0 = {}^*F_{ab}F^{ab}$ where ${}^*F_{ab}$ is the Hodge dual of F_{ab}. Equivalently, in 2-spinor terms, $\varphi_{AB}\varphi^{AB} = 0$, which tells us that

$$\varphi^{AB} = \kappa^A \kappa^B$$

for some κ^A. It is not hard to show that the Maxwell source-free equations then imply that the flagpole direction of κ^A points along a 3-parameter family – a *congruence* – of null straight lines, which turn out to be what is called "shear-free," which means that although the lines may diverge, converge, or rotate, locally, there is no shear (or distortion) as we follow along the lines.

Although, not relevant to the discussion at the moment, it is worth noting that the study of shear-free congruences of rays in *curved* spacetimes has a considerable historical significance – where I use the term *ray* simply to mean a null (i.e., lightlike) geodesic in spacetime. In particular, the well-known Kerr solution [6, 23] of the Einstein vacuum equations for a rotating black hole possesses a shear-free ray congruence, and this played a key role in its discovery, as it did also in Newman's generalization to an electrically charged black hole [34], and also in the Robinson–Trautman gravitationally radiating exact solutions [56], among other examples. As in the case of Minkowski space $\mathbb{M}$, as described above, it is also true that for any null solution φ^{AB} of Maxwell's equations in curved spacetimes, the flagpole directions of the κ^A-spinors point along a shear-free family of rays.

A simple example of a shear-free ray congruence in $\mathbb{M}$ is obtained from any fixed choice of a ray L in $\mathbb{M}$, where the family of all rays that meet L provides a shear-free ray congruence. I refer to such a congruence as a *special Robinson congruence*, and this includes the limiting case when L is taken out to infinity, so our congruence becomes a family of parallel rays in $\mathbb{M}$. Ivor Robinson had developed ways of producing null solutions of the Maxwell's equations, starting from any given shear-free null congruence, but when applied to the special congruences just described, he found that singularities would arise along the line L itself (except in the otherwise unsatisfactory case where L is at infinity). Desiring a singularity-free Maxwell field, he provided the following ingenious trick. Consider, instead, solutions of Maxwell's equations in the *complexified* Minkowski space $\mathbb{CM}$, and displace the line L in a complex direction, so that it lies in $\mathbb{CM}$, but entirely outside its real part $\mathbb{M}$. Complex analytic solutions of Maxwell's equations, based on the complex "special Robinson congruence" defined by the *displaced L* need not now be singular within $\mathbb{M}$, and the flagpoles of the κ^A-spinors within $\mathbb{M}$ now point along an entirely *non*-singular shear-free ray congruence in $\mathbb{M}$, which I later named a (general) *Robinson congruence*.

I became highly intrigued by the geometry of general Robinson congruences, and I soon realized that one could describe them in the following way. Consider an arbitrary spacelike 3-plane E in Minkowski 4-space $\mathbb{M}$. E has the geometry of ordinary Euclidean 3-space, and each ray N of the congruence will meet E in a single point, at which we can determine the location of that ray within $\mathbb{M}$ by specifying a unit 3-vector **n** at that point, pointing in the spatial direction that is the orthogonal projection into E of the null direction of N there. Thus we have a vector field of **n**s within E to represent the Robinson congruence. After some thought I realized what the nature of this vector field must be. The **n**-vectors are tangent to the oriented circles (together with one oriented straight line) obtained by stereographic projection of a family of oriented Clifford parallels on a 3-sphere. See Figure 7.1, in Section 1.1, for a picture of this configuration, and reference [53] for a detailed derivation. The large arrow at the top right indicates the direction in which the configuration appears to move with the speed of light by continually re-assembling itself in that direction, as E moves by parallel displacement into the future.

By examining this configuration, and counting the number of degrees of freedom that such configurations have, I realized that the space of Robinson congruences must be six-dimensional. Moreover, it was reasonably clear to me that by its very mode of construction, this space ought to have a *complex structure*, and so must be, in a natural way, a complex 3-manifold. Within this space would lie the space of special Robinson congruences, each of which would be determined by a single ray (namely, L). The space of rays in $\mathbb{M}$ is 5-real-dimensional, and it divides the space of general Robinson congruences into two halves, namely, those with a right-handed twist and those with a left-handed twist. The complex 3-space of Robinson congruences, which came to be known as "projective twistor space" appeared to be just what I believed was needed, where the "real" part of the space (representing light rays in $\mathbb{M}$, or their limits at infinity) would, like the "real" equator of the Riemann sphere described at the end of Section 1.2, divide the entire space into two halves. This, indeed appeared to be exactly the kind of thing that I was looking for!

2.2 Twistors in Terms of 2-Spinors

To be more explicit about things, and to understand precisely how the space of Robinson congruences does indeed provide a compact complex 3-manifold divided in two by the real 5-space of special Robinson congruences, let us turn again to the relativistic 2-spinor formalism of Section 1.2. We shall see how this allows us to provide a very neat description of individual rays in $\mathbb{M}$. In Section 2.4, we see how this generalizes to describe general Robinson congruences. The physical interpretation in terms of relativistic angular momentum of massless particles will emerge in Section 2.5.

Consider some ray Z in $\mathbb{M}$, and let us assign a *strength* to this ray in the form of a null 4-momentum covector p_a, where the vector p^a points along Z at each of its points, parallel-propagated along Z. In fact, let us go a little further than this by assigning a (dual, conjugate) 2-spinor $\pi_{A'}$, parallel-propagated along Z, where

$$p_a = \overline{\pi}_A \pi_{A'}$$

so that in addition to having $\pi_{A'}$'s flagpole pointing along Z, we also have $\pi_{A'}$'s flag *plane* (and spinor sign) assigned to Z, and which is to be parallel-propagated along it. This will be referred to as a *spinor scaling* for the ray Z.

We need to choose a spacetime origin point O within $\mathbb{M}$, so that any point X of $\mathbb{M}$ can be labeled by a position vector x^a ($= x^{AA'}$) at O. Then if X is any point on the ray Z, we can define a 2-spinor ω^A by the equation,

$$\omega^A = ix^{AA'}\pi_{A'}$$

and we find that ω^A remains unchanged if X is replaced by any other point on the ray Z, such a point having a position vector of the form

$$x^{AA'} + k\pi^{A'}\overline{\pi}^A$$

where k is any real number (since $\pi^{A'}\pi_{A'} = 0$). The pair $(\omega^A, \pi_{A'})$, serves to identify the ray Z, together with a spinor scaling for Z.

The 2-spinors ω^A and $\pi_{A'}$ are the *spinor parts* (with respect to the origin O) of the *twistor* Z^α, which represents the spinor-scaled ray Z, and often one simply writes

$$Z^\alpha = (\omega^A, \pi_{A'}).$$

However, for a *ray*, there is a particular equation that must hold between the spinor parts, namely,

$$\omega^A \overline{\pi}_A + \pi_{A'}\overline{\omega}^{A'} = 0$$

which follows from the fact that the vector x^a is *real*, so that $x^{AB'}$ has the *hermitian* property $\overline{x^{AB'}} = x^{AB'}$. The above equation can be rewritten as

$$Z^\alpha \overline{Z}_\alpha = 0$$

where $\overline{Z}_\alpha$, the complex conjugate of Z^α

$$\overline{Z}_\alpha = (\overline{\pi}_A, \overline{\omega}^{A'}),$$

(and note the reverse order of the spinor parts) is a dual twistor. When $Z^\alpha \overline{Z}_\alpha = 0$, we refer to Z^α as a *null* twistor, so it is that the null twistors represent (spinor-scaled) rays in $\mathbb{M}$ – or rays at $\mathbb{M}$'s infinity.

The above equation

$$\omega^A = ix^{AA'}\pi_{A'}$$

is referred to as the *incidence relation* between the spacetime point X and the twistor $Z^\alpha = (\omega^A, \pi_{A'})$. We may also be interested in this incidence relation when X is allowed to be a complex point. Likewise, for a dual twistor

$$W_\alpha = (\lambda_A, \mu^{A'}),$$

incidence with a (possibly complex) point X is expressed as

$$\mu^{A'} = -ix^{AA'}\lambda_A.$$

It is useful to get a picture of the geometrical role of the 2-spinor ω^A, in addition to $\pi_{A'}$, in the case of a general null twistor $Z^\alpha = (\omega^A, \pi_{A'})$. Figure 7.4 shows this, where O is the origin and the point Q is the intersection of the

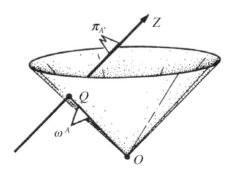

Figure 7.4 The flagpole directions of the spinor parts of a general null twistor $Z^\alpha = (\omega^A, \pi_{A'})$ are depicted, where Q is the intersection of the ray Z with the light cone of the origin O.

ray Z with the light cone of O. The null vector $\overrightarrow{OQ}$ has index form q^a and is proportional to the flagpole of ω^A where

$$q^{AA'} = \omega^A \overline{\omega}^{A'} (i \overline{\omega}^{B'} \pi_{B'})^{-1}.$$

This expression fails only when $\overline{\omega}^{B'} \pi_{B'} = 0$ (but holding in a certain limiting sense) which occurs when the ray Z lies in a null hyperplane through O, and the point Q lies at infinity.

2.3 Minkowski Space Compactified, Complexified, and Its Conformal Symmetry

At this juncture it would be helpful to clarify the nature of "infinity," with regard to Minkowski space $\mathbb{M}$. We recall that when a ray L is characterized in terms of the null congruence of rays that intersect L, we were led to consider the ray congruences that consist entirely of *parallel* rays, arising when L is moved out to infinity. There is a whole 2-sphere's-worth of such systems of parallel rays, one for each null direction. Thus the family of limiting rays L at infinity generates a kind of "light cone at infinity," frequently denoted by the script letter $\mathscr{I}$ (and pronounced "scri"). We can regard $\mathscr{I}$ as being the identification of $\mathbb{M}$'s future conformal boundary $\mathscr{I}^+$ with its past conformal boundary $\mathscr{I}^-$ (see [39, 53]). This identification also incorporates the single point i (the *vertex* of $\mathscr{I}$), which is the identification of the three points i^-, i^0, and i^+, respectively representing past infinity, spacelike infinity, and future infinity (see Figure 7.5). This provides us with the picture of *compactified Minkowski space* $\mathbb{M}^\#$ (which turns out to have topology $S^1 \times S^3$) where Figure 7.5(a) indicates the future and past null boundaries of $\mathbb{M}$, and Figure 7.5(b) shows how these two conformal boundaries $\mathscr{I}^+$ and $\mathscr{I}^-$ are to be identified as $\mathscr{I}$, where future and past endpoints of any ray in $\mathbb{M}$ are identified. This provides us with the highly symmetrical compact Lorentzian-conformal manifold $\mathbb{M}^\#$. Every ray within $\mathbb{M}$ is compactified by a single point to become a topological circle.

The global symmetry group of $\mathbb{M}^\#$ is the 15-parameter symmetry group that is frequently referred to as the *conformal group* of flat four-dimensional spacetime. This group has four connected components since it allows for reversals of time and space orientations. I shall be concerned here only with the *connected component* of the identity, referring to this group as $C(1,3)$.

Another way of understanding $\mathbb{M}^\#$ is that it represents the family of generator lines of the null cone $\mathcal{K}$ of the origin $O^{2,4}$ (i.e., of entire rays through $O^{2,4}$) in the pseudo-Minkowskian 6-space $\mathbb{M}^{2,4}$, whose signature is

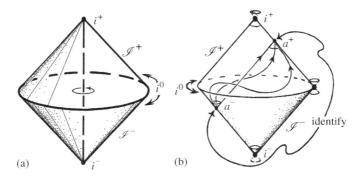

Figure 7.5 (a) A conformal picture indicating how Minkowski space $\mathbb{M}$ acquires its future null boundary $\mathscr{I}^+$, a null 3-surface supplying future endpoints to rays in $\mathbb{M}$ and, similarly, a past null boundary $\mathscr{I}^-$ supplying past endpoints to rays in $\mathbb{M}$. There are also three other conformal boundary points i^+, i^-, and i^0 denoting future, past, and spacelike infinity, respectively. (b) To complete the picture of compactified Minkowski space $\mathbb{M}^\#$, we must identify $\mathscr{I}^+$, with $\mathscr{I}^-$, so that the future endpoint a^+ of any ray in $\mathbb{M}$ is identified with its past endpoint a^-. Also the three points i^+, i^-, and i^0 must be identified.

$(++---)$. See [53]. If we consider these generator lines to be *oriented* (or as being null half-lines, starting at $O^{2,4}$), then we get a 2-fold cover $\mathbb{M}^{\#2}$ of $\mathbb{M}^\#$, since an action of the pseudo-orthogonal group $SO(2,4)$ on $\mathcal{K}$ can continuously rotate any particular one of its oriented generators into itself but with opposite orientation (i.e., reflected in the origin $O^{2,4}$). Thus, the family of oriented rays through $O^{2,4}$ provides us with a realization of the 2-fold cover $\mathbb{M}^{\#2}$ of $\mathbb{M}^\#$.

The symmetry group of the vector space $\mathbb{T}$ of twistors $Z^\alpha = (\omega^A, \pi_{A'})$ goes a step further than this. The pseudo-hermitian form $||\mathbf{Z}|| = Z^\alpha \overline{Z}_\alpha$ $(\omega^A \overline{\pi}_A + \pi_{A'} \overline{\omega}^{A'})$ has split signature $(++--)$, so the group of (complex-) linear transformations of $\mathbb{T}$ that preserve the norm is the pseudo-unitary group $SU(2,2)$. It is, indeed, one of Cartan's specific local isomorphisms (see [53]) that $SO(2,4)$ is *locally* isomorphic to $SU(2,2)$, the latter being a 2-fold cover of the former. This tells us that this $SU(2,2)$ actually acts on a 2-fold cover $\mathbb{M}^{\#4}$ of $\mathbb{M}^{\#2}$. The space $\mathbb{M}^{\#4}$ is therefore a 4-fold cover of $\mathbb{M}^\#$. Topologically, we can understand such an n-fold cover $\mathbb{M}^{\#n}$ of $\mathbb{M}^\#$ as obtained simply by "unwrapping" the S^1 of $\mathbb{M}$'s topology $S^1 \times S^3$ to the required degree n.

In fact, this strange-looking 4-fold cover $\mathbb{M}^{\#4}$ of compactified Minkowski space can be understood explicitly in terms of the geometrical representation of a null twistor Z^α in Minkowski space terms. We recall that a null twistor describes not just a ray Z in $\mathbb{M}$, but also a spinor scaling, defined by $\pi_{A'}$, assigned to the null direction at each point of the ray Z, where we think of this spinor scaling as parallel-transported along the ray Z within $\mathbb{M}$. Now, we saw in Section 1.2 (see Figure 7.3) that a 2-spinor has a $U(1)$ phase that is geometrically described by a null flag half-plane, where if the flag plane is rotated about the flagpole through 2π, the spinor changes sign. However, when we think of this flag half-plane as being parallel-transported all the way from $\mathscr{I}^-$ to $\mathscr{I}^+$ along Z, we find that if we were to try to match $\mathscr{I}^+$ directly to $\mathscr{I}^-$, as indicated in Figure 7.5(b), then we would find a discrepancy of a rotation through π, that is, the flags would point in the opposite directions from one another across $\mathscr{I}$. (This geometry is explained explicitly in [53, Section 9.4; see particularly Figures 9–11].) Since a rotation of a spinor flag-plane by 2π results in a change of sign for the spinor, we need 4π to get it back to its original value. The rotation through π that we find when we pass across $\mathscr{I}$ represents a discrepancy of i in the geometrical description of a null twistor in the space $\mathbb{M}^\#$. Moreover, the problem is not removed if we consider the flag-plane interpretation within just the 2-fold cover $\mathbb{M}^{\#2}$, since we still have a sign discrepancy. Only when we pass to the 4-fold cover $\mathbb{M}^{\#4}$ do we get a fully consistent picture of a null twistor – and, indeed, of a nonnull twistor (see [53]), whose interpretation we turn to next.

2.4 The Basic Twistor Spaces

Let us now consider how to represent a *non*-null twistor Z^α in a geometrical way. It is best to think in terms of the family of null twistors Y^α that are *orthogonal* to Z^α in the sense that

$$Z^\alpha \overline{Y}_\alpha = 0$$

(or, equivalently $Y^\alpha \overline{Z}_\alpha = 0$). If Z^α were a null twistor – where Y^α is *given* as a null twistor – these respectively representing rays Z and Y, then this vanishing of their scalar product asserts that these rays *intersect* (perhaps at infinity). Accordingly, if Z is fixed, then this condition on Y tells us that the Y belongs to the special Robinson congruence defined by the ray Z. Now, let Z be a fixed *non*-null twistor (but where Y remains null). Then the congruence of Y-rays subject to orthogonality with Z will provide a *general* Robinson congruence. See [53] for details.

As noted in Section 2.3, the space $\mathbb{T}$ of all twistors Z^α is a four-dimensional complex vector space, with pseudo-hermitian scalar product ($Z^\alpha \overline{Y}_\alpha$) of split signature ($++--$). Geometrical notions are often best expressed in terms of the *projective* twistor space $\mathbb{PT}$ of twistors up to proportionality, this being a complex projective 3-space $\mathbb{CP}^3$. This compact complex manifold $\mathbb{PT}$ – or, more strictly, in accordance with the above discussion, the $\mathbb{CP}^3$ of *dual* projective twistors $\mathbb{PT}^*$ – can indeed be identified with the space of Robinson congruences referred to above. The *dual* twistor space $\mathbb{T}^*$ is identified with the *complex conjugate* space $\overline{\mathbb{T}}$ of $\mathbb{T}$ via this pseudo-hermitian structure. The points of the dual projective space $\mathbb{PT}^*$ represent the complex projective *planes* within $\mathbb{PT}$. The complex projective *lines* within $\mathbb{PT}$ correspond to points of the complexified compactified Minkowski space $\mathbb{CM}^\#$.

Whereas, generally speaking, it is the projective twistor space $\mathbb{PT}$ that is useful to us if we are thinking of geometrical matters, the space $\mathbb{T}$ is appropriate if we are concerned with the *algebra* of twistors. For a nonzero twistor Z^α, we can have three algebraic alternatives. These are:

- $Z^\alpha \overline{Z}_\alpha > 0$, for a *positive* or *right-handed twistor* Z^α, belonging to the space $\mathbb{T}^+$,
- $Z^\alpha \overline{Z}_\alpha < 0$, for a *negative* or *left-handed twistor* Z^α, belonging to the space $\mathbb{T}^-$,
- $Z^\alpha \overline{Z}_\alpha = 0$, for a *null* Z^α, belonging to the space $\mathbb{N}$.

The entire twistor space $\mathbb{T}$ is the disjoint union of the three parts $\mathbb{T}^+$, $\mathbb{T}^-$, and $\mathbb{N}$, as is its projective version $\mathbb{PT}$ the disjoint union of the three parts $\mathbb{PT}^+$, $\mathbb{PT}^-$, and $\mathbb{PN}$ (see Figure 7.6).

Each point of $\mathbb{PT}$ represents a 1-dimensional vector subspace of $\mathbb{T}$. The points of $\mathbb{CM}^\#$ are thus described by 2-complex-dimensional subspaces of $\mathbb{T}$ (complex lines in $\mathbb{PT}$). The points of $\mathbb{M}^\#$ are described by 2-complex-dimensional subspaces in $\mathbb{N}$, that is, by complex lines in $\mathbb{PN}$. Robinson congruences are represented by the intersections of complex projective planes in $\mathbb{PT}$ with $\mathbb{PN}$. In the case of a special Robinson congruence of rays meeting a particular ray L, the complex plane in $\mathbb{PT}$ has contact with $\mathbb{PN}$ at a point representing the ray L in $\mathbb{M}^\#$.

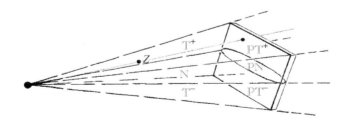

Figure 7.6 The way that the various parts of twistor space $\mathbb{T}$ relate to their various projective counterparts of $\mathbb{PT}$.

2.5 Helicity and Relativistic Angular Momentum

It is the space $\mathbb{PN}$ that has the most direct physical interpretation, since its points correspond to worldlines of free classical massless particles, which we can think of as the classical histories of (pointlike) photons in free motion, though possibly at infinity, as a limiting case in Minkowski space $\mathbb{M}$; see Figure 7.7. As stated above, not only are the points of *complexified* Minkowski space $\mathbb{CM}$ represented as (complex projective) lines in $\mathbb{PT}$, but so also are all the points of the complexified *compactified* Minkowski space $\mathbb{CM}^{\#}$. Those lines that lie in $\mathbb{PN}$ represent points of the *real* spacetime $\mathbb{M}$ (possibly at infinity), but since these lines are still complex projective lines, they are indeed *Riemann spheres*, in accordance with the ambitions put forward in Section 1.2; see Figure 7.7.

In Figure 7.8 this picture is extended to include a physical interpretation of nonnull twistors, where points of $\mathbb{PT}^+$ and $\mathbb{PT}^-$ are represented, in Minkowski space, as though they are light rays with a twist about them. This is schematic, but indeed these points can be regarded as representing massless particles with spin. In relativistic physics, if a massless particle has a *nonzero spin*, the "spin-axis" must be directed parallel or anti-parallel to the particle's velocity. We say that the particle has a *helicity s*, that can be positive or negative (or zero, for a spinless massless particle). If $s \neq 0$, then the particle's spacetime trajectory is not precisely defined (in a relativistically invariant way) as a worldline, but can be specified in terms of its 4-momentum p_a and 6-angular momentum M^{ab} about some chosen spacetime origin point **O**. These must be subject to

$$p_a p^a = 0, \quad p_0 > 0, \quad M^{(ab)} = 0, \quad \frac{1}{2}\varepsilon_{abcd} p^b M^{cd} = s p_a$$

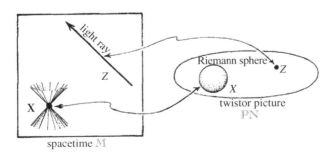

Figure 7.7 The most immediate part of the twistor correspondence: a ray Z in Minkowski space $\mathbb{M}$ corresponds to a point in $\mathbb{PN}$; a point **x** of $\mathbb{M}$ corresponds to a Riemann sphere X in $\mathbb{PN}$.

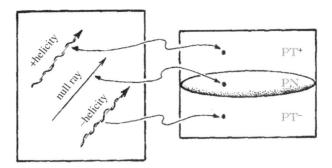

Figure 7.8 Classical massless particles with positive (right-handed) helicity can be represented as points of $\mathbb{PT}^+$ and those with negative (left-handed) helicity as points of $\mathbb{PT}^-$.

(curved or square brackets around indices respectively denoting symmetric or antisymmetric parts), where $\varepsilon_{abcd} = \varepsilon_{[abcd]}$ is the Levi-Civita tensor fixed by its component value $\varepsilon_{0123} = 1$ in a right-handed orthonormal Minkowskian frame (with time-axis basis vector δ_0^a, so p_0 is the particle's energy in units where the speed of light $c = 1$). Note that $\frac{1}{2}\varepsilon_{abcd}p^b M^{cd} = {}^*M_{ab}$ is the Hodge dual of M^{ab}, so the second displayed equation becomes ${}^*M_{ab}p^b = sp_a$. The connection between these quantities and twistor theory is that if

$$Z^\alpha = (\omega^A, \pi_{A'})$$

then we can make the interpretation

$$p_{AA'} = \pi_{A'}\bar{\pi}_A, \quad M^{AA'BB'} = i\omega^{(A}\bar{\pi}^{B)}\varepsilon^{A'B'} - i\bar{\omega}^{(A'}\pi^{B')}\varepsilon^{AB},$$

and all the above conditions are automatically satisfied, provided that $\pi_{A'} \neq 0$. Conversely, the twistor Z^α (with $\pi_{A'} \neq 0$) is determined, uniquely up to a phase multiplier $e^{i\theta}$, by p_a and M^{ab}, subject to these conditions. The helicity s finds the very simple (and fundamental) expression

$$2s = \omega^A \bar{\pi}_A + \pi_{A'}\bar{\omega}^{A'}$$
$$= Z^\alpha \bar{Z}_\alpha.$$

There is, however, the subtlety referred to above that in this interpretation of a nonnull twistor, when the helicity s is nonzero, there is no actual worldline that can describe the particle's location in a relativistically invariant way, the worldline being, in a sense, "spread out" in accordance with the configuration depicted in Figure 7.1, as we shall see at the end of Section 2.5. This issue is an important undercurrent to the application of twistor ideas in general relativity as discussed in Sections 4.3 and 4.4.

2.6 Description under Shift of Origin

Under a displacement of the origin O to a new point Q of $\mathbb{M}$,

$$O \mapsto Q,$$

where the position vector $\overrightarrow{OQ}$ is (in abstract-index form) q^a, the spinor parts of the twistor $Z^\alpha = (\omega^A, \pi_{A'})$ undergo

$$\omega^A \mapsto \omega^A - iq^{AA'}\pi_{A'}, \quad \pi_{A'} \mapsto \pi_{A'}.$$

For a dual twistor $W_\alpha = (\lambda_A, \mu^{A'})$, we correspondingly have

$$\lambda_A \mapsto \lambda_A, \quad \mu^{A'} \mapsto \mu^{A'} + iq^{AA'}\lambda_A.$$

This turns out to be consistent with the standard transformation of M^{ab} (and p_a) under origin change, where the position vector x^a of a spacetime point X correspondingly undergoes

$$x^a \mapsto x^a - q^a.$$

At this juncture, it is worth pointing out that whereas there may be a temptation to identify a twistor, as represented as a pair of 2-spinors $(\omega^A, \pi_{A'})$, with the 2-spinor form of a Dirac 4-spinor (see [3, 8]), which it superficially resembles, this would be entirely inappropriate. It is certainly true that the behavior of twistors and Dirac 4-spinors is basically the same under Lorentz transformations (leaving the origin fixed), as is defined by their spinor-index structures. But the above behavior of a twistor under shift of origin (in effect, under translation) has no place in Dirac's electron theory. As we have seen in Section 2.3, twistors provide the basic (finite-dimensional) representation space for the (restricted) conformal group $C(1,3)$, whereas Dirac spinor fields provide an infinite-dimensional representation space for the (restricted) Poincaré group. The normal Dirac equation describes a massive particle, and is not conformally invariant. The Dirac–Weyl equation for a massless neutrino is, however, and its relation to twistor theory is contained in the discussion given in Section 3.3. Massive particles can also be handled with twistor theory, but such descriptions normally require more than one twistor. See [22, 43, 53, 54, 55]; see also the remarks given at the end of Section 4.4.

There is a connection between the above direct physical interpretation of a twistor in terms of angular momentum – particularly a nonnull twistor Z^α – and the *Robinson congruence* defined by Z^α. This congruence is provided by the family of rays defined by the null (dual) twistors $W_\alpha = (\lambda_A, \mu^A)$, satisfying

$$Z^\alpha W_\alpha = 0.$$

To see the connection with angular momentum, let us examine this relation at an arbitrary point Q of $\mathbb{M}$, where we now take Q as a (variable) origin point. We are interested in the ray W of the congruence which passes through Q. With respect to Q, as origin, W_α then takes the form

$$W_\alpha = (\lambda_A, 0)$$

($\mu^{A'}$ being zero, since W_α is now incident with the origin point Q; see Section 2.2, and also Figure 7.4 in complex-conjugate form). Accordingly, the relation $Z^\alpha W_\alpha = 0$ now becomes

$$\omega^A \lambda_A = 0,$$

at the point Q. This tells us that the flagpole direction of ω^A is the same as that of λ^A, namely, the direction of the ray W. Thus, the angular momentum M^{ab} of the spinning massless particle determined by Z^α has a structure that is characterized by the flagpole directions of its two spinor parts with respect to Q. We may refer back to Figure 7.1 to see the curious spatial geometry of all this, where the flagpole directions of ω^A (small arrows in Figure 7.1) twist around in this complicated (Robinson congruence) way, while that of π_A simply points in the direction of motion of the configuration (large arrow at the top right of Figure 7.1). It may perhaps be mentioned that the choice of letters "π" and "ω" comes from the normal usage of "p" for momentum, and "ω" for angular momentum.

3 Fields, Quantization, and Curved Spacetime

3.1 Twistor Quantization Rules

Up to this point, we have been considering twistor theory only in relation to classical physics in flat spacetime geometry. *Quantum* twistor theory – and, indeed, as we shall be seeing later (in Part 4), also spacetime *curvature* – involves considering twistors (and dual twistors) as noncommuting operators, satisfying certain *commutation* laws:

$$Z^\alpha \overline{Z}_\beta - \overline{Z}_\beta Z^\alpha = \hbar \delta^\alpha_\beta$$

and, as far as our current considerations go,

$$Z^\alpha Z^\beta - Z^\beta Z^\alpha = 0, \quad \overline{Z}_\alpha \overline{Z}_\beta - \overline{Z}_\beta \overline{Z}_\alpha = 0$$

[41, 53]. Now, the twistors are taken to be *linear operators* generating a noncommutative algebra $\mathbb{A}$, whose elements are taken to be acting on an

appropriate quantum "ket-space" $|\ldots\rangle$ of some kind [10], but it is best not to be specific about this, just now. We could alternatively think of our operators as *dual* twistors, subject to the commutation laws

$$W_\alpha \overline{W}^\beta - \overline{W}^\beta W_\alpha = -\hbar \delta_\alpha^\beta$$

and

$$W_\alpha W_\beta - W_\beta W_\alpha = 0, \quad \overline{W}^\alpha \overline{W}^\beta - \overline{W}^\beta \overline{W}^\alpha = 0$$

which is the same thing as before, but with $\overline{Z}_\alpha$ relabelled as W_α.

These commutation laws are *almost* implied by the standard quantum commutators for 4-position and 4-momentum

$$p_a x^b - x^b p_a = i\hbar \delta_a^b, \quad x^a x^b - x^b x^a = 0, \quad p_a p_b - p_b p_a = 0,$$

but there appears to be an additional input related to the issue of helicity. By direct calculation, we may verify that the twistor commutation laws reproduce exactly the (more complicated-looking) standard commutation laws for the p_a and M^{ab} as defined in Section 2.5, which arise from their roles as translation and Lorentz-rotation generators of the Poincaré group (see [53]). In this calculation, there is no factor-ordering ambiguity in the expressions for p_a and M^{ab} in terms of the spinor parts of Z^α and $\overline{Z}_\alpha$ (owing to the symmetry brackets). Yet, the calculation for the helicity s (writing the operator as **s**) yields

$$\mathbf{s} = \frac{1}{4}(Z^\alpha \overline{Z}_\alpha + \overline{Z}_\alpha Z^\alpha).$$

3.2 Twistor Wave Functions

If we are to express wave functions for massless particles in twistor terms, to be in accordance with standard quantum-mechanical procedures we need functions of Z^α that are "independent of $\overline{Z}_\beta$." This means "annihilated by $\frac{\partial}{\partial \overline{Z}_\beta}$," that is, holomorphic in Z^α (Cauchy–Riemann equations). Thus, a twistor wave function (in the Z^α-description) is holomorphic in Z^α, and the operators representing Z^α and $\overline{Z}_\alpha$ act as

$$Z^\alpha \rightsquigarrow Z^\alpha \times, \quad \overline{Z}_\alpha \rightsquigarrow -\hbar \frac{\partial}{\partial Z^\alpha}.$$

Alternatively, we could be thinking of functions of $\overline{Z}_\alpha$ that are "independent of Z^β," that is, *anti*-holomorphic in Z^α. Here it would be better to rename $\overline{Z}_\alpha$ as W_α and consider functions *holomorphic* in W_α. Accordingly, in the *dual* twistor W_α-description, a wave function must be holomorphic in W_α and we have the operators representing $\overline{W}^\alpha$ and W_α, again satisfying the required commutation relations, but now with:

$$\overline{W}^\alpha \rightsquigarrow \hbar \frac{\partial}{\partial W_\alpha}, \quad W_\alpha \rightsquigarrow W_\alpha \times .$$

To ask that our wave function describe a (massless) particle of *definite helicity*, we need to put it into an eigenstate of the *helicity operator* **s**, which, by the above, is

$$\mathbf{s} = -\frac{1}{2}\hbar \left(Z^\alpha \frac{\partial}{\partial Z^\alpha} + 2 \right)$$

in the Z^α-description and

$$\mathbf{s} = \frac{1}{2}\hbar \left(W_\alpha \frac{\partial}{\partial W_\alpha} + 2 \right)$$

in the W_α-description. These are simply displaced *Euler homogeneity operators*

$$\Upsilon = Z^\alpha \frac{\partial}{\partial Z^\alpha}, \quad \tilde{\Upsilon} = W_\alpha \frac{\partial}{\partial W_\alpha},$$

so a helicity eigenstate, with eigenvalue s, in the Z^α-description requires a holomorphic twistor wave function $f(Z^\alpha)$ that is homogeneous of degree

$$n = -2s - 2,$$

where I henceforth adopt $\hbar = 1$. Then $2s$ is an integer (odd for a fermion and even order for a boson). In the W_α-description, the dual twistor wave function $\tilde{f}(W_\alpha)$ is homogeneous of degree $\tilde{n}$ where

$$\tilde{n} = 2s - 2.$$

3.3 Twistor Generation of Massless Fields and Wave Functions

In ordinary flat spacetime terms, the position-space wave function of a massless particle of helicity $2s$ [8, 9, 39] satisfies a *field equation*, this being expressible in the 2-spinor form

$$\nabla^{AA'}\psi_{AB\ldots E} = 0, \quad \Box\psi = 0, \quad \text{or} \quad \nabla^{AA'}\tilde{\psi}_{A'B'\ldots E'} = 0,$$

for the integer $2s$ satisfying $s < 0$, $s = 0$, or $s > 0$, respectively, these equations having been already considered in Section 1.3, but where the scalar case $s = 0$ is included also, involving the D'Alembertian

$$\Box = \nabla_a \nabla^a.$$

We have *total symmetry* for each of the $|2s|$-index quantities

$$\psi_{AB...E} = \psi_{(AB...E)} \quad \text{and} \quad \widetilde{\psi}_{A'B'...E'} = \widetilde{\psi}_{(A'B'...E')}.$$

What is the connection between the holomorphic twistor wave function $f(Z^\alpha)$, or dual twistor wave function $\tilde{f}(W_\alpha)$, with these spacetime equations? In most direct terms this is given by contour integrals [22, 41, 42], generalizing earlier expressions found by Whittaker [62] and Bateman [4, 5]:[1]

$$\psi_{AB...E}(x) = k \oint_{\omega=ix\cdot\pi} \frac{\partial}{\partial \omega^A} \frac{\partial}{\partial \omega^B} \cdots \frac{\partial}{\partial \omega^E} f(\omega, \pi)\, \delta\pi, \qquad \text{if } s \leqslant 0;$$

$$\widetilde{\psi}_{A'B'...E'}(x) = k' \oint_{\omega=ix\cdot\pi} \pi_{A'} \pi_{B'} \cdots \pi_{E'} f(\omega, \pi)\, \delta\pi, \qquad \text{if } s \geqslant 0.$$

Here $AB \ldots E$ or $A'B' \ldots E'$ are $|2s|$ in number, and the 1-form $\delta\pi$ is

$$\delta\pi = \varepsilon^{F'G'} \pi_{F'} d\pi_{G'},$$

and where k and k' are suitable constants. I have taken the liberty of writing x^a, ω^A, and $\pi_{A'}$ without their abstract indices in places here, and using boldface upright type instead. The *contour*, for these integrals, lies within the Riemann sphere, in $\mathbb{PT}$, of twistors $Z^\alpha = (\omega^A, \pi_{A'})$ satisfying the incidence relation $\omega^A = ix^{AA'}\pi_{A'}$ (written $\boldsymbol{\omega} = i\boldsymbol{x} \cdot \boldsymbol{\pi}$, below the integral sign), which removes the ω^A-dependence and introduces x^a-dependence, and then the contour integration itself removes the $\pi_{A'}$-dependence, leaving us with just x^a-dependence. Satisfaction of the field equations is an immediate consequence of these holomorphic expressions. The 2-form $d\pi_{0'} \wedge d\pi_{1'} = \frac{1}{2}d\delta\pi$ is sometimes more appropriate to use, rather than $\delta\pi$, the contour then being two-dimensional, lying in $\mathbb{T}$ rather than $\mathbb{PT}$. In the *dual* twistor description, we have corresponding expressions.

See Figure 7.9 for the geometrical set-up, where "X" is the Riemann sphere representing the (possibly complex) point labeled x^a. In this particular picture, X represents a complex point lying in a region of $\mathbb{CM}$ referred to as the forward tube $\mathbb{CM}^+$, this being given by complex position vectors whose imaginary parts are past-pointing timelike. In twistor terms such points are represented by lines lying entirely in $\mathbb{PT}^+$, and we see this depicted in Figure 7.9. This particular

---

[1] The upper one of these two displayed integral expressions was put forward by Lane Hughston [22], as a complement to the lower one, which I had found earlier [41, 42]. The significance of having both ways of doing it was not recognized immediately, but it was later realized that both are required for the complete picture.

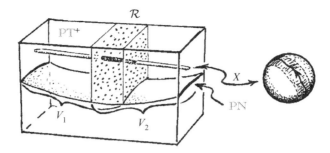

Figure 7.9 The geometry relevant to the twistor contour integral for a wave function. The regions Q_1 and Q_2 of the text are the respective complements, within $\mathbb{PT}^+$, of the depicted regions $\mathcal{V}_2$ and $\mathcal{V}_1$. Here, the open sets $\mathcal{V}_1$, $\mathcal{V}_2$ provide a 2-set open covering of $\mathbb{PT}^+$ and $\mathcal{R}$ is their intersection.

arrangement is important for quantum mechanics, because it is a convenient characterization of the important requirement, for a wave function, of energy positivity (see [58] and see also the comments at the end of Section 1.3).

3.4 Singularity Structure for Twistor Wave Functions

For such expressions to provide nonzero answers, the function f must have appropriate singularities. The situation of specific interest to us here is the case of a wave function for a free massless particle, although these formulas can also be used under many other circumstances, such as for real solutions of Maxwell's equations in particular domains. Real solutions can clearly be obtained from the complex ones described here, by taking the real part, the equations to be satisfied being linear. Completely general solutions of the equations are obtained in this way provided that they are analytic. In fact, precursors of these equations were found long ago, for the Laplace equation by Edmund Whittaker [62] in 1903, and for the wave equation (in 1904) by Harry Bateman [4] who later generalized it for Maxwell's equations in the 1930s (see [5]).

As remarked at the end of Section 3.3, for a wave function, we require complex solutions of positive frequency, and here is where the important early motivation for twistor theory referred to at the end of Section 1.3 was, in a sense, finally satisfied. But, as initially presented, this was only in a way that seemed somewhat odd. Eventually this apparent oddness was reinterpreted as something remarkably "natural" when properly understood, with potentially deep implications.

Let us see how this works. First, we take note of the fact, already noted in Section 3.3, that the family of points of $\mathbb{CM}$ that constitute the subregion

$\mathbb{CM}^+$ known as the forward tube, corresponds to the family of lines that lie entirely in $\mathbb{PT}^+$. A complex function ψ, defined on $\mathbb{M}$, which extends smoothly to a holomorphic function throughout $\mathbb{CM}^+$ is indeed of positive frequency and conversely, *positive frequency* is a key requirement for a wave function [58]. Thus, for our twistor wave function f, we require "regularity" of an appropriate sort throughout the region $\mathbb{PT}^+$. Yet it would be far too restrictive to demand holomorphicity for f over the whole of $\mathbb{PT}^+$ and, in any case, such a function would simply give the answer *zero* when contour integrated. What we seem to need is a function with two separated regions of singularity on each Riemann sphere (complex projective line) that corresponds to a point in $\mathbb{CM}^+$, that is, to a projective line in $\mathbb{PT}^+$, since then we could obtain a nontrivial answer to the contour integration, the contour being a closed loop on the Riemann sphere that separates the two regions of singularity on the sphere. The situation is depicted on the right-hand side of Figure 7.9. This is achieved if the singularities of f are constrained to lie in two disjoint regions Q_1 and Q_2 (each region being closed in $\mathbb{PT}^+$) so our contour integrations can take place within the holomorphic region $\mathcal{R}$ between them (Figure 7.9). Our twistor wave function f is thus taken to be holomorphic throughout the (open) region

$$\mathcal{R} = \mathbb{PT}^+ - (Q_1 \cup Q_2).$$

This appears to be a somewhat odd requirement for the twistor description of such a fundamental thing as a massless particle's wave function. Moreover, the region $\mathcal{R}$ is very far from being invariant under the holomorphic motions of $\mathbb{PT}^+$, some of these representing the nonreflective Poincaré (inhomogeneous Lorentz) motions of Minkowski space $\mathbb{M}$. Any particular choice of the region $\mathcal{R}$ clearly cannot take precedence over any other such choice obtained from the original one by such a motion, so there is clearly much nonuniqueness involved in the choices of $\mathcal{R}$ and f in this description. This difficulty looms large if we try to add two twistor wave functions which might have incompatible singularity structures. Linearity is, after all, a central feature of quantum mechanics as we currently understand that subject, so how are we to deal with this problem?

3.5 Čech Cohomology

The resolution of these puzzling features leads us to an understanding of what kind of an entity a twistor wave function actually "is." This lies in the notion of *Čech sheaf cohomology*. It is not appropriate that we go into much detail, here, but some indication of the issues involved will be of importance for us. What we find is that the twistor wave function f is not really to be viewed as being "just a function" in the ordinary sense, but as representing an element of "first cohomology" (actually first *sheaf* cohomology). I shall call such an entity a 1-function. An ordinary function, in this terminology, would be a 0-function. There are also higher-order entities referred to as 2-functions, 3-functions, and so on, but we shall not need to consider these here.

An important aspect of 1-functions (or of n-functions, where $n > 0$) is that they are *nonlocal* entities in an essential way (a feature of twistor theory which appears to reflect aspects of nonlocality that occur in quantum mechanics). A good intuitive way of appreciating the idea of a 1-function is to contemplate the "impossible tribar" depicted in Figure 7.10. Here we have a picture that for each local region, there is an interpretation provided, of a three-dimensional structure that is unambiguous, except for an uncertainty as to its distance from the viewer's eye. As we follow around the triangular shape, our interpretation remains consistent (though with this mild-seeming ambiguity) until we return to our starting point, only to find that it has actually become *in*consistent! The element of first cohomology that is expressed by the picture is a measure of this *global* inconsistency [47].

How might we assign such a measure to the degree of this impossibility? I shall not go into full details here, but the idea is to regard the object under consideration – here the tribar – as being built up from a number of regions

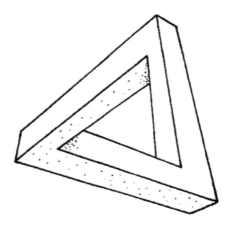

Figure 7.10 An impossible "tribar," as an illustration of the notion of (first) cohomology. There is an unambiguous interpretation of each local part, except for an ambiguity as to the distance from the viewer's eye, but globally this ambiguity leads to a nonlocal inconsistency. The measure of this inconsistency is an element of first cohomology. Twistor wave functions exhibit a similar feature, where the rigidity of analytic continuation replaces the rigidity of a material body.

(open sets) which together cover the whole object, but which are "locally trivial" in some appropriate topological (or differential) sense. In the case of the tribar, we might have a local picture of each vertex, say $\mathcal{V}_1, \mathcal{V}_2, \mathcal{V}_3$, where the three pictures overlap pairwise in smaller open regions $\mathcal{V}_i \cap \mathcal{V}_j$, somewhere along each relevant arm of the tribar, so that taken together they provide a picture of the entire tribar. On each overlap region $\mathcal{V}_i \cap \mathcal{V}_j$, we require some numerical measure F_{ij} which describes the ratio of the displacement from the eye that needs to be made for the pictures to be considered to match, and since we need an additive measure we take F_{ij} to be the logarithm of this ratio, and accordingly the F_{ij} are antisymmetric ($F_{ji} = -F_{ij}$), since the ratio goes to its reciprocal when taken in the opposite order. The triple (F_{12}, F_{23}, F_{31}) is taken modulo the particular triples of the form ($H_1 - H_2, H_2 - H_3, H_3 - H_1$), where H_i refers to the freedom that is inherent in the interpretation of each particular vertex picture $\mathcal{V}_i$. The resulting algebraic notion gives us the required cohomology element, describing the degree of impossibility in the figure. This notion is what I am calling a 1-function. For further issues see [21].

This is just to give a little flavor of what sort of an entity a 1-function actually is. More specifically, in the context of twistor theory, we are concerned with complex spaces and holomorphic functions on them. Thus, in the case of a twistor wave function there is the important subtlety, in that the global "impossibility" arises from the "rigidity" of *holomorphic functions* rather than that of the solid structures conjured up by the local parts of Figure 7.10. But let us be a bit more general here, and imagine some complex manifold $\mathcal{K}$. We shall need a locally finite open covering $\mathcal{C} = (\mathcal{V}_1, \mathcal{V}_2, \mathcal{V}_3, \dots)$, of $\mathcal{K}$. To define a 1-function f, with respect to $\mathcal{C}$, we assign a holomorphic function f_{ij} on each nonempty pairwise intersection:

$$f_{ij} = -f_{ji} \quad \text{is holomorphic on} \quad \mathcal{V}_i \cap \mathcal{V}_j,$$

and on each nonempty triple intersection:

$$f_{ij} + f_{jk} + f_{ki} = 0 \quad \text{on} \quad \mathcal{V}_i \cap \mathcal{V}_j \cap \mathcal{V}_k,$$

where the collection $\{f_{ij}\}$ is taken modulo corresponding collections of the form $\{h_i - h_j\}$, where each

$$h_i \quad \text{is holomorphic on} \quad \mathcal{V}_i,$$

so that two 1-functions are considered to be equal if the difference between their $\{f_{ij}\}$ representations is of the form $\{h_i - h_j\}$. This defines a 1-function *with respect to the particular covering* $\mathcal{C}$. For the full definition, we would have to take the direct limit for finer and finer coverings. Fortunately, in the case of complex manifolds, as is being considered here, we are assured that provided that the sets $\mathcal{V}_i$ are of suitable type (e.g., Stein spaces; see [13]) then we gain nothing from taking such a limit, and the 1-function concept is already with us. Nevertheless, in order to add two 1-functions defined by different coverings, we do need to take their common refinement in order to perform this operation, which can be a little complicated in practice.

In the case of main interest here, namely $\mathcal{K} = \mathbb{PT}^+$, it will be adequate for our immediate purposes here simply to take a 2-set covering of $\mathbb{PT}^+$, namely $\mathcal{C} = (\mathcal{V}_1, \mathcal{V}_2)$, with open sets given by the *complements* within $\mathbb{PT}^+$ of the respective singularity regions Q_2 and Q_1. Then we have our required covering $\mathcal{C}$ (not actually with Stein spaces, but that is not of great importance here)

$$\mathbb{PT}^+ = \mathcal{V}_1 \cup \mathcal{V}_2, \mathcal{R} = \mathcal{V}_1 \cap \mathcal{V}_2;$$

(see Figure 7.9), just as we had earlier. The family $\{f_{ij}\}$ consists of the single twistor function f, which, by an abuse of notation I may identify with the 1-function it determines.

This interpretation in terms of first cohomology finally fully realized the motivation described in the final paragraph of Section 1.3. The division described there, of the Riemann sphere into southern and northern hemispheres by its equator (representing the real numbers), where positive-frequency functions extend holomorphically into the southern hemisphere and negative frequency ones into the northern hemisphere, is precisely reflected by the division of $\mathbb{PT}$ into $\mathbb{PT}^+$ and $\mathbb{PT}^-$ by the "equatorial" $\mathbb{PN}$. The only essential difference, apart from the increase of dimensionality from the Riemann sphere $\mathbb{CP}^1$ to projective twistor space $\mathbb{CP}^3$, is that the holomorphic functions (0-functions) on the Riemann sphere are replaced by holomorphic first cohomology elements (1-functions) on projective twistor space.

In the cases of homogeneity 0 or 2 (left-handed electromagnetism or left-handed linearized gravity, respectively), there are generalizations of the twistor-space contour-integral expressions that allow one to view the 1-function nature of a twistor function in a different light, in which non-linearities of general relativity and particle physics begin to play a significant role. To appreciate this, let us return to the *general* Čech descriptions given earlier, where a locally finite open covering $\mathcal{C} = (\mathcal{V}_1, \mathcal{V}_2, \mathcal{V}_3, \dots)$ of a complex space $\mathcal{K}$ was considered. In the specification of a 1-function f, in relation to this covering, we required a family of holomorphic functions $\{f_{ij}\}$ defined on the nonempty overlaps $\mathcal{V}_i \cap \mathcal{V}_j$. Here, the functions are entirely passive, being just "painted on" the space $\mathcal{K}$. However, we can consider a somewhat more active role for such a 1-function f, such as (a) specifying the generation of a *bundle* above $\mathcal{K}$, or (b) using f to specify the generation of a *deformation* of $\mathcal{K}$

itself. In each case, the rules (see [13]) defining a 1-function are exactly what is needed to fulfill this purpose. However, in each case, this specification by a 1-function would only be as an infinitesimal generator of the bundle or deformed space (except for an abelian group in case (a)) because of nonlinearities. Nevertheless, the general idea expressed in (a) and (b) still holds true; it is just that the linear nature of a 1-function ceases to hold. In effect, we have a kind of "nonlinear 1-function."

It was in 1977 that Richard Ward introduced the procedure indicated in case (a) above, first in the situation provided by the (left-handed) Maxwell's equations, which allowed interactions of the field with charged particles to be considered. Almost immediately afterward he showed how this procedure could be generalized to the (left-handed) Yang–Mills equations [60]. This turned out to have considerable importance in the theory of integrable systems (see, for example, [21, 30]). Shortly before all this, in 1976, procedure (b) had been introduced [44], to provide a twistorial representation of all conformally complex–Riemannian 4-manifolds which are anti-self-dual (i.e., $\tilde{\Psi}_{A'B'C'D'} = 0$; see the end of Section 1.3). When an additional simple condition is imposed, this provides not only a (complex) metric but automatically generates the general anti-self-dual solution of the Einstein vacuum equations, either without [44] or with a cosmological constant Λ [61].

3.6 Infinity Twistors and Einstein's Equations

It is a fairly straightforward procedure to generate the desired deformed twistor spaces satisfying the required conditions ensuring satisfaction of the Einstein Λ-vacuum equations (Λ being the cosmological constant, $\Lambda = 0$ allowed). Basically, what is required is to match appropriate portions of (nonprojective) twistor space, while preserving the Euler operator

$$\Upsilon = Z^\alpha \frac{\partial}{\partial Z^\alpha}$$

and the 2-form

$$\Theta = I_{\alpha\beta} dZ^\alpha \wedge dZ^\beta$$

where the antisymmetrical Λ-*infinity twistor* $I_{\alpha\beta}$ (and its dual $I^{\alpha\beta}$) are given by

$$I_{\alpha\beta} = \begin{pmatrix} \frac{\Lambda}{6}\varepsilon_{AB} & 0 \\ 0 & \varepsilon^{A'B'} \end{pmatrix}, \quad I^{\alpha\beta} = \begin{pmatrix} \varepsilon^{AB} & 0 \\ 0 & \frac{\Lambda}{6}\varepsilon_{A'B'} \end{pmatrix}.$$

We see that $I^{\alpha\beta}$ and $I_{\alpha\beta}$ are both complex conjugates and *duals* of one another:

$$I_{\alpha\beta} = \overline{I^{\alpha\beta}}, \quad I^{\alpha\beta} = \overline{I_{\alpha\beta}},$$

$$I_{\alpha\beta} = \frac{1}{2}\varepsilon_{\alpha\beta\rho\sigma}I^{\rho\sigma}, \quad I^{\alpha\beta} = \frac{1}{2}\varepsilon^{\alpha\beta\rho\sigma}I_{\rho\sigma},$$

where $\varepsilon_{\alpha\beta\rho\sigma}$ and $\varepsilon^{\alpha\beta\rho\sigma}$ are Levi-Civita twistors, fixed by their anti-symmetry and $\varepsilon_{0123} = 1 = \varepsilon^{0123}$ in standard twistor coordinates. The preservation of Υ and Θ on the overlaps where $\mathcal{V}_i$ as matched to $\mathcal{V}_j$, is ensured, if we shift infinitesimally along the vector field

$$I^{\alpha\beta}\frac{\partial f_{ij}}{\partial Z^\alpha}\frac{\partial}{\partial Z^\beta},$$

where each f_{ij} has homogeneity degree 2 (i.e., $\Upsilon f_{ij} = 2f_{ij}$, which corresponds to helicity 2). We can imagine exponentiating these infinitesimal deformations to obtain a finite one. In the case of a 2-set covering, we can achieve this explicitly by exponentiating the single function f_{12}, but with larger numbers of sets, we can encounter difficulties in satisfying the required condition on triple overlaps. A simpler procedure for satisfying the required condition of preserving Υ and Θ is to use generating functions (see [44]). This has particular relevance to the procedures of Section 4.4.

It is, however, not at all a direct matter to obtain the (complex) curved spacetime $\mathcal{M}^{\mathbb{C}}$ from the deformed twistor space $\mathcal{T}$, according to this construction. The points of $\mathcal{M}^{\mathbb{C}}$ correspond to "lines" in $\mathbb{P}\mathcal{T}$, that are completed Riemann spheres, stretching across from one patch to the other, or perhaps several patches if the covering involves more than two. These Riemann spheres are not easy to locate, in a general way, since they are determined by the global requirement that they be compact holomorphic curves within $\mathbb{P}\mathcal{T}$ of spherical topology (and belonging to the correct topological family). The very existence of these "lines," as I shall call them, together with the fact that they belong to a 4-parameter family of such lines (provided that the deformation from $\mathbb{P}\mathbb{T}^+$, or from some other appropriate part of $\mathbb{P}\mathbb{T}$, is not too drastic), depends upon key theorems by Kodaira and Kodaira–Spencer (see [24, 25]). The space whose points represent these lines is the required complex 4-manifold $\mathcal{M}^{\mathbb{C}}$. Its complex conformal structure comes about from the simple fact that meeting lines in $\mathbb{P}\mathcal{T}$ correspond to null separated points in $\mathcal{M}^{\mathbb{C}}$, and the definition of its metric scaling comes about through use of the form Θ. With this complex metric, the complex 4-manifold automatically satisfies the Einstein Λ-vacuum equations, and the construction provides the general anti-self-dual solution. This procedure has become known as the "nonlinear graviton construction" [44, 61].[2] It has found numerous applications in differential geometry [21, 30].

At this point, it is worth emphasizing an essential but unusual feature of the nonlinear graviton construction. This is that the "curvature" in the deformed twistor space (encoding $\mathcal{M}^{\mathbb{C}}$'s actual curvature) is not local, in the sense that a small-enough neighbourhood of any point in the deformed twistor space is identical in structure with that of ordinary flat twistor space $\mathbb{T}$ (with the given Λ assigned to it). The "curvature" in the deformed twistor space is a *nonlocal* feature of the space $\mathbb{P}$, but in the construction of the "spacetime" manifold $\mathcal{M}^{\mathbb{C}}$, we consequently find genuine local curvature in the normal sense (Riemann curvature, Weyl curvature).

If we are to regard twistor theory as providing an overall approach to basic physics, however, then we must face up to the fundamental obstruction to progress that has confronted this programme for some four decades, namely, what has become known as the *googly problem* (an apposite term borrowed from the game of cricket, for a ball bowled with a right-handed spin about its direction of motion, but bowled with an action that would appear to be delivering a left-handed spin). This comes about from the very nature of the construction. The points of $\mathbb{P}\mathcal{T}$ have an interpretation within $\mathcal{M}^{\mathbb{C}}$ as what are called "α-surfaces" (totally null self-dual complex 2-surfaces) [36], and there would have to be a 1-complex-parameter family of such surfaces through each point of $\mathcal{M}^{\mathbb{C}}$ (corresponding to the 1-parameter family of points on each line of $\mathbb{P}\mathcal{T}$). This would imply $\tilde{\Psi}_{A'B'C'D'} = 0$, that is, $\mathcal{M}^{\mathbb{C}}$ being conformally anti-self-dual. Thus, it is the very existence of *points* of the space $\mathbb{P}\mathcal{T}$ that implies the existence of these troublesome α-surfaces. This illustrates the fundamental underlying difficulty of the googly problem.

Nevertheless, it is clear that if twistor theory is to have any hope of providing a basis for fundamental physics, there needs to be a way around this problem. Many ideas for addressing it have been made over the years, often resorting to examining the twistor structure at infinity, where the geometry is simpler than that at finite regions (see, for example, [48]), but none has been able to achieve very much. The most successful approach has been that of ambitwistors [26, 27], that is based on complex null geodesics, modeled on twistor and dual twistor pairs (Z^α, W_β), subject to $Z^\alpha W_\alpha = 0$. This enables complex-Riemannian 4-manifolds to be studied in relation to

---

[2] The historical point must be made here that a key input to the development of the nonlinear graviton construction was the introduction, in 1976, by Ezra T. Newman, of his notion of the $\mathcal{H}$-space, for an asymptotically flat spacetime, as described initially in [32]; for more detail, see [33]. For some fascinating new developments of these ideas, relating $\mathcal{H}$-space to equations of motion, see [33].

twistor-type ideas, and the Einstein vacuum equations to be examined in this light [26]. But it does not follow the twistor route of "nonlinearizing" the 1-function description of quantum wave functions, where left- and right-handed helicities can be combined together to describe gravitational interactions and classical spacetimes in a way that directly relates to twistor wave functions.

It should be mentioned that even without progress on the googly problem twistor ideas have found many significant applications, mainly in pure mathematics (most particularly in certain areas of differential geometry, representation theory, and integrable systems (see, in particular, [21, 30]), but, relatively recently, also in physics, where great simplifications to calculations in high-energy scatterings have been obtained. The main impetus to this relatively recent work came from Edward Witten [63], who introduced several novel ideas, partly based on earlier work in publications by others [14, 15, 20, 31, 35, 40, 51, 57]), and which subsequently stimulated a great deal of further activity, for example [1, 2, 16, 17, 18, 19, 28, 29], which allowed twistor methods to help in enormously reducing the calculations of scattering amplitudes at very high energies (where all particles involved could be considered to be massless). Nevertheless, despite all this impressive work, relatively little of the deeper results of twistor theory has been incorporated, and it is my view that these are needed for twistor theory to realize its aims to become an underlying scheme for fundamental physics generally. In Part 4, I outline what I now believe to be the appropriate direction for progress to be made in this regard.

4 Palatial Twistor Theory

4.1 Basic Ideas of Palatial Twistor Theory

As has just been remarked upon (in Section 3.6), it is the seeming need to have an unambiguous notion of a *point* in twistor space that appears to drive us inevitably to this incomplete and lopsided approach to spacetime in twistor theory, that an unresolved googly problem presents us with. It is fortunate, therefore, that there is a novel approach to generalizing the nonlinear graviton construction, so that both helicities can be accommodated within the same general framework, and that classical conformal Lorentzian spacetimes should also come under the same umbrella. To understand the basic idea, let us first return to the procedure that we considered earlier, in the nonlinear graviton construction of Section 3.6, where to produce a suitably deformed twistor space, we "glue together" pieces of complex manifold, preserving the complex structure from patch to patch. The notion of "complex structure" can be encapsulated in terms of the algebra of holomorphic functions on each patch – or, technically, the "sheaf" of such functions, where we require holomorphicity throughout a small neighbourhood of each point. Now, we saw from the above that the basic conundrum was the existence of the actual *points* in each patch, since it was the interpretation of the points in $\mathbb{P}\mathcal{T}$ that gave rise to the unwanted α-surfaces. Thus, it would seem, we somehow need to find a way of matching the sheaves of *algebras* of holomorphic functions from one patch to another, without actually having "patches" that, in the ordinary sense, would consist of individual points. This will not do as it stands, however, because the function algebras already "know" the points, this being true so long as the algebras are, like the algebra of holomorphic functions on a region of twistor space, *commutative*.

This suggests that we take, instead, a holomorphic algebra that is *non-commutative*![3] As we have seen in Section 3.1, there is a natural non-commutative algebra in twistor theory, namely, that generated (via complex linear combinations and products – these basic operations being allowed to be repeated many times – and the taking of appropriate limits) from the operators (see Section 3.2):

$$Z^\alpha \times \quad \text{and} \quad -\frac{\partial}{\partial Z^\alpha}.$$

I shall refer to this algebra as $\mathbb{A}$, the basic quantum twistor algebra for Minkowski space $\mathbb{M}$. The idea is that, in some sense, we could patch together two (or more) "subregions" of the algebra $\mathbb{A}$, analogous to the $\mathcal{V}_1$ and $\mathcal{V}_2$ of the nonlinear graviton construction, that, in an appropriate sense "cover" the entire algebraic structure of interest. This "patched" algebra $\mathcal{A}$ is then to represent some open portion of the complex(ified) spacetime $\mathcal{M}^{\mathbb{C}}$.

The essential idea is that the algebra $\mathbb{A}$ can be thought of as a system of complex linear operators acting on holomorphic functions defined locally on some complex space which, initially, we think of as twistor space $\mathbb{T}$. In Dirac's quantum-mechanical terminology [10], $\mathbb{T}$ is a "ket" space for the algebra $\mathbb{A}$ of quantum operators. We are to think of $\mathbb{A}$ as an abstract algebra that is not dependent upon this particular realization. For example, the same could also be thought of as the space of complex linear operators acting on holomorphic functions on the *dual* twistor space $\mathbb{T}^*$, where the respective operators above would (as displayed in Section 3.2) now be

---

[3] I am very grateful to Michael Atiyah for making me aware of this important requirement, in a brief conversation in 2013 (which happened to be at an occasion in Buckingham Palace – hence the name).

$$\frac{\partial}{\partial W_\alpha} \quad \text{and} \quad W_\alpha \times$$

which satisfy the same commutation rules as before. Thus, in Dirac's terminology, $\mathbb{T}^*$ would be an alternative ket space for $\mathbb{A}$.

A quantum-mechanical way of thinking about this would be to assert that the commuting complex parameters Z^0, Z^1, Z^2, and Z^3 constitute what we may call a *complete set of commuting operators*, in the sense that they, together with their respective partial derivatives $\frac{\partial}{\partial Z^0}$, $\frac{\partial}{\partial Z^1}$, $\frac{\partial}{\partial Z^2}$, and $\frac{\partial}{\partial Z^3}$, generate, in an appropriate sense, the algebra $\mathbb{A}$. A different representation of the *same* algebra $\mathbb{A}$ would be obtained if we chose, instead, the complete set of commuting operators W_0, W_1, W_2, and W_3, where the formal component replacements $W_\alpha \rightsquigarrow -\frac{\partial}{\partial Z^\alpha}$ and, accordingly, $\frac{\partial}{\partial W_\alpha} \rightsquigarrow Z^\alpha$ are made. In this sense, $\mathbb{T}$ and $\mathbb{T}^*$ would provide alternative ket spaces for the same algebra $\mathbb{A}$. There would be many other possible choices of ket space for the same algebra $\mathbb{A}$.

In order to obtain something analogous to the nonlinear graviton construction we would seem to have to involve something analogous to a connected open subset of such a ket space $\mathbb{T}$. The general idea would be to patch together various different such "ket patches," by analogy with the patching together of different open regions in a locally finite covering $\mathcal{C} = (\mathcal{V}_1, \mathcal{V}_2, \mathcal{V}_3, \dots)$, if we are to build up a nontrivial (complex) manifold. In an appropriate sense, the corresponding algebras $\mathbb{A}_1, \mathbb{A}_2, \mathbb{A}_3, \dots$ would have to "agree" on overlaps (locally isomorphic on the overlaps, in some suitable sense), but the ket spaces would generally differ from patch to patch, so that the "patched-up" algebra $\mathcal{A}$ that we end up with would have no overall ket space, and would therefore differ from $\mathbb{A}$, though agreeing with it in a *local* sense (see [49] and [50] for earlier tentative descriptions of this idea).

However, there are various thorny issues that need to be faced, with regard to this sort of "patching" if we are asking for a duly rigorous picture. We might, for example, consider some subregion χ of twistor space $\mathbb{T}$, which we propose to use as a ket-space "patch." The operator $\exp(A^\alpha \frac{\partial}{\partial Z^\alpha})$ for constant A^α, in particular, would appear to be harmless enough, but it could present us with difficulties, as a candidate for membership of the algebra whose ket space is to be χ. A problematic issue is that a holomorphic function f, defined on χ, whose analytic continuation from inside χ to a point displaced by the vector A^α to somewhere outside χ, where this analytic continuation of f becomes *singular*, would exclude $\exp(A^\alpha \frac{\partial}{\partial Z^\alpha})$ from membership of the algebra, since its action on f would be singular. On the other hand, the operation of multiplication by $\exp(-A^\alpha W_\alpha)$, on a ket space that is any subregion of $\mathbb{T}^*$

would be completely harmless. Such issues need to be better understood for a properly rigorous picture of this intended procedure to be obtained.

It is clear from all this that there is a considerable vagueness in this proposal, as put forward above. Most particularly, we do not have a clear notion of topological issues, such as "local" and "open set," when it comes to these algebras. These difficult issues are not properly resolved as things stand, and in the following sections I shall adopt a policy of providing explicit procedures for the needed patching without worrying about topological issues, complete rigour, and so forth, leaving such matters to later consideration. Nevertheless, I believe that we can go some way toward addressing these topological matters if we, in a sense, "stay close to the space $\mathbb{P}\mathcal{N}$," where $\mathbb{P}\mathcal{N}$ is the space whose points represent null geodesics in the spacetime $\mathcal{M}$ that is intended that we are describing in twistor terms. This will be explored in the next section.

4.2 The Spaces of Momentum-Scaled and Spinor-Scaled Rays

In accordance with this, let us indeed explore the ray-space $\mathbb{P}\mathcal{N}$, whose points represent individual rays in a spacetime $\mathcal{M}$, where $\mathcal{M}$ is taken to be a smooth time-oriented Lorentzian globally hyperbolic 4-manifold. Thus $\mathbb{P}\mathcal{N}$ plays the same role for $\mathcal{M}$ as does $\mathbb{P}\mathbb{N}$ for Minkowski space $\mathbb{M}$. The global hyperbolicity of $\mathcal{M}$ (equivalent to $\mathcal{M}$ containing some global spacelike 3-surface which can act as a Cauchy hypersurface for physical fields within $\mathcal{M}$; see [11]) ensures that $\mathbb{P}\mathcal{N}$ is Hausdorff and that $\mathcal{M}$ contains no causal anomalies such as closed or "almost closed" rays [11]. We are to imagine that our proposed "palatial space" $\mathbb{P}\mathcal{T}$ can, in some appropriate sense, be viewed as being some sort of "extension" of $\mathbb{P}\mathcal{N}$, to include nonnull twistors, though not in any direct sense as an ordinary manifold.

In addition, we shall be interested not only in the 5-manifold $\mathbb{P}\mathcal{N}$ of rays in $\mathcal{M}$, but also in the 6-manifold $\mathbb{p}\mathcal{N}$ of *momentum-scaled* rays in $\mathcal{M}$. Thus, each point of $\mathbb{p}\mathcal{N}$ represents a ray γ together with a momentum covector p_a which is parallel-propagated along γ, the future-null vector p^a being tangent to γ. We shall be interested also in the smooth Hausdorff 7-manifold $\mathcal{N}$ representing the *spinor-scaled* rays γ, where, in addition to the momentum scaling p_a, we attach a spinor phase provided by the flag plane of $\pi_{A'}$, where

$$p_a = \overline{\pi}_A \pi_{A'},$$

providing a spinor scaling for γ (see Section 2.2), these requirements being all conformally invariant (see [53, Chapter 7]). Importantly, this is just the arrangement of spaces required for the procedure of *geometric quantization* as described (explicitly, in relation to twistor theory) in Nicholas Woodhouse's book [65], and we shall be seeing the significance for us of some of the ideas of geometric quantization in the next section, Section 4.3. Explicitly, the 6-space $\mathbb{p}\mathcal{N}$ is a *symplectic* manifold, and $\mathcal{N}$ is a *circle bundle* over $\mathbb{p}\mathcal{N}$, the circles given by the phase freedom in $\pi_{A'}$.

I shall write the symplectic 2-form Σ of the space $\mathbb{p}\mathcal{N}$ as

$$\Sigma = i \mathrm{d} Z^\alpha \wedge \mathrm{d}\overline{Z}_\alpha$$

in anticipation of a role for twistor notation for $\mathcal{N}$, where we take note of the fact that in the case of flat twistor space $\mathbb{T}$ we then find

$$\Sigma = \mathrm{d} p_a \wedge \mathrm{d} x^a$$

(by a brief calculation in the notation of Section 2.2 for twistors in $\mathbb{M}$). This is the standard symplectic 2-form for the cotangent bundle of a manifold, where in this case we are thinking of the cotangent bundle of $\mathbb{M}$, symplectically reduced by the Hamiltonian $p_a p^a$ where we take $p_a p^a = 0$, so that the expression above indeed gives us Σ as the symplectic structure of $\mathbb{p}\mathbb{N}$ (i.e., of $\mathbb{p}\mathcal{N}$ when $\mathcal{M} = \mathbb{M}$). See [11, 27, 53] for details. This suggests that the *twistor* expression above, for the 2-form Σ might perhaps also have meaning in the case of general Lorentzian-conformal 4-manifolds $\mathcal{M}$. Indeed, a certain justification for this twistorial expression for Σ is given in [53, Section 7.4; see Figure 7.4, particularly].)

Up to this point, we have been fixing attention on what will be regarded as the region

$$Z^\alpha \overline{Z}_\alpha = 0$$

in other words, the actual geometrically defined circle bundle $\mathcal{N}$ over the symplectic $\mathbb{p}\mathcal{N}$. We are now going to try to think of this region as somehow extendible as a manifold to where $Z^\alpha \overline{Z}_\alpha \neq 0$, but there cannot in general be any *unique* way of doing this. Nevertheless, there are indeed many ways of doing it locally. The picture I am presenting is that we apply the ordinary notion of a locally finite open covering to $\mathbb{P}\mathcal{N}$, this extending to $\mathcal{N}$, by virtue of the spinor scaling, as described above. Let $(\mathbb{P}\mathcal{N}_1, \mathbb{P}\mathcal{N}_2, \dots)$ be such an open covering of $\mathbb{P}\mathcal{N}$ extending, accordingly, to a locally finite open covering $(\mathcal{N}_1, \mathcal{N}_2, \dots)$ of $\mathcal{N} - \{0\}$, where we shall require that in some sense – as yet to be specified more precisely – the individual spaces $\mathcal{N}_k$, together with their intersections $\mathcal{N}_i \cap \mathcal{N}_j$, are appropriately "simple" enough that the requirements below can be satisfied.

Whereas the patchings of these $\mathcal{N}$-regions are to be *pointwise*, in the ordinary sense, this cannot be expected to hold for the proposed "extensions" to $Z^\alpha \overline{Z}_\alpha \neq 0$. Instead, it will be that there is a (quantum) twistor *algebra* $\mathbb{A}_j$

defined for each patch $\mathcal{N}_j$, and it is these algebras that we shall require to match on the overlaps between patches. A key issue is that we cannot necessarily expect the ket spaces for these algebras will match. For if they did, we would have a matching of actual spaces, which is just what we wish to avoid, for consistency with the aspirations expressed in Section 4.1.

We shall be seeing shortly (in Section 4.3) how the procedures of geometric quantization allow us, in a general way, to construct and match the required algebras, but before proceeding to this, we shall need a little more about how twistor ideas and notation can be used to describe the needed structure of these patches. For this, I shall ask that the geometric structure of $\mathcal{N}$ be *analytic*. This may not be essential, but it makes descriptions easier. Analyticity allows us to *complexify* $\mathcal{N}$ to a complex 7-manifold $\mathbb{C}\mathcal{N}$ (a complex "thickening" of $\mathcal{N}$ which contains $\mathcal{N}$ as a real submanifold). Correspondingly, there will also be complex manifolds $\mathbb{C}P\mathcal{N}$ and $\mathbb{C}p\mathcal{N}$, where $\mathbb{C}p\mathcal{N}$ is complex-symplectic and where $\mathbb{C}\mathcal{N}$ may be regarded as a holomorphic bundle over $\mathbb{C}p\mathcal{N}$ whose fibre is an open annular region containing the unit circle in the complex (Wessel) plane.

We are to choose the open regions $\mathbb{C}P\mathcal{N}_j$ small (i.e., "simple") enough so that there can be no obstruction to mapping each $\mathbb{C}\mathcal{N}_j$ holomorphically to a subregion of the 7-complex-dimensional complexification $\mathbb{C}\mathbb{N}$ of $\mathbb{N}$, in a way that preserves the "bundle over symplectic structure" of each (there being no local notion of "curvature" for these local structures which could prevent this). This will allow us to use the ordinary twistor descriptions of Part 2 to describe the geometry and algebra of each $\mathbb{C}\mathcal{N}_j$, as inherited from standard twistor space $\mathbb{T}$. This mapping will be far from unique, of course, but the very freedom that is allowed by this nonuniqueness is an important issue for the construction.

Let us fix attention on one particular $\mathbb{C}\mathcal{N}_j$ and consider two alternative such twistor descriptions, which I denote by

$$(Z^\alpha, W_\alpha) \quad \text{and} \quad (\mathcal{Z}^\alpha, \mathcal{W}_\alpha)$$

We take Z^α and W_α to be independent twistors and dual twistors and, likewise, we take $\mathcal{Z}^\alpha$ and $\mathcal{W}_\alpha$ to be independent. The idea is that whereas the descriptions to be given below should then lead us to the description of a *complex* spacetime $\mathcal{M}^\mathbb{C}$, we can then specialize them, in a way that will be described in Section 4.4, so as to describe *real*-Lorentzian spacetimes $\mathcal{M}$, by being able to revert to the required Hermiticity:

$$W_\alpha = \overline{Z}_\alpha \quad \text{and} \quad \mathcal{W}_\alpha = \overline{\mathcal{Z}}_\alpha.$$

In the general picture, we are to regard one of these twistor descriptions (Z^α, W_α) as holding on one patch, and the other description $(\mathcal{Z}^\alpha, \mathcal{W}_\alpha)$ as holding on another which overlaps it. We are interested in the quantum twistor algebras in each system and how to match these algebras from patch to patch.

4.3 A Palatial Role for Geometric Quantization

The one piece of clear geometry that we do wish to match between the two systems is the symplectic structure given by Σ,

$$\Sigma = i\,\mathrm{d}Z^\alpha \wedge \mathrm{d}W_\alpha = i\,\mathrm{d}\mathcal{Z}^\alpha \wedge \mathrm{d}\mathcal{W}_\alpha,$$

but even this is geometrically meaningful only on the region $\mathbb{C}\mathcal{N}$. Outside this region, there is an arbitrariness involved in the extensions away from $\mathbb{C}\mathcal{N}$. Nevertheless, I shall demand that the above relation continues to be maintained outside $\mathbb{C}\mathcal{N}$, this still allows a very considerable freedom in the extension. In the picture that I am presenting, it will be legitimate to regard the 8-complex-dimensional space of pairs (Z^α, W_α) to be identifiable, in some local region of each, with the 8-complex-dimensional space of pairs $(\mathcal{Z}^\alpha, \mathcal{W}_\alpha)$, but any such identification is not to be regarded as meaningful in itself. It is only a means to an end, namely, to identify the allowable quantum twistor algebras that are to be abstracted from this framework.

Here is where the procedures of geometric quantization come into play – though, technically, it is only the more primitive procedure of geometric "prequantization" that will be called upon here. The idea will be that we require a *bundle-connection* on $\mathbb{C}\mathcal{N}$, determined by a 1-form Φ, whose curvature is the 2-form Σ (apart from the factor of i introduced here). The various possible such connections will give the different possible realizations of the quantum twistor algebra. These different possibilities come about from the different ways that the 2-form Σ is expressed as the exterior derivative of a 1-form Φ (here with a factor of i):

$$\Sigma = i\,\mathrm{d}\Phi.$$

The bundle connection is then given, in the (Z^α, W_α) system by

$$\left(\frac{\partial}{\partial Z^\alpha} + P_\alpha, \quad \frac{\partial}{\partial W_\alpha} + Q^\alpha \right)$$

where

$$\Phi = P_\alpha \mathrm{d}Z^\alpha + Q^\alpha \mathrm{d}W_\alpha.$$

Here,

$$P_\alpha = -\frac{\partial E}{\partial Z^\alpha} \quad \text{and} \quad Q^\alpha = \frac{\partial F}{\partial W_\alpha}, \quad \text{where} \quad E + F = Z^\alpha W_\alpha,$$

E and F being (holomorphic) functions of Z^α and W_α.

Using this bundle connection, we can realize the operations of the noncommutative quantum twistor algebra $\mathbb{A}$. In the description given at the beginnings of Section 3.1 and Section 3.2 (with $\hbar = 1$), we get the canonical case where the commutation relations can be written:

$$[Z^\alpha, Z^\beta] = 0, \quad \left[-\frac{\partial}{\partial Z^\alpha}, -\frac{\partial}{\partial Z^\beta}\right] = 0, \quad \left[Z^\alpha, -\frac{\partial}{\partial Z^\beta}\right] = \delta^\alpha_\beta$$

and we can directly realize these in our bundle connection if we take $P_\alpha = 0$, $Q^\alpha = Z^\alpha$ (i.e., $E = 0$, $F = Z^\alpha W_\alpha$), so that our connection is represented by

$$\left(\frac{\partial}{\partial Z^\alpha}, \frac{\partial}{\partial W_\alpha} + Z^\alpha\right).$$

Here we take our ket space to be given by holomorphic functions of Z^α, where W_α is to be taken constant, so that the first term in the above gives us $\frac{\partial}{\partial Z^\alpha}$ and the second gives us Z^α. The same algebra, but taken with functions of the *dual* twistors as the ket space, would be obtained by taking $P_\alpha = -W_\alpha$, $Q^\alpha = 0$ (i.e., $E = Z^\alpha W_\alpha$, $F = 0$), so that our connection is represented by

$$\left(\frac{\partial}{\partial Z^\alpha} - W_\alpha, \frac{\partial}{\partial W_\alpha}\right),$$

where the commutation relations can be written

$$\left[\frac{\partial}{\partial W_\alpha}, \frac{\partial}{\partial W_\beta}\right] = 0, \quad [W_\alpha, W_\beta] = 0, \quad \left[\frac{\partial}{\partial W_\alpha}, W_\beta\right] = \delta^\alpha_\beta.$$

Now, we can take our ket space to be given by holomorphic functions of *dual* twistors W_α, where we take Z^α to be constant.

Clearly, by making other ways of splitting $Z^\alpha W_\alpha$ into a sum of functions E and F, we can arrive at many different realizations of the algebra $\mathbb{A}$, at least in some local sense. Yet, the meaning of the word *local* in this context is clearly something that also needs attention, but, as indicated in Section 4.1, I am here leaving aside the thorny issues as are raised by topology and related matters, concentrating primarily on purely formal matters. Nevertheless, such topological issues must be dealt with at some stage, in order to address the important matters corresponding to those treated in Section 3.6. In relation to this, we need to recall that the considerations of this section, following the initial paragraph, have been expressed entirely in the (Z^α, W_α) system, whereas in order to express the relations in different patches, we need to relate the (Z^α, W_α) system of one patch to the $(\mathcal{Z}^\alpha, \mathcal{W}_\alpha)$ system of another. As things stand, in addition to topological matters, the actual carrying out of the transformations involved would be likely to get exceedingly complicated in practical calculations and generally unilluminating. In the following section I shall show how these twistor transformations between patches can be greatly facilitated by means of generating functions.

4.4 Palatial Generating Functions and Einstein's Equations

We have seen in Section 4.3 that although the patching together of different regions is to be understood in terms of the quantum twistor algebras assigned to the various regions, the construction of these algebras can apparently be achieved in a direct geometrical way. Most particularly, the relation between one patch and another patch overlapping it can be described in terms of respective complex symplectic structures that we may assign to each patch as defined by the different (Z^α, W_α) and $(\mathcal{Z}^\alpha, \mathcal{W}_\alpha)$ systems with the same symplectic 2-form Σ. We want the Σs to match from patch to patch, but the 1-forms Φ would be expected to differ, as would the bundle connection that Φ defines. This connection provides the needed algebra $\mathbb{A}_j$ for the jth patch and the *specific* (Z^α, W_α) or $(\mathcal{Z}^\alpha, \mathcal{W}_\alpha)$ system that is used to define this algebra, like a coordinate system in ordinary manifold construction, would not be taken to have significance.

There will be no loss of generality if we adopt the convention that we choose the ket-space description for that algebra to be the space of holomorphic functions in the Z^α variables rather than the W_α variables (and the $\mathcal{Z}^\alpha$ variables rather than the $\mathcal{W}_\alpha$ variables, etc.). In this way, we can regard the choice of Φ in the (Z^α, W_α) system as actually determining the quantum twistor algebra, together with its ket space. But the ket space itself would not be part of the palatial structure, nor would Φ, and certainly not the specific (Z^α, W_α), since these structures are not determined by the algebra. All we ask is that the algebras agree from patch to patch, being the same on the overlaps.

Yet, it needs to be pointed out that, in all this, I am being deliberately a bit vague in using terms like "the same as" rather than "isomorphic to," for the reason that I am not at all sure what the precise term should be. The global natures of the algebras might be different in situations where in some more local sense the structure of the algebras ought to be judged as "the same." Again, this raises an issue of rigor that I am leaving aside in this article.

The most explicit way to relate one (Z^α, W_α) system to another one $(\mathcal{Z}^\alpha, \mathcal{W}_\alpha)$ while preserving Σ is by means of a generating function G, which chooses one set of half the variables from one system and the other half from the other one:

$$G(Z^\alpha, \mathcal{W}_\alpha).$$

The remaining variables are then provided by

$$\mathcal{Z}^\alpha = \frac{\partial G}{\partial \mathcal{W}_\alpha} \quad \text{and} \quad W_\alpha = \frac{\partial G}{\partial Z^\alpha}$$

which by direct calculation immediately gives the required

$$\Sigma = i\,\mathrm{d}Z^\alpha \wedge \mathrm{d}W_\alpha = i\,\mathrm{d}\mathcal{Z}^\alpha \wedge \mathrm{d}\mathcal{W}_\alpha$$

For the purposes of palatial twistor theory, we require this to have a *total* homogeneity of 2 in all its variables. In terms of Euler homogeneity operators, this can be expressed as

$$\left(Z^\alpha \frac{\partial}{\partial Z^\alpha} + \mathcal{W}^\alpha \frac{\partial}{\partial \mathcal{W}^\alpha} \right) G = 2G$$

from which, we obtain

$$G = \frac{1}{2}(Z^\alpha W_\alpha + \mathcal{Z}^\alpha \mathcal{W}_\alpha).$$

Having an explicit expression for the patching, in terms of the given generating function, and therefore of the twistor algebras and their respective ket spaces, we can envisage piecing together an entire quantum twistor algebra $\mathcal{A}$ for a complex spacetime $\mathcal{M}^\mathbb{C}$, as was envisaged in Section 4.1. But how are we to identify the *points* in $\mathcal{M}^\mathbb{C}$? From general considerations, it can be seen that the points ought to correspond to completely commutative four-dimensional subalgebras of $\mathcal{A}$, although this does not appear to be quite a sufficient characterization. Yet, in this kind of way, the notion of points being null separated would have a simple interpretation in terms of subalgebras so that the identification of $\mathcal{M}^\mathbb{C}$ as a complex-conformal manifold ought to be derived from the structure of the algebra. Clearly much clarification is needed, and not the least of these problems would appear to be a need for theorems analogous to those of Kodaira and Kodaira–Spencer [24, 25] that were so important for the original nonlinear graviton construction of Section 3.6.

Even if all this works out well, there would remain at least four important issues for this programme to have importance for physics: (1) How do we incorporate the conformal scaling that will actually provide a spacetime metric? (2) How do we express Einstein's equations for this metric? (3) How do we ensure that we obtain a real-Lorentzian spacetime $\mathcal{M}$, rather than just a complex one $\mathcal{M}^\mathbb{C}$? (4) How do we generalize this to apply to the Ward construction for Yang–Mills fields [60], and hence to particle physics?

It is perhaps very remarkable that it appears to be possible to address all three of (1), (2), and (3) with a single generating function of a particular type, which I shall come to very shortly. With regard to (4), I see no reason why these palatial procedures should not apply also to the Ward construction and perhaps give some new insights into particle physics. However for this to be realistic, one needs to understand the proper way to introduce mass. Earlier proposals for this involved functions of several twistors [43, 54, 55], and it might be well worthwhile to reopen these discussions in the light of the above palatial ideas. It would be interesting to see whether this leads to any new insights.

To end this article I provide a method, using a curious kind of generating function, which preserves not only the 2-form Σ, but also another 2-form

$$\Pi = dZ^\alpha \wedge dZ^\beta I_{\alpha\beta} + dW_\alpha \wedge dW_\beta I^{\alpha\beta}$$

where $I_{\alpha\beta}$ and $I^{\alpha\beta}$ are the Λ-infinity twistors introduced in Section 3.6. From general considerations of twistor theory, it can be inferred that the preservation of Π defines for us not only a metric scaling (over and above the conformal structure that is inherent in the twistor formalism), but also the Einstein Λ-vacuum equations (Ricci scalar equals 4Λ), although I do not have a direct argument for this that can be presented succinctly here. Bearing this in mind, if we can find a generating function that preserves *both* Σ and Π, then our palatial procedure should provide us with the general $\mathcal{M}^\mathbb{C}$ satisfying the Λ-vacuum equations. Somewhat remarkably this can be achieved by a generating function

$$\Gamma(\lambda Z^0, W_1, Z^3, \lambda W_2, ; \mathcal{W}_0, \lambda \mathcal{Z}^1, \lambda \mathcal{W}_3, \mathcal{Z}^2),$$

which is homogeneous of total degree 2, where

$$\lambda = i\sqrt{\frac{\Lambda}{6}}$$

and we impose the symmetry that Γ is unchanged if

> the first and second entries are interchanged, together with the third and fourth being interchanged

or else

> the fifth and sixth entries are interchanged, together with the seventh and eighth being interchanged.

This should give us a general $\mathcal{M}^\mathbb{C}$ satisfying the Λ-vacuum equations. If, in addition, we impose the Hermiticity relation that Γ becomes its complex conjugate under the complete reversal of the order of the first four entries, together with the complete reversal of the order of the last four entries, then we should obtain a real Lorentzian $\mathcal{M}$ satisfying the Λ-vacuum equations. If desired, we can use Γ first to obtain the required relation between the (Z^α, W_α)

and ($\mathcal{Z}^\alpha, \mathcal{W}_\alpha$) systems and then reconstruct the original type of generating function G used earlier, from the relation $G = \frac{1}{2}(Z^\alpha W_\alpha + \mathcal{Z}^\alpha \mathcal{W}_\alpha)$ and proceed as before.

It should be reiterated that although ideas from quantum mechanics are crucially incorporated into this construction, this is still just *classical* Einstein general relativity. There is no role here for the Planck length, for example. To extend these ideas for a *quantum* gravity theory, it would appear that the commutation rules of Section 3.1 should be extended to include

$$Z^\alpha Z^\beta - Z^\beta Z^\alpha + W_\rho W_\sigma \varepsilon^{\alpha\beta\rho\sigma} = \varsigma I^{\alpha\beta},$$

ς being some constant connected with the Planck length (and we may note that the above formula is actually symmetric with respect to interchange of Z^α with W_α because of an identity involving the Levi-Civita twistor $\varepsilon^{\alpha\beta\rho\sigma}$).

References

[1] N. Arkani-Hamed, F. Cachazo, C. Cheung, and J. Kaplan, *The s-matrix in twistor space*, J. High Energy Phys. 110 (2010)

[2] N. Arkani-Hamed, A. Hodges, and J. Trnka, *Positive amplitudes in the amplituhedron*, Preprint, https://arxiv.org/abs/1412.8478v1

[3] W. L. Bade and H. Jehle, *An introduction to spinors*, Rev. Mod. Phy. 25 (1953) 714–28

[4] H. Bateman, *The solution of partial differential equations by means of definite integrals*, Proc. London Math. Soc. s2-1 (1904) 451–58

[5] H. Bateman, *Partial Differential Equations of Mathematical Physics*, Dover (1944)

[6] R. H. Boyer and R. W. Lindquist, *Maximal analytic extension of the Kerr metric*, J. Math. Phys. 8 (1967) 265–81

[7] É. Cartan, *The Theory of Spinors*, Hermann (1966)

[8] E. M. Corson, *Introduction to Tensors, Spinors and Relativistic Wave Equations*, Blackie (1953)

[9] P. A. M. Dirac, *Relativistic wave equations*, Proc. R. Soc. London, Ser. A 155 (1936) 447–59

[10] P. A. M. Dirac, *The Principles of Quantum Mechanics*, 3rd ed., Clarendon Press (1947)

[11] R. Geroch, *Domain of dependence*, J. Math. Phys. 11 (1970) 437–49

[12] B. Greene, *The Elegant Universe; Superstrings, Hidden Dimensions, and the Quest for the Ultimate Theory*, Jonathan Cape (1999)

[13] R. C. Gunning and R. Rossi, *Analytic Functions of Several Complex Variables*, Prentice Hall (1965)

[14] A. Hodges, *Twistor diagrams*, Phys. A 114 (1982) 157–75

[15] A. Hodges, *A twistor approach to the regularization of divergences*, Proc. R. Soc. A 397 (1985) 341–74

[16] A Hodges, *Scattering amplitudes for eight gauge fields*, Preprint, https://arxiv.org/abs/hep-th/0603101v1

[17] A Hodges, *Twistor diagrams for all tree amplitudes in gauge theory: a helicity-independent formalism*, Preprint, https://arxiv.org/abs/hep-th/0512336v2

[18] A Hodges, *Eliminating spurious poles from gauge-theoretic amplitudes*, https://arxiv.org/abs/0905.1473v1

[19] A. Hodges, *Theory with a twistor*, Nat. Phys. 9 (2013) 205–206

[20] A. Hodges and S. Huggett, *Twistor diagrams*, Surv. High Energy Phys. 1 (1980) 333–53

[21] S. A. Huggett, L. J. Mason, K. P. Tod, S. T. Tsou, and N. M. J. Woodhouse, eds., *The Geometric Universe: Science, Geometry, and the Work of Roger Penrose*, Oxford University Press (1998)

[22] L. P. Hughston, *Twistors and Particles*, Lecture Notes in Physics 97, Springer (1979)

[23] R. P. Kerr, *Gravitational field of a spinning mass as an example of algebraically special metrics*, Phys. Rev. Lett., 11 (1963) 237–38

[24] K. Kodaira, *On stability of compact submanifolds of complex manifolds*, Amer. J. Math. 85 (1963) 79–94

[25] K. Kodaira and D. C. Spencer, *On deformations of complex analytic structures*, I, Ann. Math. 67 (1958) 328–401, 403–66

[26] C. R. LeBrun, *Ambi-twistors and Einstein's equations*, Class. Quant. Grav. 2 (1985) 555–63

[27] C. R. LeBrun, *Twistors, Ambitwistors, and Conformal Gravity*, London Mathematical Society Lecture Note Series, Cambridge University Press (1990)

[28] L. Mason and D. Skinner, *Dual superconformal invariance, momentum twistors and grassmannians*, Preprint, https://arxiv.org/abs/0905.1473v1

[29] L. Mason and D. Skinner, *Scattering amplitudes and BCFW recursion in twistor space*, Preprint, https://arxiv.org/abs/0903.2083v3

[30] L. J. Mason and N. M. J. Woodhouse, *Integrability, Self-Duality, and Twistor Theory*, Oxford University Press (1996)

[31] V. P. Nair, *A current algebra for some gauge theory amplitudes*, Phy. Lett. B 214 (1988) 215–18

[32] E. T. Newman, *Heaven and its properties*, Gen. Rel. Grav. 7 (1976) 107–11

[33] E. T. Newman, *Deformed twistor space and h-space*, in D. E. Lerner and P. D. Sommers, eds., *Complex Manifold Techniques in Theoretical Physics*, Pitman (1979) 154–65

[34] E. T. Newman, E. Couch, K. Chinnapared, A. Exton, A. Prakash, and R. Torrence, *Metric of a rotating, charged mass*, J. Math. Phys. 6 (1965) 918–19

[35] S. J. Parke and T. R. Taylor, *Amplitude for n-gluon scattering*, Phys. Rev. Lett. 56 (1986) 2459–60
[36] W. T. Payne, *Elementary spinor theory*, Amer. J. Phys. 20 (1952) 253–62
[37] R. Penrose, *The apparent shape of a relativistically moving sphere*, Math. Proc. Camb. Philos. Soc. 55 (1959) 137
[38] R. Penrose, *A spinor approach to general relativity*, Ann. Phys. 10 (1960) 171–201
[39] R. Penrose, *Zero rest-mass fields including gravitation: asymptotic behaviour*, Proc. R. Soc. London, Ser. A. 284 (1965) 159–203
[40] R. Penrose, *Twistor algebra*, J. Math. Phys. 8 (1967) 345–66
[41] R. Penrose, *Twistor quantisation and curved space-time*, Int. J. Theoretical Phys. 1 (1968) 61–99
[42] R. Penrose, *Solutions of the zero-rest-mass equations*, J. Math. Phys. 10 (1969) 38–39
[43] R. Penrose, *Twistors and particles: An outline*, in *Quantum Theory and the Structures of Time and Space: Proceedings, Feldafing Conference, July 1974*, (1974) 129–45
[44] R. Penrose, *Nonlinear gravitons and curved twistor theory*, Gen. Rel. Grav. 7 (1976) 31–52
[45] R. Penrose, *Null hypersurface initial data for classical fields of arbitrary spin and for general relativity*, Gen. Rel. Grav. 12 (1980) 225–64
[46] R. Penrose, *On the origins of twistor theory*, in W. Rindler and A. Trautman, eds., *Gravitation and Geometry: A Volume in Honour of I. Robinson*, Bibliopolis (1987)
[47] R. Penrose, *On the cohomology of impossible figures (la cohomologie des figures impossibles)*, Structural Topol. 17 (1991) 11–16
[48] R. Penrose, *A new angle on the googly graviton*, in L. J. Mason, L. P. Hughston, P. Z. Kobak, and K. Pulverer, eds., *Further Advances in Twistor Theory, vol. III, Curved Twistor Spaces*, Research Notes in Mathematics 424, Chapman and Hall (2001) 264–69
[49] R. Penrose, *Palatial twistor theory and the twistor googly problem*, Philos. Trans. R. Soc. A 373 (2015)
[50] R. Penrose, *Towards an Objective Physics of Bell Nonlocality: Palatial Twistor Theory*, Cambridge University Press (2016)
[51] R. Penrose and M. A. H. MacCallum, *Twistor theory: An approach to the quantisation of fields and space-time*, Phys. Rep. 6 (1973) 241–315
[52] R. Penrose and W. Rindler, *Spinors and Space Time, vol. 1, Two-Spinor Calculus and Relativistic Fields*, Cambridge Monographs on Mathematical Physics, Cambridge University Press (1984)
[53] R. Penrose and W. Rindler, *Spinors and Space Time, vol. 2, Spinor and Twistor Methods in Space-Time Geometry*, Cambridge Monographs on Mathematical Physics, Cambridge University Press (1986)
[54] Z. Perjés, *Introduction to twistor particle theory*, in H. D. Doebner and T. D. Palev, eds., *Twistor Geometry and Non-Linear Systems*, Springer (1982)
[55] Z. Perjés and G. A. J. Sparling, *The twistor structure of hadrons*, in L. P. Hughston and R. S. Ward, eds., *Advances in Twistor Theory*, Pitman (1979)

[56] I. Robinson and A. Trautman, *Some spherical gravitational waves in general relativity*, Proc. R. Soc. London, Ser. A 265 (1962) 463–73
[57] W. T. Shaw and L. P. Hughston, *Twistors and Strings*, London Mathematical Society Lecture Note Series, Cambridge University Press (1990)
[58] R. F. Streater and A. S. Wightman, *PCT, Spin and Statistics, and All That*, Princeton University Press (2000)
[59] B. L. van der Waerden, *Spinoranalyse*, Nachr. Ges. Wiss. Göttingen Math.-Phys. (1959) 100–109
[60] R. S. Ward, *On self-dual gauge fields*, Phy. Lett. A 61 (1977) 81–82
[61] R. S. Ward, *Self-dual space-times with cosmological constant*, Commun. Math. Phys. 78 (1980) 1–17
[62] E. T. Whittaker, *On the partial differential equations of mathematical physics*, Math. Ann. 57 (1903) 333–55
[63] E. Witten, *Perturbative gauge theory as a string theory in twistor space*, Commun. Math. Phys. 252 (2004) 189–258
[64] L. Witten, *Invariants of general relativity and the classification of spaces*, Phys. Rev. 113 (1959) 357–62
[65] N. M. J. Woodhouse, *Geometric Quantization*, Oxford Mathematical Monographs, Clarendon Press (1997)

Roger Penrose
Mathematical Institute, University of Oxford
rpenroad@gmail.com

8
Quantum Geometry of Space

Muxin Han*

Contents

1	Gravity, Geometry, and Quantization	373
2	Quantum Configuration Space	378
3	Cylindrical Functions on Quantum Configuration Space	382
4	Holonomy-Flux Algebra and Representation	383
5	Spin-Network Basis of Hilbert Space	388
6	Geometrical Operators	394
7	Gauss Constraint, Gauge Invariance, and Closure	397
8	Conclusion and Outlook	399
References		402

1 Gravity, Geometry, and Quantization

One of the most important lessons from Einstein's general relativity is that the gravitational field is the curved geometry of spacetime. The dynamics of gravity, governed by Einstein's field equations, describes how four-dimensional spacetime geometries are determined by various matter distributions or boundary conditions. Four-dimensional spacetime geometries from Einstein's equations may be understood as dynamics of 3d space geometries (or, namely,

---

* The author would like to thank Gabriel Catren for the invitation to contribute to this volume and Carlo Rovelli for the recommendation. The author acknowledges support from US National Science Foundation through grant PHY-1602867, and Start-up Grant at Florida Atlantic University, USA.

histories of 3d geometries). Thus Einstein's field equations may be equivalently understood as the determination of the dynamics of 3d space geometry.

As is standard in classical field theory, Einstein's equations determine a trajectory in the phase space of gravity. Since Einstein's equations govern the dynamics of 3d space geometry, it is not hard to imagine that each point of the gravity phase space determines a geometry of 3d space. This point of view is important because it makes clear how a quantum theory can be developed for 3d space geometries. Indeed since the phase space of gravity relates to 3d space geometries, the quantization of gravity phase space makes the quantization of 3d space geometries.

The program of loop quantum gravity (LQG) [1, 2, 3, 4] provides a concrete realization of the above idea, and achieves the quantum theory of space geometry. LQG is an attempt toward a nonperturbative and background independent quantum theory of gravity in 2+1, 3+1, and higher spacetime dimensions [2, 3, 4, 5, 6]. In this chapter we mainly focus on 3+1 spacetime dimensions. LQG is originated by the canonical formulation of classical gravity in 4d as a dynamical theory of gauge connections [1]. In this formalism, the phase space $\mathcal{P}$ of gravity has a similar structure as an SU(2) gauge theory. But the canonical variables represent the geometry on 3d spatial slices. The quantization of the phase space $\mathcal{P}$ is similar to the quantization of nonabelian gauge theory. It has been well understood in LQG (see, e.g., [4, 7, 8]), and leads to the Hilbert space $\mathcal{H}_{LQG}$, shown to be the *unique* representation of the operator algebra quantizing the phase space $\mathcal{P}$ [9]. Promoting the canonical variables to operators on $\mathcal{H}_{LQG}$ quantizes the 3d spatial geometry. Many geometrical quantities are represented as (self-adjoint) operators on $\mathcal{H}_{LQG}$ (e.g., [10, 11, 12, 13, 14, 15, 16]). Two of the most important examples are the area operator $\hat{A}r_S$ and the volume operator $\hat{V}_\mathcal{R}$. Both of them have discrete spectra (eigenvalues), which implies that in quantum geometry the area and volume are fundamentally discrete at Planck scale. The area and volume operators share the same set of eigenstates in $\mathcal{H}_{LQG}$, which are known as *spin-network* states. The area and volume eigenvalues are understood as the quanta carried by the spin-networks at Planck scale.

The following in this section summarizes the ideas and results of the phase space and quantization in LQG, including the emergence of quantum space geometry. The systematic discussion of the theory is given in the main body of this chapter.

The phase space $\mathcal{P}$ underlying LQG is derived from the formulation of classical gravity using connection variables. The phase space $\mathcal{P}$ can be obtained via a 3+1 decomposition of the Holst action [17]

$$S_{\text{Holst}}\left[e^I, \omega^{IJ}\right] = \frac{1}{8\pi G_N} \int_{M_4} e^I \wedge e^J \wedge \left(*F + \frac{1}{\beta}F\right)_{IJ} \quad (1.1)$$

where e_μ^I is the 4d tetrad and $F_{\mu\nu}^{IJ}$ is the curvature of the $so(1,3)$-valued connection ω_μ^{IJ} ($\mu, I = 0, \cdots, 3$). $\beta \in \mathbb{R}$ is the Barbero–Immirzi parameter, a free parameter in LQG. The variational principle of S_{Holst} gives vacuum Einstein's equations, and shows S_{Holst} is on-shell equivalent to the Einstein–Hilbert action of gravity.

The 3+1 decomposition and Hamiltonian analysis of S_{Holst} leads to the phase space $\mathcal{P}$ of 4d gravity. The canonical conjugate variables in $\mathcal{P}$ are the Ashtekar–Barbero connection $A_a^i = \Gamma_a^i + \gamma K_a^i$ and densitized triad $E_i^a = \sqrt{\det q} \, e_i^a$ on the spatial slices M_3. Here e_i^a ($a, i = 1, 2, 3$) is the triad on M_3, which determines the metric $q_{ab} = e_a^i e_b^i$ and the spin connection Γ_a^i. K_a^i relates to the extrinsic curvature K_{ab} of $M_3 \hookrightarrow M_4$ by $K_{ab} = K_{(a}^i e_{b)}^i$. In contrast to the $so(1,3)$-valued connection ω_μ^{IJ} in 4d, A_a^i is a spatial connection in 3d with gauge group SU(2). The breaking of the gauge group from the Lagrangian S_{Holst} to the Hamiltonian formulation is due to the 3+1 decomposition of spacetime, together with an internal partial gauge fixing (usually called "time gauge" in the literature). The detailed derivation of the canonical conjugate variables can be found, for example, in [17, 18]. The symplectic structure of the phase space $\mathcal{P}$ gives the Poisson brackets

$$\{A_a^i(x), E_j^b(x')\} = 8\pi G_N \beta \, \delta_a^b \delta_j^i \delta^{(3)}(x, x'),$$
$$\{A_a^i(x), A_b^j(x')\} = \{E_i^a(x), E_j^b(x')\} = 0. \tag{1.2}$$

Therefore $A_a^i(x)$ is the position variable of gravity, and $E_j^b(x')$ is the momentum variable.

In the quantization of phase space $\mathcal{P}$ [4, 7, 8], the wave function can be understood as a function $\psi(A_a^i)$ of the connection field on M_3. More precisely, the wave functions are functions ψ of SU(2) holonomies $h_e(A) = P \exp \int_e A$ along a number of oriented edges (analytic curves) $e_1, \ldots, e_N \subset M_3$:

$$\psi = \psi\left(h_{e_1}(A), \ldots, h_{e_N}(A)\right). \tag{1.3}$$

The edges $e_1, \ldots, e_N$ form a graph (a network) γ. A general graph γ consists of a finite set of oriented edges denoted by $E(\gamma)$ and a set of vertices $V(\gamma)$. The vertices in γ are the sources and targets of the edges $e \in E(\gamma)$. A wave function (1.3) is defined upon a choice of graph γ, and depends on only a finite number of degrees of freedom. Thus ψ in equation (1.3) is referred to as a *cylindrical function* of LQG. Obviously, the full infinite number of degrees of freedom of gravity is achieved by putting together all possible choices of γ.[1] All possible

---

[1] The LQG Hilbert space is the completion of the union of all possible cylindrical functions, modulo some equivalence relations. In simple language, the equivalence relations include the cylindrical functions of a small graph into the cylindrical functions of a larger graph. The integration of cylindrical functions and all operators have to respect the equivalence relations, which is known as cylindrical consistency.

cylindrical functions of the type equation (1.3) by considering all possible γ form a Hilbert space of L^2-type,

$$\mathcal{H}_{LQG} = L^2(\bar{\mathcal{A}}/\bar{\mathcal{G}}, \mathrm{d}\mu_{AL}). \qquad (1.4)$$

$\mathcal{H}_{LQG}$ is the Hilbert space of LQG and carries the representation of quantum geometry on the spatial manifolds M_3. The configuration space $\bar{\mathcal{A}}/\bar{\mathcal{G}}$ is the space of all SU(2) connection fields (including certain nonsmooth and distributional connection fields) over the spatial slices M_3, modulo the gauge transformations [19]. $\mathrm{d}\mu_{AL}$ denotes the Ashtekar–Lewandowski measure on $\bar{\mathcal{A}}/\bar{\mathcal{G}}$ [20]. Importantly, the LQG quantization of the gravity phase space $\mathcal{P}$ is systematic and mathematically rigorous. The formalism is even *unique*. It is proved in [9] that $\mathcal{H}_{LQG}$ is the unique representation of the quantization of $\mathcal{P}$, provided that the theory is diffeomorphism invariant.

There is a useful orthonormal basis in $\mathcal{H}_{LQG}$ which is called *spin-network* basis. The spin-network basis can be constructed by the following observation: Consider the simplest graph consisting only of a single edge e; the associated cylindrical functions $\psi = \psi(h_e)$ is a function on the group SU(2). More precisely ψ belongs to the space $L^2(\mathrm{SU}(2), \mathrm{d}\mu_H)$ where $\mathrm{d}\mu_H$ is the Haar measure. An orthogonal basis in $L^2(\mathrm{SU}(2), \mathrm{d}\mu_H)$ is given by the matrix elements $R^j_{mn}(h_e)$ of all SU(2) irreducible representations labeled by the spins j

$$R^j_{mn}(h_e) = \langle j, m | h_e | j, n \rangle. \qquad (1.5)$$

Thus $\psi(h_e)$ can be written as a linear combination of $R^j_{mn}(h_e)$: $\psi = \sum_{j=0}^{\infty} \sum_{m,n=-j}^{j} c^j_{mn} R^j_{mn}$. A general cylindrical function $\psi(h_{e_1}, \cdots, h_{e_N})$ on a closed graph γ can be decomposed in the same way at each entry h_{e_i}. The basis to make the decomposition is a product of $R^{j_e}_{m_e n_e}(h_e)$ over all $e \in E(\gamma)$. However in order to preserve the gauge invariance at each vertex $v \in V(\gamma)$, an invariant tensor I_v has to be inserted and contract the m_e, n_e from the adjacent e's. Therefore we have the (gauge invariant) spin-network basis $T_{\gamma, \{j_e\}, \{I_v\}}$ for decomposing arbitrary $\psi(h_{e_1}, \cdots, h_{e_N})$

$$T_{\gamma, \{j_e\}, \{I_v\}}(h_{e_1}, \cdots, h_{e_N}) = \sum_{m_e, n_e} \prod_{e \in E(\gamma)} R^{j_e}_{m_e n_e}(h_e) \prod_{v \in V(\gamma)} (I_v)^{\{j_e\}}_{\{m_e, n_e\}}, \qquad (1.6)$$

where $I_v^{\{j_e\}} \in \mathrm{Inv}_{\mathrm{SU}(2)}(\otimes_e V_{j_e})$ is an invariant tensor in SU(2) tensor representation $\otimes_e V_{j_e}$ with j_e on adjacent edges. I_v is often called an *intertwiner* of SU(2).

The spin-network state is defined as a triple $|\gamma, \{j_e\}, \{I_v\}\rangle$ containing the graph γ, spins j_e associated to all edges, and intertwiners I_v associated to all vertices. The vertex in a closed graph is not univalent. The spin-network state is related to the spin-network function $T_{\gamma,\{j_e\},\{I_v\}}(h_{e_1},\cdots,h_{e_N})$ by

$$\langle h_{e_1},\cdots,h_{e_N}|\gamma,\{j_e\},\{I_v\}\rangle = T_{\gamma,\{j_e\},\{I_v\}}(h_{e_1},\cdots,h_{e_N}). \quad (1.7)$$

The spin-networks have been firstly invented by Penrose, motivated by the twistor theory [21]. The above displays that spin-networks are naturally applied in LQG. Moreover, there are also extensive applications of spin-networks to topological invariants of manifolds of three and four dimensions (see, e.g., [22]).

The spin-network basis is a complete orthogonal basis in the LQG Hilbert space $\mathcal{H}_{LQG}$. It also has nice interpretations in terms of the 3d quantum geometry on the spatial manifolds M_3. Recall that $E_i^a(x)$ is related to the spatial triad and metric. A large class of geometrical quantities on M_3 have been represented as the self-adjoint operators acting on $\mathcal{H}_{LQG}$, constructed using the derivative $\hat{E}_i^a(x) = -8\pi i\beta\ell_P^2 \frac{\delta}{\delta A_a^i(x)}$ (see, e.g., [10, 11, 12, 13, 14, 15, 16]). Among the class of geometrical operators, two important ones are the area operator $\hat{Ar}_S$ of a 2-surface S, and the volume operator $\hat{V}_R$ of a 3d region R. It turns out that the spin-network basis is simultaneously the eigenbasis of both $\hat{Ar}_S$ and $\hat{V}_R$. For S cuts transversely a number edges in γ and R encloses a number of vertices

$$\hat{Ar}_S\big|\gamma,\{j_e\},\{I_v\}\big\rangle = 8\pi\beta\ell_P^2 \sum_{S\cap e\neq\emptyset} \sqrt{j_e(j_e+1)}\big|\gamma,\{j_e\},\{I_v\}\big\rangle;$$

$$\hat{V}_R\big|\gamma,\{j_e\},\{I_v\}\big\rangle = \sum_{v\in R} V_v(I_v)\big|\gamma,\{j_e\},\{I_v\}\big\rangle. \quad (1.8)$$

The eigenvalue of the area operator is given by a sum over all cut edges weighted by $\sqrt{j(j+1)}$ according to the spin labels carried by the edges. The spin labels j_e are thus interpreted as the quanta of the area elements transverse to the edges. The eigenvalue of the volume operator is a sum of local contributions $V_v(I_v)$ at $v \in \mathcal{R}$, which is determined by the intertwiner I_v at v [23, 24]. Geometrically, one may imagine that there is a tiny polyhedron (infinitesimal from the macroscopic point of view) enclosing a single vertex v (Figure 8.1). The polyhedron faces are all transverse to the edges adjacent to v. The intertwiner I_v is a quantum parameterization of the shapes of the tiny polyhedron [25, 26], while the js on the edges adjacent to v give the quantum areas of the tiny polyhedron faces. In this way, a spin-network with n vertices encodes a discretized geometry made by gluing n polyhedra. Given that the graph Γ is sufficiently refined, the discrete geometry approximates the smooth geometry arbitrarily well.

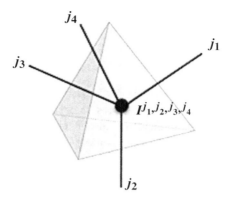

Figure 8.1 A quantum tetrahedron located at a spin-network vertex.

A generic state in $\mathcal{H}_{LQG}$ is a linear combination of spin-network states $|\psi\rangle = \sum_s c_s |s\rangle$, $s \equiv (\gamma, \{j_e\}, \{I_v\})$, which is interpreted as a quantum superposition of geometries. The coefficient c_s is the quantum probability amplitude of the corresponding quantum geometry s.

2 Quantum Configuration Space

In quantum mechanics, the kinematical Hilbert space is $L^2(\mathbf{R}^3, d^3x)$, where the simple $\mathbf{R}^3$ is the classical configuration space of a free particle which has finite degrees of freedom, and d^3x is the Lebesgue measure on $\mathbf{R}^3$. In quantum field theory, it is expected that the kinematical Hilbert space is also the L^2 space on the configuration space of the field, which is infinite-dimensional, with respect to some Borel measure naturally defined. However, it is often hard to define a concrete Borel measure on the classical configuration space, since the integral theory on infinite-dimensional spaces is involved. Thus the intuitive expectation should be modified, and the concept of quantum configuration space should be introduced as a suitable enlargement of the classical configuration space so that an infinite-dimensional measure, often called cylindrical measure, can be well defined on it. The example of a scalar field can be found in the references [3, 27]. For quantum gravity, it should be emphasized that the construction of the quantum configuration space must be background independent. Fortunately, general relativity has been reformulated as a dynamical theory of $SU(2)$ connections, which is greatly helpful for this development.

The classical configuration space for the gravitational field, which is denoted by $\mathcal{A}$, is a collection of the $su(2)$-valued connection 1-form fields smoothly distributed on Σ. The idea of the construction of the quantum configuration space is due to the concept of holonomy.

Definition 2.1 Given a smooth SU(2) connection field A_a^i and an analytic curve c with the parameter $t \in [0, 1]$ supported on a compact subset (compact support) of Σ, the corresponding holonomy is defined by the solution of the parallel transport equation

$$\frac{d}{dt}A(c,t) = -\left(A_a^i \dot{c}^a \tau_i\right) A(c,t), \tag{2.1}$$

with the initial value $A(c,0) = 1$, where $\dot{c}^a$ is the tangent vector of the curve and $\tau_i \in su(2)$ constitute an orthonormal basis with respect to the Killing–Cartan metric $\eta(\xi, \zeta) := -2\text{Tr}(\xi\zeta)$, which satisfy $[\tau_i, \tau_j] = \epsilon^k{}_{ij} \tau_k$ and are fixed once for all. Thus the holonomy is an element in SU(2), which can be expressed as

$$A(c) = \mathcal{P} \exp\left(-\int_0^1 [A_a^i \dot{c}^a \tau_i]\, dt\right), \tag{2.2}$$

where $A(c) \in SU(2)$ and $\mathcal{P}$ is a path-ordering operator along the curve c.

The definition can be well extended to the case of piecewise analytic curves via the relation:

$$A(c_1 \circ c_2) = A(c_1)A(c_2), \tag{2.3}$$

where $\circ$ stands for the composition of two curves. It is easy to see that a holonomy is invariant under reparameterizations and covariant under changes of the orientation, that is,

$$A(c^{-1}) = A(c)^{-1}. \tag{2.4}$$

So one can formulate the properties of holonomy in terms of equivalence classes of curves.

Definition 2.2 Two analytic curves c and c' are said to be equivalent if and only if they have the same source $s(c)$ (beginning point) and the same target $t(c)$ (endpoint), and the holonomies of the two curves are equal to each other, that is, $A(c) = A(c') \ \forall A \in \mathcal{A}$. An equivalence class of analytic curves is defined to be an edge, and a piecewise analytic path is an composition of edges.

To summarize, the holonomy is actually defined on the set $\mathcal{P}$ of piecewise analytic paths with compact supports. The two properties (2.3) and (2.4) mean that each connection in $\mathcal{A}$ is a homomorphism from $\mathcal{P}$, which is a so-called groupoid by definition [28], to our compact gauge group SU(2). Note that the internal gauge transformations and spatial diffeomorphisms act covariantly on a holonomy as

$$A(e) \mapsto g(t(e))^{-1} A(e) g(s(e)) \quad \text{and} \quad A(e) \mapsto A(\varphi \circ e), \qquad (2.5)$$

for any SU(2)-valued function $g(x)$ on Σ and a spatial diffeomorphism φ. The above discussion is for classical smooth connections in $\mathcal{A}$. The quantum configuration space for loop quantum gravity can be constructed by extending the concept of holonomy, since its definition does not depend on an extra background. One thus obtains the quantum configuration space $\overline{\mathcal{A}}$ of loop quantum gravity as the following.

Definition 2.3 The quantum configuration space $\overline{\mathcal{A}}$ is a collection of all quantum connections A, which are algebraic homomorphism maps without any continuity assumption from the collection $\mathcal{P}$ of piecewise analytic paths with compact supports on Σ to the gauge group $SU(2)$, that is, $\overline{\mathcal{A}} := \text{Hom}(\mathcal{P}, SU(2))$.[2] Thus for any $A \in \overline{\mathcal{A}}$ and edge e in $\mathcal{P}$,

$$A(e_1 \circ e_2) = A(e_1) A(e_2) \quad \text{and} \quad A(e^{-1}) = A(e)^{-1}.$$

The transformations of quantum connections under internal gauge transformations and diffeomorphisms are defined by equation (2.5).

The above discussion on the smooth connections shows that the classical configuration space $\mathcal{A}$ can be understood as a subset of the quantum configuration space $\overline{\mathcal{A}}$. Moreover, the Giles theorem [29] shows precisely that a smooth connection can be recovered from its holonomies by varying the length and location of the paths.

On the other hand, it was shown in [4, 28] that the quantum configuration space $\overline{\mathcal{A}}$ can be constructed via a projective limit technique and admits a naturally defined topology. To make the discussion precise, we begin with a few definitions:

Definition 2.4

1. A finite set $\{e_1, ..., e_N\}$ of edges is said to be independent if the edges e_i can only intersect each other at their sources $s(e_i)$ or targets $t(e_i)$.
2. A finite graph is a collection of a finite set $\{e_1, ..., e_N\}$ of independent edges and their vertices, that is, their sources $s(e_i)$ and targets $t(e_i)$. We denote by

---

[2] It is easy to see that the definition of $\overline{\mathcal{A}}$ does not depend on the choice of a local section in the SU(2)-bundle, since the internal gauge transformations leave $\overline{\mathcal{A}}$ invariant.

$E(\gamma)$ and $V(\gamma)$, respectively, the sets of independent edges and vertices of a given finite graph γ. And N_γ is the number of elements in $E(\gamma)$.
3. A subgroupoid $\alpha(\gamma) \subset \mathcal{P}$ can be generated from γ by identifying $V(\gamma)$ as the set of objects and all $e \in E(\gamma)$ together with their inverses and finite compositions as the set of homomorphisms. This kind of subgoupoid in $\mathcal{P}$ is called tame subgroupoid. $\alpha(\gamma)$ is independent of the orientation of γ, so the graph γ can be recovered from the tame subgroupoid α up to the orientations on the edges. We will also denote by N_α the number of elements in $E(\gamma)$ where γ is recovered by the tame subgroupoid α.
4. $\mathcal{L}$ denotes the set of all tame subgroupoids in $\mathcal{P}$.

One can equip $\mathcal{L}$ with a partial order relation $<$,[3] defined by $\alpha < \alpha'$ if and only if α is a subgroupoid in α'. Obviously, for any two tame subgroupoids $\alpha \equiv \alpha(\gamma)$ and $\alpha' \equiv \alpha(\gamma')$ in $\mathcal{L}$, there exists $\alpha'' \equiv \alpha(\gamma'') \in \mathcal{L}$ such that $\alpha, \alpha' < \alpha''$, where $\gamma'' \equiv \gamma \cup \gamma'$. Define $X_\alpha \equiv Hom(\alpha, SU(2))$ as the set of all homomorphisms from the subgroupoid $\alpha(\gamma)$ to the group $SU(2)$. Note that an element $A_\alpha \in X_\alpha$ ($\alpha = \alpha(\gamma)$) is completely determined by the $SU(2)$ group elements $A(e)$ where $e \in E(\gamma)$, so that one has a bijection $\lambda: X_{\alpha(\gamma)} \to SU(2)^{N_\gamma}$, which induces a topology on $X_{\alpha(\gamma)}$ such that λ is a topological homomorphism. For any pair $\alpha < \alpha'$, one can define a surjective projection map $P_{\alpha'\alpha}$ from $X_{\alpha'}$ to X_α by restricting the domain of the map $A_{\alpha'}$ from α' to the subgroupoid α, and these projections satisfy the consistency condition $P_{\alpha'\alpha} \circ P_{\alpha''\alpha'} = P_{\alpha''\alpha}$. Thus a projective family $\{X_\alpha, P_{\alpha'\alpha}\}_{\alpha<\alpha'}$ is obtained by the above constructions. Then the projective limit $\lim_\alpha(X_\alpha)$ is naturally obtained.

Definition 2.5 The projective limit $\lim_\alpha(X_\alpha)$ of the projective family $\{X_\alpha, P_{\alpha'\alpha}\}_{\alpha<\alpha'}$ is a subset of the direct product space $X_\infty := \prod_{\alpha \in \mathcal{L}} X_\alpha$ defined by

$$\lim_\alpha(X_\alpha) := \{\{A_\alpha\}_{\alpha \in \mathcal{L}} | P_{\alpha'\alpha} A_{\alpha'} = A_\alpha, \forall \alpha < \alpha'\}.$$

Note that the projection $P_{\alpha'\alpha}$ is surjective and continuous with respect to the topology of X_α. One can equip the direct product space X_∞ with the so-called Tychonov topology. Since any X_α is a compact Hausdorff space, by Tychonov theorem X_∞ is also a compact Hausdorff space. One then can prove that the projective limit, $\lim_\alpha(X_\alpha)$, is a closed subset in X_∞ and hence a compact Hausdorff space with respect to the topology induced from X_∞.

---

[3] A partial order on $\mathcal{L}$ is a relation, which is reflective ($\alpha < \alpha$), symmetric ($\alpha < \alpha'$, $\alpha' < \alpha$ $\Rightarrow \alpha' = \alpha$) and transitive ($\alpha < \alpha'$, $\alpha' < \alpha'' \Rightarrow \alpha' < \alpha''$). Note that not all pairs in $\mathcal{L}$ need to be related.

At last, one can find the relation between the projective limit and the prior constructed quantum configuration space $\overline{\mathcal{A}}$. As one might expect, there is a bijection Φ between $\overline{\mathcal{A}}$ and $\lim_\alpha(X_\alpha)$ [4]:

$$\Phi: \overline{\mathcal{A}} \to \lim_\alpha(X_\alpha);$$
$$A \mapsto \{A|_\alpha\}_{\alpha \in \mathcal{L}},$$

where $A|_\alpha$ means the restriction of the domain of the map $A \in \overline{\mathcal{A}} = \mathrm{Hom}(\mathcal{P}, SU(2))$. As a result, the quantum configuration space is identified with the projective limit space and hence can be equipped with the topology. In conclusion, the quantum configuration space $\overline{\mathcal{A}}$ is constructed to be a compact Hausdorff topological space.

By the Giles theorem [29], the space of smooth connections can be viewed as a dense subspace in the quantum configuration space $\overline{\mathcal{A}}$. The extension to $\overline{\mathcal{A}}$ includes nonsmooth and distributional connections. The extension from smooth configurations to distributional configurations (quantum configurations) is standard in quantum field theory (e.g., free scalar field). There have been earlier studies on constructive quantum field theory which demonstrate this common feature [3, 27], which inspired the construction of LQG.

3 Cylindrical Functions on Quantum Configuration Space

Given the projective family $\{X_\alpha, P_{\alpha'\alpha}\}_{\alpha < \alpha'}$, the cylindrical functions on its projective limit $\overline{\mathcal{A}}$ are well defined as follows.

Definition 3.1 Let $C(X_\alpha)$ be the set of all continuous complex functions on X_α, two functions $f_\alpha \in C(X_\alpha)$ and $f_{\alpha'} \in C(X_{\alpha'})$ are said to be equivalent or cylindrically consistent, denoted by $f_\alpha \sim f_{\alpha'}$, if and only if $P^*_{\alpha''\alpha} f_\alpha = P^*_{\alpha''\alpha'} f_{\alpha'}$, $\forall \alpha'' > \alpha, \alpha'$, where $P^*_{\alpha''\alpha}$ denotes the pullback map induced from $P_{\alpha''\alpha}$. Then the space $Cyl(\overline{\mathcal{A}})$ of cylindrical functions on the projective limit $\overline{\mathcal{A}}$ is defined to be the space of equivalent classes $[f]$, that is,

$$Cyl(\overline{\mathcal{A}}) := [\cup_\alpha C(X_\alpha)] / \sim .$$

One can then easily prove the following proposition by definition.

Proposition 3.2 *All continuous functions f_α on X_α are automatically cylindrical since each of them can generate an equivalent class $[f_\alpha]$ via the pullback map $P^*_{\alpha'\alpha}$ for all $\alpha' > \alpha$, and the dependence of $P^*_{\alpha'\alpha} f_\alpha$ on the groups associated to the edges in α' but not in α is trivial, that is, by the definition of the pullback map,*

$$(P^*_{\alpha'\alpha} f_\alpha)(A(e_1), ..., A(e_{N_\alpha}), ..., A(e_{N_{\alpha'}})) = f_\alpha(A(e_1), ..., A(e_{N_\alpha})), \quad (3.1)$$

where N_α denotes the number of independent edges in the graph recovered from the groupoid α. On the other hand, by definition, given a cylindrical function $f \in Cyl(\overline{\mathcal{A}})$ there exists a suitable groupoid α such that $f = [f_\alpha]$, so one can identify f with f_α. Moreover, given two cylindrical functions f, $f' \in Cyl(\overline{\mathcal{A}})$, by definition of cylindrical functions and the property of the projection map, there exists a common groupoid α and f_α, $f'_\alpha \in C(X_\alpha)$ such that $f = [f_\alpha]$ and $f' = [f'_\alpha]$.

Given f, $f' \in Cyl(\overline{\mathcal{A}})$, there exists a graph α such that $f = [f_\alpha]$, and $f' = [f'_\alpha]$. Then the following operations are well defined

$$f + f' := [f_\alpha + f'_\alpha], \quad ff' := [f_\alpha f'_\alpha], \quad zf := [zf_\alpha], \quad \bar{f} := [\bar{f}_\alpha],$$

where $z \in \mathbf{C}$ and $\bar{f}$ denotes the complex conjugate. So we construct $Cyl(\overline{\mathcal{A}})$ as an abelian $*$-algebra. In addition, there is a unital element in the algebra because $Cyl(\overline{\mathcal{A}})$ contains constant functions. Moreover, we can define the sup-norm for $f = [f_\alpha]$ by

$$\|f\| := \sup_{A_\alpha \in X_\alpha} |f_\alpha(A_\alpha)|, \quad (3.2)$$

which satisfies the C^*-property $\|f \bar{f}\| = \|f\|^2$. Then $\overline{Cyl(\overline{\mathcal{A}})}$ is a unital abelian C^*-algebra, after the completion with respect to the norm.

From the theory of C^*-algebras, it is known that a unital abelian C^*-algebra is identical to the space of continuous functions on its spectrum space via an isometric isomorphism, the so-called Gelfand transformation (see, e.g., [4]). So we have the following theorem [30, 31], which finishes this section:

Theorem 3.3

(1) The space $Cyl(\overline{\mathcal{A}})$ has the structure of a unital abelian C^*-algebra after completion with respect to the sup-norm.
(2) The quantum configuration space $\overline{\mathcal{A}}$ is the spectrum space of the completed $\overline{Cyl(\overline{\mathcal{A}})}$ such that $\overline{Cyl(\overline{\mathcal{A}})}$ is identical to the space $C(\overline{\mathcal{A}})$ of continuous functions on $\overline{\mathcal{A}}$.

4 Holonomy-Flux Algebra and Representation

In analogy with the quantization procedure of Section 3.1, in this subsection we would like to perform the background-independent construction of an algebraic quantum field theory for general relativity. First we construct the algebra of classical observables. Taking into account the future quantum analogs, we define the algebra of classical observables $\mathfrak{P}$ as the Poisson $*$-subalgebra generated by the functions of holonomies (cylindrical functions) and the fluxes of triad fields smeared on some 2-surface. Namely, one can define the classical algebra in analogy with geometric quantization in the finite-dimensional phase space case by the so-called classical Ashtekar–Corichi–Zapata holonomy-flux $*$-algebra as follows [9].

Definition 4.1 The classical Ashtekar–Corichi–Zapata holonomy-flux $*$-algebra is defined to be a vector space $\mathfrak{P}_{ACZ} := Cyl(\overline{\mathcal{A}}) \times \mathcal{V}^{\mathbf{C}}(\overline{\mathcal{A}})$, where $\mathcal{V}^{\mathbf{C}}(\overline{\mathcal{A}})$ is the vector space of algebraic vector fields spanned by the vector fields $\psi Y_f(S)$ $\psi \in Cyl(\overline{\mathcal{A}})$, and their commutators. Here the smeared flux vector field $Y_f(S)$ is defined by acting on any cylindrical function:

$$Y_f(S)\psi := \{\int_S \eta_{abc} \widetilde{P}_i^c f^i, \psi\},$$

for any $su(2)$-valued function f^i with compact support on S and where ψ are cylindrical functions on $\overline{\mathcal{A}}$. We equip $\mathfrak{P}_{ACZ}$ with the structure of a $*$-Lie algebra as follows :
(1) Lie bracket $\{,\} : \mathfrak{P}_{ACZ} \times \mathfrak{P}_{ACZ} \to \mathfrak{P}_{ACZ}$ is defined by

$$\{(\psi, Y), (\psi', Y')\} := (Y \circ \psi' - Y' \circ \psi, [Y, Y']),$$

for all $(\psi, Y), (\psi', Y') \in \mathfrak{P}_{ACZ}$ with $\psi, \psi' \in Cyl(\overline{\mathcal{A}})$ and $Y, Y' \in \mathcal{V}^{\mathbf{C}}(\overline{\mathcal{A}})$.
(2) Involution: $p \mapsto \bar{p}$ $\forall$ $p \in \mathfrak{P}_{ACZ}$ is defined by means of the complex conjugate of cylindrical functions and vector fields, that is, $\bar{p} := (\overline{\psi}, \overline{Y})$ $\forall$ $p = (\psi, Y) \in \mathfrak{P}_{ACZ}$, where $\overline{Y}\psi := \overline{Y\overline{\psi}}$.
(3) $\mathfrak{P}_{ACZ}$ admits a natural action of $Cyl(\overline{\mathcal{A}})$ by

$$\psi' \circ (\psi, Y) := (\psi'\psi, \psi'Y),$$

which gives $\mathfrak{P}_{ACZ}$ a module structure.

Note that the action of the flux vector field $Y_f(S)$ can be expressed explicitly on any cylindrical function $\psi_\gamma \in C^1(X_{\alpha(\gamma)})$ via a suitable regularization [4]:

$$Y_f(S)\psi_\gamma = \{\int_S \eta_{abc} \widetilde{P}_i^c f^i, \psi_\gamma\},$$

$$= \sum_{v \in V(\gamma) \cap S} \sum_{e \ at \ v} \frac{\kappa(S, e)}{2} f^i(v) X_i^{(e,v)} \psi_\gamma,$$

where $X_i^{(e,v)}$ is the left(right) invariant vector field $L^{(\tau_i)}(R^{(\tau_i)})$ of the group associated with the edge e if v is the source(target) of edge e where by definition

$$L^{(\tau_i)}\psi(A(e)) := \frac{d}{dt}|_{t=0}\psi(A(e)\exp(t\tau_i)),$$
$$R^{(\tau_i)}\psi(A(e)) := \frac{d}{dt}|_{t=0}\psi(\exp(-t\tau_i)A(e)),$$

and

$$\kappa(S,e) = \begin{cases} 0, & \text{if } e \cap S = \emptyset, \text{ or } e \text{ lies in } S; \\ 1, & \text{if } e \text{ lies above } S \text{ and } e \cap S = p; \\ -1, & \text{if } e \text{ lies below } S \text{ and } e \cap S = p. \end{cases}$$

Since the surface S is oriented with normal n_a, "above" means $n_a \dot{e}^a|_p > 0$, and "below" means $n_a \dot{e}^a|_p < 0$, where $\dot{e}^a|_p$ is the tangent vector of e at p. And one should consider $e \cap S$ contained in the set $V(\gamma)$ and some edges are written as the union of elementary edges which either lie in S, or intersect S at their source or target. On the other hand, from the commutation relations for the left(right) invariant vector fields, one can see that the commutators between flux vector fields do not necessarily vanish when $S \cap S' \neq \emptyset$. This unusual property is the classical origin of the noncommutativity of quantum Riemannian structures [32].

The classical Ashtekar–Corichi–Zapata holonomy-flux ∗-algebra serves as a classical algebra of elementary observables in our dynamical system of gauge fields. Then one can construct the quantum algebra of elementary observables from $\mathfrak{P}_{ACZ}$ in analogy with Definition 3.1.2.

Definition 4.2 ([9]) The abstract free algebra $F(\mathfrak{P}_{ACZ})$ of the classical ∗-algebra is defined by the formal direct sum of finite sequences of classical observables $(p_1, ..., p_n)$ with $p_k \in \mathfrak{P}_{ACZ}$, where the operations of multiplication and involution are defined as

$$(p_1, ..., p_n) \cdot (p'_1, ..., p'_m) := (p_1, ..., p_n, p'_1, ..., p'_m),$$
$$(p_1, ..., p_n)^* := (\bar{p}_n, ..., \bar{p}_1).$$

A 2-sided ideal $\mathfrak{J}$ can be generated by the following elements,

$$(p+p') - (p) - (p'), \quad (zp) - z(p),$$
$$[(p), (p')] - i\hbar(\{p, p'\}),$$
$$((\psi, 0), p) - (\psi \circ p),$$

where the canonical commutation bracket is defined by

$$[(p), (p')] := (p) \cdot (p') - (p') \cdot (p).$$

Note that the ideal $\mathfrak{J}$ is preserved by the involution $*$, and the last set of generators in the ideal $\mathfrak{J}$ cancels the overcompleteness generated from the module structure of $\mathfrak{P}_{ACZ}$.

The quantum holonomy-flux $*$-algebra is defined by the quotient $*$-algebra $\mathfrak{A} = F(\mathfrak{P}_{ACZ})/\mathfrak{J}$, which contains the unital element $1 := ((1,0))$. Note that a sup-norm has been defined by equation (3.2) for the abelian sub-$*$-algebra $Cyl(\overline{\mathcal{A}})$ in $\mathfrak{A}$.

For simplicity, we denote the one element sequences (equivalence classes) $\widehat{((\psi,0))}$ and $\widehat{((0,Y))}$ $\forall\, \psi \in Cyl(\overline{\mathcal{A}})$, $Y \in \mathcal{V}^C(\overline{\mathcal{A}})$ in $\mathfrak{A}$ by $\hat{\psi}$ and $\hat{Y}$, respectively, where the "hat" denotes the equivalence class with respect to the quotient. In particular, for all cylindrical functions $\hat{\psi}$ and flux vector fields $\hat{Y}_f(S)$,

$$\hat{\psi}^* = \hat{\bar{\psi}} \quad \text{and} \quad \hat{Y}_f(S)^* = \hat{Y}_f(S).$$

It can be seen that the free algebra $F(\mathfrak{P}_{ACZ})$ is greatly simplified after the quotient, and every element of the quantum algebra $\mathfrak{A}$ can be written as a finite linear combination of elements of the form

$$\hat{\psi},$$
$$\hat{\psi}_1 \cdot \hat{Y}_{f_{11}}(S_{11}),$$
$$\hat{\psi}_2 \cdot \hat{Y}_{f_{21}}(S_{21}) \cdot \hat{Y}_{f_{22}}(S_{22}),$$
$$\ldots$$
$$\hat{\psi}_k \cdot \hat{Y}_{f_{k1}}(S_{k1}) \cdot \hat{Y}_{f_{k2}}(S_{k2}) \cdot \ldots \cdot \hat{Y}_{f_{kk}}(S_{kk}),$$
$$\ldots$$

Moreover, given a cylindrical function ψ and a flux vector field $Y_f(S)$, one has the following relation from the commutation relation:

$$\hat{Y}_f(S) \cdot \hat{\psi} = i\hbar \widehat{Y_f(S)\psi} + \hat{\psi} \cdot \hat{Y}_f(S). \tag{4.1}$$

Then the kinematical Hilbert space $\mathcal{H}_{LQG}$ can be properly obtained via the GNS-construction for the unital $*$-algebra $\mathfrak{A}$ in the same way as in Definition 3.1.3. By the GNS-construction, a positive linear functional, that is, a state ω_{kin}, on $\mathfrak{A}$ defines a cyclic representation $(\mathcal{H}_{LQG}, \pi_{kin}, \Omega_{kin})$ for $\mathfrak{A}$. In the case of the quantum holonomy-flux $*$-algebra, the state with both Yang–Mills gauge invariance and diffeomorphism invariance is defined for any $\psi_\gamma \in Cyl(\overline{\mathcal{A}})$ and nonvanishing flux vector field $Y_f(S) \in \mathcal{V}^C(\overline{\mathcal{A}})$ as [9]

$$\omega_{kin}(\hat{\psi}_\gamma) := \int_{SU(2)^{N_\gamma}} \prod_{e \in E(\gamma)} d\mu_H(A(e))\psi_\gamma(\{A(e)\}_{e \in E(\gamma)}),$$

$$\omega_{kin}(\hat{a} \cdot \hat{Y}_f(S)) := 0, \quad \forall \hat{a} \in \mathfrak{A},$$

where $d\mu_H$ is the Haar measure on the compact group SU(2) and N_γ is the number of elements in $E(\gamma)$. This ω_{kin} is called Ashtekar–Isham–Lewandowski state. The null space $\mathfrak{N}_{kin} \in \mathfrak{A}$ with respect to ω_{kin} is defined as $\mathfrak{N}_{kin} := \{\hat{a} \in \mathfrak{A} | \omega_{kin}(\hat{a}^* \cdot \hat{a}) = 0\}$, which is a left ideal. Then a quotient map can be defined as

$$[.]: \mathfrak{A} \to \mathfrak{A}/\mathfrak{N}_{kin};$$
$$\hat{a} \mapsto [\hat{a}] := \{\hat{a} + \hat{b} | \hat{b} \in \mathfrak{N}_{kin}\}.$$

The GNS-representation for $\mathfrak{A}$ with respect to ω_{kin} is a representation map: $\pi_{kin}: \mathfrak{A} \to \mathcal{L}(\mathcal{H}_{LQG})$ such that $\pi_{kin}(\hat{a} \cdot \hat{b}) = \pi_{kin}(\hat{a})\pi_{kin}(\hat{b})$, where $\mathcal{H}_{LQG} := \langle \mathfrak{A}/\mathfrak{N}_{kin} \rangle = \langle Cyl(\overline{\mathcal{A}}) \rangle$ by straightforward verification and the $\langle \cdot \rangle$ denotes the completion with respect to the natural equipped inner product on $\mathcal{H}_{LQG}$,

$$< [\hat{a}] | [\hat{b}] >_{kin} := \omega_{kin}(\hat{a}^* \cdot \hat{b}).$$

To show how this inner product works, given any two cylindrical functions $\psi = [\psi_\alpha], \psi' = [\psi'_{\alpha'}] \in Cyl(\overline{\mathcal{A}})$, the inner product between them is expressed as

$$< [\hat{\psi}] | [\hat{\psi}'] >_{kin} := \int_{X_{\alpha''}} (P^*_{\alpha''\alpha} \overline{\psi_\alpha})(P^*_{\alpha''\alpha'} \psi'_{\alpha'}) d\mu_{\alpha''}, \quad (4.2)$$

for any groupoid α'' containing both α and α'. The measure $d\mu_\alpha$ on X_α is defined by the pullback of the product Haar measure $d\mu_H^{N_\alpha}$ on the product group $SU(2)^{N_\alpha}$ via the identification bijection between X_α and $SU(2)^{N_\alpha}$, where N_α is the number of maximal analytic edges generating α. In addition, a nice result shows that given such a family of measures $\{\mu_\alpha\}_{\alpha \in \mathcal{L}}$, a probability measure μ is uniquely well defined on the quantum configuration space $\overline{\mathcal{A}}$ [30], such that the kinematical Hilbert space $\mathcal{H}_{LQG}$ coincides with the collection of the square-integrable functions with respect to the measure μ on the quantum configuration space, that is, $\mathcal{H}_{LQG} = L^2(\overline{\mathcal{A}}, d\mu)$, just as we expected at the beginning of our construction.

The representation map π_{kin} is defined by

$$\pi_{kin}(\hat{a})[\hat{b}] := [\hat{a} \cdot \hat{b}], \quad \forall \hat{a} \in \mathfrak{A}, \text{ and } [\hat{b}] \in \mathcal{H}_{LQG}.$$

Note that $\pi_{kin}(\hat{a})$ is an unbounded operator in general. It is easy to verify that

$$\pi_{kin}(\hat{Y}_f(S))[\hat{\psi}] = i\hbar [\widehat{Y_f(S)\psi}]$$

via equation (4.1), which gives the canonical momentum operator. In the following context, we denote the operator $\pi_{kin}(\hat{Y}_f(S))$ on $\mathcal{H}_{LQG}$ by $\hat{P}_f(S)$, and just denote the elements $[\hat{\psi}]$ in $\mathcal{H}_{LQG}$ by ψ for simplicity.

Moreover, since $\Omega_{kin} := 1$ is a cyclic vector in $\mathcal{H}_{LQG}$, the positive linear functional which we begin with can be expressed as

$$\omega_{kin}(\hat{a}) = <\Omega_{kin}|\pi_{kin}(\hat{a})\Omega_{kin}>_{kin}.$$

Thus the Ashtekar–Isham–Lewandowski state ω_{kin} on $\mathfrak{A}$ is equivalent to a cyclic representation $(\mathcal{H}_{LQG}, \pi_{kin}, \Omega_{kin})$ for $\mathfrak{A}$, which is the Ashtekar–Isham–Lewandowski representation for the quantum holonomy-flux $*$-algebra of a background independent gauge field theory. One thus obtains the kinematical representation of loop quantum gravity via the construction of an algebraic quantum field theory. It is important to note that the Ashtekar–Isham–Lewandowski state is the unique state on the quantum holonomy-flux $*$-algebra $\mathfrak{A}$ that is invariant under internal gauge transformations and spatial diffeomorphisms,[4] which are both automorphisms α_g and α_φ on $\mathfrak{A}$ and can be verified that $\omega_{kin} \circ \alpha_g = \omega_{kin}$ and $\omega_{kin} \circ \alpha_\varphi = \omega_{kin}$. So these gauge transformations are represented as unitary transformations on $\mathcal{H}_{LQG}$, while the cyclic vector Ω_{kin}, representing a "no-geometry vacuum" state, is the unique state in $\mathcal{H}_{LQG}$ invariant under internal gauge transformations and spatial diffeomorphisms. This is a very crucial uniqueness theorem for the canonical quantization of gauge field theory [9]:

Theorem 4.3 *There exists exactly one Yang–Mills gauge invariant and spatial diffeomorphism invariant state (positive linear functional) on the quantum holonomy-flux $*$-algebra. In other words, there exists a unique Yang–Mills gauge invariant and spatial diffeomorphism invariant cyclic representation for the quantum holonomy-flux $*$-algebra, which is called Ashtekar–Isham–Lewandowski representation. Moreover, this representation is irreducible with respect to an exponential version of the quantum holonomy-flux $*$-algebra (defined in [33]), which is analogous to the Weyl algebra.*

Hence we have finished the construction of the kinematical Hilbert space for a background independent gauge field theory and represented the quantum holonomy-flux $*$-algebra on it. Then following the general program presented in the last subsection, we should impose the constraints as operators on the kinematical Hilbert space since we are dealing with a gauge system.

5 Spin-Network Basis of Hilbert Space

The kinematical Hilbert space $\mathcal{H}_{LQG}$ for loop quantum gravity has been well defined. In this subsection, it will be shown that $\mathcal{H}_{LQG}$ can be decomposed into

---

[4] The proof of this conclusion depends on the compact support property of the smear functions f^i.

the orthogonal direct sum of one-dimensional subspaces and find a basis, called spin-network basis, in the Hilbert space, which consists of uncountably infinite elements. So the kinematic Hilbert space is nonseparable. In the following, we will show the decomposition in three steps.

- Spin-network decomposition on a single edge

  Given a graph consisting of only one edge e, naturally associated with a group $SU(2) = X_{\alpha(e)}$, the elements of $X_{\alpha(e)}$ are the quantum connections that take nontrivial values on e. Then we consider the decomposition of the Hilbert space $\mathcal{H}_{\alpha(e)} = L^2(X_{\alpha(e)}, d\mu_{\alpha(e)}) \simeq L^2(SU(2), d\mu_H)$, which is nothing but the space of square integrable functions on the compact group $SU(2)$ with the natural L^2 inner product. It is natural to define several operators on $\mathcal{H}_{\alpha(e)}$. First, the so-called configuration operator $\hat{f}(A(e))$ whose operation on any ψ in a dense domain of $L^2(SU(2), d\mu_H)$ is nothing but multiplication by the function $f(A(e))$, that is,

  $$\hat{f}(A(e))\psi(A(e)) := f(A(e))\psi(A(e)),$$

  where $A(e) \in SU(2)$. Second, given any vector $\xi \in su(2)$, it generates a left invariant vector field $L^{(\xi)}$ and a right invariant vector field $R^{(\xi)}$ on $SU(2)$ by

  $$L^{(\xi)}\psi(A(e)) := \frac{d}{dt}|_{t=0}\psi(A(e)\exp(t\xi)),$$
  $$R^{(\xi)}\psi(A(e)) := \frac{d}{dt}|_{t=0}\psi(\exp(-t\xi)A(e)),$$

  for any function $\psi \in C^1(SU(2))$. Then one can define the so-called momentum operators on the single edge by

  $$\hat{J}_i^{(L)} = iL^{(\tau_i)} \quad \text{and} \quad \hat{J}_i^{(R)} = iR^{(\tau_i)},$$

  where the generators $\tau_i \in su(2)$ constitute an orthonormal basis with respect to the Killing–Cartan metric. The momentum operators have the well-known commutation relations of the angular momentum operators in quantum mechanics:

  $$[\hat{J}_i^{(L)}, \hat{J}_j^{(L)}] = i\epsilon_{ij}^{k}\hat{J}_k^{(L)}, \ [\hat{J}_i^{(R)}, \hat{J}_j^{(R)}] = i\epsilon_{ij}^{k}\hat{J}_k^{(R)}, \ [\hat{J}_i^{(L)}, \hat{J}_j^{(R)}] = 0.$$

  Third, the Casimir operator on $\mathcal{H}_e$ can be expressed as

  $$\hat{J}^2 := \delta^{ij}\hat{J}_i^{(L)}\hat{J}_j^{(L)} = \delta^{ij}\hat{J}_i^{(R)}\hat{J}_j^{(R)}. \tag{5.1}$$

  The decomposition of $\mathcal{H}_e = L^2(SU(2), d\mu_H)$ is provided by the following Peter–Weyl Theorem:

Theorem 5.1 (**Peter–Weyl**) *Given a compact group G, the function space $L^2(G, d\mu_H)$ can be decomposed as an orthogonal direct sum of finite-dimensional Hilbert spaces, and the matrix elements of the equivalence classes of finite-dimensional irreducible representations of G form an orthogonal basis in $L^2(G, d\mu_H)$.*

Note that a finite-dimensional irreducible representation of G can be regarded as a matrix-valued function on G. Then the matrix elements are functions on G. Using this theorem, one can find the decomposition of the Hilbert space:

$$L^2(SU(2), d\mu_H) = \oplus_j [\mathcal{H}_j \otimes \mathcal{H}_j^*],$$

where j, labeling irreducible representations of $SU(2)$, are half integers, $\mathcal{H}_j$ denotes the carrier space of the j-representation of dimension $2j+1$, and $\mathcal{H}_j^*$ is its dual space. The basis $\{\mathbf{e}_m^j \otimes \mathbf{e}_n^{j*}\}$ in $\mathcal{H}_j \otimes \mathcal{H}_j^*$ maps a group element $g \in SU(2)$ to a matrix $\{\pi_{mn}^j(g)\}$, where $m, n = -j, ..., j$. Thus the space $\mathcal{H}_j \otimes \mathcal{H}_j^*$ is spanned by the matrix element functions π_{mn}^j of equivalent j-representations. Moreover, the spin-network basis can be defined.

Proposition 5.2 *The system of spin-network functions on $\mathcal{H}_{\alpha(e)}$, consisting of matrix elements $\{\pi_{mn}^j\}$ in finite-dimensional irreducible representations labeled by half-integers $\{j\}$, satisfies*

$$\hat{J}^2 \pi_{mn}^j = j(j+1)\pi_{mn}^j, \quad \hat{J}_3^{(L)} \pi_{mn}^j = m\pi_{mn}^j, \quad \hat{J}_3^{(R)} \pi_{mn}^j = n\pi_{mn}^j,$$

where j is called angular momentum quantum number and $m, n = -j, ..., j$ magnetic quantum number. The normalized functions $\{\sqrt{2j+1}\pi_{mn}^j\}$ form an orthonormal basis in $\mathcal{H}_{\alpha(e)}$ by the above theorem and

$$\int_{\overline{\mathcal{A}_e}} \overline{\pi_{m'n'}^{j'}} \pi_{mn}^j d\mu_e = \frac{1}{2j+1} \delta^{j'j} \delta_{m'm} \delta_{n'n},$$

which is called the spin-network basis on $\mathcal{H}_{\alpha(e)}$. So the Hilbert space on a single edge has been decomposed into one-dimensional subspaces.

Note that the system of operators $\{\hat{J}^2, \hat{J}_3^{(R)}, \hat{J}_3^{(L)}\}$ forms a complete set of commuting operators in $\mathcal{H}_{\alpha(e)}$. There is a cyclic "vacuum state" in the Hilbert space, which is the $(j=0)$-representation $\Omega_{\alpha(e)} = \pi^{j=0} = 1$, representing that there is no geometry on the edge.

- Spin-network decomposition on a finite graph

  Given a groupoid α generated by a graph γ with N oriented edges e_i and

M vertices, one can define the configuration operators on the corresponding Hilbert space $\mathcal{H}_\alpha = L^2(X_\alpha, d\mu_\alpha) \simeq L^2(SU(2)^N, d\mu_H^N)$ by

$$\hat{f}(A(e_i))\psi(A(e_1), ..., A(e_N)) := f(A(e_i))\psi(A(e_1), ..., A(e_N)).$$

The momentum operators $\hat{J}_i^{(e,v)}$ associated with an edge e connecting a vertex v are defined as

$$\hat{J}_i^{(e,v)} := (1 \otimes ... \otimes \hat{J}_i \otimes ... \otimes 1),$$

where $\hat{J}_i = \hat{J}_i^{(L)}$ if $v = s(e)$ and $\hat{J}_i = \hat{J}_i^{(R)}$ if $v = t(e)$, so $\hat{J}_i^{(e,v)} = iX_i^{(e,v)}$. Note that $\hat{J}_i^{(e,v)}$ only acts nontrivially on the Hilbert space associated with the edge e. Then one can define a vertex operator associated with the vertex v in analogy with the total angular momentum operator via

$$[\hat{J}^v]^2 := \delta^{ij}\hat{J}_i^v \hat{J}_j^v,$$

where

$$\hat{J}_i^v := \sum_{e \text{ at } v} \hat{J}_i^{(e,v)}.$$

Obviously, $\mathcal{H}_\alpha$ can be firstly decomposed by the representations on each edge e of α as

$$\mathcal{H}_\alpha = \otimes_e \mathcal{H}_{\alpha(e)} = \otimes_e [\oplus_j (\mathcal{H}_j^e \otimes \mathcal{H}_j^{e*})] = \oplus_\mathbf{j} [\otimes_e (\mathcal{H}_j^e \otimes \mathcal{H}_j^{e*})]$$
$$= \oplus_\mathbf{j} [\otimes_v (\mathcal{H}_{\mathbf{j}(s)}^{v=s(e)} \otimes \mathcal{H}_{\mathbf{j}(t)}^{v=t(e)})],$$

where $\mathbf{j} := (j_1, ..., j_N)$ assigns to each edge an irreducible representation of $SU(2)$. In the fourth step the Hilbert spaces associated with the edges are allocated to the vertices where these edges meet so that for each vertex v,

$$\mathcal{H}_{\mathbf{j}(s)}^{v=s(e)} \equiv \otimes_{s(e)=v} \mathcal{H}_j^e \quad \text{and} \quad \mathcal{H}_{\mathbf{j}(t)}^{v=t(e)} \equiv \otimes_{t(e)=v} \mathcal{H}_j^{e*}.$$

The group of gauge transformations $g(v) \in SU(2)$ at each vertex is reducibly represented on the Hilbert space $\mathcal{H}_{\mathbf{j}(s)}^{v=s(e)} \otimes \mathcal{H}_{\mathbf{j}(t)}^{v=t(e)}$ in a natural way. So this Hilbert space can be decomposed as a direct sum of irreducible representation spaces via the Clebsch–Gordon decomposition:

$$\mathcal{H}_{\mathbf{j}(s)}^{v=s(e)} \otimes \mathcal{H}_{\mathbf{j}(t)}^{v=t(e)} = \oplus_l \mathcal{H}_{\mathbf{j}(v),l}^v.$$

As a result, $\mathcal{H}_\alpha$ can be further decomposed as

$$\mathcal{H}_\alpha = \oplus_\mathbf{j}[\otimes_v (\oplus_l \mathcal{H}_{\mathbf{j}(v),l}^v)] = \oplus_\mathbf{j}[\oplus_\mathbf{l}(\otimes_v \mathcal{H}_{\mathbf{j}(v),l}^v)] \equiv \oplus_\mathbf{j}[\oplus_\mathbf{l}\mathcal{H}_{\alpha,\mathbf{j},\mathbf{l}}]. \quad (5.2)$$

It can also be viewed as the eigenvector space decomposition of the commuting operators $[\hat{J}^v]^2$ (with eigenvalues $l(l+1)$) and $[\hat{J}^e]^2 \equiv \delta^{ij}\hat{J}_i^e \hat{J}_j^e$. Note that $\mathbf{l} := (l_1, ..., l_M)$ assigns to each vertex (objective) of α an irreducible representation of SU(2). One may also enlarge the set of commuting operators to further refine the decomposition of the Hilbert space. Note that the subspace of $\mathcal{H}_\alpha$ with $\mathbf{l} = 0$ is Yang–Mills gauge invariant, since the representation of gauge transformations is trivial.

- Spin-network decomposition of $\mathcal{H}_{LQG}$

  Since $\mathcal{H}_{LQG}$ has the structure $\mathcal{H}_{LQG} = \langle \bigcup_{\alpha \in \mathcal{L}} \mathcal{H}_\alpha \rangle$, one may consider constructing it as a direct sum of $\mathcal{H}_\alpha$ by canceling some overlapping components. The construction is precisely described as a theorem below.

Theorem 5.3 *Consider assignments $\mathbf{j} = (j_1, ..., j_N)$ to the edges of any groupoid $\alpha \in \mathcal{L}$ and assignments $\mathbf{l} = (l_1, ..., l_M)$ to the vertices. The edge representation j is nontrivial on each edge, and the vertex representation l is nontrivial at each spurious[5] vertex, unless it is the base point of a close analytic loop. Let $\mathcal{H}'_\alpha$ be the Hilbert space composed by the subspaces $\mathcal{H}_{\alpha,\mathbf{j},\mathbf{l}}$ (assigned the above conditions) according to equation (5.2). Then $\mathcal{H}_{LQG}$ can be decomposed as the direct sum of the Hilbert spaces $\mathcal{H}'_\alpha$, that is,*

$$\mathcal{H}_{LQG} = \oplus_{\alpha \in \mathcal{L}} \mathcal{H}'_\alpha \oplus \mathbf{C}.$$

Proof: Since the representation on each edge is nontrivial, by definition of the inner product, it is easy to see that $\mathcal{H}'_\alpha$ and $\mathcal{H}'_{\alpha'}$ are mutual orthogonal if one of the groupoids α and α' has at least an edge e more than the other due to

$$\int_{\overline{\mathcal{A}}_e} \pi^j_{mn} d\mu_e = \int_{\overline{\mathcal{A}}_e} 1 \cdot \pi^j_{mn} d\mu_e = 0$$

for any $j \neq 0$. Now consider the case of the spurious vertex. The Hilbert space $\mathcal{H}^e_j \otimes \mathcal{H}^{e*}_j$ is assigned to an edge e with j-representation in a graph. Inserting a vertex v into the edge, one obtains two edges e_1 and e_2 split by v both with j-representations, which belong to a different graph. By the decomposition of the corresponding Hilbert space,

$$\mathcal{H}^{e_1}_j \otimes \mathcal{H}^{e_1*}_j \otimes \mathcal{H}^{e_2}_j \otimes \mathcal{H}^{e_2*}_j = \mathcal{H}^{e_1}_j \otimes (\oplus_{l=0...2j} \mathcal{H}^v_l) \otimes \mathcal{H}^{e_2*}_j,$$

the subspace for all $l \neq 0$ are orthogonal to the space $\mathcal{H}^e_j \otimes \mathcal{H}^{e*}_j$, while the subspace for $l = 0$ coincides with $\mathcal{H}^e_j \otimes \mathcal{H}^{e*}_j$ since $\mathcal{H}^v_{l=0} = \mathbf{C}$ and $A(e) = A(e_1)A(e_2)$. This completes the proof.

Since there are uncountably many graphs on Σ, the kinematical Hilbert $\mathcal{H}_{LQG}$ is nonseparable. We denote the spin-network basis in $\mathcal{H}_{LQG}$ by

---

[5] A vertex v is spurious if it is bivalent and $e \circ e'$ is itself an analytic edge with e, e' meeting at v.

Π_s, $s = (\gamma(s), \mathbf{j}_s, \mathbf{m}_s, \mathbf{n}_s)$ and vacuum $\Omega_{kin} \equiv \Pi_0 = 1$, where

$$\Pi_s := \prod_{e \in E(\gamma(s))} \sqrt{2j_e + 1} \pi^{j_e}_{m_e n_e} \quad (j_e \neq 0), \tag{5.3}$$

which form an orthonormal basis with the relation $< \Pi_s | \Pi_{s'} >_{kin} = \delta_{ss'}$. And $Cyl_\gamma(\overline{\mathcal{A}}) \subset Cyl(\overline{\mathcal{A}})$ denotes the linear span of the spin-network functions Π_s for $\gamma(s) = \gamma$. ∎

The spin-network basis can be used to construct the so-called spin-network representation of loop quantum gravity.

Definition 5.4 The spin-network representation is a vector space $\widetilde{\mathcal{H}}$ of complex valued functions

$$\widetilde{\Psi} : S \to \mathbf{C}; \ s \mapsto \widetilde{\Psi}(s),$$

where S is the set of the labels s for the spin-network states. $\widetilde{\mathcal{H}}$ is equipped with the scalar product

$$< \widetilde{\Psi}, \widetilde{\Psi}' > := \sum_{s \in S} \overline{\widetilde{\Psi}(s)} \widetilde{\Psi}(s)'$$

between square summable functions.

The relation between the Hilbert spaces $\widetilde{\mathcal{H}}$ and $\mathcal{H}_{LQG}$ is clarified by the following proposition:

Proposition 5.5 *The spin-network transformation*

$$T : \mathcal{H}_{LQG} \to \widetilde{\mathcal{H}}; \ \Psi \mapsto \widetilde{\Psi}(s) := < \Pi_s, \Psi >_{kin}$$

is a unitary transformation with inverse

$$T^{-1}\Psi = \sum_{s \in S} \widetilde{\Psi}(s) \Pi_s.$$

Thus the connection representation and the spin-network representation are "Fourier transforms" of each other, where the role of the kernel of the integral is played by the spin-network basis. Note that, in the gauge invariant Hilbert space of loop quantum gravity which we will define later, the Fourier transform with respect to the gauge invariant spin-network basis is the so-called loop transform, which leads to the unitary equivalent loop representation of the theory [34, 35, 36].

To conclude this subsection, we show the explicit representation of elementary observables on the kinematical Hilbert space $\mathcal{H}_{LQG}$. The action of the canonical momentum operator $\hat{P}_f(S)$ on differentiable cylindrical functions $\psi_\gamma \in Cyl_\gamma(\overline{\mathcal{A}})$ can be expressed as

$$\hat{P}_f(S)\psi_\gamma(\{A(e)\}_{e\in E(\gamma)})$$
$$= \frac{\hbar}{2} \sum_{v\in V(\gamma)\cap S} f^i(v)[\sum_{e\ at\ v} \kappa(S,e)\hat{J}_i^{(e,v)}]\psi_\gamma(\{A(e)\}_{e\in E(\gamma)})$$
$$= \frac{\hbar}{2} \sum_{v\in V(\gamma)\cap S} f^i(v)[\hat{J}_{i(u)}^{(S,v)} - \hat{J}_{i(d)}^{(S,v)}]\psi_\gamma(\{A(e)\}_{e\in E(\gamma)}), \quad (5.4)$$

where

$$\hat{J}_{i(u)}^{(S,v)} \equiv \hat{J}_i^{(e_1,v)} + ... + \hat{J}_i^{(e_u,v)},$$
$$\hat{J}_{i(d)}^{(S,v)} \equiv \hat{J}_i^{(e_{u+1},v)} + ... + \hat{J}_i^{(e_{u+d},v)}, \quad (5.5)$$

for the edges $e_1,...,e_u$ lying above S and $e_{u+1},...,e_{u+d}$ lying below S. And it was proved that the operator $\hat{P}_f(S)$ is essentially self-adjoint on $\mathcal{H}_{LQG}$ [4]. On the other hand, it is obvious how to construct configuration operators by spin-network functions:

$$\hat{\Pi}_s\psi_\gamma(\{A(e)\}_{e\in E(\gamma)}) := \Pi_s(\{A(e)\}_{e\in E(\gamma(s))})\psi_\gamma(\{A(e)\}_{e\in E(\gamma)}).$$

Note that $\hat{\Pi}_s$ may change the graph, that is, $\hat{\Pi}_s: Cyl_\gamma(\overline{\mathcal{A}}) \to Cyl_{\gamma\cup\gamma(s)}(\overline{\mathcal{A}})$. So far, the elementary operators of quantum kinematics have been well defined on $\mathcal{H}_{LQG}$.

6 Geometrical Operators

The well-established quantum kinematics of loop quantum gravity is now in the same status as Riemannian geometry before the appearance of general relativity and Einstein's equations, giving general relativity a mathematical foundation and offering a place to live to Einstein's equations. Instead of classical geometric quantities, such as length, area, and volume, the quantities in quantum geometry are operators on the kinematical Hilbert space $\mathcal{H}_{LQG}$, and their spectra serve as the possible values of the quantities in measurements. So far, the kinematical quantum geometric operators constructed properly in loop quantum gravity include the length operator [13, 14, 15], the area operator [10, 11], the volume operator [10, 12], and the $\hat{Q}$ operator [16].

First, we define the area operator with respect to a 2-surface S by the elementary operators. Given a closed 2-surface or a surface S with boundary, we can divide it into a large number N of small area cells S_I. Taking into account the classical expression of an area, we set the area of the 2-surface to be the limit of the Riemannian sum

$$\mathbf{Ar}_S := \lim_{N\to\infty} [\mathbf{Ar}_S]_N = \lim_{N\to\infty} \kappa\beta \sum_{I=1}^{N} \sqrt{P_i(S_I)P_j(S_I)\delta^{ij}}.$$

Then one can unambiguously obtain a quantum area operator from the canonical momentum operators $\hat{P}_i(S)$ smeared by constant functions. Given a cylindrical function $\psi_\gamma \in Cyl_\gamma(\overline{\mathcal{A}})$ which has second-order derivatives, the action of the area operator on ψ_γ is defined in the limit by requiring that each area cell contains at most only one intersecting point v of the graph γ and S as

$$\hat{\mathbf{Ar}}_S \psi_\gamma := \lim_{N\to\infty} [\hat{\mathbf{Ar}}_S]_N \psi_\gamma = \lim_{N\to\infty} \kappa\beta \sum_{I=1}^{N} \sqrt{\hat{P}_i(S_I)\hat{P}_j(S_I)\delta^{ij}} \, \psi_\gamma.$$

The regulator N is easy to remove, since the result of the operation of the operator $\hat{P}_i(S_I)$ does not change when S_I shrinks to a point. Since the refinement of the partition does not affect the result of the action of $[\hat{\mathbf{Ar}}_S]_N$ on ψ_γ, the limit area operator $\hat{\mathbf{Ar}}_S$, which is shown to be self-adjoint [11], is well defined on $\mathcal{H}_{LQG}$ and takes the following explicit expression:

$$\hat{\mathbf{Ar}}_S \psi_\gamma = 4\pi\beta\ell_p^2 \sum_{v \in V(\gamma \cap S)} \sqrt{(\hat{J}_{i(u)}^{(S,v)} - \hat{J}_{i(d)}^{(S,v)})(\hat{J}_{j(u)}^{(S,v)} - \hat{J}_{j(d)}^{(S,v)})\delta^{ij}} \, \psi_\gamma,$$

where $\hat{J}_{i(u)}^{(S,v)}$ and $\hat{J}_{i(d)}^{(S,v)}$ have been defined in equation (5.5). It turns out that the finite linear combinations of spin-network basis in $\mathcal{H}_{LQG}$ diagonalize $\hat{\mathbf{Ar}}_S$ with eigenvalues given by finite sums,

$$a_S = 4\pi\beta\ell_p^2 \sum_v \sqrt{2j_v^{(u)}(j_v^{(u)}+1) + 2j_v^{(d)}(j_v^{(d)}+1) - j_v^{(u+d)}(j_v^{(u+d)}+1)}, \tag{6.1}$$

where $j^{(u)}, j^{(d)}$ and $j^{(u+d)}$ are arbitrary half-integers subject to the standard condition

$$j^{(u+d)} \in \{|j^{(u)} - j^{(d)}|, |j^{(u)} - j^{(d)}| + 1, ..., j^{(u)} + j^{(d)}\}. \tag{6.2}$$

Hence the spectrum of the area operator is fundamentally pure discrete, while its continuum approximation becomes excellent exponentially rapidly for large eigenvalues. However, at the fundamental level, the area is discrete and so is the quantum geometry. One can see that the eigenvalues of $\hat{\mathbf{Ar}}_S$ do not vanish even in the case where only one edge intersects the surface at a single point, whence the quantum geometry is distributional.

The volume operator in LQG is defined as follows. Given a region R with a fixed coordinate system $\{x^a\}_{a=1,2,3}$ in it, one can introduce a partition of R in the following way. Divide R into small volume cells C such that each cell C is a cube with coordinate volume less than ϵ and two different cells only share the points on their boundaries. In each cell C, we introduce three 2-surfaces $s = (S^1, S^2, S^3)$ such that x^a is constant on the surface S^a. We denote this partition (C, s) as $\mathcal{P}_\epsilon$. Then the volume of the region R can be expressed classically as

$$V_R^s = \lim_{\epsilon \to 0} \sum_C \sqrt{|q_{C,s}|},$$

where

$$q_{C,s} = \frac{(\kappa \beta)^3}{3!} \epsilon^{ijk} \underline{\eta}_{abc} P_i(S^a) P_j(S^b) P_k(S^c).$$

This motivates us to define the volume operator by naively changing $P_i(S^a)$ to $\hat{P}_i(S^a)$:

$$\hat{V}_R^s = \lim_{\epsilon \to 0} \sum_C \sqrt{|\hat{q}_{C,s}|},$$

$$\hat{q}_{C,s} = \frac{(\kappa \beta)^3}{3!} \epsilon^{ijk} \underline{\eta}_{abc} \hat{P}_i(S^a) \hat{P}_j(S^b) \hat{P}_k(S^c).$$

Note that, given any cylindrical function $\psi_\gamma \in Cyl_\gamma(\overline{\mathcal{A}})$, we require the vertices of the graph γ to be at the intersecting points of the triples of 2-surfaces $s = (S^1, S^2, S^3)$ in the corresponding cells. Thus the limit operator will trivially exist for the same reason in the case of the area operator. However, the volume operator defined here depends on the choice of orientations for the triples of surfaces $s = (S^1, S^2, S^3)$, or essentially, the choice of coordinate systems. So it is not uniquely defined. Since the resulting operators have the correct semi-classical limit for all choices of $s = (S^1, S^2, S^3)$, one settles up the problem by averaging different operators labeled by different s [12]. The process of averaging removes the freedom in defining the volume operator up to an overall constant κ_0. The resulting self-adjoint operator acts on any cylindrical function $\psi_\gamma \in Cyl_\gamma(\overline{\mathcal{A}})$ as

$$\hat{V}_R \psi_\gamma = \kappa_0 \sum_{v \in V(\alpha)} \sqrt{|\hat{q}_{v,\gamma}|} \psi_\gamma,$$

where

$$\hat{q}_{v,\gamma} = (8\pi \beta \ell_p^2)^3 \frac{1}{48} \sum_{e,e',e'' \text{ at } v} \epsilon^{ijk} \epsilon(e, e', e'') \hat{J}_i^{(e,v)} \hat{J}_j^{(e',v)} \hat{J}_k^{(e'',v)},$$

here $\epsilon(e, e', e'') \equiv \mathrm{sgn}(\epsilon_{abc}\dot{e}^a \dot{e}'^b \dot{e}''^c)|_v$ where $\dot{e}^a$ is the tangent vector of edge e and ϵ_{abc} is the orientation of Σ. The only unsatisfactory point in the present volume operator is the ambiguity in the choice of κ_0. However, fortunately, the most recent discussion shows that the overall undetermined constant κ_0 can be fixed to be $\sqrt{6}$ by the consistency check between the volume operator and the triad operator [37, 38].

7 Gauss Constraint, Gauge Invariance, and Closure

The Hilbert space $\mathcal{H}_{LQG}$ contains states which are not invariant under SU(2) gauge transformations. It is obvious from the expression of spin-network states Π_s in equation (5.3) that each Π_s is not invariant under gauge transformations $A(e) \mapsto g(t(e))^{-1}A(e)g(s(e))$. In order to project down to a subspace $\mathcal{H}^G \subset \mathcal{H}_{LQG}$ of gauge invariant states we have to solve the quantum Gauss constraint.

The classical Gauss constraint is expressed in terms of the canonical variables and a smear function $\mathrm{L}^i(x)$ as

$$\mathcal{G}(\Lambda) = \int_\Sigma d^3 x \, \Lambda^i D_a E^a_i = -\int_\Sigma d^3 x \, E^a_i D_a \Lambda^i \equiv -P(D\Lambda),$$

where $D_a \Lambda^i = \partial_a \Lambda^i + \epsilon^i{}_{jk} A^j_a \Lambda^k$. The Hamiltonian flow of the Gauss constraint generates the SU(2) gauge transformations of the canonical variables.

As in the situation of the triad flux, the Gauss constraint can be defined as cylindrically consistent vector fields $Y_{D\Lambda}$ on $\overline{\mathcal{A}}$, which act on any cylindrical function $f_\gamma \in Cyl_\gamma(\overline{\mathcal{A}})$ by

$$Y_{D\Lambda} \circ f_\gamma(\{A(e)\}_{e \in E(\gamma)}) := \{-P(D\Lambda), f_\gamma(\{A(e)\}_{e \in E(\gamma)})\}.$$

Then the Gauss constraint operator can be defined in analogy with the momentum operator, which acts on f_γ as

$$\hat{\mathcal{G}}(\Lambda) f_\gamma(\{A(e)\}_{e \in E(\gamma)}) := i\hbar Y_{D\Lambda} f_\gamma(\{A(e)\}_{e \in E(\gamma)})$$
$$= \hbar \sum_{v \in V(\gamma)} [\Lambda^i(v) \hat{J}^v_i] f(\{A(e)\}_{e \in E(\gamma)}), \quad (7.1)$$

which is the generator of SU(2) gauge transformations on $Cyl_\gamma(\overline{\mathcal{A}})$. The kernel of the operator is easily obtained in terms of the spin-network decomposition, which is the internal gauge invariant Hilbert space:

$$\mathcal{H}^G = \oplus_{\alpha, \mathbf{j}} \mathcal{H}'_{\alpha, \mathbf{j}, \mathbf{l}=0} \oplus \mathbb{C}.$$

The gauge invariant spin-network basis T_s, $s = (\gamma(s), \mathbf{j}_s, \mathbf{i}_s)$ in $\mathcal{H}^G$ is obtained as

$$T_{s=(\gamma,\mathbf{j},\mathbf{i})} = \bigotimes_{v \in V(\gamma)} i_v \bigotimes_{e \in E(\gamma)} \pi^{j_e}(A(e)), \quad (j_e \neq 0),$$

which assigns a nontrivial spin representation j on each edge and an invariant tensor i (intertwiner) at each vertex. We denote the vector space of finite linear combinations of the vacuum state and gauge invariant spin-network states $Cyl(\overline{\mathcal{A} \propto \mathcal{G}})$, which is dense in $\mathcal{H}^G$. And $Cyl_\gamma(\overline{\mathcal{A} \propto \mathcal{G}}) \subset Cyl(\overline{\mathcal{A} \propto \mathcal{G}})$ denotes the linear span of the gauge invariant spin-network functions T_s for $\gamma(s) = \gamma$. All SU(2) gauge invariant operators are well defined on $\mathcal{H}^G$.

At each vertex v, the quantum Gauss constraint equation (7.1) reduces to

$$\sum_{e \text{ at } v} \widehat{\vec{\mathcal{J}}^{(e)}} |i_v\rangle = 0, \tag{7.2}$$

where $|i_v\rangle$ is the invariant tensor located at v. The above constraint equation is the quantum analog of a classical constraint equation in describing polyhedron geometries by using area vectors. Consider a classical flat polyhedron with n faces. We define the area vector $\vec{J}_f$ for each face f by requiring that $|\vec{J}_f|$ is the area of the face and that $\vec{J}_f/|\vec{J}_f|$ is the unit normal vector to f. Because all the faces of the polyhedron form a closed surface, all the area vectors satisfy a closure constraint

$$\sum_f \vec{J}_f = 0. \tag{7.3}$$

It is clear that equation (7.2) is the quantum analog of the closure constraint, by promoting each $\vec{J}_f$ to be operators, and understanding that each polyhedron face f is dual to an edge e (recall Figure 8.1). The quantization condition imposed by LQG requires that the components of $\vec{J}_f$ are not commutative but satisfy $[\hat{J}_f^k, \hat{J}_f^l] = i\beta \ell_P^2 \varepsilon^{klm} \hat{J}_f^m$.

The set of classical data $\vec{J}_f$ subject to the closure constraint determines uniquely the polyhedron geometry, which is known as the Minkowski Theorem [39]. The quantum constraint equation (7.2) implies that the invariant tensor $i_v \in \text{Inv}(\otimes_{e \text{ at } v} \mathcal{H}_{j_e})$ quantizes the space of shapes of a polyhedron with given face areas $8\pi\beta\ell_P^2\sqrt{j_e(j_e+1)}$ (see, e.g., [25, 26, 40] for discussions about quantizing the space of polyhedron shapes). Namely, i_v is a quantum number parameterizing the shapes of a quantum polyhedron with a given set of face areas.

By the above geometrical interpretation of the invariant tensor i_v, a gauge invariant spin-network state T_s can be understood as a quantum gluing of a set of quantum polyhedra, each of which encloses a vertex v of the graph γ. Each edge e of γ connecting 2 vertices is dual to a face shared by a pair of polyhedra. For each quantum polyhedron, the spins j_e parameterize face areas, and the invariant tensor i_v parameterizes the polyhedron shape. Quantum gluing a pair of polyhedra only requires the matching of face areas j_e at the interface, but does not require that the shapes of the pair of faces also match. The semiclassical approximation of this type of discrete geometry is known as twisted geometry [41, 42, 43, 44].[6]

8 Conclusion and Outlook

In this chapter we have obtained the quantum geometry of 3d space from a quantization of the phase space of 4d gravity. The formulation of quantum geometry, including the Hilbert space and the operators, was obtained from the representation theory of the holonomy-flux $*$-algebra of LQG. The geometrical quantities of the space, including areas and volumes, were promoted to operators of the theory. The area and volume operators are featured with their discrete spectra, which implies that the area and volume only take discrete values at the Planck scale.

After understanding the quantum geometry of space, a step further is to understand the quantum geometry of spacetime in four dimensions. As it was mentioned at the beginning of the chapter, the spacetime geometry can be viewed as a dynamical history of space geometries. Quantum spacetime is thus the dynamics of quantum space geometries. If we take the conventional point of view that the space geometry is the 3d spatial metric h_{ab}, while the spacetime geometry is the 4d metric $g_{\mu\nu}$, then the quantum spacetime may be described by a path integral on a 4d-manifold (with boundary)

$$Z = \int_{h_{ab}^{out}}^{h_{ab}^{in}} Dg_{\mu\nu}\, e^{i S[g_{\mu\nu}]}. \tag{8.1}$$

The boundary of the path integral is given by two spatial slices $\Sigma^{out}, \Sigma^{in}$, whose spatial geometries $h_{ab}^{out}, h_{ab}^{in}$ are the boundary data of the path integral (Figure 8.2). The above path integral sums all spacetime geometries $g_{\mu\nu}$ developing the histories from h_{ab}^{in} to h_{ab}^{out}. Z, as a function of the boundary data, is understood as a transition amplitude between two spatial geometries. This transition amplitude is the quantum spacetime.

---

[6] The twisted geometry mentioned here is made by flat polyhedra. It has been generalized in [45, 46] to define twisted geometries using constant curvature tetrahedra.

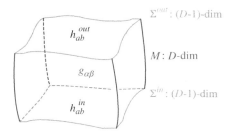

Figure 8.2 D-metric as a history of $(D-1)$-metrics.

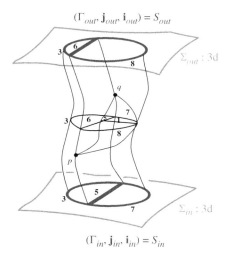

Figure 8.3 An example of a spinfoam as an evolution history of spin-networks. A link in the spin-network evolves and creates a surface in the spinfoam, and a node in the spin-network evolves and creates an edge in the spinfoam. If we imagine that the spinfoam is embedded in a 4-manifold, any hypersurface transverse to the spinfoam edges intersects the spinfoam and gives a spin-network as the intersection.

In LQG, 3d geometries are replaced by the quantum 3d geometries described in the main part of the chapter. The data h_{ab} should be replaced by its quantum analog, namely, the spin-network $s = (\gamma, \mathbf{j}, \mathbf{i})$. The quantum spacetime geometry in 4d is a transition amplitude obtained by summing over all the histories of spin-networks. Histories of spin-networks are known as *spinfoams* (Figure 8.3). Spinfoams are based on cellular decompositions of a 4-manifold whose boundary is given by spin-network graphs. In LQG, the amplitude Z is known as the *spinfoam amplitude*. There have been many recent developments of spinfoam models in which the spinfoam amplitude can be computed (e.g., [47, 49, 50, 51, 52, 53, 54, 118]). The semiclassical limit of the amplitude has been shown to reproduce all simplicial geometries and the discrete Einstein–Hilbert action (Regge action) in 4d [55, 56, 57, 58, 59]. Notions of classical spacetime such as distance, locality, and causality can be recovered in the semiclassical analysis. The formalism has been extended to include a cosmological constant (e.g., [60, 61, 62, 63, 64, 65, 66]) and matter (e.g., [67, 68, 69]). The continuum limit of the spinfoam can be taken at the semiclassical level, which reproduces 4d smooth spacetimes and the Einstein's equations [70]. The spinfoam is viewed as a promising formalism given its success in reproducing correct semiclassical behavior. The semiclassical reconstruction of spacetime geometries realizes spinfoam LQG as a program of emergent spacetimes.

The spinfoam amplitudes are expected to solve the quantum Hamiltonian constraint in LQG [71, 72]. The Hamiltonian constraint operator has been extensively studied in LQG (e.g., [73, 74, 75]; see [4] for a review). As a feature of a background independent and diffeomorphism invariant theory, the Hamiltonian of gravity is a linear combination of constraints. The time evolution of gravity has no distinction with the gauge transformations (diffeomorphisms), which is known as "the problem of time." The quantum dynamics of LQG is governed by the quantum Hamiltonian operator which should generate the diffeomorphisms in the time direction. The Dirac observables in principle should be gauge invariant, that is, commute with the Hamiltonian constraint and other constraint operators. At first sight, it seems highly nontrivial to construct the Dirac observable for any physical quantity, given the complexity of the Hamiltonian constraint. The area and volume operators discussed above do not commute with the Hamiltonian constraint. However there is a "relational formalism" which can upgrade any non-gauge-invariant kinematical geometric operator to a physical Dirac operator [34, 76, 77]. This formalism is expected to provide all Dirac operators in LQG.

The dynamical theory of quantum geometries is not yet complete at the moment. One of the important open research directions is to understand whether the discrete set up of quantum spacetime (spinfoam) is fundamental, or there should be a continuum theory behind the spinfoam formulation, where the spinfoam would only be an effective theory. There have been efforts to formulate the continuum theory as a "Group Field Theory," a nonlocal field theory defined on a group manifold which generates spinfoams from its perturbative expansion (see, e.g., [78], and references therein). There are also proposals to renormalize the spinfoam models, and approach the continuum limit at the quantum level (e.g., [79]).

Another important open research direction is to extract physical predictions from the formalism of quantum geometry. There have been recent results by applying LQG to black holes and cosmology (e.g., [80, 81, 82]). These results indicate that the quantum geometry is nonsingular even at the classical singularity. The geometry near the classical singularity has large quantum fluctuations. But the quantum geometry, including the Hilbert space and the geometrical operators, is well defined.

Recently the formalism of quantum geometry has been applied to study holographic bulk-boundary dualities, and it has been shown that it is related to the holographic tensor network program (the program studying bulk geometries emerging from quantum states of boundary many-body systems) [83, 84, 85, 86]. It is shown that the quantum geometry reproduces the Ryu–Takayanagi formula of holographic entanglement entropy of the boundary theory. These results open a new direction of research to understand holographic dualities from quantum gravity as the first principle.

References

[1] A. Ashtekar, *New variables for classical and quantum gravity*, Phys. Rev. Lett. 57 (1986) 2244–47

[2] M. Han, W. Huang, and Y. Ma, *Fundamental structure of loop quantum gravity*, Int. J. Mod. Phys. D16 (2007) 1397–1474

[3] A. Ashtekar and J. Lewandowski, *Background independent quantum gravity: A status report*, Class. Quant. Grav. 21 (2004) R53

[4] T. Thiemann, *Modern Canonical Quantum General Relativity*, Cambridge University Press (2007)

[5] C. Rovelli and F. Vidotto, *Covariant Loop Quantum Gravity: An Elementary Introduction to Quantum Gravity and Spinfoam Theory*, Cambridge Monographs on Mathematical Physics, Cambridge University Press (2014)

[6] N. Bodendorfer, T. Thiemann, and A. Thurn, *New Variables for Classical and Quantum Gravity in All Dimensions III. Quantum Theory*, Class. Quant. Grav. 30 (2013) 045003

[7] C. Rovelli and L. Smolin, *Knot Theory and Quantum Gravity*, Phy. Rev. Lett. 61 (1988) 1155–58

[8] A. Ashtekar and C. J. Isham, *Representations of the holonomy algebras of gravity and non-Abelian gauge theories*, Class. Quant. Grav. 9 (1992) 1433–68

[9] J. Lewandowski, A. Okolow, H. Sahlmann, and T. Thiemann, *Uniqueness of diffeomorphism invariant states on holonomy-flux algebras*, Commun. Math. Phys. 267 (2006) 703–33

[10] C. Rovelli and L. Smolin, *Discreteness of area and volume in quantum gravity*, Nucl. Phys. B 442 (1995) 593–619

[11] A. Ashtekar and J. Lewandowski, *Quantum theory of geometry. 1: Area operators*, Class. Quant. Grav. 14 (1997) A55–82

[12] A. Ashtekar and J. Lewandowski, *Quantum theory of geometry. 2. Volume operators*, Adv. Theor. Math. Phys. 1 (1998) 388–429

[13] T. Thiemann, *A length operator for canonical quantum gravity*, J. Math. Phys. 39 (1998) 3372–92

[14] E. Bianchi, *The length operator in loop quantum gravity*, Nucl. Phys. B807 (2009) 591–624

[15] Y. Ma, C. Soo, and J. Yang, *New length operator for loop quantum gravity*, Phys. Rev. D81 (2010) 124026

[16] Y. Ma and Y. Ling, *The Q-hat operator for canonical quantum gravity*, Phys. Rev. D62 (2000) 104021

[17] S. Holst, *Barbero's Hamiltonian derived from a generalized Hilbert–Palatini action*, Phys. Rev. D53 (1996) 5966–69

[18] J. F. Barbero, *Real Ashtekar variables for Lorentzian signature space times*, Phys. Rev. D51 (1995) 5507–10

[19] D. Marolf and J. M. Mourao, *On the support of the Ashtekar–Lewandowski measure*, Commun. Math. Phys. 170 (1995) 583–606

[20] A. Ashtekar and J. Lewandowski, *Representation theory of analytic holonomy C^* algebras*, Preprint, arXiv:gr-qc/9311010

[21] R. Penrose, *Angular momentum: An approach to combinatorial spacetime*, in T. Bastin, ed., Quantum Theory and Beyond, Cambridge University Press (1971) 151–180

[22] L. Kauffman and R. Baadhio, *Quantum Topology*, Series on Knots and Everything, IOP (1993)

[23] J. Brunnemann and D. Rideout, *Properties of the volume operator in loop quantum gravity. I. Results*, Class. Quant. Grav. 25 (2008) 065001

[24] E. Bianchi and H. M. Haggard, *Discreteness of the volume of space from Bohr–Sommerfeld quantization*, Phys. Rev. Lett. 107 (2011) 011301

[25] E. Bianchi, P. Dona, and S. Speziale, *Polyhedra in loop quantum gravity*, Phys. Rev. D83 (2011) 044035

[26] F. Conrady and L. Freidel, *Quantum geometry from phase space reduction*, J. Math. Phys. 50 (2009) 123510

[27] A. Ashtekar, J. Lewandowski, D. Marolf, J. Mourao, and T. Thiemann, *A manifestly gauge invariant approach to quantum theories of gauge fields*, in Geometry of Constrained Dynamical Systems. Proceedings, Conference, Cambridge, UK, June 15–18, 1994, Elsevier (1994) 60–86

[28] J. M. Velhinho, *A groupoid approach to spaces of generalized connections*, J. Geom. Phys. 41 (2002) 166–80

[29] R. Giles, *The reconstruction of gauge potentials from Wilson loops*, Phys. Rev. D24 (1981) 2160

[30] A. Ashtekar and J. Lewandowski, *Projective techniques and functional integration for gauge theories*, J. Math. Phys. 36 (1995) 2170–91

[31] A. Ashtekar and J. Lewandowski, *Differential geometry on the space of connections via graphs and projective limits*, J. Geom. Phys. 17 (1995) 191–230
[32] A. Ashtekar, A. Corichi, and J. A. Zapata, *Quantum theory of geometry III: Noncommutativity of Riemannian structures*, Class. Quant. Grav. 15 (1998) 2955–72
[33] H. Sahlmann and T. Thiemann, *Irreducibility of the Ashtekar–Isham–Lewandowski representation*, Class. Quant. Grav. 23 (2006) 4453–72
[34] C. Rovelli, *Quantum Gravity*, Cambridge University Press (2004)
[35] R. Gambini and J. Pullin, *Loops, Knots, Gauge Theories and Quantum Gravity*, Cambridge Monographs on Mathematical Physics, Cambridge University Press (2000)
[36] C. Rovelli and L. Smolin, *Loop space representation of quantum general relativity*, Nucl. Phys. B331 (1990) 80–152
[37] K. Giesel and T. Thiemann, *Consistency check on volume and triad operator quantisation in loop quantum gravity. I*, Class. Quant. Grav. 23 (2006) 5667–92
[38] K. Giesel and T. Thiemann, *Consistency check on volume and triad operator quantisation in loop quantum gravity. II*, Class. Quant. Grav. 23 (2006) 5693–5772
[39] H. Minkowski, *Ausgewählte Arbeiten zur Zahlentheorie und zur Geometrie, vol. 12, Teubner-Archiv zur Mathematik*, Springer (1989)
[40] L. Freidel, K. Krasnov, and E. R. Livine, *Holomorphic factorization for a quantum tetrahedron*, Commun. Math. Phys. 297 (2010) 45–93
[41] L. Freidel and S. Speziale, *Twisted geometries: A geometric parametrisation of SU(2) phase space*, Phys. Rev. D82 (2010) 084040
[42] M. Langvik and S. Speziale, *Twisted geometries, twistors and conformal transformations*, Phys. Rev. D94 (2016), 024050
[43] C. Rovelli and S. Speziale, *On the geometry of loop quantum gravity on a graph*, Phys. Rev. D82 (2010) 044018
[44] H. M. Haggard, C. Rovelli, W. Wieland, and F. Vidotto, *Spin connection of twisted geometry*, Phys. Rev. D87 (2013) 024038
[45] H. M. Haggard, M. Han, and A. Riello, *Encoding curved tetrahedra in face holonomies: phase space of shapes from group-valued moment maps*, Annales Henri Poincare 17 (2016) 2001–48
[46] M. Han and Z. Huang, *SU(2) flat connection on Riemann surface and twisted geometry with cosmological Constant*, Preprint, arXiv:1610.01246
[47] J. Engle, E. Livine, R. Pereira, and C. Rovelli, *LQG vertex with finite Immirzi parameter*, Nucl. Phys. B799 (2008) 136–49
[48] L. Freidel and K. Krasnov, *A new spin foam model for 4d gravity*, Class. Quant. Grav. 25 (2008) 125018
[49] M. Han and T. Thiemann, *Commuting simplicity and closure constraints for 4D spin foam models*, Class. Quant. Grav. 30 (2013) 235024
[50] M. Dupuis and E. R. Livine, *Holomorphic simplicity constraints for 4d spinfoam models*, Class. Quant. Grav. 28 (2011) 215022
[51] E. R. Livine and S. Speziale, *A new spinfoam vertex for quantum gravity*, Phys. Rev. D76 (2007) 084028
[52] Y. Ding, M. Han, and C. Rovelli, *Generalized spinfoams*, Phys. Rev. D83 (2011) 124020

[53] W. Kaminski, M. Kisielowski, and J. Lewandowski, *Spin-foams for all loop quantum gravity*, Class. Quant. Grav. 27 (2010) 095006 Erratum: Class. Quant. Grav. 29 (2012) 049502
[54] B. Bahr, B. Dittrich, F. Hellmann, and W. Kaminski, *Holonomy spin foam models: Definition and coarse graining*, Phys. Rev. D87 (2013) 044048
[55] J. W. Barrett, R. Dowdall, W. J. Fairbairn, F. Hellmann, and R. Pereira, *Lorentzian spin foam amplitudes: Graphical calculus and asymptotics*, Class. Quant. Grav. 27 (2010) 165009
[56] F. Conrady and L. Freidel, *On the semiclassical limit of 4d spin foam models*, Phys. Rev. D78 (2008) 104023
[57] M. Han and M. Zhang, *Asymptotics of spinfoam amplitude on simplicial manifold: Lorentzian theory*, Class. Quant. Grav. 30 (2013) 165012
[58] M. Han and T. Krajewski, *Path integral representation of Lorentzian spinfoam model, asymptotics, and simplicial geometries*, Class. Quant. Grav. 31 (2014) 015009
[59] M. Han, *Covariant loop quantum gravity, low energy perturbation theory, and Einstein gravity with high curvature UV corrections*, Phys. Rev. D89 (2014) 124001
[60] H. M. Haggard, M. Han, W. Kaminski, and A. Riello, *SL(2,C) Chern–Simons theory, a non-planar graph operator, and 4D loop quantum gravity with a cosmological constant: Semiclassical geometry*, Nucl. Phys. B900 (2015) 1–79
[61] H. M. Haggard, M. Han, W. Kaminski, and A. Riello, *Four-dimensional quantum gravity with a cosmological constant from three-dimensional holomorphic blocks*, Phys. Lett. B752 (2016) 258–62
[62] H. M. Haggard, M. Han, W. Kaminski, and A. Riello, *SL(2,C) Chern–Simons theory, flat connections, and four-dimensional quantum geometry*, Preprint, arXiv:1512.07690
[63] M. Han, *4d quantum geometry from 3d supersymmetric gauge theory and holomorphic block*, JHEP 01 (2016) 065
[64] M. Han, *4-dimensional spin-foam model with quantum Lorentz group*, J. Math. Phys. 52 (2011) 072501
[65] W. J. Fairbairn and C. Meusburger, *Quantum deformation of two four-dimensional spin foam models*, J. Math. Phys. 53 (2012) 022501
[66] K. Noui and P. Roche, *Cosmological deformation of Lorentzian spin foam models*, Class. Quant. Grav. 20 (2003) 3175–3214
[67] E. Bianchi, M. Han, C. Rovelli, W. Wieland, E. Magliaro, and C. Perini, *Spinfoam fermions*, Class. Quant. Grav. 30 (2013) 235023
[68] M. Han and C. Rovelli, *Spin-foam fermions: PCT symmetry, Dirac determinant, and correlation functions*, Class. Quant. Grav. 30 (2013) 075007
[69] D. Oriti and H. Pfeiffer, *A spin foam model for pure gauge theory coupled to quantum gravity*, Phys. Rev. D66 (2002) 124010
[70] M. Han, *Einstein equation from covariant loop quantum gravity in semiclassical continuum limit*, Preprint, arXiv:1705.09030
[71] E. Alesci, T. Thiemann, and A. Zipfel, *Linking covariant and canonical LQG: New solutions to the Euclidean scalar constraint*, Phys. Rev. D86 (2012) 024017

[72] M. Han and T. Thiemann, *On the relation between operator constraint –, master constraint –, reduced phase space –, and path integral quantisation*, Class. Quant. Grav. 27 (2010) 225019
[73] T. Thiemann, *Quantum spin dynamics (QSD)*, Class. Quant. Grav. 15 (1998) 839–73
[74] J. Lewandowski and H. Sahlmann, *Symmetric scalar constraint for loop quantum gravity*, Phys. Rev. D91 (2015) 044022
[75] J. Yang and Y. Ma, *New Hamiltonian constraint operator for loop quantum gravity*, Phys. Lett. B751 (2015) 343–47
[76] C. Rovelli, *Relational quantum mechanics*, Int. J. Theor. Phys. 35 (1996) 1637–78
[77] T. Thiemann, *Reduced phase space quantization and Dirac observables*, Class. Quant. Grav. 23 (2006) 1163–80
[78] D. Oriti, *Group field theory as the 2nd quantization of loop quantum gravity*, Class. Quant. Grav. 33 (2016) 085005
[79] B. Dittrich, *The continuum limit of loop quantum gravity – a framework for solving the theory*, Preprint, arXiv:1409.1450
[80] M. Christodoulou, C. Rovelli, S. Speziale, and I. Vilensky, *Planck star tunneling time: An astrophysically relevant observable from background-free quantum gravity*, Phys. Rev. D94 (2016) 084035
[81] M. Han and M. Zhang, *Spinfoams near a classical curvature singularity*, Phys. Rev. D94 (2016) 104075
[82] I. Agullo, A. Ashtekar, and B. Gupt, *Phenomenology with fluctuating quantum geometries in loop quantum cosmology*, Preprint, arXiv:1611.09810
[83] M. Han and L.-Y. Hung, *Loop quantum gravity, exact holographic mapping, and holographic entanglement entropy*, Phys. Rev. D95 (2017) 024011
[84] G. Chirco, D. Oriti, and M. Zhang, *Group field theory and tensor networks: Towards a Ryu–Takayanagi formula in full quantum gravity*, Preprint, arXiv:1701.01383
[85] L. Smolin, *Holographic relations in loop quantum gravity*, Preprint, arXiv:1608.02932
[86] M. Han and S. Huang, *Discrete gravity on random tensor network and holographic Rényi entropy*, Preprint, arXiv:1705.01964

Muxin Han
Department of Physics, Florida Atlantic University
and Institut für Quantengravitation, Universität
Erlangen-Nürnberg, Germany
hanm@fau.edu

9
Stringy Geometry and Emergent Space

Marcos Mariño*

Contents

1	Introduction	407
2	Stringy Geometry	409
3	Strings from Gauge Theories and Emergent Space	418
4	Conclusions and Open Problems	426
References		427

1 Introduction

String theory appeared 40 years ago as a tentative, unified theory of all particles and interactions. It is still not clear if the description of nature provided by string theory will be vindicated by experiments in the near future, although in view of the absence of supersymmetry signals at the LHC, one should not be too optimistic. However, string theory seems to have the rare virtue of being *a* consistent theory of quantum gravity. It might not be the right theory of actually existing gravity, but it contains propagating gravitons and black hole solutions which can be studied in the quantum regime. The Bekenstein–Hawking entropy of many of these black holes can be correctly accounted for in terms of microstates. It has also been conjectured that in some cases,

---

* I would like to thank Gabriel Catren for the invitation to contribute to this volume and Marcelo Alé for his injections of healthy skepticism. This work is supported in part by the Swiss National Science Foundation, subsidies 200020-149226, 200021-156995, and by NCCR 51NF40-141869 "The Mathematics of Physics" (SwissMAP).

string theories are dual to gauge theories, therefore providing the first concrete realization of holographic dualities. Last, but not least, string theory has had a continuing impact on mathematics, which goes well beyond the traditional patterns of interaction between physics and mathematics. Although one might be skeptical of string theory models of particle physics, string theory has found in the meantime interesting applications in condensed matter theory and in hadronic physics.

As a theory of quantum gravity, string theory proposes a new picture of the underlying structure of spacetime, and some of the most successful applications of string theory in mathematics have involved "stringy" versions of geometry, with deep relations to enumerative geometry. The modern theory of enumerative invariants has grown out of the challenges (and the answers) proposed by string theory. More recently, holography has suggested that, in string theory, spacetime should be regarded as an "emergent" phenomenon. In this short article I will review some of these developments, and I will try to give some concrete examples in which string theory has modified or challenged our notions of space (insightful articles on this subject include [14, 19, 21]).

After a summary of string theory (hopefully for nonspecialists), I discuss how geometry can be deformed to take into account the extended nature of the string. This leads to the notion of "stringy" geometry put forward in, for example, topological string theory. As I will emphasize, string geometry should also take into account string interactions, and this paves the way to a discussion of nonperturbative aspects of the theory. Our best guide for such a nonperturbative formulation is the string/gauge theory correspondence. In this framework, the space in which strings propagate seems to be an emergent notion, akin to a macroscopic description in thermodynamics. "Classical" geometry appears as an effective description at large distances. It becomes "stringy" and "brany" as we go to shorter distances and increase the strength of string interactions, and might finally "dissolve" into gauge-theoretic degrees of freedom as we reach the Planck scale. I illustrate these considerations in one particular example of a string/gauge duality, in which precise calculations seem to vindicate this picture of spacetime.

I should haste to say that my view of this subject is very partial, and this article does not pretend to be exhaustive or make justice to the enormous amount of work done on this problem. It should be read as an essay on the subject, rather than as an expository article or review.

2 Stringy Geometry

2.1 From Particles to Strings

In order to understand some of the implications of string theory for physical conceptions of space, it is necessary to present some of its ingredients, and in particular to explain in which sense string theory is a generalization (or more precisely, a deformation) of theories of point particles. Our presentation will be necessarily sketchy and incomplete. Modern textbooks on superstring theory include [15, 20], and the summary presented below is based on [17], which discusses the interaction of string theory with modern mathematics.

In a classical theory of point particles, the fundamental ingredient is the *trajectory* of the particle in a given spacetime, which is typically represented by a differentiable, Riemannian manifold X. This trajectory can be represented by an application

$$\begin{aligned} x: I &\to X, \\ \tau &\mapsto x(\tau), \end{aligned} \tag{2.1}$$

where $I \subset \mathbb{R}$ is an interval, and $\tau \in I$ is the time parameterizing the trajectory. This specification of the trajectory provides only the kinematical data. Determining the dynamics requires as well an *action functional* $S(x(\tau))$. For example, for a (nonrelativistic) free particle one can take,

$$S(x(\tau)) = \int d\tau \, G_{\mu\nu}(x(\tau))\dot{x}^{\mu}(\tau)\dot{x}^{\nu}(\tau), \tag{2.2}$$

where $G_{\mu\nu}$ is the metric of X. The classical equations of motion can be derived from the variational principle

$$\frac{\delta S}{\delta x} = 0, \tag{2.3}$$

and in the case of the free particle this means that the motion occurs along *geodesics* of the Riemannian manifold X.

The quantization of the theory is appropriately done by using Feynman's path integral formalism. In this formalism, the quantum mechanical propagator (in Euclidean signature) is obtained by integrating the weight $exp(-\frac{1}{\hbar}S(x))$ over all possible trajectories $x(\tau)$ with fixed boundary conditions,

$$x(\tau_0) = x_i, \quad x(\tau_1) = x_f, \tag{2.4}$$

where τ_0, τ_1 are the endpoints of the interval. This is written, formally, as

$$K(x_i, x_f) = \int Dx(\tau) \, e^{-\frac{1}{\hbar}S(x(\tau))}. \tag{2.5}$$

The weight $\exp(-\frac{1}{\hbar}S(x))$ might be regarded as a probability distribution for paths, and the propagator is closely related to the probability that a quantum particle which starts at x_i at $\tau = \tau_0$ is detected at x_f at $\tau = \tau_1$. One can also consider *periodic* trajectories, that is, maps of the form

$$x: \mathbb{S}^1 \to X \tag{2.6}$$

where the circle has length β. This means that the map x is defined on a *closed* one-manifold. As is well known, the path integral with these boundary conditions,

$$Z(\beta) = \int Dx(\tau)\, e^{-\frac{1}{\hbar}S(x(\tau))} \tag{2.7}$$

gives the partition function at temperature $kT = \beta^{-1}$ (where k is the Boltzmann constant), which describes the properties of a particle in a thermal bath at this temperature.

In string theory, point-particles are replaced by 1-dimensional objects. Classically, the embedding of such an object in a spacetime manifold is described by a map

$$\begin{aligned} x: I \times S &\to X, \\ (\tau,\sigma) &\mapsto x(\tau,\sigma) \end{aligned} \tag{2.8}$$

where $\sigma \in S \subset \mathbb{R}$ parameterizes now the string. The dynamics is specified again by a classical action $S(x)$, which in the simplest case takes the form

$$S(x,h) = \frac{1}{\ell_s^2} \int d\tau d\sigma\, \sqrt{h}\, G_{\mu\nu} h^{ab} \partial_a x^\mu \partial_b x^\nu. \tag{2.9}$$

Here, h is a metric on the 2-dimensional manifold $\Sigma = I \times S$, and it plays the role of an auxiliary field. The quantity ℓ_s, which has the dimensions of a length, sets the scale of this 1-dimensional extended object and for this reason it is called the *length of the string*. It is convenient to parameterize $S = [0, \pi]$. A *closed string* is a loop with no free ends, and in this case the appropriate boundary condition is

$$x(\tau,0) = x(\tau,\pi), \qquad \tau \in I. \tag{2.10}$$

A freely propagating closed string, as it evolves in time, spans the surface with the topology of a cylinder (see Figure 9.1).

There are two types of string interaction: in a *splitting* process one single string splits into two, and in a *joining* process two strings merge into one. As in quantum physics, the strength of such an interaction is measured by a constant g_{st} called the *string coupling constant*. For closed strings, the basic process of

Stringy Geometry and Emergent Space

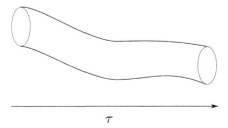

Figure 9.1 A closed string propagating in time.

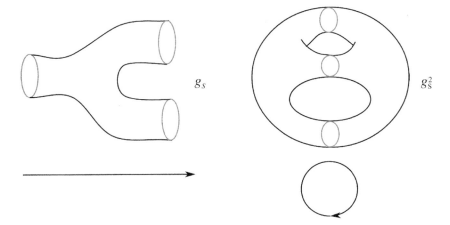

Figure 9.2 A closed string interaction takes place when one single string splits into two, as shown in the left-hand side of the figure. Such a process is weighted by a power of g_s. In the right-hand side we show a periodic configuration in which a closed string splits and then joins again before coming back to itself, spanning a Riemann surface of genus 2. Since there are two string interactions involved, this process has a weight g_s^2.

joining or splitting is described by a "pair of pants" diagram as in the left-hand side of Figure 9.2, and it has a single power of g_{st} associated to it. When this process takes place n times, it has the factor g_{st}^n.

As in the theory of point particles, periodic configurations generalizing (2.6) play a very important role. For a closed string, a periodic configuration is described by a single string which, after various processes of splitting and joining, comes back to itself. This process produces a *closed, orientable Riemann surface* Σ_g, and it is easy to see that it has the weight g_{st}^{2g-2}, where g is the genus of the resulting Riemann surface. In the right-hand side of

Figure 9.2 we show a periodic configuration in which a closed string evolves, splits into two strings which merge back to a single closed string, and this string goes back to the starting point. The time evolution produces a Riemann surface of genus 2, and since there were two string interactions (one splitting and one joining) the whole process has a factor g_{st}^2 associated to it.

We conclude that, in string theory, a periodic map is just a map from a closed Riemann surface Σ_g to the spacetime X:

$$x : \Sigma_g \to X. \tag{2.11}$$

The Riemann surface Σ_g is called the *worldsheet* of the string, while the manifold X is called its *target space*.

The quantization of a theory of strings is rather delicate. Formally, one considers a path integral over all possible configurations of the fields, as in (2.5), with the appropriate boundary conditions. For simplicity we will consider periodic boundary conditions, that is, the analog of (2.7). Since we have to consider all possible configurations of the string, we have to take into account all possible splitting/joining processes, and this means that we should sum over all possible genera for the Riemann surfaces spanned by the string. In the computation of Z one considers *disconnected* Riemann surfaces, but in the so-called free energy, defined by $F = \log Z$, one should sum only over connected Riemann surfaces, labeled by the genus g. This means that the free energy is given by a formal infinite series over the different genera

$$F = \sum_{g=0}^{\infty} g_{\text{st}}^{2g-2} F_g, \tag{2.12}$$

which is sometimes called the genus expansion of the string free energy. To calculate F_g for a fixed genus g, we should integrate over all metrics on Σ_g and all configurations of maps x. The space of all metrics on Σ_g, after taking into account the relevant symmetries, turns out to be equivalent to the moduli space of Riemann surfaces $\overline{M}_g$ constructed in algebraic geometry, and one has

$$F_g = \int_{\overline{M}_g} Dh_{ab} \, Dx \, e^{-S(x,h)}. \tag{2.13}$$

Note that, in (2.13), the integration over the metric has been in fact reduced to an integration over the $3g - 3$ complex moduli parameterizing $\overline{M}_g$. It is important to note that the "total string amplitude" F in (2.12) depends on two parameters: the string length ℓ_s (which appears in the string action) and the string coupling constant g_{st}. In particular, if we regard string theory as a 2-dimensional quantum field theory, we see from (2.9) that the squared string length ℓ_s^2 plays the same role as $\hbar$. When $\ell_s \to 0$, the length of the string vanishes and we recover a theory of point particles. Of course, the coupling constant g_{st} can be also regarded as a quantum parameter, and in the point-particle limit of string theory it becomes a standard coupling constant governing quantum interactions of particles. The existence of these two quantum parameters makes it clear that string theory is a consistent deformation of a quantum theory of point particles.

The structure we have described is roughly speaking the so-called *bosonic string* with target space X, and it can be easily generalized after we identify its key ingredients. On one hand, the map x can be regarded as a 2-dimensional quantum field described by the action (2.9). This field theory has the property of being invariant under the full conformal group in two dimensions, that is, it is an example of a 2-dimensional *conformal field theory* (CFT). On the other hand, we also introduced a metric in the 2-dimensional surface where this field lives. The combination of these two ingredients is a particular example of a *2-dimensional conformal field theory coupled to 2-dimensional gravity*. In more abstract terms, a string theory is just such a system, and depending on the conformal field theory involved – the so-called matter content of the string theory – we will have different string theories. For example, in *supersymmetric* string theories the conformal field theory is supersymmetric, that is, there are extra fermionic fields as well as a symmetry exchanging the bosonic and the fermionic fields. In some cases these theories lead to supersymmetry in spacetime, and the resulting theories are called *superstring theories*.

One surprising aspect of string theories is that they only make sense as quantum theories if one imposes constraints on the field content. For example, the bosonic string is consistent only if the target space has 26 dimensions, that is, if the field x has 26 components. Superstring theories require X to have 10 dimensions. If string theory is regarded as a model of the real world, the target space X should be identified as the physical spacetime, and the operators of the CFT give rise to quantum fields propagating on X. We conclude that string theory models require extra physical dimensions. It is widely believed that consistent theories of strings have to be superstring theories, since the bosonic string is unstable (it has a tachyonic state in its spectrum). There are five different consistent superstring theories, which are called type I, type IIA, type IIB and the two different heterotic strings. In this article we will not need the details of how these theories are constructed (see for example [20]). One crucial property of all these string theories is that they contain, in its perturbative spectrum, interacting gravitons, which at large distances are described by General Relativity or modifications thereof. Therefore, these theories provide in principle a framework to study the quantum regime of interacting gravitons. In addition, they lead naturally to gauge interactions, so they have all the necessary ingredients to describe particles and the forces among them, including gravity. One can go further and assume that X is the product of a tiny, compact 6-dimensional manifold K, and a four-dimensional Minkowski spacetime. In this way one finds unified theories in four dimensions, reasonably close to supersymmetric versions of the Standard Model.

There are strong indications that the five, apparently different, string theories are in reality different descriptions (in different regimes) of the same underlying theory, which is sometimes called M-theory. Unfortunately, we do not have a basic description of this fundamental theory, although we know some of its ingredients. For example, this theory contains, on top of strings or extended 1-dimensional objects, other extended objects like *membranes* (2-dimensional objects) and also five-branes.

In addition to the "physical" string theories that we have described, there are other, simplified models of string theory which play an important role in mathematical physics. One of them is "noncritical" (super)string theory, in which the string propagates on a space of dimension less than or equal to 1. We should note that, in string theory, the dimension of spacetime is given by the so-called central charge c of the corresponding CFT, and a noncritical string is simply a CFT with $c \leq 1$ coupled to 2-dimensional gravity, in the way we have sketched above.

Another important simplified model is *topological string theory*, first introduced by Witten. In this case, one considers a two-dimensional *topological* field theory, that is, a 2-dimensional QFT whose correlation functions are independent of the metric, and couples the resulting theory (which is in particular conformal) to 2-dimensional gravity. Topological string theory is the physical counterpart of the modern theory of enumerative invariants (i.e., Gromov–Witten theory and their close cousins). It offers a precise arena in which one can explore "stringy" versions of space.

2.2 Strings and Geometry

As we have just seen, string theory can be understood as a *deformation* of a theory of point particles, in which the deformation parameter is the string length ℓ_s. When this parameter goes to zero, string theories become theories of point particles. For example, superstring theories become, in the point-particle limit, supergravity theories.

The natural geometric arena for a theory of point particles are differentiable manifolds. Since string theory is a deformation of a theory of point particles, it is natural to suppose that it will lead to a natural deformation of the

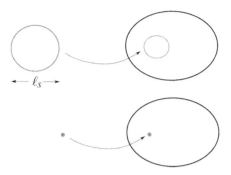

Figure 9.3 In string theory, a point in a manifold should be regarded as an embedded string in the limit in which the string length goes to zero size, $\ell_s \to 0$.

usual structures appearing in the theory of manifolds. A precise and general formulation of these deformed structures has not been constructed yet, but a simplified version thereof has been studied in detail in the context of topological string theory.

The basic idea of "stringy" geometry is that, in string theory, a manifold X comes equipped with all possible maps of Riemann surfaces to X (satisfying some additional conditions, like for example holomorphy). In this framework, a *point* in X should be regarded as a limiting case of such maps, when the length of the string ℓ_s goes to zero size (see Figure 9.3).

In order to elaborate on this, let us consider one of the most important examples of "stringy" geometry (see [4, 13] for comprehensive references on these topics). As we mentioned above, there is a simplified version of string theory, called topological string theory, which can be studied on a special type of spaces called Calabi–Yau manifolds. Technically, a Calabi–Yau manifold is a complex, Kähler manifold which is Ricci-flat. In string theory, Calabi–Yau manifolds of real dimension equal to six (also called Calabi–Yau threefolds) play an important rôle. In some models, they provide the six additional dimensions that are needed in order to go from our spacetime of four dimensions to the 10-dimensional spacetime of the superstring. However, one can also study the "stringy" geometry of these manifolds from a more mathematical point of view. Let X be a compact, Calabi–Yau threefold. If $\omega \in H^{1,1}(X)$ is the Kähler form of X, it is customary to introduce a *complexified* Kähler form as

$$J = \omega + iB, \qquad B \in H^{1,1}(X), \tag{2.14}$$

where B is sometimes called the B-field. Let $\{S_a\}$ be a basis of the two-homology of $H_2(X,\mathbb{Z})$. The "complexified" sizes or Kähler parameters of the S_a are defined as

$$t_a = \int_{S_a} J, \quad a = 1, \cdots, h^{1,1}(X), \qquad (2.15)$$

and they have dimension of an area. Positivity of the metric requires that $\operatorname{Re} t_a > 0$.

In topological string theory, the quantum theory of embedded strings on a Calabi–Yau X reduces to a study of holomorphic maps from the worldsheet of the string Σ_g to X, and a one-loop calculation of the small fluctuations around it. This is due to the fact that the underlying sigma model describing topological string theory is a topological field theory of the cohomological type, and for these theories the semiclassical approximation to the path integral gives the exact result. When one considers Riemann surfaces of genus zero, the free energy reduces to a sum over the different holomorphic embeddings, and it has the structure

$$F_0(t_a) = \frac{1}{6\ell_s^6} \sum_{a,b,c} C_{abc} t_a t_b t_c + \sum_{Q \in H_2(X,\mathbb{Z})} N_{0,Q} e^{-\int_Q J/\ell_s^2}. \qquad (2.16)$$

In this expression, C_{abc} is the intersection number of the three homology classes

$$C_{abc} = S_a \cap S_b \cap S_c, \qquad (2.17)$$

and the sum over Q is over two-homology classes. The numbers $N_{0,Q}$ appearing in this expression are the *Gromov–Witten invariants* of the Calabi–Yau X for genus zero and for the class Q. These invariants have an enumerative interpretation as an appropriate counting of holomorphic curves of genus zero in the homology class Q. Physically, the Gromov–Witten invariants are due to "periodic" strings, holomorphically embedded in X. An important property of (2.16) is that, when $\ell_s \to 0$, the contribution from the Gromov–Witten invariants is exponentially suppressed. In this limit, the genus zero free energy is dominated by the first term in (2.16), which involves a purely classical property of the manifold X, namely, its triple intersection numbers C_{abc}. When ℓ_s is not zero, this point-particle or "classical" quantity gets a very rich, infinite series of "stringy corrections" which contain information about the enumerative geometry of X.

The above example is particularly important since it has led to a new branch of mathematics, Gromov–Witten theory and its refinements and generalizations, which has been directly motivated by string theory. However, the physical principle behind this example is at work in many other situations: quantities computed in string theory are typically given by a point-particle quantity (involving the "classical" or Riemannian geometry of the spacetime manifold), together with various corrections coming from the extended nature of the string. These corrections involve a generalized notion of geometry in which the manifold has to be enriched with an additional, "stringy" structure.

The "stringy" structure of the geometry that we have just described has led to many interesting qualifications to the classical notions of space and distance. Note that, if L is the characteristic scale of the manifold where the strings propagate (i.e., the "size" of the manifold), the dimensionless parameter measuring the importance of string corrections is

$$\frac{L}{\ell_s}. \qquad (2.18)$$

When the size of the manifold is much larger than the size of the string, stringy corrections are negligible. However, as the size of the manifold decreases, these corrections become more and more important and classical geometric intuition is no longer reliable. For example, in a compact Calabi–Yau threefold X, one can use the corrections involving embedded strings to define the "stringy volume" of cycles of even dimension. One of these cycles is a six-cycle, that is, the manifold X itself. In a classical geometry with a single Kähler parameter, if we decrease the volume of X to zero, all lower dimensional cycles inside X will squeeze to zero volume, too. However, if we take into account stringy corrections, one can find examples in which this is not the case, and the ambient manifold can have zero "stringy" volume while lower dimensional cycles still have a nonzero "stringy" volume [9, 10]. This phenomenon occurs in the regime in which the characteristic size of the manifold is comparable to the length of the string, so classical geometry receives important corrections.

2.3 Quantum Strings

Most of the studies of changing notions of space and distance in string theory have focused on "stringy" corrections in *classical* string theory, in which the string coupling constant vanishes and spacetime string interactions are negligible. In the classical limit, only the first term in (2.12) is kept, corresponding to Riemann surfaces of genus zero, that is, with the topology of a sphere. For example, in (2.16) we have only considered genus zero Gromov–Witten invariants. This is clearly not the whole story, and one should consider interacting strings. In considering the string free energy (2.12), one should sum over all possible genera of the Riemann surfaces. We should also expect modifications in the geometry of spacetime due to a nonzero string coupling constant, and not only to a nonzero string length. We will refer to the resulting geometry as a "quantum" geometry, to distinguish it from the "stringy" geometry analyzed above.

The effects of interacting strings on the geometry have been much less studied, due to various limitations. Perhaps the most important one is the fact that the genus expansion appearing in string theory (like in the formal infinite sum in (2.12)) is asymptotic, and leads to a divergent series. This is a typical feature of quantum perturbation theories, and is usually solved by considering an appropriate nonperturbative definition of the theory. The problem is that, in string theory, there is no such a definition in most of the situations. This makes it difficult to use the string perturbation series as a guide to quantum geometry.

Although we do not have general, nonperturbative definitions of superstring theory, we have some indications about the ingredients that enter into such a definition. One solid hint that has emerged in recent years is that, at the nonperturbative level, string theory requires extended objects of higher dimensions, like membranes and more generally p-branes (i.e., extended objects with p dimensions). This suggests that a quantum geometry taking into account interactions between strings will involve Riemann surfaces of higher genus, as it follows from string perturbation theory, but also embedded, extended objects of higher dimensions in the spacetime manifold. For example, periodic configurations of membranes will lead to embedded 3-dimensional configurations in the target manifold X. M-theory, which underlies the different superstring theories, is known to contain membranes and five-branes. These objects lead to new corrections in the computation of physical amplitudes. In many cases, these corrections have a nonanalytic dependence on the string coupling, that is, they involve small exponentials of the form

$$e^{-1/g_{st}}, \quad (2.19)$$

which are invisible in the perturbative framework. We will revisit these issues in the light of the string/gauge theory correspondence in the next section.

3 Strings from Gauge Theories and Emergent Space

3.1 The String/Gauge Theory Correspondence

As we have mentioned, string theory contains interacting gravitons in its perturbative spectrum, and it provides in principle a framework to study the quantum regime of these interacting gravitons. However, string theory is not defined nonperturbatively, and this has limited the applications of string theory to quantum gravity. In order to fully assess the implications of string theory, one needs a nonperturbative definition of the theory.

The search for a nonperturbative definition of string theory is almost as old as string theory itself. In the late 1980s, it was found that, in some situations, noncritical strings can be described nonperturbatively in terms of quantum gauge theories in zero dimension (also known as matrix integrals) and in one dimension (also known as matrix quantum mechanics); see [6, 8] for a review of those developments. These gauge theories in low dimension have a $U(N)$ symmetry, and it was discovered by 't Hooft [22] that the observables of any $U(N)$ gauge theory can be organized in a $1/N$ expansion which looks very much like a genus expansion of a string theory. In the nonperturbative descriptions of noncritical strings, the perturbative genus expansion of string theory is in fact identified to the $1/N$ expansion of the corresponding gauge theory. This means, in particular, that the string free energy (2.12), which as we saw is a formal (and in fact divergent) power series expansion, can be obtained as the asymptotic expansion of the free energy for the "dual" gauge theory in the $1/N$ expansion.

These discoveries, remarkable as they were, involved string theories in no more than one dimension, and did not provide a clear picture of how gravitational physics could emerge from the "dual" degrees of freedom. However, in 1998, Juan Maldacena postulated an exact relationship between various superstring theories in 10 dimensions and supersymmetric gauge theories in lower dimensions [16]. The most celebrated example is the correspondence between $N = 4$ supersymmetric Yang–Mills theory in four dimensions, and the type IIB superstring theory on the manifold $AdS_5 \times \mathbb{S}^5$, where AdS_n denotes the n-dimensional Anti de Sitter space (a maximally symmetric space with negative cosmological constant). Maldacena's correspondence is based on a study of extended objects in superstring theory called D-branes. These objects support gauge-theoretic degrees of freedom, and at the same time they can be effectively described by a curved background involving AdS spaces. In a suitable scaling limit, both descriptions are equivalent, and this leads to a duality between a superstring AdS background, and the gauge theory living in the D-branes. A similar duality can be obtained by considering membranes in M-theory.

The conjecture of Maldacena has triggered an enormous literature, and has many interesting implications. This conjecture, if true, provides a nonperturbative definition of string theory, and therefore a nonperturbative definition of a quantum theory of gravity. In this definition the fundamental degrees of freedom of string theory are not gravitational, but rather gauge-theoretic.

This correspondence has been compared to the relationship between a thermodynamic system, like a gas or a fluid, and its microscopic description in terms of elementary constituents, like atoms or molecules. It suggests that strings, and their spacetime physics, are not fundamental phenomena, but rather "effective" or "emergent" descriptions of very different degrees of freedom. In particular, the 10-dimensional spacetime of superstring theory "emerges" from the quantum dynamics of a gauge theory in lower dimensions (see [5, 14, 21] for discussions of emergence in string theory and in the gauge/string correspondence). It is believed that the original proposal in [16] can be extended to a more general correspondence between a gravity theory in an AdS space of $d+1$ dimensions, and a CFT in d dimensions living on the boundary of AdS. Finding and studying examples of such a correspondence remains a very active research area.

3.2 Emergent Space: An Example

In order to understand a little bit more quantitatively the emergence of space in the string/gauge theory correspondence, it is useful to look at a concrete example. I will focus on the duality between the gauge theory known as ABJM theory, and type IIA superstring on an AdS_4 background. ABJM theory is built upon Chern–Simons gauge theory, which is defined by the action

$$S(A) = -\frac{\mathrm{i}k}{4\pi} \int \mathrm{d}^3 x \, \mathrm{Tr}\left(A \wedge \mathrm{d}A + \frac{2\mathrm{i}}{3} A^3 \right). \tag{3.1}$$

Here, A is a $U(N)$ gauge connection, and k is the so-called Chern–Simons level. It has to be an integer number, in order to prevent the appearance of gauge anomalies. In ABJM theory one considers two different $U(N)$ connections with Chern–Simons actions, so that the gauge group is $U(N) \times U(N)$, but with opposite levels k and $-k$. In addition, one considers a supersymmetric extension of the theory, with additional matter fields in the bifundamental representation of $U(N) \times U(N)$, and a coupling between the two gauge groups. The resulting theory is a superconformal field theory depending on two integer parameters: the rank N appearing in the gauge group, and the level k. Note that k plays the role of the inverse gauge coupling, in the sense that ABJM theory is weakly coupled when k is large. As in many other gauge theories, ABJM theory admits a 't Hooft or $1/N$ expansion, in which N and k are taken to be large, while the quotient

$$\lambda = \frac{N}{k}, \tag{3.2}$$

also called 't Hooft parameter, is kept fixed. In particular, it can be shown, order by order in perturbation theory, that the free energy F of ABJM theory on a three-manifold, defined as the logarithm of the partition function Z, has the following asymptotic $1/N$ expansion:

$$F = \log Z \sim \sum_{g=0}^{\infty} F_g(\lambda) N^{2-2g}. \qquad (3.3)$$

If we compare this expansion to (2.12), we find that it has precisely the structure of a genus expansion in (super)string theory. Similar $1/N$ expansions hold for other observables. At large N but λ fixed, the main contribution comes from the first term, and one has

$$F \approx F_0(\lambda) N^2, \qquad N \to \infty. \qquad (3.4)$$

This is called the *planar approximation* to the gauge theory, in which only a subset of diagrams are kept [22].

According to [1], ABJM theory is dual to type IIA superstring theory on the manifold

$$AdS_4 \times \mathbb{CP}^3, \qquad (3.5)$$

where $\mathbb{CP}^3$ is the projective space in three complex dimensions. The manifold (3.5) has one single scale, the common "radius" L of AdS_4 and $\mathbb{CP}^3$. Therefore, the type IIA superstring on (3.5) is characterized by two dimensionless parameters. One of them is the quotient L/ℓ_s, while the other is the string coupling constant g_{st}. According to the string/gauge theory duality of [1], there is a dictionary between the two parameters of the superstring theory and the two parameters of the gauge theory. The dictionary reads as follows:

$$g_{st}^2 = \frac{\sqrt{32\pi^2 \lambda}}{k^2}, \qquad \lambda = \frac{1}{32\pi^2} \left(\frac{L}{\ell_s}\right)^4. \qquad (3.6)$$

According to this dictionary, the regime in which the scale of the spacetime L is large in comparison to the string length, that is, the point-particle approximation to string theory, corresponds to the regime in which λ, the 't Hooft parameter, is large. In addition, the regime in which the string is weakly coupled, that is, the string coupling constant g_{st} is small, corresponds to the regime in which k is large. Since λ is fixed, this corresponds to the planar limit of the gauge theory, in which N is large. The 2-dimensional parameter space of the theories is shown in Figure 9.4. The gauge/string correspondence also postulates that the $1/N$ expansion of the gauge theory corresponds to the genus expansion of string theory. This means in particular that the free energy

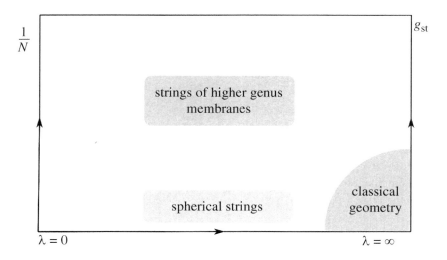

Figure 9.4 The 2-dimensional parameter space of superstring theory on the manifold (3.5) and its relation to the gauge theory parameter space. The description of the theory in terms of classical, Riemannian geometry corresponds to the point-particle limit of noninteracting strings. This is the regime of the gauge theory in which λ and N are very large. Away from this regime, the classical picture is modified by "stringy," "quantum," and "brany" corrections.

in (2.12) corresponds, order by order, to the expansion of the free energy (3.3). The three-manifold on which (3.3) is computed is precisely the *boundary* of the AdS$_4$ geometry appearing in (3.5).

What are the implications of this duality for the structure of spacetime? According to the dictionary (3.6), the Riemannian geometry of the space (3.5) corresponds to a very particular regime of the dual gauge theory, namely, the regime in which λ is large (so we are in the point-particle limit) and k is large (so that strings do not interact). In particular, in this regime, N, the rank of the gauge group, is very large. This suggests that the standard notions of Riemannian geometry appear as emergent phenomena in the limit of a large number of constituents in the gauge theory.

Let us now suppose that we decrease slightly the value of λ, but we keep N very large. This is equivalent to going away from the point-particle limit, and to making manifest the "stringy" nature of the geometry. Explicit calculations indicate that, indeed, away from this limit, the gauge theory excitations organize themselves into embedded, spherical curves in the spacetime (3.5), similar to what happens for topological strings in a Calabi–Yau manifold. Consequently, the free energy is given by the result obtained from General Relativity on (3.5), together with an infinite number of corrections coming from these embedded curves [7].

If we now increase the value of the string coupling constant, curves of higher genera start contributing. From the point of view of the gauge theory, they come from the subleading corrections in the $1/N$ expansion (3.3). As we mentioned above, the resulting genus expansion in string theory is badly divergent, but now it makes sense as the asymptotic expansion of a well-defined quantity, namely, the free energy of the gauge theory.

It turns out that, as g_{st} is increased, the corrections to Riemannian geometry are not only due to strings of all genera, but also to extended objects of two spatial dimensions, that is, to *membranes*. These configurations are invisible in string perturbation theory, since they have the nonanalytic dependence on the string coupling constant in (2.19). Nevertheless, we expect such corrections in the nonperturbative picture of type IIA superstring theory given by M-theory [2]. In fact, it can be shown that, if we consider only the corrections to classical geometry due to embedded strings, the resulting answer is inconsistent, and both membranes and strings have to be included [11]. The detailed analysis of this example of gauge/string correspondence shows that the emergent space of string theory is not just the standard arena of Riemannian geometry, but a "stringy" and "brany" space.

As g_{st} becomes large, the contribution of membranes becomes more and more important, and the description based on fundamental strings is no longer appropriate. Rather, we should find a formulation of the theory in which membranes and fundamental strings are on equal footing. Such a formulation is provided, in principle, by M-theory. At large distances, M-theory is described by 11-dimensional supergravity [23]. Therefore, a remarkable aspect of the transition from type IIA string to M-theory is that the space itself *changes*: as the string interactions becomes stronger, and membranes become more and more important, an eleventh dimension appears. In this particular example, the spacetime background for M-theory in the long-distance regime is

$$\text{AdS}_4 \times \left(\mathbb{S}^7/\mathbb{Z}_k\right). \tag{3.7}$$

A 7-dimensional sphere (modded by a discrete group $\mathbb{Z}_k$), of radius L, has grown out of the 6-dimensional projective space $\mathbb{CP}^3$. The parameter k, which as we noted in (3.6) is related to the value of the string coupling constant, is now purely geometric, since it defines the modding of the seven-sphere by a discrete group. In M-theory, the fundamental length is no longer the string length, but the Planck length ℓ_p, and the dimensionless parameter L/ℓ_p is now related to N and k by

$$\left(\frac{L}{\ell_p}\right)^6 = 32\pi^2 kN. \tag{3.8}$$

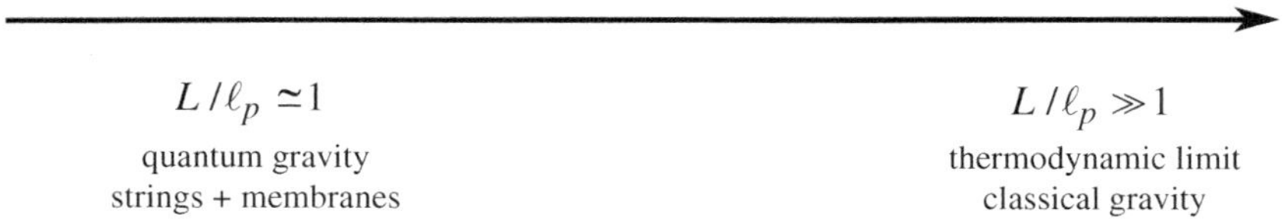

$L/\ell_p \simeq 1$ $L/\ell_p \gg 1$

quantum gravity   thermodynamic limit
strings + membranes   classical gravity

Figure 9.5 The parameter space of M-theory on the manifold (3.7). In the long-distance regime $L/\ell_p \gg 1$, the theory is well described by classical Riemannian geometry (i.e., classical gravity). This corresponds to the large N limit of the gauge theory at fixed gauge coupling (the "thermodynamic limit" of the gauge theory). When L is of the order of the Planck length, quantum gravity effects become important, as well as the contributions of extended objects like strings and membranes.

This relation holds at large N. As we will see later on, it receives corrections. The long-distance regime of M-theory corresponds to $L/\ell_p \gg 1$. As L/ℓ_p becomes smaller, the description in terms of 11-dimensional supergravity is less and less reliable, and should be replaced by a more fundamental description of M-theory (in the same way that supergravity is replaced by string theory when L/ℓ_s becomes smaller). Although we know some properties of this description (for example, it contains membranes and five-branes) we do not know its precise formulation.

The transition from (3.5) to (3.7) supports the view that, in string/M-theory, space is a not an ultimate entity, but rather an emergent or "effective" description. From the point of view of the gauge/string correspondence, both type IIA superstring on (3.5) and 11-dimensional supergravity on (3.7) have to be considered as "effective" descriptions of two different regimes (or "phases") of the gauge theory. The type IIA string description is convenient if we want to describe the gauge theory in the 't Hooft expansion, in which N and k are large and we keep the ratio N/k fixed. Eleven-dimensional supergravity is useful if we want to describe the gauge theory in the regime in which k is fixed (but not necessarily large) and N is large. This is a sort of "thermodynamical" regime of the gauge theory, different from the 't Hooft regime. Note that, when N is large and k is fixed but not large, the geometry of the type IIA string is 10-dimensional, but it is very "quantum" (since higher genus strings are not suppressed). However, this "quantum" 10-dimensional geometry can be effectively described (as long as N is still large) by the classical geometry (3.7) in 11-dimensional supergravity.

In both the type IIA string and the M-theory picture, classical or Riemannian geometry appears as a limiting regime, and the corrections based on strings and membranes become important as we move slightly away from it. Let us focus on the M-theory picture, which is in many ways simpler, since its parameter space is 1-dimensional (see Figure 9.5). The corrections to Riemannian geometry on the space (3.7) become important when L, the size of the target space, is of the order of the Planck length ℓ_p. These include quantum gravity corrections, as well as corrections due to extended objects like strings and membranes (on equal footing). The structure of such corrections can be made very precise, since in the case of ABJM theory, the full expansion of the free energy at large N and fixed k can be obtained from a gauge theory calculation. It has the form (see [12, 18] for reviews)

$$F(N,k) = -\frac{1}{384\pi^2 k}\zeta^{3/2} + \frac{1}{6}\log\left[\frac{\pi^3 k^3}{\zeta^{3/2}}\right] + A(k) + \sum_{n=1}^{\infty} c_{n+1} k^n \zeta^{-3n/2}$$
$$+ O\left(e^{-\sqrt{N/k}}, e^{-\sqrt{kN}}\right). \tag{3.9}$$

In this equation,

$$\zeta = 32\pi^2 k \left(N - \frac{k}{24} - \frac{1}{3k}\right), \tag{3.10}$$

$A(k)$ is a known function of k, and c_{n+1} are calculable coefficients. It is natural to identify

$$\zeta = \left(\frac{L}{\ell_p}\right)^6, \tag{3.11}$$

which at large N agrees with the dictionary (3.8), but it incorporates some subleading corrections in $1/N$. If this identification is made, then the first term in (3.9) is precisely the contribution to the free energy obtained from classical (super)gravity, that is, from classical Riemannian geometry. The remaining terms in the first line are quantum corrections in supergravity: the logarithmic term is a one-loop correction, while the nth term in the series,

$$\zeta^{-3n/2} = \left(\frac{\ell_p}{L}\right)^{9n}, \tag{3.12}$$

has precisely the form of an $(n+1)$-loop correction in a quantum theory of gravity in 11 dimensions. Finally, the terms in the second line (which can be computed explicitly) are nonperturbative corrections at large N. The corrections which go like

$$e^{-\sqrt{N/k}} \sim e^{-(L/\ell_s)^2} \tag{3.13}$$

are due to "stringy" effects. They can be obtained from perturbative string theory, and they are similar to the corrections due to holomorphic curves appearing in (2.16). The corrections which go like

$$e^{-\sqrt{kN}} \sim e^{-(L/\ell_p)^3} \sim e^{-1/g_{st}} \tag{3.14}$$

are due to membrane instantons. They are nonperturbative in the string coupling constant, and they are invisible in string perturbation theory. Clearly, when N is very large, the free energy is dominated by the contribution of classical Riemannian geometry. As N becomes small, that is, as we enter the Planckian regime, both quantum corrections to classical gravity (in the first line of (3.9)) as well as corrections due to strings and branes (in the second line of (3.9)) become more and more important. The "classical" Riemannian geometry of the 11-dimensional space (3.7) provides an accurate description only when N is large.

It is interesting to note that, since N is the rank of a gauge group, it can only take positive integer values. According to the dictionary (3.8), this seems to imply that the scale L should be quantized in units of the Planck length. Therefore, as N becomes small, we might encounter an intrinsic discreteness in space, inherited from the gauge theory realization. Unfortunately, not much is known about the ultra-Planckian regime of the correspondence between gauge theory and string/M-theory. Perhaps, in this regime, the spacetime of string/M-theory, after being substantially corrected by string and membrane excitations, finally "dissolves" into gauge-theoretic degrees of freedom. However, recent calculations show that, in the case of ABJM theory and its string/M-theory dual, it is possible to resum the expansion in (3.9) in such a way that the free energy of the gauge theory makes sense for arbitrary complex values of N. Therefore, at least in this concrete example, one might be able to avoid the discretization of space at Planckian lengths [3].

4 Conclusions and Open Problems

In this article I have summarized some of the ways in which string theory qualifies or modifies our notions of space. In going from pointlike particles to extended strings, the framework of classical or Riemannian geometry gets modified in ways which in some cases can be made mathematically precise. In Gromov–Witten theory or quantum cohomology, classical topological and geometrical notions have to be enlarged in order to take into account the structure of embedded strings in spacetime. One finds in this way a "stringy" geometry which incorporates the corrections due to the finite length of the strings. However, a full string theory geometry should also be "quantum," that is, it should incorporate the string interactions and the effects of a finite string coupling constant. This requires a nonperturbative formulation of string theory.

A useful framework to address the full quantum nature of the strings is the string/gauge theory correspondence. In these dualities, space can be regarded as an "effective" or emergent description of a completely different system, namely, a gauge theory. The emergent space comes equipped with corrections to the classical geometry due to embedded strings, as expected from perturbative string theory, but also to membranes and other extended objects, as expected from M-theory and nonperturbative string theory. An emergent notion of space also appears when we consider the relation between type IIA superstring and 11-dimensional supergravity, in which an eleventh dimension appears dynamically in order to describe the strongly coupled regime of the superstring. From the point of view of string/gauge theory dualities, both descriptions, with their different spacetimes, should be regarded as different "effective" pictures for two different regimes of the underlying gauge theory.

In most of our discussion of the string/gauge theory duality, we have assumed that the gauge theory side has a conceptual and ontological primacy, since it provides the "microscopic" description of "macroscopic" gravity theories. This is a reasonable assumption, since the gauge theory is a relatively well-defined object, and it can be used to provide a nonperturbative description of the elusive string/M-theory side of the correspondence. However, there might still be a description of M-theory/string theory in which spacetime recovers a more fundamental status, and the string/gauge duality would then become an equivalence of two different theories, rather than a hierarchy along the lines of the microscopic/macroscopic divide (see [5] for a related discussion).

In order to answer this and related questions, we need a better understanding of the way in which the gauge theory encodes the spacetime of the dual string/M-theory, and of the behavior of the gauge/string correspondence at small distances. In spite of all the work done on holography, this remains a relatively unexplored subject. The lessons extracted from this line of work might provide important hints about a possible fundamental formulation of M-theory, and, hopefully, also about the nature of actually existing spacetime.

References

[1] O. Aharony, O. Bergman, D. L. Jafferis, and J. Maldacena, *N=6 superconformal Chern-Simons-matter theories, M2-branes and their gravity duals*, JHEP 0810 (2008) 091

[2] K. Becker, M. Becker, and A. Strominger, *Five-branes, membranes and nonperturbative string theory*, Nucl. Phys. B 456 (1995) 130

[3] S. Codesido, A. Grassi, and M. Mariño, *Exact results in $\mathcal{N} = 8$ Chern-Simons-matter theories and quantum geometry*, JHEP 1507 (2015) 001

[4] D. A. Cox and S. Katz, *Mirror Symmetry and Algebraic Geometry*, American Mathematical Society (1999)

[5] D. Dieks, J. van Dongen, and S. de Haro, *Emergence in holographic scenarios for gravity*, arXiv:1501.04278 [hep-th]

[6] P. Di Francesco, P. H. Ginsparg, and J. Zinn-Justin, *2-D gravity and random matrices*, Phys. Rept. 254 (1995) 1–133

[7] N. Drukker, M. Mariño, and P. Putrov, *From weak to strong coupling in ABJM theory*, Commun. Math. Phys. 306 (2011) 511–563

[8] P. H. Ginsparg and G. W. Moore, *Lectures on 2-D gravity and 2-D string theory*, Preprint, arXiv:hep-th/9304011

[9] B. R. Greene and Y. Kanter, *Small volumes in compactified string theory*, Nucl. Phys. B 497 (1997) 127–145

[10] B. R. Greene and C. I. Lazaroiu, *Collapsing D-branes in Calabi–Yau moduli space. 1*, Nucl. Phys. B 604 (2001) 181–255

[11] Y. Hatsuda, S. Moriyama, and K. Okuyama, *Instanton effects in ABJM theory from Fermi gas approach*, JHEP 1301 (2013) 158

[12] Y. Hatsuda, S. Moriyama, and K. Okuyama, *Exact instanton expansion of the ABJM partition function*, PTEP 2015 (2015) 11B104

[13] K. Hori, S. Katz, A. Klemm, R. Pandharipande, R. Thomas, C. Vafa, R. Vakil, and E. Zaslow, *Mirror Symmetry*, Clay Mathematics Monographs, American Mathematical Society (2003)

[14] G. T. Horowitz, *Spacetime in string theory*, New J. Phys. 7 (2005) 201

[15] E. Kiritsis, *String Theory in a Nutshell*, Princeton University Press (2007)

[16] J. M. Maldacena, *The large N limit of superconformal field theories and supergravity*, Int. J. Theor. Phys. 38 (1999) 1113–1133

[17] M. Mariño, *La teoria delle stringhe*, in C. Bartocci and P. Odifredi, eds., *La Matematica*, vol. IV, Einaudi (2010)

[18] M. Mariño, *Localization at large N in Chern–Simons–matter theories*. Unpublished manuscript.

[19] E. Martinec, *Evolving notions of geometry in string theory*, Found. Phys. 43 (2013) 156–173

[20] J. Polchinski, *Superstring Theory*, Cambridge University Press (2005)

[21] N. Seiberg, *Emergent spacetime*, Preprint, arXiv:hep-th/0601234

[22] G. 't Hooft, *A planar diagram theory for strong interactions*, Nucl. Phys. B 72 (1974) 461–473

[23] E. Witten, *String theory dynamics in various dimensions*, Nucl. Phys. B 443 (1995) 85–126

Marcos Mariño
Département de Physique Théorique et Section de Mathématiques,
Université de Genève
marcos.marino@unige.ch